THE OXFORD HANDBOOK OF

MUSIC AND THE BODY

THE OXFORD HANDBOOK OF

MUSIC AND THE BODY

Edited by

YOUN KIM

and

SANDER L. GILMAN

OXFORD
UNIVERSITY PRESS

OXFORD
UNIVERSITY PRESS

Oxford University Press is a department of the University of Oxford. It furthers the University's objective of excellence in research, scholarship, and education by publishing worldwide. Oxford is a registered trade mark of Oxford University Press in the UK and certain other countries.

Published in the United States of America by Oxford University Press
198 Madison Avenue, New York, NY 10016, United States of America.

Library of Congress Cataloging-in-Publication Data
Names: Kim, Youn. | Gilman, Sander L.
Title: The Oxford handbook of music and the body / edited by Youn Kim and Sander L. Gilman.
Description: New York, NY: Oxford University Press, 2019. |
Includes bibliographical references and index.
Identifiers: LCCN 2018037866 | ISBN 9780190636234 (hardcover)
Subjects: LCSH: Music—Psychological aspects. | Music—Physiological aspects. |
Music—Performance—Psychological aspects. | Music—Performance—Physiological aspects. |
Human body in music.
Classification: LCC ML3830.O88 2019 | DDC 781.1—dc23
LC record available at https://lccn.loc.gov/2018037866

1 3 5 7 9 8 6 4 2

Printed by Sheridan Books, Inc., United States of America

Contents

PART III EMBODIED MIND, EMBODIED RHYTHM

PART IV MUSIC AND THE DISABLED AND SEXUAL BODY

PART V MUSIC AS MEDICINE

PART VI MUSIC AND THE MULTIMODAL BODY

Acknowledgments

Our Handbook is the product of a truly collaborative endeavor across disciplines and continents. We thank Anna-Lise Santella, the music editor of the Oxford Handbooks, for her support and commitment to this complex project. We are also grateful to Carla Calandra, development editor at OUP, for her invaluable help in moving the Handbook to publication. The Handbook also owed a great deal in its final stages to Jennifer van der Grinten, whose meticulous attention to detail improved every draft. We are also grateful to all the participants of our conference Music and the Body held at the University of Hong Kong in 2012. The project to bring together the specialized perspectives into a broader context had its point of departure from that international meeting. We are especially grateful to all colleagues, readers, and reviewers for their suggestions and comments on our initial proposal and chapters. We thank our colleagues, friends, and families for their support. Finally, but by no means least, we thank the contributors for their contributions, commitment and belief in this project, and patience during the long process of putting this volume together.

This work was supported by the Seoul National University Research Grant in 2016.

About the Editors

Youn Kim is associate professor of music at the University of Hong Kong. Her research interests include history of music theory, psychology of music, history of listening, and in particular, the interrelationship between music theory and the science of the mind. She has published articles on these subjects and is currently writing a book on body and force in music.

Sander L. Gilman is a distinguished professor of the Liberal Arts and Sciences as well as professor of psychiatry at Emory University. A cultural and literary historian, he is the author or editor of over ninety books. His *Stand Up Straight! A History of Posture* appeared with Reaktion Press (London) in 2018. He is the author of the basic study of the visual stereotyping of the mentally ill, *Seeing the Insane*, published by John Wiley and Sons in 1982 (reprinted: 1996 and 2014) as well as the standard study *Jewish Self-Hatred*, the title of his Johns Hopkins University Press monograph of 1986, which is still in print. For twenty-five years he was a member of the humanities and medical faculties at Cornell University, where he held the Goldwin Smith Professorship of Humane Studies. For six years he held the Henry R. Luce Distinguished Service Professorship of the Liberal Arts in Human Biology at the University of Chicago. For four years he was a distinguished professor of the Liberal Arts and Medicine at the University of Illinois at Chicago and recently as the Alliance Professor of History at the Ludwig Maximilian University, Munich (2017–2018). He has been a visiting professor at numerous universities in North America, South Africa, the United Kingdom, Germany, Israel, China, and New Zealand. He was president of the Modern Language Association in 1995. He has been awarded a Doctor of Laws (*honoris causa*) at the University of Toronto in 1997, elected an honorary professor of the Free University in Berlin (2000) and an honorary member of the American Psychoanalytic Association (2007), and made a fellow of the American Academy of Arts and Sciences (2016).

Contributors

Michael B. Bakan, professor of ethnomusicology and head of the World Music Ensembles Program in the College of Music at Florida State University

Hanne Blank, Denison University

Marco Donnarumma, independent scholar

H. M. Evans, principal of Trevelyan College and professor of philosophy of music in the Department of Music at Durham University

Yayoi U. Everett, professor of music at the University of Illinois at Chicago

Marina Gilman, speech language pathologist at the Emory Voice Center of Emory University

Sander L. Gilman, distinguished professor of the Liberal Arts and Sciences, professor of psychiatry, and director of the Program in Psychoanalysis and the Health Sciences Humanities Initiative at Emory University

Evelyn Glennie, Solo Percussionist

Rolf Inge Godøy, professor of musicology at the University of Oslo

Stefan Sunandan Honisch, associate fellow of the Centre for Studies in Religion and Society at the University of Victoria.

Blake Howe, assistant professor of musicology at Louisiana State University

Linda Hutcheon, professor emeritus in the Department of English and the Centre for Comparative Literature at the University of Toronto

Michael Hutcheon, professor of medicine at the University of Toronto

Shersten Johnson, professor of music at the University of St. Thomas

James Kennaway, senior research fellow in the Department of Humanities at the University of Roehampton

Youn Kim, associate professor of music at the University of Hong Kong

Hedy Law, assistant professor of musicology at the University of British Columbia

Paul Lennard, director of Neuroscience and Behavioral Biology Department at Emory University

Xuejing Lu, postdoctoral research fellow at the Chinese Academy of Sciences

Eugene Montague, assistant professor of music at George Washington University

Mia Nakamura, associate professor of communication design science at Kyushu University

Jay Schulkin, research professor of obstetrics and gynecology at the University of Washington's School of Medicine

Daniel B. Stevens, director of the English Language Institute at the University of Delaware

Byron Suber, senior lecturer of performing and media arts at Cornell University

Atau Tanaka, professor of media computing at Goldsmiths, University of London

William Forde Thompson, professor of psychology at Macquarie University

Sandra E. Trehub, professor emeritus at the Infant and Child Studies Centre of the University of Toronto

CONTEXTUALIZING MUSIC AND THE BODY

An Introduction

YOUN KIM AND SANDER L. GILMAN

IN the 1994 film adaptation of Mary Shelley's *Frankenstein*, the monster learns to speak, sense, perceive, think, and even play a flute. When he encounters his creator, Dr. Victor Frankenstein, the creature asks about his musicality: "Did you know I knew how to play this? In which part of me did this knowledge reside? In these hands? In this mind? In this heart? . . . Who am I?" The creature was not the only one captivated by such questions. The precise seat of musicality in the human body has fascinated many for generations. In the early nineteenth century, phrenologists were intrigued by Franz Josef Haydn's skull, whereas subsequent generations revered the hands of Franz Liszt and Nicolò Paganini as divine musical relics. Today, the idea of a "music room" in the brain appears to be prevalent. Scientists scan musicians' brains to reveal their musical minds (Spitzer 2005). All these engaging cases bear on one of the themes of the present Handbook: music is created, performed, and appreciated by the human body.

Still, we can continue to ask: Can one's musicality be relegated to any single part of the body? Imagining the hands, ears, and brains as the body parts most immediately related to performing and listening to music is possible, but they do not function independently. The entire body works as the subject of musical activities. How about composing or mentally rehearsing a melody, which may appear to be "purely intellectual"? How is the body related to these musical activities? For a long time, composition and imagination have been considered to be the results of "a mysterious and solipsistic process that emerges unpredictably from deep within an individual" (Thompson 2009, 210). They were associated with intuition, inspiration, and insight—as matters of mentality, spirituality, and the psyche detached from the body and soma. Nonetheless, can we separate the musical mind from the body? The questions concerning how the body as a whole is related to musical activities, thus, invite us to rethink the old binary oppositions of "mind versus body." We compose and imagine while making conscious and unconscious choices within individual and social spheres. In doing so, the biophysical and cognitive as well as stylistic and social constraints influence, if not limit, how we create and imagine music. In other words, our musical thoughts are grounded in the body, thereby enabling and shaping music.

In the 2004 TV adaptation of *Frankenstein*, yet another interesting scene highlighted the relationship between music and the body. Upon hearing music for the first time—a blind man's violin performance of Johann Sebastian Bach, Sonatas and Partitas for Solo Violin, BWV 1001, Adagio—, the monster suddenly bursts into tears. In fact, the first stage adaptation of the novel, Richard Brinsley Peake's *Presumption* (1823), also contains a stage direction in which the monster physically responds to music (Cox 1993). These scenes—which did not appear in the original novel—point to music's power over the body. One can easily think of experiencing chills, shivers, or frissons when listening to certain musical passages. Music can induce not only these temporal physiological changes but also lasting effects on the human body. Thus, as much as the body is the subject of music-making and appreciation, music also influences and changes the body.

The link between music and the body seems self-evident. Nonetheless, only recently have music and the body been "brought back in" discussion (Frank 1990) and explicitly investigated from various perspectives. The study of music and the body intensified with the shift in the humanities—which has traditionally tended to shy away from the corporeal—and the development of the new technologies of looking into the body in sciences. The topic provides a converging platform for the humanistic sciences and scientific humanities to interact. After all, our body is "bio-socio-psycho-cultural," and music is what we do with such a body in the spaces and within the meanings attendant to it. Music and the body can be subjected to various interpretations and understandings, and the relationship between the two is even more multifaceted.

This chapter aims to provide a background for the rich and complex field of music and the body. It initially discusses multiple definitions of the body and raises a set of fundamental questions in the general context of body studies. The chapter then contextualizes the topic with regard to music and identifies six different yet related aspects of music and the body. The subsequent chapters—each discussing the significance of the body in music from specific perspectives—are configured accordingly.

The Body and Music

The past decade has seen an explosion of scholarly works on the body in virtually every field in the humanities and social sciences in ways that have paralleled changes and innovations in the biomedical sciences. The more we learn about the body the more we need theoretical models that provide alternative readings not just to the biomedical body but also, as John O'Neill argued in 1985 in his seminal work *Five Bodies*, to all of the different meanings, senses, and maps of the body with which we live. He sees the human body as "an intelligent and critical resource in the anthropomorphic production of those small and large orders that underlie our social, political and economic institutions" ([1985] 2004, 16). With the burgeoning field of "body studies," we are now talking not merely about the collection of in-depth yet isolated and independent studies, but about the all-embracing attempt to investigate the human body from diversified yet comprehensive perspectives. The present volume shows how these multiple definitions of the body and the equivalently complicated definitions of music interact and provides a set of case studies for complexity in understanding human experience, creativity, and interaction. Writing about music and the body means providing an understanding of the interrelationships between two ambiguous categories. Yet, as we shall see, such ambiguities are not necessarily erased in writing about their intersection.

Even thinking about the body as an organic reality experienced both from the perspective of an objective observer as well as from the very subjective position of embodiment has shifted over the recent past. In the nineteenth century the intellectual debate was which of the "higher senses," sight and sound, was most accurate in capturing the complexity of the world. The question was which provided the best map: for some it meant a map that was more abstract; for some a map that was more concrete (Chua 1999). The biomedical model shifted from "listening," which had been dominant from at the very beginning of the nineteenth century, to the world of "sight." Indeed, the biomedical sciences moved from the study of the whole organism to the cell in the late nineteenth century with the publication of the German physiologist Rudolf Virchow's 1858 monograph *Cell Pathology*. In his volume there is an image of the complete human body, but most of his plates reflect the new and exciting cellular level of the body, visualized through the application of coal tar dyes developed in nineteenth-century chemical laboratories. By the mid-twentieth century, molecular biology, in fields as diverse as genetics and psychiatry, began to move the study of the body to ever more complex (if ever less holistic) levels of analysis and explanation. The teaching of gross anatomy in the course of the twentieth century, which sought to teach the entire body, gave way to the study of systems in medicine. The older models of listening to the body in the late eighteenth and early nineteenth century, developed first by Leopold Auenbrugger (percussing the chest) and René Laennec (who developed auscultation and the technology of the stethoscope), were rooted in a detailed understanding of gross anatomy. Wilhelm Röntgen's discovery of X-rays in 1895 moved the emphasis from listening to the body to seeing inside of the body, echoing a core debate in nineteenth-century aesthetics as to which was the more sophisticated sense: hearing or seeing. But in the medical sciences both relied on gross anatomy in comprehending the body. The mid-twentieth century moved the focus from the cellular level to that of molecules and atoms. This refocusing provided new ways of seeing the body: computerized tomography (CT scan), which used X-rays to look at a now three-dimensional body, and functional neuroimaging (fMRI), which relied on reading the body through electronic scans of blood flow. (The irony is that the fMRI is significantly about "sound," as radiofrequency pulse is used to accentuate nuclei to higher magnetization levels. Likewise, Thorsten Wiesel and David Hubel's Nobel Prize–winning work [1981] on the neuroscience of the visual system was done in the 1960s via the sonification of neural firings.) Thus the living brain came to be visible in its functions just as the neurobiology of the brain became more and more transparent. Seeing seems to have won out at the moment, but listening to the body remains a powerful way, as we shall see in this volume, that the "psy disciplines," to use Nicolas Rose's term (1996), imagine their undertaking. Yet listening and viewing are central to understanding the complex relationship of ideas and representations of the body to the equally complex meanings attached to music (Benjamin 1968).

All of these innovations were coupled with complex models of understanding not only how data was to be generated but also how it would be interpreted. These models appeared both in the social sciences, often including social and economic data ascribed to the individual. One key example was the revelation that while high blood pressure in the African American community had been ascribed solely to genetics, the fact that the population of West Africa from which the ancestors of African Americans had sprung had much lower rates of high blood pressure led researchers to examine stress as a factor in the expression of genetic materials (Cooper et al. 2005). Models from the humanities, including models taken from women's history, examined how data, which recorded normal, healthy female experiences such as childbirth and menopause, were interpreted within medicine as pathological. The approaches and methods of a greater number of often conflicting models of understanding the body drew the interpretative sciences in the humanities and the so-called soft social sciences closer and closer into the world of the descriptive sciences (such as the so-called hard sciences such as genetics and evolutionary biology) so that in the end body studies came to be a merger of both (Gatherer 1995). It was the quality of the body experienced and observed that defined the type and focus of such studies. Each lays claim to its own methodology and each methodology to its own definition of evidence (Gould 1980).

Yet none of the changes, advancements, or alterations in our understanding of the body is free from the core methodological question: what manifestations do we examine to provide evidence for our reading of the body? Medical readings have usually departed from pathology, and we have had complex systems of pathology reading in and onto the body since the ancients. Entire arenas of body studies, such as disability studies, have questioned the pathological reading of a range of bodies earlier seen as ill or deviant. Central among these rereadings, as the philosopher Hilary Putnam argued in the 1970s, are those that rely on aesthetic criteria rather than scientific evidence (Conant and Żegleń 2001, 14–24). Is an abstract visual abstraction of the body in musical performance through opera or dance (in very different ways) as vital to our understanding of music in and of the body as our acoustic reception of music as ears and brains that listen? Today the study of bodily representations in their aesthetic mode is a key dimension of body studies. Yet anatomical questions also remain foregrounded. Which is the "real" body?

The interface between music and the body is multidimensional and multidisciplinary, involving biological, social, cultural, historical, ideological, psychological, and philosophical dimensions. Each of these fields has focused on different aspects of the relationship between music and the body. We believe that it is necessary to consider the issues in a broader context. By uniting in this Handbook approaches from very different contexts, we hope to highlight the complexity of music and the body and encourage further interdisciplinary research.

Music and the Body: Lines of Inquiry

In the context of music, we can contemplate on various background factors behind the recent surge and formation of music and the body as a field. One such factor is the conceptual expansion from a rather narrow approach centering on scores, composers, and the "work" toward a broader consideration embracing performers, listeners, and societies (see Goehr 1992). If the previously advocated "structural listening" implies "the almost total exclusion of human physical presence" (Dell'Antonio 2004, 8), the new comprehensive view considers music as neither a text nor an artifact but rather as a process and a phenomenon (see Cook 2001). Such a proposition naturally calls attention to the bodily dimension of musical activities. This dimension not only involves the most apparently somatic activity of performing but also incorporates composing, listening, and even the notion of music itself. Music is conceptualized not only as a noun but also as a verb: "musicking" or "to music" denotes "to take part, in any capacity, in a musical performance, whether by performing, by listening, by rehearsing or practicing, by providing material for performance (what is called composing), or by dancing" (Small 1998, 3).

Increasing sensitivity toward different types of musical media is also pertinent to the relationship between music and the body. Now, more than at any time in history, we have diverse opportunities to experience music in all genres and periods, via recordings (Philip 2004; Borio 2016). The development of technology has made music increasingly portable and accessible. Hence, it is generally considered an object rather than an activity. Conjointly, acousmatic, which means hearing without seeing the source of the sound, became the dominant mode of listening. Although the term "acousmatic" is attributed to the ancient Greek Pythagorean teaching (see Schaeffer 2017), it is not long since we began to listen without seeing. In this mode of separating sight and hearing, music is experienced without "that intangible sense of the music taking shape before us" (Johnson 2002, 197). Thus, the source of the sound, the musicking body, can evade our attention. The prevalence of the acousmatic mode of listening seemingly points to the removal of the sonic body. The notion of the acousmatic itself, however, ironically highlights the significance of the body in musical experience. We become aware of the body by consciously realizing its absence. Pierre Shaeffer's discussion of *musique concrète* (Shaeffer 2017), Michel Chion's theory of film sound (Chion 1994), and even popular TV shows like *The Voice*—which feature blind auditions for singing competitions—explore the imagination and potential extension of the unseen musicking body. Here, the body is projected even in the acousmatic setting (Kane 2014; see also Leppert 1995).

In conjunction with the body's gaining prominence in intellectual discourse, the development and availability of new imaging technologies in neurology and neuroscience have been instrumental in such an ideological expansion. Emphasis has been redirected to previously marginalized aspects of experience. The notion of the musical mind is broadened to include not only the biological brain and cognition but also emotions, feelings, trance, and the human psyche in a wider sense (Cochrane et al. 2013). Along with this, putting the body in contexts demonstrates yet another interesting aspect of music and the body. Music can inspire transcendental—the so-called out-of-body experiences—in deep trance states (see Rouget 1985; Becker 2004).

The body produces and experiences music in a given time and space. Therefore, the environment must be considered in approaching the relationship between music and the body. Such ecological perspectives in the psychology and sociology of music as well as ethnomusicology all converge on the significance of the environment and the body's interaction with it (see Clarke 2011; DeNora 2011; Feld 1993).

This is not a mere reflection of an intellectual trend. To quote the musicologist Susan McClary, "It is not . . . for the sake of following fads that questions about the body need to be brought to music, for the body has always been there. Rather, the current interrogation of the body in the academy finally bestows legitimacy on a topic that ought never to have been banished: the constant interaction between music and bodies" (1995, 87). Such legitimacy seems today to be given, yet when the competing notions of both "body" and "music" in the critical literature are examined, one notes the constant attempts to delegitimize various notions of each, highlighting the "authentic" or "natural" or "given" assumptions that underlie one or the other. One classic example is when the noted historian William H. McNeill claimed that dance music reflects the intuitive imperative of the beating heart translated into movement (1995, 2).

The Oxford Handbook of Music and the Body is the outcome of a multidisciplinary effort to explore the diversified perspectives on music and the body. We first view the performing and listening body as an object from anatomical, physiological, kinesiological, and psychological perspectives. However, even the most physiological act of hearing cannot be completely objective. To paraphrase the vision psychologist Daniel Simons, we "hear the world as it isn't" (Chabris and Simons 2011). Thus, it is equally important to incorporate the body, which experiences music from a subjective, first-person perspective while moving dynamically in its physical and cultural surroundings. We are interested in the history as well as the science of such claims. The volume will serve as a handbook for both students in a wide range of fields (from anthropology to human medicine and from musicology to psychology) and specialists, who need broader orientations to questions of music and the body.

In mapping out this new field, we aim at bringing together the questions previously separated by the traditional dichotomies of the sciences/humanities, nature/culture, mind/body, and activity/object. We consider various modalities of the music–body relationship (such as hearing, seeing, and kinesthetics), the subjects of the action (i.e., composers, performers, and listeners), and the different genres and cultures of music. This Handbook is organized around the following lines of inquiries. However, one chapter often deals with more than a single theme and thus can be grouped under different sectional headings. Readers are invited to listen to the rich and resonant dialogues created by diverse perspectives across chapters and sections, which span a wide and interlocking range of topics.

Moving and Performing Body

"In the beginning was movement"—with these words, Maxine Sheets-Johnstone indicates the importance of the moving body living in an animate world in her interdisciplinary study on the kinetic phenomena (2011, 331). The moving body is central to the understanding of music. Particularly, musicking and dancing bodies share certain features. Music and dance—being nonlinguistic (nonverbal) forms of communication—manifest the inseparability of mind and body: in both, "we observe mind at work through movements of bodies in space and time" (Blacking 1977, 18).

Dancing and music are inseparable in many cultures and were probably integrated together in the prehistoric time. Recent evolutionary theories consider music and dance as a coalition signaling system (Hagen and Bryant 2003). What is then the musicality of the moving body, more specifically, that of the dancing body? Byron Suber's chapter discusses the fundamental issues on the musicking and dancing body in the specific context of Western concert dance, which assumes visual as well as aural consumption as a theatrical form. As movement implies temporality, the moving body is also the foundation of our aesthetic judgment of music. Cognitive expectation plays a critical role in organizing human actions and arousing emotions, and Jay Schulkin's chapter connects music, movement, and thinking in providing a kinesthetic/anticipatory description of musical sensibility.

The body not only responds to but also produces music. Music-making, particularly performance, has been relatively overlooked compared with composition and listening. Traditionally, in the communication model proposed by Claude Elwood Shannon and Warren Weaver (1949), performance is a mediator or a "transmitter" that connects the "information source" (composer) and the "destination" (listener) of the musical communication system (see also Meyer 1961). However, it is also possible to conceptualize the relationship between musical works and performance conversely: "*performance does not exist in order to present musical works, but rather, musical works exist in order to give performers something to perform*" (Small 1998, 8; emphasis in original). Above all, we experience music mainly through performance: "The experience of live or recorded performance is the primary form of music's existence, not just the reflection of notated text" (Cook 2014, 1; see also Cook 2007; Rink 2005; Le Guin 2005). Attention to the somatic and tactile dimension of musical activities also leads to growth in studying various types of physical gestures and bodily movements for producing sound, facilitating performance, and communicating among peer performers and the audience (Davidson 2012; Gritten and King 2006, 2011).

The body as an instrument, therefore, will be the groundwork for understanding the relationship between music and the body. All parts of the human body can produce sounds (and music). People in various cultures have been making music by stomping, clapping, patting or hitting body parts, and snapping fingers.[1] Nonetheless, the most immediate musical instrument would be the voice. The emergence of music itself usually begins with the discussion of various vocalizations including "protolanguage," "protomusic," or "musilanguage" (Brown 1999; Tomlinson 2015; Mithen

2007). Focusing on voice, Marina Gilman's chapter presents the scientific and practical understanding of the human body as the instrument, discussing the anatomy and physiology as well as the performance and pedagogy of the voice in the context of the entire body.

The body as a sound-producing instrument can be extended beyond one's actual body. As Maurice Merleau-Ponty affirmed, "Those actions in which I habitually engage incorporate their instruments into themselves and make them play a part in the original structure of my own body" (2002, 104). When a person makes music with an instrument, the body mediates between the instrument and the world and acts on both by developing its own complex motor capacity. The use of instruments transforms the motor areas of our brain. Musicians generally speak of the inseparability of their bodies and instruments. In the words of the pianist Claudio Arrau, you have to "feel the unity of the instrument with your body" (cited in Arx 2014, 77). Musicians' instruments are, in fact, incorporated into their brain maps after years of practice (see, for example, Gaser and Schlaug 2003; Haueisen and Knösche 2001). Coincidentally, the oldest surviving musical instrument, the Geißenklösterle flute, dating perhaps to 40,000 years ago, is made of a bone, a part of the animal body (Higham et al. 2012). Instruments extend somatic boundaries, not only figuratively but also literally. Recent developments in technology further expand the limits of the musicking body and reconfigure the interaction between the body and instruments (see Haraway 1991; Hayles 1999; Ihde 2015, 1990). Atau Tanaka and Marco Donnarumma's chapter examines the evolvement of the idea of the body as musical instruments in a cross-disciplinary context, leading up to the examples of "posthuman body" in contemporary experimental musical practice.

Musical Brain, Psyche, and Beyond

Studies on music and the brain have drastically increased over the last two decades with the development of new techniques, such as fMRI, position emission tomography (PET), electroencephalography (EEG), and magnetoencephalography (MEG). Despite some of the apparent limitations, such as the noise of the machines or the restrictions of bodily movements, these techniques enabled real-time exploration of a living body's brain activities for the first time in history. Such neuroscientific studies elucidate brain responses to music in listening, as well as the neural activities in imagining and performing music. Others provide case studies of individuals with neurological disorders. Often collectively called the "cognitive neuroscience of music," these studies illuminate our insight into the musical brain. What is the underlying brain mechanism in music perception and cognition? The pursuit of this question illuminates the nature of the human brain. In Stefan Koelsch's words, "Listening to music, and music making, engages [*sic*] a large array of psychological processes, including perception and multimodal integration attention, learning and memory, syntactic processing and processing of meaning information, action, emotion, and social cognition" (2012, x). In this view, musicality is located in the brain and many scholars have paid attention to the topic of the musical brain. As Oliver Sacks confirmed, anatomists would have difficulty identifying the brain of artists, writers, or scientists, but they would recognize the brain of musicians (2007, 94). Considerable research is conducted on the brain, taking music as "an ideal tool to investigate human psychology and the working of the human brain" (Koelsch 2012, 2). Paul Lennard's chapter in the present volume not only examines how music changes the brain but also elucidates the connection between music and language by comparing nonmusicians' holistic and musician's syntax-oriented listening from a neuroscientific perspective. A related yet different approach to music and the body can be taken in the context of psychoanalysis. In addition to causing functional and morphological changes to the brain and providing therapeutic effects to the body, music and listening play critical roles in revealing and understanding the psyche, the inner self. Sander L. Gilman's chapter examines the multiple links among music, unconscious thoughts, listening, and healing, and sheds yet a different light on music and the body in the context of psychanalytic theories. Such concerns for the brain and the psyche should not exclude cultural and social factors. For instance, mirror neuron theory proposes that humans can feel emotions by the neural mechanism of observing and simulating others' action in the brain (Molnar-Szakacs et al. 2006). This theory deals with the intersubjective and interpersonal brain and body. Human brains produce and consume music through interactions. Thus, the two approaches of the brain as a biological and cultural entity must converge (Pitts-Taylor 1995). Mia Nakamura's chapter offers one such example of this convergent approach to the musical brain, exploring an intersection between sociology and neuroscience. Musical experience is viewed as not only a cognitive but also an expressive and aesthetic experience that includes affective responses to music (see Hodges 2016).

Embodied Mind, Embodied Rhythm

Brain science locates the control center of the body in the central nervous system, consisting of the brain and the spinal cord. It proposes that all bodily actions and movements are regulated by the brain. Therefore, the mind is equated with the brain and has control over the body. The idea of embodied cognition, proposed by the philosopher Mark Johnson and the linguist George Lakoff, overturns this relationship between the mind and the body by emphasizing the significance of kinesthetic action with regard to thinking: the body influences thoughts (Lakoff and Johnson 1999; Johnson 1990). Our thoughts are grounded in "the phenomenological experience of our bodies in action" (Gibbs 2005, 3). The invisible and disembodied psyche, the mind, is intertwined with the embodied soma, the body. Unlike Western tradition, where mind and thoughts are viewed as formless and body-less, we argue that the mind is embodied, that it is grounded in the body. In response to these arguments in cognitive linguistics and psychology, physical and musical gestures have been developed as the fundamental ideas of music production and listening (Zbikowski 2006, 2002).

In the present Handbook, Rolf Inge Godøy's chapter addresses embodied music cognition in the context of the mind–body issue. All activities of the mind, such as perception and cognition, are grounded in the phenomenological bodily experience. Moreover, the physical movement of the body shapes these mental activities and phenomena (Gallagher 2006). Thus, the relationship between music and the body goes beyond the most immediate dimension of the body as executing music. In listening to Robert Schumann's *Kreisleriana*, Roland Barthes famously says, "I actually hear no note, no theme, no contour, no grammar, no meaning, nothing which would permit me to reconstruct an intelligible structure of the work. No, what I hear are blows: I hear what beats in the body, what beats the body, or better: I hear this body that beats" (1986, 299). We are hearing the body *within* the music. As Nicholas Cook aptly validated, the body is "encoded within the sound" as the "virtual body of the music" (2016, 191; see also Cox 2016).

Music resounds in our body, which also echoes music. We synchronize our bodies to the rhythm of the external world or to that of other bodies. Traditional Western classical concerts demand that the audience be silent and motionless while listening to the music. However, this tradition is a rather peculiar and historically recent phenomenon. As the writer Philip Ball puts it, "Music is often said to find a short cut to the heart, but the right sort of music takes a quick route to the legs" (2012, 208). This capacity to synchronize with other rhythmic cycles and coordinate rhythmic movements is believed to have played a critical role in the survival of mankind (see Bücher 1899, on the subject of work rhythm; see also Thaut 2005; Bispham 2006). While some zoomusicologists take a biosemiotic approach, proposing the concept of animal music (Keller 2012; Martinelli 2010), others would claim that only humans have music (Nattiez 1990). Music is indeed a special way of gaining insight into humankind and the body. The capacity of entrainment can enlighten the ensuing debate on this issue (Patel et al. 2009).

The chapters in this section illumine the different aspects of embodied rhythm and mind. Eugene Montague's chapter identifies and examines the different types of entrainment through a case study of J. S. Bach's prelude. Daniel B. Steven's chapter also deals with Western classical keyboard music, Frederick Chopin's etude and Leopold Godowsky's paraphrase-study of that etude, but with the focus on expressive timing in the act of performance and how it is related to the complex and dynamic gestures of the performer's body. Shersten Johnson's chapter explores the multimodal embodiments of rhythm in selected operatic scenes and how they affect the experience of the dramas. The human body, which is important for understanding the perception and production of rhythm, crucially highlights the very notion of music and musicality. In examining the relationships among music, the body, and freedom in seventeenth- and eighteenth-century French music, Hedy Law's chapter demonstrates how the seemingly transcendental concepts like freedom can be addressed through the context of embodiment. Bacchus, the focus of inquiry here, presents a close link to the next section by engendering the idea of the extraordinary body.

Music and the Disabled and Sexual Body

The human body is a biological, "natural" entity, but this body is highly susceptible to social and cultural influences. Disability and sexuality are both qualities of the body that are subject to the medical definition but also social constructs. How does a body defined by the normative standards of any given society work in the production of music?

There has been a growing and sustained interest in the physical dimension of the human body in the humanities, as demonstrated in the studies of such culturally and historically conceptualized notions as health, disorder, disability, disease, sensuality, and medicine (including pioneering works like Bordo 1993; Butler 1990; Foucault 1978). This interest has paralleled and merged with work undertaken on theories of representation of the body and its pitfalls that grew out of continental philosophy in the 1960s. The debate between a "soft" and a "hard" history of science, which rejects the image for the material, has led to a new materialism in the history of medicine (Gilman 2015, xvvi–xxi). Only the lived reality of the experience embodied in the materiality of the world is taken as "real." Recently this merger has become apparent in disability studies in music, which includes both the representation of disabilities and a focus on the disabled performer, observer, and/or composer, rather than on the interaction between the cultural assumptions of disability (positive or negative) and the lived experience of the disabled performer.

The interest in the relationship between music and disability has rapidly and intensively arisen during the first decade of the twentieth-first century (see Lerner and Straus 2006; Straus 2011; and Lubet 2011). The chapters in this section provide a multidisciplinary perspective on the relationship between music and disability. Michael B. Bakan's chapter explores how musical experience mediates bodily existence through the ethnographical studies of two individuals. Blake Howe's chapter presents various formulations of music as the agent of healing and rehabilitating disabled bodies. Stefan Honisch's chapter explores blindness and deafness in the context of competitive music performance, where disability and virtuosity converge with each other. In Linda Hutcheon and Michael Hutcheon's chapter, the singing, acting body on the stage is foregrounded, and the representations of the differently abled and sexual bodies are examined in the operatic context. Hanne Blank's chapter focuses on nineteenth-century operas and the experience of the sublime, enabled by the operatic presence of sexuality and dis/ability. In our discussion, we include the embodied experiences of individual practitioners of music in approaching music and the individualized body. Such an attempt includes the chapter based on the interview with Evelyn Glennie. The most recent rethinking of disability redefines disability on a scale of "human variability" that understands the difficulties facing the disabled as resulting from the inflexibility of social institutions rather than from impairment. The feminist disability studies scholar Rosemarie Garland-Thomson has defined "disability as a way of being in an environment" produced by the "discrepancy between body and world, between that which is expected and that which is" (2012, 342). Not only is disability the disjuncture between the world and the individual, but also, as all who age become disabled, "disability is thus inherent in our being: What we call disability is perhaps the essential characteristic of being human" (342). This discussion eventually engenders a more general question regarding listening as defined by social institutions with the whole body.

Music as Medicine

The relationship between music and medicine has never ceased to be discussed. This is not surprising, given that music is commonly an integral part of healing rituals in several cultures. When we think of music and medicine, the first thing that comes to mind is music therapy, the standardized practice, and application of clinical music strategies in medicine. Nonetheless, professional activities for applying music *as* medicine are not the only link between the two. Above all, the topic of music and medicine presents an interesting connection between the mind and the body, which were separated and opposed in the traditional conceptualization. The notion of pneuma proposed by the pneumatic school of medicine as some sort of vitalistic fluid, for example, may avail in connecting the mind and the body.

In addition to the obvious connection of music as medicine, the two are interrelated on many different levels. Since the 1990s, the liaisons between them have been discussed in a wide historical, social, and cross-cultural context. In addition to practicing music therapists, neuroscientists, and psychologists, scholars in history, philosophy, cultural studies, anthropology, and musicology have been active in this interdisciplinary discussion (see Koen et al. 2011; Gouk 2000; Horden 2000). James Kennaway's chapter examines music and medicine against the historical background of embodiment and takes a medical approach to explain the effect of music. H. M. Evans's chapter investigates the healing potential of music by looking into musical imagination and hallucinosis and thus addresses the question of why music-as-therapy can be effective.

Music and the Multimodal Body

Our experience of music is mostly and naturally multimodal, frequently involving hearing, seeing, and often other sensory modalities. The unimodal, sightless, and audio-only mode of music listening was far from normal before World War I (see Cook 2016, 1998; Katz 2012). This section approaches music and the body from a holistic perspective. Instead of discussing each mode of perception independently, the chapters in this section investigate the issue across modalities and disciplinary boundaries. Xuejing Lu and William Forde Thompson focus on cross-modality and spatial representation, discussing the spatial representation of musical pitch and how this spatial conceptualization of music is bodily grounded and represented through gestures (see Baily 1985 for the discussion of the spatiomotor thinking in performance). Today's "new multimedia" (Cook 2016, 188), such as film and YouTube, further highlight the multimodality of musical experience with internet technology. Thus, they even more increase the need for cross-modal approaches in the study of music and the body. Yayoi Everett's chapter takes the experience of opera in its remediated form as film as a case and examines how audiovisual congruence induces the multimodal embodied experience of music. The multimodality of musical experience is more evident among infants and young children. Sandra Trehub's chapter investigates the multimodal music of early childhood in the ecological context of home and school beyond the laboratory settings (for a review of the aging body and music, see Halpern and Bartlett 2002). Youn Kim's chapter examines the historical emergence of the holistic conceptualization of listening by engaging in various discourses on seeing, hearing, and kinesthetics in the late nineteenth and early twentieth centuries. The relationship between music and the body is historically contingent.

Going back to the story of Frankenstein, the questions that the monster asked are not really about the main bodily organs of musicality. Asking about the relationship between music and the body ultimately led to the ontological question of "Who am I?" A similar scene is found in Jean-Philippe Rameau's acte de ballet *Pigmalion* (1748), set to a libretto by Sylvain Ballot de Sauvot. The sculptor Pigmalion creates a beautiful statue, which becomes vivified, sings, and dances. Just like Frankenstein's monster, the enlivened statue, Galathée, begins her song with a series of questions: "Que vois-je? Où suis-je? Et qu'est-ce que je pense? D'où viennent ces mouvements?" (What do I see? Where am I? And what do I think? From where come these movements?).[2] Both Galathée and Frankenstein's creature are linking their musicalities with their very existence. Questions about musicking bodies answer to "who I am" as an individual and further to "who we are" as a species. We have been doing music for such a long time in the evolutionary history. *Homo musicus*, the man who makes music, is believed to be even older than *Homo sapiens*, the man who thinks and reasons (Kirnarskaya 2009; see also Tomlinson 2015; Mithen 2007). This intimate connection between music and the body reminds us of the poetic question posed by William Butler Yeats in 1927: "O body swayed to music, O brightening glance, / How can we know the dancer from the dance?" As we embark on our journey to explore music and the body, we may likewise ask, "How can we know the musicking body from music?"

Notes

1. Bobby McFerrin's body percussion is one of such recent examples and historically we can also think of the extraordinary success of the French flatulist *Le Pétomane*, Joseph Pujol (1857–1945).
2. We thank Alexander Rehding for bringing our attention to this parallelism.

References

Arx, Victoria A. von. 2014. *Piano Lessons with Claudio Arrau: A Guide to His Philosophy and Techniques*. New York: Oxford University Press.

Baily, John. 1985. "Music Structure and Human Movement." In *Musical Structure and Cognition*, edited by Peter Howell, Ian Cross, and Robert West, 237–258. London: Academic Press.

Ball, Philip. 2012. *The Music Instinct: How Music Works and Why We Can't Do Without It*. Repr. ed. Oxford: Oxford University Press.

Barthes, Roland. 1986. *The Responsibility of Forms: Critical Essays on Music, Art, and Representation*. Translated by Richard Howard. Oxford: Blackwell.

Becker, Judith. 2004. *Deep Listeners: Music, Emotion, and Trancing*. Bloomington: Indiana University Press.

Benjamin, Walter. 1968. *Illuminations: Essays and Reflections*. Edited by Hannah Arendt. Translated by Harry Zohn. New York: Schocken Books.

Bispham, John. 2006. "Rhythm in Music: What Is It? Who Has It? And Why?" *Music Perception: An Interdisciplinary Journal* 24 (2): 125–134. https://doi.org/10.1525/mp.2006.24.2.125.

Blacking, John, ed. 1977. *The Anthropology of the Body*. New York: Academic Press.

Bordo, Susan. 1993. *Unbearable Weight: Feminism, Western Culture, and the Body*. Berkeley: University of California Press.

Borio, Gianmario, ed. 2016. *Musical Listening in the Age of Technological Reproduction*. Burlington, VT: Routledge.

Brown, Steven. 1999. "The 'Musilanguage' Model of Music Evolution." In *The Origins of Music*, edited by N. L. Wallin, B. Merker, and S. Brown, 271–301. Cambridge, MA: MIT Press.

Bücher, Karl. 1899. *Arbeit und Rhythmus*. Leipzig: Teubner.

Butler, Judith. 1990. *Gender Trouble: Feminism and the Subversion of Identity*. New York: Routledge.

Chabris, Christopher, and Daniel Simons. 2011. *The Invisible Gorilla: How Our Intuitions Deceive Us*. Repr. ed. New York: Harmony.

Chion, Michel. 1994. *Audio-Vision: Sound on Screen*. Translated by Claudia Gorbman. 14th ed. New York: Columbia University Press.

Chua, Daniel. 1999. *Absolute Music and the Construction of Meaning*. New York: Cambridge University Press.

Clarke, Eric. 2011. *Ways of Listening: An Ecological Approach to the Perception of Musical Meaning*. Repr. ed. Oxford University Press.

Cochrane, Tom, Bernardino Fantini, Klaus R. Scherer, and Kristen Jafflin. 2013. *The Emotional Power of Music*. 1st ed. Oxford: Oxford University Press.

Conant, James, and Ursula M. Żegleń, eds. 2001. *Hilary Putnam: Pragmatism and Realism*. 1st ed. New York: Routledge.

Cook, Nicholas. 1998. "The Domestic Gesamtkunstwerk, or Record Sleeves and Reception." In *Composition, Performance, Reception: Studies in the Creative Process*, edited by Wyndham Thomas, 105–117. Aldershot, UK: Ashgate.

Cook, Nicholas. 2001. "Between Process and Product: Music and/as Performance." *Music Theory Online* 7 (2). http://www.mtosmt.org/issues/mto.01.7.2/mto.01.7.2.cook.html.

Cook, Nicholas. 2007. *Music, Performance, Meaning: Selected Essays*. Aldershot, UK: Ashgate.

Cook, Nicholas. 2014. *Beyond the Score: Music as Performance*. New York: Oxford University Press.

Cook, Nicholas. 2016. "Seeing Sounds, Hearing Images: Listening Outside the Modernist Box." In *Musical Listening in the Age of Technological Reproduction*, edited by Gianmario Borio, 185–202. New York: Routledge.

Cooper, Richard S., Katharina Wolf-Maier, Amy Luke, Adebowale Adeyemo, José R. Banegas, Terrence Forrester, et al. 2005. "An International Comparative Study of Blood Pressure in Populations of European vs. African Descent." *BMC Medicine* 3 (January): 2. https://doi.org/10.1186/1741-7015-3-2.

Cox, Arnie. 2016. *Music and Embodied Cognition: Listening, Moving, Feeling, and Thinking*. Bloomington: Indiana University Press.

Cox, Jeffrey N. 1993. *Seven Gothic Dramas, 1789–1825: 1789–1825*. Athens: Ohio University Press.

Davidson, Jane W. 2012. "Bodily Movement and Facial Actions in Expressive Musical Performance by Solo and Duo Instrumentalists: Two Distinctive Case Studies." *Psychology of Music* 40 (5): 595–633. https://doi.org/10.1177/0305735612449896.

Dell'Antonio, Andrew. 2004. *Beyond Structural Listening? Postmodern Modes of Hearing*. Berkeley: University of California Press.

DeNora, Tia. 2011. *Music-in-Action: Selected Essays in Sonic Ecology*. New York: Routledge.

Feld, Steven. 1993. "From Ethnomusicology to Eco-Muse-Ecology." *The Soundscape Newsletter*, 1993.

Foucault, Michel. 1978. *The History of Sexuality, Vol. 1: An Introduction*. New York: Vintage Books.

Frank, Arthur W. 1990. "Bringing Bodies Back in: A Decade Review." *Theory, Culture, and Society* 7 (1): 131–162.

Gallagher, Shaun. 2006. *How the Body Shapes the Mind*. Oxford: Clarendon Press.

Garland-Thomson, Rosemarie. 2012. "'The Case for Conserving Disability.'" *Journal of Bioethical Inquiry* 9 (3): 339–355.

Gaser, Christian, and Gottfried Schlaug. 2003. "Brain Structures Differ between Musicians and Non-Musicians." *Journal of Neuroscience* 23 (27): 9240–9245.

Gatherer, Derek. 1995. "The Evolution of Music—A Comparison of Darwinian and Dialectical Methods." *Journal of Social and Evolutionary Systems* 20 (1): 75–92.

Gibbs, Raymond W., Jr. 2005. *Embodiment and Cognitive Science*. New York: Cambridge University Press.

Gilman, Sander L. 2015. *Illness and Image: Case Studies in the Medical Humanities*. New Brunswick, NJ: Transaction.

Goehr, Lydia. 1992. *The Imaginary Museum of Musical Works: An Essay in the Philosophy of Music*. Oxford: Clarendon Press.

Gouk, Penelope, ed. 2000. *Musical Healing in Cultural Contexts*. 1st ed. Aldershot, UK; Ashgate.

Gould, Stephen Jay. 1980. *The Panda's Thumb: More Reflections in Natural History*. New York: W. W. Norton.

Gritten, Anthony, and Elaine King, eds. 2006. *Music and Gesture*. Burlington, VT: Ashgate.

Gritten, Anthony, and Elaine King. 2011. *New Perspectives on Music and Gesture*. SEMPRE Studies in the Psychology of Music. Burlington, VT: Ashgate.

Hagen, Edward H., and Gregory A. Bryant. 2003. "Music and Dance as a Coalition Signaling System." *Human Nature* 14 (1): 21–51. https://doi.org/10.1007/s12110-003-1015-z.

Halpern, Andrea R., and James C. Bartlett. 2002. "Aging and Memory for Music: A Review." *Psychomusicology: A Journal of Research in Music Cognition* 18 (1–2): 10–27. https://doi.org/10.1037/h0094054.

Haraway, Donna. 1991. "A Cyborg Manifesto: Science, Technology, and Socialist—Feminism in the Late Twentieth Century." In *Simians, Cyborgs and Women: The Reinvention of Nature*, 149–181. New York: Routledge.

Haueisen, J., and T. R. Knösche. 2001. "Involuntary Motor Activity in Pianists Evoked by Music Perception." *Journal of Cognitive Neuroscience* 13 (6): 786–792. https://doi.org/10.1162/08989290152541449.

Hayles, N. Katherine. 1999. *How We Became Posthuman: Virtual Bodies in Cybernetics, Literature, and Informatics*. Chicago: University of Chicago Press.

Higham, Thomas, Laura Basell, Roger Jacobi, Rachel Wood, Christopher Bronk Ramsey, et al. 2012. "Testing Models for the Beginnings of the Aurignacian and the Advent of Figurative Art and Music: The Radiocarbon Chronology of Geißenklösterle." *Journal of Human Evolution* 62 (6): 664–676. https://doi.org/10.1016/j.jhevol.2012.03.003.

Hodges, Donald. 2016. "The Neuroaesthetics of Music." In *Oxford Handbook of Music Psychology*, 2nd ed., edited by Susan Hallam, Ian Cross, and Michael H. Thaut, 247–262. Oxford: Oxford University Press.

Horden, Peregrine, ed. 2000. *Music as Medicine: The History of Music Therapy since Antiquity*. 1st ed. Brookfield, VT: Routledge.

Ihde, Don. 1990. *Technology and the Lifeworld: From Garden to Earth*. Bloomington: Indiana University Press.

Ihde, Don. 2015. *Acoustic Technics*. Lanham, MD: Lexington Books.

Johnson, Mark. 1990. *The Body in the Mind: The Bodily Basis of Meaning, Imagination, and Reason*. 1st ed. Chicago: University of Chicago Press.

Johnson, Peter. 2002. "The Legacy of Recordings." In *Musical Performance: A Guide to Understanding*, edited by John Rink, 197–211. Cambridge: Cambridge University Press.

Kane, Brian. 2014. *Sound Unseen: Acousmatic Sound in Theory and Practice*. 1st ed. New York: Oxford University Press.

Katz, Mark. 2012. "Sound Recording Introduction." In *Music, Sound, and Technology in America*, edited by Timothy D. Taylor, Mark Katz, and Tony Grajeda, 11–28. Durham, NC: Duke University Press.

Keller, Marcello Sorce. 2012. "Zoomusicology and Ethnomusicology: A Marriage to Celebrate in Heaven." *Yearbook for Traditional Music* 44: 166–183. https://doi.org/10.5921/yeartradmusi.44.0166.

Kirnarskaya, Dina. 2009. *The Natural Musician: On Abilities, Giftedness, and Talent*. Translated by Mark H. Teeter. 1st ed. New York: Oxford University Press.

Koelsch, Stefan. 2012. *Brain and Music*. Hoboken, NJ: Wiley-Blackwell.

Koen, Benjamin, Jacqueline Lloyd, Gregory Barz, and Karen Brummel-Smith, eds. 2011. *The Oxford Handbook of Medical Ethnomusicology*. Repr. ed. New York: Oxford University Press.

Le Guin, Elisabeth. 2005. *Boccherini's Body: An Essay in Carnal Musicology*. Berkeley: University of California Press.

Lakoff, George, and Mark Johnson. 1999. *Philosophy in the Flesh: The Embodied Mind and Its Challenge to Western Thought*. New York: Basic Books.

Leppert, Richard. 1995. *The Sight of Sound: Music, Representation, and the History of the Body*. Berkeley: University of California Press.

Lerner, Neil, and Joseph Straus, eds. 2006. *Sounding Off: Theorizing Disability in Music*. New York: Routledge.

Lubet, Alex. 2011. *Music, Disability, and Society*. Philadelphia, PA: Temple University Press.

Martinelli, Dario. 2010. *A Critical Companion to Zoosemiotics: People, Paths, Ideas*. Dordrecht, Netherlands: Springer.

McClary, Susan. 1995. "Music, the Pythagoreans, and the Body." In *Choreographing History*, edited by Susan Leigh Foste, 82–104. Bloomington: Indiana University Press.

McNeill, William Hardy. 1995. *Keeping Together in Time: Dance and Drill in Human History*. Cambridge, MA: Harvard University Press.

Merleau-Ponty, Maurice. 2002. *Phenomenology of Perception*. New York: Routledge.

Meyer, Leonard B. 1961. *Emotion and Meaning in Music*. Chicago: University of Chicago Press.

Mithen, Steven. 2007. *The Singing Neanderthals: The Origins of Music, Language, Mind, and Body*. Cambridge, MA: Harvard University Press.

Molnar-Szakacs, Istvan, and Katie Overy. 2006. "Music and Mirror Neurons: From Motion to 'e'motion." *Social Cognitive and Affective Neuroscience* 1 (3): 235–241. https://doi.org/10.1093/scan/nsl029.

Nattiez, Jean-Jacques. 1990. *Music and Discourse*. Translated by Carolyn Abbate. 3rd ed. Princeton, NJ: Princeton University Press.

O'Neill, John. 2004. *Five Bodies: Re-Figuring Relationships*. Thousand Oaks, CA: SAGE.

Patel, Aniruddh D., John R. Iversen, Micah R. Bregman, and Irena Schulz. 2009. "Experimental Evidence for Synchronization to a Musical Beat in a Nonhuman Animal." *Current Biology: CB* 19 (10): 827–830. https://doi.org/10.1016/j.cub.2009.03.038.

Philip, Robert. 2004. *Performing Music in the Age of Recording*. New Haven, CT: Yale University Press.

Pitts-Taylor, Victoria. 1995. "Social Brains, Embodiment and Neuro-Interactionism." In *Choreographing History*, 171–182. Bloomington: Indiana University Press.

Rink, John, ed. 2005. *The Practice of Performance: Studies in Musical Interpretation*. Cambridge and New York: Cambridge University Press.

Rose, Nicolas. 1996. *Inventing Our Selves: Psychology, Power and Personhood*. Cambridge: Cambridge University Press.

Rouget, Gilbert. 1985. *Music and Trance: A Theory of the Relations between Music and Possession*. Chicago: University of Chicago Press.

Sacks, Oliver. 2007. *Musicophilia: Tales of Music and the Brain*. 1st ed. New York: Knopf.

Schaeffer, Pierre. 2017. *Treatise on Musical Objects: An Essay across Disciplines*. Berkeley: University of California Press.

Sheets-Johnstone, Maxine. 2011. *The Primacy of Movement*. Amsterdam and Philadelphia, PA: John Benjamins.

Shannon, Claude Elwood, and Warren Weaver. 1949. *The Mathematical Theory of Communication*. Urbana: University of Illinois Press.

Small, Christopher. 1998. *Musicking: The Meanings of Performing and Listening*. Middletown, CT: Wesleyan University Press.

Spitzer, Manfred. 2005. *Musik im Kopf: Hören, Musizieren, Verstehen und Erleben im neuronalen Netzwerk*, Vol. 9. Stuttgart: Schattauer.

Straus, Joseph N. 2011. *Extraordinary Measures: Disability in Music*. 1st ed. New York: Oxford University Press.

Thaut, Michael. 2005. *Rhythm, Music, and the Brain: Scientific Foundations and Clinical Applications*. New York: Routledge.

Thompson, William Forde. 2009. *Music, Thought, and Feeling: Understanding the Psychology of Music*. New York: Oxford University Press.

Tomlinson, Gary. 2015. *A Million Years of Music: The Emergence of Human Modernity*. New York: Zone Books.

Zbikowski, Lawrence. 2002. *Conceptualizing Music: Cognitive Structure, Theory, and Analysis*. Oxford: Oxford University Press.

Zbikowski, Lawrence. 2006. "The Cognitive Tango: Music, Mind and Brain." In *The Artful Mind: Cognitive Science and the Riddle of Human Creativity*, edited by Mark Turner, 115–132. New York: Oxford University Press.

PART I

THE MOVING AND PERFORMING BODY

CHAPTER 1

MUSICALITIES AND THE MOVING BODY IN WESTERN CONCERT DANCE

BYRON SUBER

Musicality and Dance

In the field of Western concert dance, what does it mean to say a dancer or a choreographer is musical? In the context of dance, being musical is part of the lexicon of dance analysis. The term "musicality" is not being used metaphorically but rather as a description of one of the many common facilities attributed to a dancer or choreographer. For most viewers, musicality is understood as an innate ability. Other examples of characteristics attributed to "inborn" proclivities are "good lines" (meaning the physical geometric lines of the body), or spatial intelligence, or a good memory for choreography. Innate gifts or talents are arguably learned capacities (Mauss 1973). But in dance, if learned at all, it is assumed to be an ability acquired outside of formal dance training. A dancer will usually only profess they are musical when they have been told as much by someone else. It is rare that a dancer or choreographer is self-aware of their own musicality. (An obnoxious exception to this rule is discussed toward the end of this chapter.) In the context of the perception or reception of dance, a viewer stating that a dancer is musical can be described as evidence of an aesthetic or learned experience. The viewer is most often understanding the musicality of the dancer or choreographer as something the artist has always possessed. Musicality comes across as an element the artist is embedding in the event of an embodied performance.

Rather than embark on a polemic about nature or nurture, this chapter focuses on the premise that the reading, in the form of a perceptual moment, is an assumption of musicality as innate. In other words, I am deciphering that moment when a viewer of a dancing body perceives a confluence of physical movement with the sound of some music that is playing, and the ways in which that viewer understands the musicality to be something intrinsic to that individual dancer. (Confluence of movement and music can also take the form of dissonance (Buck-Morss 1977). (This topic is discussed to some extent later in the chapter.)

To ask more precise questions, what exactly is it about the physical movement that appears to be musical? Or rather, how exactly is the movement musical? Or, to propose a slightly more nuanced distinction, what specific moment in the physicality is perceived as confluent with what part of the music? Or, from the perspective of the music, what moment in the sound or music is being highlighted with the expression of physicality? And lastly, what musicality is being brought forward in the movement, a musicality that is not heard in the music but seen in the movement?

There are multiple ways of being musical or reading musicality when moving, or creating movement. This can be discerned or made evident in a casual social dance movement or in movement made specifically for presentation on a stage. In this sense, discerning movement and musicality is a form of communication and therefore can be understood as a form of learning. But what information is being transmitted and what or how is knowledge being formed in the act of audience reception? Thinking in terms of transmission and reception, it is helpful to think of any single moment within a viewer's comprehension of a confluence of movement and musicality as transmitting multiple messages on multiple wavelengths to multiple receivers (or audience members) with multiple possible interpretations.

To build even further on multiplicity by adding yet another level of production and perception, in a step removed from the group dynamics of social dance, or the audience-based reception of concert dance, striving for a particular musicality can be integral in the improvement of dance technique in the studio. In all these contexts, but most literally and intentionally in the studio, knowledge is transmitted. Knowledge formation by visual and audio observation, and the audience's simultaneous absorption of the information contained in physical movement and musicality differs from knowledge gained from linguistic forms of communication. Dance and musicality as a site of knowledge formation is therefore extralinguistic. Within the many nonlinguistic modes of communication in dance and musicality, there are various routes and mechanisms that while related, are distinctly different in schema and outcome.

Making and Understanding Dance

Sometimes movement is planned ahead of time, sometimes it is created spontaneously on the bodies of dancers in rehearsal, and sometimes it is improvised in rehearsal or performance. But dance as an embodied practice, the work of putting dance onto physical bodies, begins in the studio, through experimentation and repetition in a classroom (the process of training dancers in any particular style or dance technique) or the teaching and acquisition of choreography in the rehearsal room. These two sites of learning are different but related spaces of knowledge formation. And in a more extended relational context, these pedagogic methods are related but quite separate from the processes of knowledge formation that can be achieved by both performers and audiences in the context of a live concert on a stage. For dancers, in any context, whether or not there is a goal of vocational expertise, knowledge is acquired, albeit through different paths. And knowledge formation that occurs for the performers in the concert form is related to, but different from, processes of knowledge acquisition for the audience.

To complicate these questions further, there are also countless ways to define the concept of music, especially as it relates to dance. These differences can be seen in trajectories that veer far from traditional melody and instrumentation. For example, in what are sometimes controversial moves, the use of silence and how the body responds to silence can also be understood in terms of musicality in dance.

Lacking some parameters, this inquiry could take the form of an interminable labyrinth. The framework for this examination takes the form of a specific site of physical movement creation, and one that most often engages with music in specific and illuminating ways. The first several of the preceding statements and questions are informed by the context of a specific tradition of physical movement creation, the context of Western concert dance. This chapter focuses on musicality as it is attached to the aesthetic principles that have emerged from the context of this formal aesthetic field, a tradition and trajectory that will continue to be defined throughout the chapter.

Western Concert Dance: A Brief History

To put it simply, Western concert dance is physical movement created most often for visual consumption in the form of a staged event. The tradition has its roots in aristocratic social dance in the courts of the late Italian Renaissance (Desmond 2003; Jowitt 1988; Kirstein 1984; Copeland and Cohen 1983). The form then became a mixture of social dance interfacing with presentational dance. In this form, concert dance was first codified in seventeenth-century France under the reign of Louis XIV. Over the following several centuries, the art form, now known as classical ballet, continued to develop as a uniquely individual embodied practice and to diverge into parallel and contrasting styles and genres. Ironically, classical ballet was formulated in the neoclassical (and baroque) period for music (along with other aesthetic forms such as neoclassical art and architecture). As the genre of classical ballet continued to develop into the eighteenth and through the twentieth century, the art form crossed national boundaries that included most prominently, Russia (and eventually the Soviet Union), England, Denmark, Germany, and the United States, each nation developing and claiming its own individual style. In the twentieth century, while classical ballet continued to develop globally, including national forms in China, Korea, and Japan, the neoclassical period in ballet emerged. Again, ironically, neoclassicism in ballet emerged coterminous with twentieth-century modernist aesthetic movements in art and architecture. Also, modern dance developed as a mostly oppositional form of concert dance to that of the descendants of court ballet in the popular theater of the late nineteenth century. Some of the most notable diversions in these new dance forms were German Expressionist dance (Bergsohn and Partsch-Bergsohn 2003), American Modern dance (Manning 2004; Foulkes 2002; Burt 1998; Adshead-Lansdale 1994), and American Jazz dance (Valis Hill 2010; Sommer 1998; Malone 1996; Stearns and Stearns 1994; Giordano 1978).

At the same time, other related genres began to emerge in the realm of social dance. And as a theatrical form, in the last quarter of the twentieth century, hybrid forms such as American and European postmodern dance, German *Tanztheater* (Climenhaga 2013), and Japanese *Butoh* (Fraleigh 1999) stepped onto the global stage. Yet among this great diversity and contradiction to the world of classical ballet, certain tenets, particularly related to how the work is presented, viewed, and understood by audiences, have been maintained through many of these oppositional formations. In many ways, what is also preserved, or some may say, perpetuated, through privileged Cartesian perspectivalism, are connections to the original hierarchical structures that developed out of Italian and French aristocratic social dance. While many forms contest the structures of classical ballet, in terms of defining tenets of vision as Cartesian, components of early Western concert dance remain vestigially as a major defining principle of any style of presentational dance. The history of vision and visuality has been explored extensively in art, photography, and cinema (Foster 1998), and similar paths of inquiry are being applied to the very visual/auditory forms of Western concert dance (Lepecki 2004). Even in moments of revolutionary confrontation with those vestigial components, in a way, the original hierarchy is supported and maintained, upheld through what are sometimes revolutionary gestures of antagonism.

Dance or Not Dance

Before returning to the main topic of this chapter (the ways in which the moving body can be defined as musical) it will be helpful to define the term "dance" as it is differentiated from the broad concept of the "moving body." While some practitioners would consider any body movement a form of dance (including the movement of a camera without any visible body in sight of the camera's frame), it will be most helpful in the context of this chapter to limit our definition to dance that in some way relates to sound or music (or silence as a form or at least a component of sound or music) in an event-format where the movement is presented to an audience. From a twenty-first-century perspective, in reviewing the history of dance as an art form, it is difficult now to differentiate dance from what many understand as the common-sense opposite of dance, pedestrian movement. Postmodern dance (Banes 1987), beginning in the 1960s, in a strong move, included, and at times even focused on, pedestrian movement as a form of concert dance. For instance, Lucinda Childs work, "Carnation" (1964), a solo she created in 1964 for the groundbreaking experimentalist Judson Dance Theater in New York City, consists mainly of her manipulating everyday objects such as pink hair rollers, green sponges, and a vegetable steamer. Performed in silence, this work in some ways broke through the hierarchy of virtuosity in dance, in certain respects, democratizing the field. This work may or may not have achieved that goal for every audience member, and for some that accepted these movements as dance, and for those that did not, they may see it differently more than fifty years after the premiere. In any case, it is necessary to use a broad (yet still specifically limited) definition for the concept of dance, a concept like that of visual art, a definition that hinges on the context of where and under what circumstances the action occurs. For the purposes of this chapter, we can begin by narrowing the field by determining what is not concert dance. I begin by looking at the contrasts and similarities between what can be understood as social dance as opposed to what can be understood as concert dance (both in performance and in training or rehearsal), remembering that concert dance is defined as a specific type of presentational performance.

Social Dance and "Bad Musicalities"

Consider first an example familiar to almost anyone: the occurrence of a body moving to music, an individual dancer in a social setting such as a social or school dance or a nightclub. A social or club dancer can be understood as being musical as they relate to the clear punctuating beat of pop music (see Bakan's chapter for more on this subject). But this relationship is rarely a one-on-one match of a single, clear, and emphatic movement occurring at precisely the same moment in time as a single, clear, and emphatic musical downbeat. Usually when one notices that a dancer in a club is musical, they are referring to a perception of that dancer as moving to the music in a way that is new and unique to the viewer and specific to that particular dancer. While you may hear of a group of people being experts at synchronization, you rarely hear of a group of people being described as musical. Although the dancer in question may be executing movements that acknowledge the downbeat, the musicality resides in the ways that the dancer moves into the downbeat and away from it, and then toward the next downbeat—the quality of the dancer's musicality resides in what they do in the spaces in-between the downbeats, not necessarily on the downbeat. Those physical actions can be endlessly variable with one dancer across a stretch of music, and between any number of dancers, on any given occasion. So, it is the dancer's individuality expressed in their own specific way of relating to the music that is more often being noticed as musicality and less that the dancer has a clear ability to keep time with the downbeat of each measure. But it is important to note that not being able to acknowledge the downbeat is clearly visible to an onlooker. This will be addressed more extensively.

So what is "good" musicality? It is difficult to determine the precise nature of good musicality precisely because of the nebulous and endless variations of pushing the moving body into, against, or even away from the certainty of the downbeat in unexpected ways, a manner of individually punctuated, measured sonic tones. So perhaps we can more clearly move toward a definition of good musicality by defining the opposite: bad musicality. So then, what is a nonmusical dancer? To return to the example cited earlier, what is a nonmusical dancer in the social setting of a social or school dance or a nightclub?

Bring to mind the ubiquitous awareness of the cliché of the "bad" dancer at a wedding. Most have had the experience of watching someone who appears "to have no rhythm whatsoever." Familiar to many would be Julia Louis-Dreyfus's character Elaine Benis in the *Seinfeld* episode where, to the intentionally comic horror of everyone at a wedding, Elaine "kicked" off the dancing with what most can agree to be terrible nonmusical dancing.[1] Yet it is important to realize that Louis-Dreyfus, as an actress, developed this style of "bad" dancing consciously. The humor in this television episode resonates with the summoning of the archetype of a bad dancer, a symbol for all the bad dancers at all the weddings we have all been to at any point in time. In the *Seinfeld* scene, Elaine not only missed the musical downbeat but also had some awkwardly zealous movements that seemed to make everyone uncomfortable. That discomfort uncannily echoes in Western culture whenever social dancing occurs in any social situation. Simply stated, the "bad" dancer can be understood as being nonmusical, at least. But when using that parameter as a meter for comparison is everyone else then a "good" musical dancer? As stated above, being able to move to a downbeat is rarely a qualifier for a person to be considered gifted musically when moving to music. Consider our ever-familiar "bad" dancer: music is playing, and the playing of that music is why this person is moving. So, in some ways there is a confluence of physical movement to music, if only in the fact that the two events are intentionally occurring simultaneously in time and space. Is it just that this person is hearing the music differently than everyone else? If so, how then can we talk about this individual's musicality? What or how is that person hearing, feeling, or engaging with the experience of body movement as it coincides with music if the "bad" dancer is not moving to, with, or against the music? Perhaps, rather than being nonmusical, this "bad" dancer is simply relating to the music in an alternative way than what is comfortable for others, albeit in a manner that many may find unpleasant to watch.

This discomfort can be used to an aesthetic advantage. Many years ago, I was hired to choreograph a striptease for the actress Loretta Gross in the director Maggie Greenwald's 1989 film adaptation of a Jim Thompson novel, *The Kill-Off* (Greenwald and Thompson 1989). Loretta was a trained dancer, and I worked many hours with her to deconstruct her technique and sense of musicality to make her come across as an amateur dancer in the film. One of my main concerns was not only to get her to dance off the downbeat but also to have her approach the dance with a specific musicality that rendered her dancing unprofessional in appearance. Years later, I realized I had been heavily influenced in this effort by a subconscious channeling of Marilyn Monroe's portrayal of the character Chérie in Josh Logan's 1956 film, *Bus Stop* (Axelrod 1956). Both examples, if my unintentional homage was successful, created vulnerability in the character by having the dancing that each character performed come off as "bad" dancing. It was not an easy task to convince Loretta to commit herself as a failed dancer, to be forever captured on celluloid. For both her dancing and my choreography, in the end, our achievements could be measured by our success at appearing as failures.

Concert Dance

Now, returning to a differently contemplated, or consciously developed, form of physical movement in confluence with music, I consider choreography or dance that is created for presentation on the stage. Like a film played in a cinema, one can posit that a dance concert occurs in a predetermined place, usually a theater intended for presentation. It begins at a stated time with a group of people gathering together to witness the performance as an audience. Usually there is a separation between the audience space and the performing space, most often with a single plane separating the two spaces in an actual or suggested proscenium plane. (Many smaller concerts have audiences on three or even four sides of the performing space, but this concept of "front" or presentation to the space in which the audience inhabits is deeply embedded in the event of any dance concert as it is in Western visuality.) There is a definitive ending to the presentation at which point (more often for live performance than for film) the audience communicates its appreciation (or lack thereof) of the concert with applause. Finally, the audience departs, marking the very end of the experience. But for live concert dance, unlike dance framed for inclusion in a film, this experience is an ephemeral one, an important distinction. The experience has longevity in the minds of the audience, in what the audience learned during the viewing experience and what they retain after it. Part of what can help with retention is how the physical movement relates to music, or in some cases how the movement intentionally disregards music.

In the many genres within the field of Western concert dance, in a dancer's performance, or a choreographer's creation of movement to music, musicality is more formally complex in how it connects to a history of visuality, how it is thought about, and the ways in which it is discussed. This is not to say that a club dancer's musicality is not complex; it is just that in Western concert dance, musicality, or absence of musicality, is an area of critical concern and discernment in ways that have received more consideration than in analyses of club dancing (although that is changing to some degree with the advent of deep analyses of pop culture; see Banes 1994). With a long history of formal analysis in concert dance, musicality is, or can be, addressed in this field from an abstracted, analytic point of view specific to the aesthetics required to create products in this field.

Although it was previously suggested that merely keeping a beat is not essentially a sign of musicality, in Western concert dance as a vocation, it is rare that a dancer is unable to keep time with the downbeat of the music. In some ways contrary to this earlier statement, not being able to simply keep time with the downbeat is a characteristic that is most definitely understood as nonmusical. Why this is a nonissue for concert dance is that a dancer who cannot keep beat with the music is one who usually will not make it onto the stage of advanced or professional work. In part, it is because there is often use of unison movement, or unified movement, where dancers take musical or physical cues from other dancers. An individual's musical inability in this respect would be very visible. This inability can even result in the destruction of a dance work by offsetting the synchronicity of an entire piece.[2] So to some degree, it is safe to conclude that all advanced or professional performers in Western concert dance are musical in the simplest way one can define musicality; they can all keep a beat. Because of this, keeping a beat is rendered invalid in understanding a metric from which to judge the musicality of concert dance. Not to be too obviously general in summarizing, but all advanced or professional dancers can keep time with the music. So being called a musical dancer or choreographer in Western concert dance is a more complicated index of movement to musicality or music to movement.

Performance as Practice as Research

The following examples of Western concert dance do not constitute an exhaustive or even linear chronological historiography of dance and musicality. Rather, they represent a specific trajectory of knowledge formation that I have experienced myself, which therefore is subjectively experiential and not an objective discursive exposition. The works I cite are either works that I have seen several times or works I have actually performed on numerous occasions. Some of the works cited are pieces I have created myself, to perform myself, or to have other dancers perform for an audience. I do not cite any linguistic texts, although it can be argued that many of the works I refer to emerge from properly codified dance forms with components of movement that have been named and are put together in a way that can be compared to syntax. The word "choreograph" comes from the ancient Greek (when dance was never separate from music), meaning to dance/write, or to write with the body—ephemerally—in the air. But I stop here with linguistic comparisons and state that my response to these works is formed from knowledge gained from watching or performing these works, not writing or reading texts about them. This is an important distinction that relates precisely to my own way of hearing and moving to, or creating movement for or with music. Working under or beyond text raises a few questions that are not easily answered. What does dance, and/or music, or evident musicality in movement, do that text cannot do? What does musicality and movement affect that text cannot affect? What does a musically movement-based experience provide that text will never be able to provide? What is made clearer by watching a dance (that is considered to be musical) that can never be clarified by reading text? And what guidelines or level of analytical rigor can we access when looking at and interpreting movement while hearing or listening to music? The point here is to validate receptivity of music and dance as a different path of knowledge formation that is not just on a par with textual analysis, but at times, necessarily replacing it. That said, I continue here by writing with words about music and dance.

Musicalities of Ballet and Modern Dance

For this chapter, when I first began considering complex musicality in Western concert dance, my thoughts immediately turned to Suzanne Farrell, former prima ballerina in the New York City Ballet (NYCB; Farrell and Bentley 1990; Kirkland and Lawrence 1986; Ashley 1984; Bentley 1982). Farrell was the most renowned muse of the Russian American NYCB founder, director, and choreographer George Balanchine. Farrell had a unique way of addressing music. Particular to Farrell, her musicality has not been replicated by any other dancer performing in those same ballets. In Balanchine's work, Farrell would push movements to the very end of the musical measure forcing a quick physical transition, sometimes so quick as to render any transition imperceptible to an audience. Her sense of movement was affected by this risky relationship with almost missing the downbeat and her body moved not in response to the exigencies of staying in time, but rather by pushing far beyond it. The specificity of her musicality was but one innovative aspect of her individuality as a dancer. She also had a tendency to work physically off-balance in a way that no other dancer has ever quite been able to replicate. She made the choice to reorient the lines of her body in ways that are still anathema on most dancers' bodies but for Farrell's body, it was a part of her individually intrinsic style of expression. Along with her unique musicality and several other aspects of her performing body, Farrell stands out as a genius in the dance world. Her unique relationship to movement and musicality created a Suzanne Farrell mystique that remains unapproachable in the field of Western concert dance.

Her mentor, George Balanchine, a neoclassical ballet choreographer was also known for his unique musicality, but as a creator, not a performer of dances (Gottlieb 2010). One of his choreographic works, *Concerto Barocco* (Balanchine 1948), used excerpts of music by Johann Sebastian Bach. Some of the selections were the same excerpts that his contemporary, Paul Taylor, a modern dance choreographer, used in his contemporaneous work, *Esplanade*. As a member of the Hartford Ballet, I performed both *Esplanade* and *Concerto Barocco* in the same performance on several occasions. Audience members rarely detected that the same music was being paired with different physical movements of the dancers in each work. It was because each choreographer heard and addressed the music so differently from the other. For instance, while both Balanchine and Taylor utilized the body's negotiation with gravity, they each addressed that negotiation in vastly different ways. In both cases, each choreographer's engagement with gravitational pull brought out very different aspects of the same pieces of music. In another example, *plié*, or the bending of the knee, was used by Blanchine as a way to sometimes emphasize the downbeat while remaining in full supported suspension away from gravity. At other times, to the same music, Balanchine uses the *plié* to move toward a full *relevé* at the accent of the downbeat, emphasized by a suspended balance *en pointe*, the highest physical position made by a ballerina without leaving the ground. In a very different way, Taylor used the *plié* in sustained runs that glide the body across the floor with a strong and smooth connection into the ground. As a third example, each choreographer used counterpoint and canon in ways that reflect the counterpoint and canon in musical moments. While Balanchine's movements often revealed micro-canons in the music that would not be as apparent when played alone, Taylor focused on the more apparent musical canons that might be easily heard when listening to the music without physical accompaniment.

Taylor's work can be seen as a contrast to Balanchine's work, while Balanchine's work can be seen as an extension and development of his Russian imperial predecessor, Marius Petipa. Petipa's choreography, most often to music by Peter Ilyich Tchaikovsky in such familiar ballets as *Sleeping Beauty*, *Swan Lake*, and the *Nutcracker*, brought out musicality in less complex but still surprising ways. Lilts in the melody would be repeated in the movement but in physical enchainement that would combine surprising strings of steps (Vaganova 1969).

In all of these examples, the choreographers' works are musical by giving attention to, and affirming, apparent melodies and different relationships to marked time, but the question can be asked as to whom and how those melodies are apparent. These questions are more easily answered when the music and dance confluence is actually seen and interpreted by a viewer. It is impossible to describe adequately in text. One must see (not read) to learn. But it remains that, even if seen, there is not only one definitive answer. Differences of choreographic and viewer interpretations are limitless.

Musicality from the Inside

I have been told at numerous points in my career, first as a dancer, then as a choreographer, and again as a teacher, that I possess a very different musicality than anyone else. I have never had an intention of being different in how I hear and then physically express music. Admittedly I do work toward innovation in movement creation, but never does it occur to me to try to hear music in an innovative way. I have noticed that when I try to read the sheet music while listening the music, the geometric patterns on the page suggest something entirely different to me than when listening to a musical selection alone for any number of times. When reading sheet music, the process to me becomes so mathematical and complex and so predetermined that my choreographic process is to ask for physical acumen that is unattainable. I work best when I listen to a musical phrase, then respond to what I have just heard. I also create music entirely apart from any specific music and add different pieces of music after the choreography is completed, getting different physical responses from my dancers for each different musical configuration. I have explored this difference in a work that spanned across three nights, with a slightly different performance each night. The work was titled *The Breaking Series*. Using the music of Arvo Pärt, Olivier Messiaen, and Béla Bartók, for three performances, I changed which piece of music accompanied which set choreography for each night's performance. In terms of musicality, the dancers reacted differently each night while performing the same choreography but responding to vastly different musical accompaniment.

My own response to music usually has to do with how I hear a melody or what I hear in a melody. Sometimes it is a specific dialectical pairing of certain instruments together that produces an aural reaction in my musical sensibilities. When I try to dismantle music by looking at the way the music is written on paper, by delineating specific musical notes, or when I try to determine what exact notes are being played by any particular instrument, my ability to hear the music as a whole disappears to the point of choreographic immobilization. Additionally, my choreographic response to music can change if another musician plays the same piece of written music. I have often choreographed specific movements to a specific recording but when the same piece of music is played live, or if another recording is played, the original movement loses much of its power of confluence. This is not to say I do not enjoy and appreciate live musicians, but that is another way of choreographing and presenting music and offers a different type of knowledge formation for both the performers and the audience.

Other Musicalities: Other Politics

There are many other ways to create musical choreography. One can also attach to and reiterate clear rhythmic patterns that support, but are apart from any apparent melody. Sometimes this can come across as a fugue of movement as an additional group of layers to the layers of music. Sometimes the most apparent association is highlighting very contrasting melodies (some in the music and some in the movement). All of these methods work to democratize musical and physical spatialities. As I slip into a discussion of political metaphors, dance and musicality as a site for knowledge formation becomes less abstract and more symbolic. Movement and music each separately act as a conduit for knowledge formation, and when put together as a more complex creative form, they have an even more complex agenda to decipher. Resorting to simpler methods of expression does not necessarily result in a simpler message.

Even though I have opined that focusing on the downbeat as a method of choreography is not necessarily musical, in some instances, it is mostly, or exclusively, how a choreographer steers attention, creating a spatiality that can be read in absolutist or uncompromising ways. Comparatively, the more complexly physical movement engages with or against music, the more democratic the space becomes. A tendency toward downbeat punctuation, or in opposition to downbeat punctuation, both ends of this polarity can be understood as being deeply political or ideological. For instance, in some genres, such as competition dance, a genre of which I am readily critical, at use is an extremely codified dance vocabulary that is closer to sport than art, competition being a defining term for the genre. In competition dance, downbeat punctuation and codified or habitual tendencies of moving between the downbeats is a defining element. This codification can be quite powerful but can also lead to movement that is interpreted in ways that can be described as militaristic. Comparisons can be drawn between the affect this kind of dance promotes for an audience and a similar affect, or populist pleasure, that can be derived from nationalistic spectacles created by authoritarian or totalitarian regimes. If my assessments here appear as overindulgent criticisms, it is important to consider other contexts in which these competition dances are composed and presented. Consider what elementary, but incredibly complex, social identities are being portrayed in these formations. Also consider to what groups these performances are catering. Issues of gender, race, class, and sexual identity are explicit in these representations and must be included in a critical analysis. These connotations can be problematic for a choreographer whose goal is explicitly oppositional to the dominant sentiment of competition dance, a sentiment that I propose borders on a fascist aesthetic. Problems can arise in choreographers that only hear, or choose to only listen to the downbeat. It is an ideological problem in the ways that listeners polarize movement into moments of stillness and repetitive cadences, creating absolutist on-and-off statements about cultural movement and spatialities. And more simply, choreographing or moving only to the downbeat quite literally confirms preconceived ideas. One such idea is the concept that an audience already possesses all the knowledge it needs in terms of understanding movement and musical interpretation. This motion flattens the potential of what bodies moving to music can teach us as choreographers, performers, and audience members, both as local and as global citizens.

Returning to a more comfortable or less ideologically challenging approach, the abstract application of knowledge formation, a moving body can unlock musical gestures that would be unapparent before pairing the music with a particular movement sequence. In contrast, it is possible to be musical by going completely against the music or by pairing the music to the movement arbitrarily, as I did in my *Breaking Series* mentioned above. This was the method of many of the collaborations of choreographer Merce Cunningham when working with his life partner and artistic collaborator, the composer John Cage (Brown 2007). Sometimes the dancers in these works would never hear the music they would be performing to until the actual performance. Often, that music would be entirely different from subsequent performances of the same choreography. This was just one aspect of a game of chance used by Cunningham when creating dance. *Chance Dance*, being the nomenclature chosen by Cunningham himself, was not as arbitrary as it seemed. Cunningham and Cage shared so much of their personal and professional lives, they were bound to create moments of confluence in their aesthetic sensibilities. These confluences would merge and emerge, even without conscious planning. The movements and music would appear to be spontaneous when in fact there was a shared aesthetic methodology at work.

More dissonant a connection of dance to music is movement created to music emerging from the tradition of atonal musicality, which is one way of challenging an audience's habitus of how they hear music and then how they hear music and see dance to that music. Although apocryphal, the premiere of Vaslav Nijinksy's choreography to Igor Stravinsky's *Rite of Spring* was said to have caused riots when premiered for a 1913 Parisian audience (Eksteins 1989; Garafola 1989). Combining uncomfortable, and even revolutionary music and movement was just too much for the audience to bear.

Moving from dissonance to absence, a choreographer can reject music entirely in the creation of movement (or even stillness) to silence. John Cage created a musical piece titled *4′33″*—four minutes and thirty-three seconds of silence to be "not" played by any number of musicians sitting before various types of instruments. The choreographer Paul Taylor choreographed a staged version of this work in 1952. His work titled, *Duets*, consisted of a male and female dancer standing still on stage to silence for four minutes. In a humorous antitextual response to the work, Louis Horst published a review of the piece for the *Dance Observer* that consisted of four inches of blank space sandwiched by the time and place of the performance at the top of the page and Horst's signature below the blank space on the page.

Silence as Music

I have mentioned silence several times as a form of musicality with dance but how exactly is silence musical? Apart from the sounds that are made by instruments or voices in a traditional melodic piece of music, the duration of the music, and on a micro-level, the duration of silence between notes, combine to make the musical patterning. A pattern is not sustaining or repeating an unbroken note. It is when the monotone or repetition is broken in some way that a pattern is created. A note played continuously for four minutes and thirty-three seconds without a break or a rhythmic ground level, is more challenging for an audience to witness than silence for the same amount of time. In Cage's *4′33″*, the sounds of the room, audience members shifting in their seats, mechanical sounds from an air conditioning system or elevator moving in the building, sounds that would usually go unnoticed, and other seemingly arbitrary noises create a pattern and a type of melody. This experience of listening to silence teaches an audience to question what they understand to have been music up until that point and what they will understand as music from that point and after.

Off-Setting Movement and Musicality

For a more recent postmodern example, William Forsythe is an American choreographer who has worked mostly in Germany, only recently returning to the United States to call it his home again. His work in the last thirty years has moved classical ballet as far forward as, if not further than, anyone has moved it in its four-hundred-year history. His innovation was by way of deconstruction (borrowed from the architectural arena) rather than construction, a possible way to define Balanchine's work that allows construction to be classified as a neoclassical move. Forsythe's movement creation in relationship to music has been a large part of that innovation. Working with innovative composers such as Leslie Stuck, Thom Willems, and Gavin Byers, Forsythe helped open up what it meant to dance to music and new ways in which music for dance could be defined. My own experience of viewing Forsythe's work was that it was uncannily familiar but I could not quite see it as legible until I viewed his first work a second time. I see Forsythe's work as being offset just enough to allow you to recognize what is familiar but be totally inundated with something entirely new and unfamiliar. We need the familiar to render legible the unfamiliar. In his 1984 work *Artifact*, Forsythe worked from the very strict classical improvisations on a Bach Partita as played by the rehearsal pianist Eva Crossman-Hecht. Both the music and movements are extensions of a classical canon, stretched beyond simplistic recognition or comprehension but familiar enough to resonate with the past. For this work, Forsythe also created a text that is spoken. Paraphrasing: "Every time I step outside, I forget what I always knew. Every time I step inside, I remember what I never knew" (Forsythe 1984). Although claims to abstraction persist, for me, this text is explaining how the mind learns to accept the new without totally discarding the past and, in fact, needs recollection of the past progressions in recognizing the new. Not only does the inclusion of spoken text in this piece begin to dissolve my argument about the exclusivity of dance as a nonlinguistic art form but also, in *Artifact*, spoken text elevates the role of music beyond traditional melody and instrumentation. A discursive turn dialectically embedded in Forsythe's work replicates the extension from the familiar to the unfamiliar and teaches the audience textually, musically, and choreographically how to move forward.

Quintessential Postmodernism

At the same time as Forsythe was working in Germany, Twyla Tharp, in New York City, was defining a quintessential postmodernist approach to movement and musicality herself. Early in 1976, Tharp premiered a work titled *Push Comes to Shove*, where she combined classical ballet with jazz and modern dance performed to music that alternated between the classical music of Franz Joseph Haydn and ragtime composer Joseph Lamb. *Push Comes to Shove* is not simply a montage or collage of different music and choreographic styles, a modernist move, but rather a postmodernist contextual juxtaposition. A simple way to explain the difference between modern dance and postmodern dance is that modernism might pull differing styles of music or movement together to form a whole. The immediate adjacency of contrasting elements produces a third meaning, or shifts the meaning of each of the two elements. Using a simple visual comparison, consider the way two colors may appear to change tone when placed side by side. Stravinsky and Balanchine were fine examples of mixing jazz music/dance with classical music/dance, the jazz being "elevated" and the classical being "innovated." But this side-by-side concurrence, or even conflation of two elements (of the classical with jazz in this case), creates a product that becomes contiguous in nature. It is not a simple gesture to conflate such contrasting elements emerging from such different socially and politically charged realms, particularly when comparted to one another. Through confluence and conflation, the true (and problematic) origins of the classical and jazz, along with the social, political, and concomitant racial connotations of these juxtapositions, can be lost. An erasure of the originals through this proximal or blended placement silences the tension necessary to understand these problematic origins. Postmodernism, on the other hand, moves beyond object placement and uses the enveloped contexts from which the many elements emerge. Bringing in various elements and positioning them together so that their contexts remain included allows for a number of audience responses to occur, or various sets of knowledge to be gained. Tharp, a true postmodernist, brought full pieces of music and movement together, maintaining the original integrity of the different styles or genres. In turn, with the final product maintaining references to the original contexts of the styles and genres, an audience is allowed to include in their aesthetic experience important and necessary contextual and historical information. Allowing the viewer/listener to include contexts surrounding the various musical and dance styles or genres prevents information from blending into oblivion and encourages recognition of ideological contradictions on many levels of perception.

Restating the Obvious in an Old Modernism, or a Very Bad Boy

Later in the twentieth century, it would follow that the most obvious turn for new choreographers would be advancement within the combining of movement to the musical, an expansion of what had come before. Unfortunately, this is not the case with the populist choreographer Mark Morris. In his first appearances on the dance scene in New York City, he was deemed the "bad boy" of dance, not because of anything shocking he did in his choreography, or in the ways in which he combined movement with music, but because of his public behavior unrelated to his position as a creative artist, all too tedious to detail here.[3] But to speak to his expression of musicality, Morris was incredibly adept at creating the most literal and overtly obvious musical visualizations in his pairing of movement to music. Literal movement had for decades been anathema in modern dance (as it was in abstract art) and it would seem that a counterreaction to overtly abstract dance would have been refreshing. And in the work of many true postmodernists, it was. But Morris missed the mark. His music to movement correlations most often bred simplicity and regression. In contrast to these failures, he tirelessly pronounced his own gift toward a progressive musicality, a claim that is easily put down as a parallel to hitting every note of the downbeat, a populist, even fascistic move. This is not to say that he was not commercially successful but, in fact, explains why he was. It is difficult to determine any way that he progressed the field in a forward direction. He gave his audiences what they most expected, what they already knew, and this is not knowledge formation, only knowledge confirmation.

Making One's Own Special Music

Tere O'Conner, working today on a global scale, in his early years was a choreographer who rejected musicality created outside of the dance. It was O'Conner himself who created the music in response to his own choreography and it was his soundtracks that became the music for his performed dances. His understanding of the sounds, silences, and rhythms that he felt should accompany his choreography was informed by his own intimate knowledge of where the rhythms of the movement originated, in the ways he had asked those moving bodies to move (O'Conner 1996, 1998a, 1998b). He included alongside recognizable samples of traditional sounding Western music, sounds that could be called noise, and text, both prerecorded and spoken live. O'Conner is a true postmodernist as well, using context and cracking expectations at every turn, producing a knowledge-forming form of dance and dance/theater.

Don't . . . Stop . . . the Music

Before concluding, I briefly attend to a disturbing moment in dance and musicality, that of the butchering of music when used to accompany sports such as gymnastics, ice-skating, and synchronized swimming. In these instances, mash-ups of some familiar and some not-so-familiar pieces of music are edited together to fit the stringent requirements of sports choreography. This is not always unpleasant, and mixing music as a technique can often have very illuminating results, but it is rarely the case in these activities. The regulatory needs of the sport disregard the integrity of the original music in ways that can be quite unsettling to purists and nonpurists alike.

As stated earlier, this chapter is not meant to be a comprehensive linear historiography of Western concert dance but is rather coded by my own personal tastes and preferences. I would now like to write from my perspective as a teacher of dance technique and the ways in which musicality can be used to progress embodied performance.

Teaching Musically and Teaching Musicality

First is a question concerning the predisposed tendencies or gifted talent in artistic fields, most particularly, dance. Some bodies seem to possess inborn or innate qualities that enable virtuosity in the performance of technical feats. I am not here to question those abilities but rather to ask whether someone who is not musical can learn how to be musical. Or if someone clearly exhibits the ability of musicality, can they learn to be more musical? I have already stated there are many ways to be musical, so it would follow, if someone is musical in one way, can someone learn to be musical in multiple ways? When is musicality perceived as learned and when is it understood as uniquely innate? Musicality is not always a dancer's or choreographer's response to music. Nor is it always a particular way an audience receives the music as it intertwines with the dance. Musicality in dance also occurs in the sequestered studio or lab environment of the dance studio. In this context, an instructor can use music as a tool in the training of the dancer's body. Providing musically based exercises can challenge the dancer to achieve a particular music/body goal that opens up new ways of moving or can access certain technical abilities that were previously hidden.

Looking back to the Romantic period in French classical ballet, the music, which at times could be called light and buoyant, never came near producing the lightness of music and the resulting lifting dance movements of Denmark's nineteenth-century choreographer Augustus Bournonville. I bring Bournonville into the discussion because of his seemingly unlikely influence on the present-day training of dancers in Balanchine's school, The School of American Ballet.[4] I attended this school when I was young and my most influential teacher was Stanley Williams, an Englishman who was trained in the mid-twentieth-century version of the nineteenth-century style of Augustus Bournonville at the Royal Danish Ballet. As an adult in New York City, I also trained extensively with Wilhelm Burmann, who himself trained in Balanchine's school but adopted many of the musical strategies of the Bournonville technique as well. What I learned from these two great instructors is something I put to use every day in my own classes that I teach today. For instance, I often ask dancers to move to the music in ways that are very difficult to pin down in the usual way of relying on the emphasis on downbeat punctuation accents. I also create exercises that make it downright impossible to depend on the downbeat. The result I am looking for is that the dancers remain suspended in the movement through consistent and continuous control of the muscles rather than an overextension from movement that has an emphasis of punctuated momentum. As an example, I can look to the way I ask dancers to execute the ubiquitous movement of a *rond de jambe a terre*: the circle of the leg on the floor. There are four positions to this movement: (1) the heels touching in 1st position (with the legs rotated out as close to 180 degrees as possible), (2) then straight forward to the *tendue* front position, (3) reaching the toes in an arc along the floor to the side *tendue* position, (4) then in an arc along the floor to the *tendue* position back before returning the first position (this movement is also repeated in the opposite direction). The music for this exercise for most instructors is executed to a 3/4 meter. I ask the dancers to glide through each of these positions and to pass the positions in subsequent order for each measure. This means that the dancer must glide through five positions for every three counts, since they are moving around the circle and past the first position of that movement within the 3-count measure. This very difficult musical maneuver requires the dancer to be just ahead and just behind the beat over and over again on a continuous loop and if attempted, makes it quite difficult to habitually punctuate the downbeat in the same place over and over.

Earlier in the class, I always use music divided into 3/4 or 6/8 meter. Music counted this way encourages the physicality of the dancer to be continuous rather than polarizing to a 2/4 meter. I ask that the continuity be maintained when the subsequent exercises move into a 2/4 meter and we move back and forth between these subdivisions of the downbeat to test the dancers' bodies' habits of downbeat punctuation.

There are many other ways that I work with music to push and pull the embodied technique of my students forward by insisting the body operate beyond its conscious technical abilities. As a last example for this chapter, I ask dancers to avoid the tendency to accent the downbeat of their jumps with the landing from the jump. This tendency pulls dancers into the ground and freezes them into physical complacency, or even physical surrender. If the accent of the downbeat is at the highest point in the jump, the movement seems to be released from gravity. I say "seems to be" because while this attempt looks effortless in contrast to the heavy-handed landing into the ground on the downbeat, the dancer looks free from gravity, but in reality, the effort is immensely difficult because of the increased demands on the body in order to physically negotiate the physical weight away from gravitational pull.

In any case, working with musicality can push the body to move in ways that are beyond cerebral recognition of the embodied performance but in this effort toward somatic "thinking," the body is still learning and advancing in ways that cerebral attention does not allow. At the same time, an audience's perception of the ways in which gravity and musicality is being negotiated offers unconscious recognition of mood and intent of the choreographer and the performer.

Conclusion

To conclude, musicality in dance exists as an endless set of multiplicities that can be compared to the multiplicity of meaning that the art forms of music and dance produce individually as nonlinguistic forms of communication. Dance, like music, is its own separate art form as well as a distinctive form of knowledge that can produce forms of knowledge that no other art form can. And when put together dialectically, dance and music produce a third art form that, as a unity, produces an even more complex production of knowledge formation. In this essay, I hope to have begun an examination of the many possibilities within that range.

Notes

1. See *Elaine's Dance*. YouTube video, 0:46. Posted by "Roncagliolo," August 7, 2006, https://www.youtube.com/watch?v=5xi4O1yi6bo&noredirect=1; Spike Feresten, *Seinfeld*, "The Little Kicks," Season 8, Episode 4 (Los Angeles, NBC, October 10, 1996).
2. Synchronicity in dance as related to other paths of knowledge formation is explored extensively in a project at the Ohio State University. Scholars and scientists from several fields analyze in numerous ways a work of choreography by William Forsythe. The choreographic work is titled *One Flat Thing*, the project of multiple analyses is titled *One Flat Thing, Reproduced*. The project is detailed on the website *SYNCHRONOUSOBJECTS* (http://synchronousobjects.osu.edu).
3. Brendan Kiley, "I May Sound Cunty, but I'm Doing It Fondly," *The Stranger*, October 3, 2012, http://www.thestranger.com/seattle/i-may-sound-cunty-but-im-doing-it-fondly/Content?oid=14914165.
4. Jennifer Dunning, "Stanley Williams, 72, Teacher of Top Male Ballet Dancers," *New York Times*, October 24, 1997. See also Yule (1999).

References

Adshead-Lansdale, Janet, and June Layson. 1994. *Dance History: An Introduction*. New York: Routledge.

Ashley, Merrill. 1984. *Dancing for Balanchine*. New York: Dutton.

Balanchine, George. 1948. *Concerto Barocco*. New York; Rio de Janeiro, Teatro Municipal, October 11, 1948.

Banes, Sally. 1987. *Terpsichore in Sneakers: Post-Modern Dance*. Middletown, CT: Wesleyan University Press.

Banes, Sally. 1994. *Writing Dancing, in the Age of Postmodernism*. Hanover, NH: Wesleyan University Press.

Bentley, Toni. 1982. *Winter Season: A Dancer's Journal*. New York: Random House.

Bergsohn, Harold, and Isa Partsch-Bergsohn. 2003. *The Makers of Modern Dance in Germany: Rudolf Laban, Mary Wigman, Kurt Jooss*. Princeton, NJ: Princeton Book Company.

Brown, Carolyn. 2007. *Chance and Circumstance: Twenty Years with Cage and Cunningham*. New York: Alfred A. Knopf.

Buck-Morss, Susan. 1977. *The Origin of Negative Dialectics: Theodor W. Adorno, Walter Benjamin, and the Frankfurt Institute*. New York: The Free Press.

Burt, Ramsey. 1998. *Alien Bodies: Representations of Modernity, "Race," and Nation in Early Modern Dance*. New York: Routledge.

Childs, Lucinda. 1964. *Carnation*. New York: Judson Memorial Church.

Climenhaga, Royd, ed. 2013. *The Pina Bausch Sourcebook: The Making of Tanztheater*. New York: Routledge.

Copeland, Roger, and Marshall Cohen, eds. 1983. *What Is Dance?* Oxford: Oxford University Press.

Desmond, Jane C. 2003. *Dancing Desires, Choreographing Sexualities on and off Stage*. Madison: University of Wisconsin Press.
Eksteins, Modris. 1989. *Rites of Spring: The Great War and the Birth of the Modern Age*. New York: Houghton Mifflin.
Farrell, Suzanne, and Toni Bentley. 1990. *Holding on to the Air*. New York: Summit.
Forsythe, William. 1984. *Artifact*. Frankfurt: Frankfurt Ballet.
Foster, Hal, ed. 1998. *Vision and Visuality*. New York: New Press.
Foulkes, Julia L. 2002. *Modern Bodies: Dance and American Modernism from Martha Graham to Alvin Ailey*. Chapel Hill: University of North Carolina Press.
Fraleigh, Sondra Horton. 1999. *Dancing into Darkness: Butoh, Zen, and Japan*. Pittsburgh, PA: University of Pittsburgh Press.
Garafola, Lynn. 1989. *Diaghilev's Ballets Russes*. New York: Oxford University Press.
Giordano, Gus. 1978. *Anthology of American Jazz Dance*. Evanston, IL: Orion.
Gottlieb, Robert. 2010. *George Balanchine: The Ballet Maker (Eminent Lives)*. New York: Harper Collins.
Jowitt, Deborah. 1988. *Time and the Dancing Image*. New York: William Morrow.
Kirkland, Gelsey, and Greg Lawrence. 1986. *Dancing on My Grave*. Garden City, NY: Doubleday.
Kirstein, Lincoln. 1984. *Four Centuries of Ballet: Fifty Masterworks*. New York: Dover.
Lepecki, André. 2004. *Of the Presence of the Body, Essays on Dance and Performance Theory*. Middletown, CT: Wesleyan University Press.
Malone, Jacqui. 1996. *Steppin' on the Blues*. Chicago: University of Illinois Press.
Manning, Susan. 2004. *Modern Dance, Negro Dance: Race in Motion*. Minneapolis: University of Minnesota Press.
Mauss, Marcel. 1973. "Techniques of the Body." *Economy and Society* 2 (1): 70–88.
O'Conner, Tere. 1996. *Mother*. New York: Dance Theatre Workshop.
O'Conner, Tere. 1998a. *House*. New York: Dance Theatre Workshop.
O'Conner, Tere. 1998b. *The World Is a Missing Girl*. New York: Dance Theatre Workshop.
Sommer, Sally R. 1998. "Tap Dance." In *International Encyclopedia of Dance*, vol. 6, edited by Selma Jeanne Cohen. New York: Oxford University Press.
Stearns, Marshall, and Jean Stearns. 1994. *Jazz Dance: The Story of American Vernacular Dance*. New York: Da Capo.
Vaganova, Agrippina. 1969. *Basic Principles of Classical Ballet*. New York: Dover.
Valis Hill, Constance. 2010. *Tap Dancing America: A Cultural History*. New York: Oxford University Press.
Yule, Sean. 1999. *Suki Schorer on Balanchine Technique*. New York: Knopf.

Videography

Axelrod, George. 1956. *Bus Stop*. Los Angeles: 20th Century Fox.
Greenwald, Maggie, and Jim Thompson. 1989. *The Kill-Off*. New Jersey: Pilcher.

CHAPTER 2

MUSIC AND MOVEMENT

Expectations, Aesthetics, and Representation

JAY SCHULKIN

Introduction

Grappling with the unexpected while sustaining the familiar is a core feature that underlies human experience and inquiry in general, but which, in part, is closely tied to discussions of musical sensibilities. Leonard Meyer, the great twentieth-century music theorist, focused on the interplay of the familiar and the unfamiliar in musical experiences and musical meaning. Music is no different than problem-solving and statistical inference, as it is probabilistic and anticipatory; unconscious inferences, expectations, and the generation of habits underlie the perception and production of musical experiences.

Cognitive anticipatory expectations stand out when one considers the close link between music and dance. Music and dance evolved in the context of social contact and meaning. The human condition is linked to music and dance, and the range of human emotional expression is fundamental in this regard. Music allows us to reach out to others and expand our human experience toward and with others. This process began with song and was expanded through instruments and dance. Music and movement go together in every human society: "music to my feet," as it were. Indeed, this sense of aesthetics is a feature of the way we come prepared to interpret the world. The brain is prepared to detect movement and rhythm, both familiar and unfamiliar, with aesthetic judgment reflecting our cognitive flexibility.

This chapter begins by discussing the interplay of the familiar and unfamiliar, variation amid repetition, in musical expectations and experience. Next, the probabilistic and the anticipatory in musical aesthetic sensibility and emotional experience is presented in the context of a neural structure oriented toward adaptation and change. Art and the aesthetic judgment of both music and movement are then explored in the context of human representation. Music and movement are also linked to rhythmic generators and internal perceptions of time that underlie the cognitive formation of expectations. Next, the coevolution of music and dance amid human connection, communication, and culture is examined. Finally, music and dance are presented as expressions rich in lived and shared experience, manifestations of meaning and cognitive predilections.

Musical Expectations: Familiarity and Variation

A vast array of human pleasures underlies musical experience (Huron 2005; Merker 2005). Feelings of pleasure are, in many contexts, linked to familiarity. We enjoy the familiar; it is reassuring. With music, we listen to a song over and over; the familiar reassuring as we embrace something from the past, explore, and find something new. Our cognitive capabilities and generative creativity allow us to experience enjoyment in the music, to derive pleasure from the familiar and at the same time to take comfort in it as we explore interesting features; we can also simply enjoy the sensory state. Exposure tends to evoke enhanced preference for that to which one has been exposed, or with which one is already familiar (e.g., "oldies" music, comfort food, or a known and loved person).

Musical pleasure includes a capacity to prepare our mind to hear sound amid conceptual/auditory coherence, endless beauty in audition, and deep social symbolic meanings. We come prepared with biological prepotent neural structures to the allure of the auditory, to the musical, to song and sound rich in meaning, through which we are compelled to complete and fulfill expectations and experience pleasure (Vuust and Kringelbach 2010; Huron 2005). Music is inherently familiar to us, even if it contains some unpredictable elements. This could play a role in why it is pleasurable to us. It is like eating variations of mom's chicken soup—maybe a bit surprising, but always comforting and delightful.

Leonard Meyer focused on the interplay of the familiar and the unfamiliar in musical experiences and musical meaning. In describing music, Meyer noted, "underlying the reciprocity of play, regularity, and style, is in part the patterns of repetition" (1956, 72; see also Meyer 1967). The repetition of the familiar is of course part of the experience of music. Indonesian gamelan bands, for example, depend on the repetition of an already-familiar musical motif with minor tone variations in some form of endless, repetitive, "eternal recurrence," to borrow a phrase from Nietzsche (1927).

Meyer also pointed out that music depends on the contrast between the familiar and its variations. Variation at multiple levels of appraisal is a feature of most musical experiences. Both Mozart and Coltrane, for instance, delighted in slight deviations in expectations pervading the familiar. John Coltrane's recording of the song "My Favorite Things," written by Oscar Hammerstein and Richard Rodgers for *The Sound of Music*, relies on the audience's familiarity with the song to launch into magnificent variations, expected and unexpected, as he expands into novel uncharted territory.

We have a wide variety of musical expectations that range, as David Huron (2008) notes, from pitch proximity to the direction of the music. He ties this organizational sensibility to "heuristic listening." Scale tones set up diverse expectations amid surprises and "sweet anticipation." Indeed,

> some surprises start right from the moment a work begins. When Igor Stravinsky began his "Rite of Spring" with a solo bassoon, he violated several well-known conventions in classical music. The vast majority of Western orchestral works do not begin with a solo. Moreover, the bassoon is one of the least likely orchestral instruments to perform by itself. Finally, Stravinsky placed the instrument at the very top of its range. In other words, Stravinsky began the "Rite of Spring" in a highly unorthodox (that is, improbable) way. (Huron 2006, 270)

That shock of unexpectedness mimics the way the season of spring itself surprises us after the cold of winter.

Surprise and tension underlie human experience, but they take new and extraordinary forms in music (Meyer 1973). Indeed, such events pervade the sense of musical experience (Cuddy and Lunney 2005; Russo and Cuddy 1999; Narmour 1991). Consider, for instance, the expectations we have concerning melodic intervals (Thompson, Cuddy, and Plaus 1997). Neural capabilities underlie diverse forms of human expression, and predictability is a core feature of the brain's responses to rhythmic expectations (Vuust et al. 2009).

Indeed, expectations are a core feature of neural organization. Repetition is essential in habit formation (Peirce 1992; Dewey 1929), and small deviations underlie responses to even the smallest novelties. These factors operate in music as well as in daily life (Narmour 2008; Rozin 1976). Expectations underlie a pattern itself, and are indeed part of the pattern. A common pattern in Western music (and also Western humor) is the repetition of two similar phrases followed by a dissimilar event, the A-B-A-B-C pattern (Rozin et al. 2006). Thus, a break in expectation is part of sustaining coherence (Narmour 2008).

Meyer understood that melodies are often acting on more than one level, in which there is the familiar and the unexpected. We are anchored to frames of reference in which there are expectations, detection of deviations or gaps, and then exploration. Continuity, building on the insights of C. S. Peirce (1868), is a primary feature of expectations situated in a cultural and contextual milieu, within a "cultural style" (Meyer 1956). We forge coherence by relying on the familiar amid the unfamiliar, and it is this relationship that permeates much of our experience in music and, indeed, in most avenues of human expression.

Musical Expectations: Probability, Emotion, and Aesthetic Sensibilities

Meyer was anchored in classical pragmatism, and he placed music and its appreciation in the broader arena of problem solving, exploring the probabilistic and the anticipatory in human musical sensibility. Underlying such exploration is statistical inference within frameworks of expectations rich in meaning, problem-solving, and neural investment.

Statistics are embedded in practices in which reliability and predictive patterns are depended on and noticed when they break down (Goodman 1955). This view underlies reasoning about musical expectation, and to some extent about art more generally, in which a mixture of expectation and discrepancy is held to underlie aesthetic sensibility (Goodman 1968; Dewey 1934). The mental tools used in statistical reasoning also underlie diverse forms of musical competence, replete within a social context. They permeate the cogitative processes involved in listening to music, and motivate our response to violations of musical rules (Huron 2008; Juslin 2001; Smith and Melara 1990; Meyer 1956).

Probability judgments to assess the condition of uncertainty are at the heart of human reasoning, and while it is a mistake to exaggerate their role in aesthetic judgment, one can assume that in some contexts they do play a role. Since uncertainty is a basic feature of our existence, we have developed a variety of resources to cope with it. Moreover, as Meyer has noted, "uncertainty is anathema to humankind," and "we devise ways of reducing uncertainty both in the out-there world and in our personal lives," but "in the arts and other playful activities such as sports, games and gambling we actually relish and cultivate a considerable amount of uncertainty" (2001, 352; see also Krumhansl 2002).

The play of ideas amid the uncertainty of one's existential condition defines an aesthetics in which probability expectations are part of our everyday cognitive adaptation, in music or otherwise (Dewey 1929). The codification of well-worn habits and expectations is the grounding frame in which musical orientation takes place; it requires well-orchestrated habits, cognitively mediated in regions of the brain that underlie repetitive occurrences and their breakdown.

Cognitive systems underlie all aspects of music and aesthetic sensibility (Temperley 2001; Sloboda 2000). Cognition is often construed as detached, and since emotions are anything but detached, they are considered noncognitive by many investigators. Aesthetic experience is surely up close and personal, and therefore one can understand why many investigators construe the aesthetic experience as not "exclusively cognitive" (Elster 2000). However, aesthetics is linked to information-processing systems in the brain. Aesthetic appreciation is replete with information processing; it is not as if one side of us is doing the thinking and another is only appreciating (Kivy 2001, 1991; Davies 1994).

Adaptation is a core feature of musical sensibility (Cross 2010; Huron 2006); the biological systems are tied to auditory acuity and the prediction of auditory events.

Appraisal systems in musical pleasure and displeasure are embedded in the auditory system analysis and the larger connectivity in neural systems. Infused within are diverse forms of learning, rich in semiotics (Meyer 1973; Dewey 1929). Neural capabilities underlie musical sensibilities amid a sense of problem solving and detection; the broader theme is the creation process (Sloboda 2000). Moreover, as Dewey understood, in art and experience, problem-solving is endemic to aesthetics. Music is no different; unconscious inferences, expectations, and the generation of habits underlie the perception and production of musical experiences.

Something inherent in Darwin, who did not acknowledge a separation of cognitive systems from emotional systems, is of course the view of Meyer, and certainly Dewey, that emotional responses are embedded in diverse forms of information-processing neural systems. Indeed, John A. Sloboda, a psychologist who studies musical sensibility, has commented on Meyer's work regarding emotion: "emotion is intimately linked to the cognitive processes of generating and testing predictions about future musical events" (Sloboda 2008, 489). Violations of expectations are one reliable feature linked to emotions, and while Meyer had doubts about the accuracy of the emotions in music, physical embodiment is embedded in these expectations and their violation in musical appreciation.

It has been found that the greater the deviation from expectations about the music's harmonic characteristics, the greater the emotional response (Steinbeis, Koelsch, and Sloboda 2006). That is, neural systems come prepared to recognize harmonic relationships, syntactically organized in the diverse semantic meanings expressed in music; deviations result in greater and more diverse neural responses. The attention here becomes focused when encountering something that stands out as discrepant. Our curiosity is piqued; beholding an aesthetically pleasing or offensive object, our interest is aroused, our sensibility offended (Hebb 1949). Expectancy in information processing in the brain permeates and results in what Karl Lashley called "the serial order of behavior" (1951).

Dewey's view of learning, whether aesthetic or otherwise, is one in which the failure of an expectation initiates the process of learning, a cognitive behaviorism (1925). This view of aesthetics is explicated, for example, in Meyer's *Emotion and Meaning in Music* (1956), in which the discrepancy model of learning figures throughout, as well as in the conflict theory of human appraisals and meaning (see also Meyer 1973, 1967). In the discrepancy model, disruptions of expected events result in the recruitment of a greater number of behaviors that might reflect learning. Functionalist and cognitivist views place the behavior patterns that are generated by a central state of the brain in the context of acquiring information, which has been consistently linked to the context of musical sensibility (Huron 2005; Sloboda 2000; Meyer 1973; Berylne 1960).

Musical sensibilities exist amid the information-processing systems in the brain that underlie perception and are oriented to novelty, familiarity, and syntax. We are curious for a lot of reasons, and there is no doubt that one of these reasons is informational discrepancy in an event (Loewenstein and Lerner 2003). Discrepancy (or novelty) detection is a core motivator of brain processing in both musicians and nonmusicians

(Koelsch et al. 2000), and the recognition of musical discrepancy is expressed early in development. Expectancy effects permeate the organization of the brain and are linked to appraisal systems for gains and losses; the events are embedded in information processing. A broad-based response to discrepancy is an important behavioral adaptation, underlying aesthetic sensibility.

Novelty detectors are operative across the human experience, and in a number of regions of the human brain. Several areas of the brain, including the amygdala, the basal ganglia, and regions of the prefrontal cortex, as well as Broca's area and its analogue in the right hemisphere, are importantly involved in some forms of aesthetic judgment as well as the detection of discrepancies. Moreover, activation of the left prefrontal cortex is linked to music that generates feelings of joy and happiness, while activation of the right prefrontal cortex is linked to music that generates feelings of fear or sadness (Schmidt and Trainor 2001).

Aesthetic sensibility is much broader than simply the detection of discrepancy and the appreciation of filling in the information that is needed. Sensing the syntactical and informational content of the aesthetic object is also important for aesthetic appreciation. While expectations linked to probability judgments are only part of the story for musical aesthetic appreciation, they still seem to be elements of the musical feel for phenomena, which requires a number of information-processing systems (Repp 2005). The sense of musical expectations is embedded in appraisal systems, replete with emotions (Huron 2008; Gjerdingen 2007; Sloboda 1991; Meyer 1973).

Musical sensibility involves a neural structure oriented toward change. The sense of change is what underlies the appreciation of knowledge in music and adaptation; the sense of uncertainty is part of the interplay of the regularity of the expected with a deviation, however slight, that keeps our interest. Such is the structure of musical expectation, amid style and form; staying the same, alternatively, is surely not a feature of many musical forms, although variation within set parameters is a recurrent musical feature. Expectations and variations are thus common themes for which musical structure sets the conditions for diverse expression.

Music and Movement: Art and Aesthetics in the Context of Human Representation

Aesthetic judgment is a profound part of our life. We do not just view a masterfully rendered painting with awe. We value a person's character as a work of beauty, a beautiful soul; we enjoy the aesthetic contrast of orange carrots and green peas on our dinner plates; a majestic mountain glimpsed through a car window renders us speechless as it comes into view. Aesthetics is pervasive in everyday life; it is not just for the museum, parlor, or theater. Music, like other arts, can heighten and deepen our experience.

John Dewey, in his book *Art as Experience* (1934), stated romantically, "because experience is the fulfillment of an organism in its struggles and achievements in a world of things, it is art in germ. Even in its rudimentary forms, it contains the promise of that delightful perception which is aesthetic experience" (18). Dewey was fond of linking human activity with common threads: aesthetic experiences, problem solving, and a common currency. Art infuses life, and there is no higher or lower art. Folk music can be as elegant as Mozart, and as moving.

The work of Doris Humphrey, a twentieth-century modern dancer, in *Art of Making Dances* ([1959] 1991) for example, suggests that aesthetic sensibility evolved because elegant classificatory cognitive systems are adaptive (Rozin 1976). We come prepared with a variety of cognitive systems, information processing, inferential mechanisms, and representational capacities for depicting and making sense of the objects that we are likely to encounter (Dunbar 1998; Rozin 1998, 1976). These systems pervade our experiences and help determine how we interpret and predict events. They also underlie our problem-solving proclivities that reach into aesthetic judgment (Cross 2009; Huron 2008).

Representation of objects is an ancient human activity, and we come prepared to understand objects in our environments. Cave art is but one example of this phenomenon that we find across cultures (Mithen 1996). The desire to represent and understand the objects in our world is a fundamental human motivation. This orientation is embedded in practice, in action, in movement or dance, and in music (Peirce 1992; Hanson 1971, 1958). Such a predilection is wired in our minds/brains (Ramachandran and Hirstein 1999; Zeki, Watson, and Frackoiak 1993). Aesthetic judgment is part of our appraisal of ourselves and of others, and is about both what is attractive and what we are trying to understand. It represents the objects we eat, copy, capture, and try to control (Mithen 1996).

Kant ([1792] 1951) tied aesthetic judgment to a "free play of the imagination," of cognitive capacities and faculties interacting to determine the perception of beauty or the sublime. Kant and the free play of the imagination set the stage for the "mind to think [of] the unattainability of nature regarded as a presentation of ideas" (151). In Kant's lofty terms, the construction of aesthetics, musical or otherwise, pushes one beyond what can be represented and known toward a semblance of the sublime. This occurs through embodied rituals amid an expanded sense of sensibility through play and practice, which pervades our sense of music and movement or dance (Bellah 2011; Schulkin 2004; Johnson 1987).

In Langer's view (1957), information-processing systems are continuous with bodily representation, and they reach their zenith in aesthetic sensibility or perception. But feelings are part of the perceptual apparatus that underlies information processing. An aesthetic appreciation emerges from trying to capture nature amid cognitive sensibility, tied to adaptation and aesthetic judgment, and intimately linked to bodily sensibility (Budd 2002; Mithen 1996). Perhaps music, often construed as one of the more abstract forms of representation, demonstrates this core cognitive capacity to reach beyond sheer beauty and the sensuous to the sublime and a free play of cognitive capabilities.

In experiments, Sarah, a chimpanzee with extensive communication training, was shown movement by humans that we would judge as graceful and aesthetically pleasing. She was then shown what we would consider a more awkward movement in comparison. The chimpanzee tended to look more at the graceful performance (Premack and Premack 1983). Perhaps form matters to the visual systems of the chimpanzee. But these effects in Sarah are modest when compared to the aesthetic judgment of movement in a five-year-old human child. Children in the same experiment express a specific and forceful preference for the more graceful movement.

A sense of aesthetics is historically variable and rich when the ecological conditions are suitable (Mithen 1996; Humphrey 1973). Aesthetic judgment reflects our cognitive flexibility and our extension and use of specific cognitive mechanisms to wide domains of human expression (Rozin 1998, 1976; Mithen 1996).

Music and Movement: Expectations, Rhythmic Generators, and Internal Clocks

Rhythmic generators are a feature of cognitive systems and cephalic function across music and virtually all human activities (Kruse 2007; Parrott and Schulkin 1993; Dipert 1983). In other words, anticipatory expectations are endemic to cognitive function, musical or otherwise, and thus cognitive systems pervade musical activities. Therefore, the distinction between the cognitive and the noncognitive is undercut at all levels of the neural axis. Rhythmic expectations stand out when one considers the close link between music and dance. Indeed, music and dance usually run together, pregnant with cognitive anticipatory expectations. Consider the George M. Cohan song "I'm a Yankee Doodle Dandy," now mentally inseparable from the dancing of Jimmy Cagney, capturing a very precise sense of place and history in dance and music.

Furthermore, movement is innately connected with a sense of time. Our brains are primed to detect movement, both familiar and unfamiliar, and our perception of time and self are tied to movement. The concept of a river of movement and time also reflects our origins in water; after all, most of our body and our planetary bodies are composed of water. The composer Claude Debussy's "Water Pieces," for instance, effectively captures the great flux of existence, of movement, of feeling moved to sing, as well as fantasies of singing and moving.

Clocks are linked to the prediction of periodicity as we cope with the world. The 24-hour clock is just one of many: monthly, yearly, and seasonal. There is also the time sense that can pervade our activities; whether we are aware of it or not, timing is endemic.

Of course, timing mechanisms pervade music when we are keeping time, tempo, and so on. Time management and organization is also inherent in music, and entrained in aesthetics more generally. Indeed, entrainment to predictive core features is at the heart of nature and is tied to anticipation of events. Endogenous clocks like the 24-hour clock are used biologically to anticipate events, such as when food or other vital resources are available (Foster and Kreitzman 2004; Richter 1965), and clock-like mechanisms underlie the tempo of music.

Cognitive systems bind us together and set the conditions for the organization of action in a coherent but disruptive world; regularities are pervasive. The 24-hour rhythm is the essential rotation of our behavior and is tied to light/dark cycles. This clock is so essential that it is found in diverse regions of the body, such as the liver, as well as in regions of the brain (Moore 1992; Richter 1965). Our many body clocks are pervasive in bodily rhythms, and clocks in every organ regulate diverse information molecules such as dopamine and vasopressin. However, it is the cephalic clocks (e.g., 24-hour clocks) that are tied to music. These are the rhythms that we come prepared to perceive, entrain, and coordinate with, and with which we regulate ourselves in response to rhythmic patterns (Thaut 2003).

For a dancer, the rhythmic beat is at the heart of song. Naturally, the nervous system itself is rich in rhythmic function (Richter 1965). The neuroscientist G. Buzsáki's book *Rhythms of the Brain* (2006) provides an excellent understanding of cognitive capability. Inherent rhythmic properties ready the brain for action; there is relative rest, but not the elimination of the inherent patterns of the neural system that underlie all cellular and regulatory events. Indeed, evidence indicates that neural networks oscillate in functional networks independent of external stimuli (Simmons and Martin 2012).

Thus, our bodies are full of all sorts of regulatory and innate clock-like rhythms, the most obvious one of which is the palpable beat of the heart, but most of which are unknown and undetected. In spite of this, they surely must figure into primitive rhythmic beating, sounds, and eventually songs. These clocks are tied to the rhythms and the core themes of nature as we forge coherence toward keeping time, tracking time, and organizing ourselves with others by "keeping to the beat" internally, socially, and ecologically.

The Greek concept of the "harmony of the spheres" is merely one expression of a sentiment that flows across diverse cultural experiences, and of the ecological rhythms of events, internalized in diverse end organ systems, which are not simply passive. When it comes to neurally generated rhythms, we become anticipatory and linked to others, something essential for the sense and evolution of both music and language (Thomas 2001; Cook 1998; James 1993; Guthrie 1955).

Perhaps not surprisingly, our sense of motion, movement, and physical sensibility impacts our sense of time and space. This means that there is a bodily component to our sense of time and memory of events, which is rich in sensorimotor experiences, as well as agency and action; the experiences of music are boundless with their sense of movement. As Eric Clarke notes, "musical sound requires and inevitably involves movement" (2005).

Music and Dance: Connection, Communication, and Culture

Art is a vital source of communication, and it is therefore not surprising that rudimentary forms of aesthetic appreciation may be expressed in a variety of species, and are reflected in diverse neural structures (Brown et al. 2004; Geissman 2002; Marler 2000). Art is a feature of the human condition, and all cultures have music and dance of some sort (B. Nettl 1956; P. Nettl 1948). We are ultimately a social species. Music reinforced our social instincts by allowing us to expand our human experience toward and with others; coupled with the development of instruments for touching each other through sound, our social expression continued to swell. This process began with song and was expanded through instruments and dance. With culture and memory coded into external artifacts, music and dance coevolved in contexts of adaptation, human meaning, and social contact. The human condition is intimately linked to music and movement, which are in turn fundamental for the range of human emotional expression.

Leaping with graceful ease, gesture fills the space of Fred Astaire and Ginger Rogers, as is immediately called to mind by those who know their work. They exemplify how we are all physical in our singing, how our bodies swing to and fro to music. As Zbikowski notes, "the connection between dance and music seems completely undeniable—Astaire and Roger's movements are simply the fanfare from Kern's tune" (2012, 152). The gestures are brimming, bristling and breezing, effortless and beautiful. The music is highlighted by the dance and the dance is highlighted by the music. The dance completes the music, and the music is embedded in the dance.

Suzanne Langer understood that movement and dance are at the heart of music, and music is at the heart of movement (1953, 1957; see also Bellah 2011; Schultz 1962). Music and movement run together, just as cognitive systems are grounded in human action, understanding, and what Meyer, following Dewey, understood as "embodied." There is no Cartesian separation of a mind in a body in our actual experience.

The cognitive architecture is in part revealed in action; the dance and the song revealing gestures of cognitive structure (Zbikowski 2011). Indeed, music is tied to the organization of movement in cognitive systems within the motor system and linked to different kinds of information processing within the brain. Premotor planning and motor control are inherent in this cephalic process (Gallese 2007) between the order of behavior and expressed syntactical patterns of movements that are the products of generative processes. A neuroscientific perspective on dance ties diverse regions of the basal ganglia (e.g., putamen and diverse regions of the cortex, including the premotor and motor cortex, and the cerebellum) in the organization of dance (Brown and Parsons 2008). We distinguish between animate and less animate movement, and animacy is a core category in our cognitive arsenal for understanding the world.

Music is infused with the extension of animacy and agency, two core categories that underlie our sense of purpose, action, direction, and meaning. Expanding predilection in activity is a primary feature in our creative sensibility in reaching to aesthetic sensibility and perhaps the "neural sublime" (Richardson 2010), a capacity to search for the unbounded amid the finite and bounded (Kant [1787] 1965).

Embedded in the narratives, of which poetry is a rhythmic example, is the rumination of time as in "Burnt Norton" by T. S. Eliot (1943), in which the directions of time and the "words move, music moves" in time. The poetry or the odes and rhythmic sensibility inherent musically in Greek or Indian tales, to name just two, reflect the extension of cephalic categories of understanding and representing (Trimble 2007; Cooke 1959). The animate categories pervade the poetic seduction.

Poetry, like music, is less literal in general than language; of course, language nevertheless is ripe with metaphors that expand our conceptual horizons (Lakoff and Johnson 1999). Poetic metaphor is infused in the dances and movements and the stories. As Doris Humphrey notes, "rhythm so permeates every aspect of a human being and indeed of the known world that it might be compared to the ambience of existence, like the water in which the fish moves" (Humphrey and Pollack [1959] 1991, 104).

Within this mixture of what Meyer (1967) called "an aesthetics of stability" is a coherence of expansion into the unfamiliar and the expected. Metaphors within music expand our sensibilities. Metaphor and myth run through musical thought, as they do in all of human thought (Lakoff and Johnson 1999), through expression of mythic belief codified in movement and music (Spitzer 2004; Lévi-Strauss 1969). Much of the infusion of creativity in music and dance took place amid a moment of broader cultural change and uncertainty. The aesthetic ascended to the sublime, but even the less sublime, everyday manifestations of "singing in the rain, what a glorious day," show us that music and dance are a piece of our biology and an evolving part of our culture (Suskin 2010; Bordman 1992). What pervades is the recurrent familiar amid pockets of change and variation.

For instance, the stately Bill Robinson, an elegant man with endless dignity and grace (Suskin 2010), danced with the little sweetheart of America, Shirley Temple, in the grips of the Great Depression. African Americans in many parts of the country were not supposed to touch or mingle with whites in this manner, but there they are, moving through space, up and down a stairwell with elegant grace and genius, on cue together.

Dance is tied to our sociability. Without the ability to move together, music becomes far less social. It is the sensation of "feeling the music" that really brings people together around music (see Levi 1967; Mead 1938). Music is expressed across cultures and has been with us, evolving for as long as we as a species have been here with instruments. It follows that musical complexity is a feature of the larger cultural milieu (Burgess and Haynes 2004).

Music and Dance: Human Expression and Lived Experience

A noted musical critic of the nineteenth century, Eduard Hanslick, stated, "the essence of music is sound" in motion (cited in Clarke 2005, 4). Music and dance are endlessly intertwined, and any history of dance is in part a history of music (Sachs 1937). Diverse forms of life and ritual are expressed, and bodily rituals are a piece of nature, from the dancing of the bees to the synchronous bodily and howling rituals of wolves (von Frisch 1955).

Dance is a phylogenetically ancient expression in which no separation exists between our desire to move and our desire to reach into rich rituals of bodily expression. Ages in music, like ages in dance, are cognitive and historical constructs (B. Nettl 2005, 1956; Wiora 1965; Sachs 1937). But dance and ritual are linked in ages past and present in the context of the larger cultural milieu. For instance, within dance, a place in St. Petersburg was unique: an assembly called the Ballet Russe. There, Stravinsky, Fokine, and Ravel all produced music that broke boundaries, coupled with the elegance of movement of Nijinsky and Pavlova (Joseph 2011; Scheijen 2010; Taruskin 1996). The combination of musical and dance talent produced an extraordinary explosion of creative energy. While this assembly was unique in the breadth and quality of its output, these fusions are in fact commonplace and suggest the fundamental link between music and dance.

Martha Graham, for instance, changed the landscape of modern dance paradoxically with forms of ancient myths accompanying contemporary music. She produced a rich landscape of pieces including *Lamentation, Primitive Mysteries, American Mysteries, Clytemnestra, Appalachian Spring,* and *Dark Meadow* (see Morgan [1941] 1980). Diverse composers, including Aaron Copland, captured Americana music for Graham and for ballets. They were and are a wonder of breadth and depth. Copland was a senior colleague and early mentor of Meyer, who writes about rhythm, melody, and harmony. Copland himself produced a series of lectures (1957), which specifically note the link between music and dance.

A wonderful chain in dance runs from Hana Holm to Isadora Duncan, and to Jose Limon and Alvin Ailey, capturing diverse spirits of movement, music, and meaning (Siegel 1979). For Ailey, capturing the African American experiences of spirituality was central to his body of work, from pieces set to spirituals and church music, to the music of Ralph Vaughn Williams in "The Lark Ascending." The complicated and endlessly intricate movements of the dances of India are also infused throughout with religious sensibility. Religious dancers, as diverse as the Shakers and Whirling Dervishes, express the ecstasy of movement and the simplicity of aesthetic ascendance through movement and dance; the song matches the movement. These expanding categories of understanding, always toward the mystical and toward endless transcendental sensibility, are usually based on a narrative of travel, of being on a path.

Music expands dance, and dance expands music: both fuse as expanding landscapes of meaning. "Revelations," for instance, is a depiction of the dance and movements linked to black spiritual music and cries for freedom. Indeed, the choreography of Alvin Ailey called "Cry" conveys in movement a sense of story, toward some form of human liberation, a common theme. At the same time, musical expression was also connected to themes of liberation; Marion Anderson's "Songs of Freedom" lionized that theme in song and expression in front of the Lincoln Memorial in 1939, after she was denied the right to perform in the Daughters of the Revolution Hall (Arsenault 2009).

Our respect for individual expression and initiative, and for a narrative and metaphorical expression of categories endlessly embedded in our stories, reflects an evolutionary predilection and adaptation. These categories set the conditions for coherent action in adapting to terrains both social and ecological. These categories are not abstract, and in fact, they are set in everyday action.

We are not Cartesian machines, thinking in an abstract, divorced vacuum, nor are we random inductive machines. Musical sensibility makes this very transparent. We bring with us diverse forms of cognition that underlie what Dewey used to call "lived experiences," or what others have called "embodied cognition" (Gallagher 2005; Schulkin 2004; Lakoff and Johnson 1999; Varela, Thompson, and Rosch 1991).

Music is often full of purpose and bound to movement, whether literal or not (Zbikowski 2012). This goes from the structured ballet of Chopin-type music, with its endless practice and the expansion of musical movements, through Charles Ives's "The Fourth of July." Music is expanded and tied to meaning, stories, and living experience, running together in both long and short versions. Our sense of self and our sense of music are often rooted in our life histories or our trajectory of movement through space.

Thinking, movement, and music are understood in the context of action and transacting with others and are quite close to a pragmatist position where cognitive systems are embedded in the organization of action (Schulkin 2009; Dewey 1896; James 1890). The emphasis is on embodied and expanded cognitive systems (Noë, 2004; Barsalou 2003; Wilson 2002; Clark 1997, 1999). The sensorimotor systems are themselves knotted to cephalic machinations across all regions of the brain (Schulkin 2007; Barton 2004; Berthoz 2002; Jeannerod 1997; Dewey 1896).

In other words, cognitive systems are not just a cortical affair, but are endemic to cephalic function. Meyer, of course, understood this, with his kinesthetic/anticipatory description of musical sensibility. Movement and purpose, living a life with music rich in expectations and kinesthetic affordances, is replete with memory.

Indeed, Johnson, a philosopher, and Larson, a musicologist, have nicely depicted the common themes of music, movement, and musical trajectories and, consistent with the earlier works of Langer, the utter kinesthetic sensibility running through music. Music is tied to movement and memory with vivid imagery, such as the one cited in "Over the Rainbow," and the expansion of our horizons by metaphors, as in George Harrison's "Something in the Way She Moves" (Johnson and Larson 2003). The metaphors of movement and destination traverse the "musical landscape." Music is a force to be experienced.

An inherent tension in aesthetic sensibility, particularly in music, is the search for stability and stasis (Meyer 1967). The construction of events shapes the way action is orchestrated in music and otherwise. The shared experience with one another is a fundamental adaptation (Tomasello and Carpenter 2007; Gallagher 2005), and the cephalic innervations of bodily sensibility are a pervasive common element in our evolution (Schulkin 2004; Berthoz 2002; Damasio 1994; Merleau-Ponty 1962; James 1890).

We come prepared with an evolved brain and a set of cognitive predilections that are situated toward context, flexibility, and perceptual embodiment about objects that are conceptually rich and vital to behavioral adaptations of action, perception, and the brain. The emphasis is on the adaptive nature of these systems (Gigerenzer 2000) that run through human cephalic capability, including that of two important expressions: music and dance.

Conclusion

Aesthetic sensibilities exist amid information-processing systems in the brain that underlie perception and are oriented toward novelty, familiarity, and syntax. The interplay of the unfamiliar amid the familiar, variation amid repetition, is central to the formation of our musical expectations and the pleasure we derive from surprising deviations. The experience of music itself reflects the interaction between the performer and the one listening (Chapin et al. 2010). Indeed, we come prepared to recognize the probabilities of an event. Rooted in neural structures oriented toward adaptation and change, the probabilistic and the anticipatory elicit emotional and pleasurable experiences as well as aesthetic judgments. Musical experiences form patterns of expectations, categorical markers (Neuhaus, Knosche, and Friederici 2006), and sentiments amplified by depth and beauty (Rosen 2010). In sum, cognitive expectations about music are fundamental to our experience of it.

Music, like other arts, can heighten and deepen experience. The desire to represent and understand objects in the world is a fundamental human motivation embedded in practice, in action, in movement or dance, and in music (Peirce 1992; Hanson 1971, 1958). We come prepared to interpret objects in our environments, and aesthetic judgment is part of our appraisal of ourselves and of others, of what is attractive and what we are trying to understand. As Dewey (1934) noted, art is part of the ordinary human experience and is part of the framework in which our worlds are understood, adapted, and invented. Information-processing systems are continuous with bodily representation (Langer 1957), and aesthetic judgment of movement as well as music can therefore be understood in the context of human representation. Music and movement are also linked through rhythmic generators and internal perceptions of time, which are pervasive throughout nature and bodily systems. These natural phenomena underlie our cognitive formation of expectations.

An extension of bodily representation, dance coevolved with music, reflecting human narratives and communications of culture. Indeed, the human quest for understanding, connection, and self-expression is highlighted in music and dance, which are expressions of lived and shared experience, manifestations of meaning and cognitive predilections. Cognitive anticipatory expectations stand out when one considers the close link between music and dance. The brain is prepared to detect movement and rhythm, both familiar and unfamiliar, with aesthetic judgment reflecting our cognitive flexibility. Music and dance are two human activities (Huron 2001; Temperley 2001; Sloboda 1985) among others in which expectations are cardinal features of the organization of human action and capability.

An experimental sensibility broadens the horizon as we go from the familiar to the novel. Our sense of continuity has its roots in problem solving and in making sense of discrepancies. Memory and the diverse forms of cultural expressions provide a rich array to draw on in the expansion of musical and dance expressions as we explore the unfamiliar amid the familiar (Clarke 2005). This process began with song and was expanded through instruments and dance. Music and movement go together in every human society. The human condition is linked to music and dance, and the range of human emotional expression is fundamental in this regard.

Note

This chapter was adapted with permission from a book by Jay Schulkin titled, *Reflections on the Musical Mind* (Princeton, NJ: Princeton University Press, 2013).

References

Arsenault, Raymond. 2009. *The Sound of Freedom: Marian Anderson, the Lincoln Memorial, and the Concert That Awakened America.* New York: Bloomsbury.

Barsalou, Lawrence W. 2003. "Abstraction in Perceptual Symbol Systems." *Philosophical Transactions of the Royal Society of London B: Biological Sciences* 358: 1177–1187.

Barton, R. A. 2004. "Binocularity and Brain Evolution in Primates." *Proceedings of the National Academy of Sciences of the United States* 101: 10113–10115.

Bellah, Robert Neelly. 2011. *Religion in Human Evolution: From the Paleolithic to the Axial Age.* Cambridge, MA: Belknap.

Berthoz, A. 2002. *The Brain's Sense of Movement.* Cambridge MA: Harvard University Press.

Berylne, D. E. 1960. *Conflict, Arousal and Curiosity.* New York: McGraw-Hill.

Bordman, Gerald Martin. 1992. *American Musical Theatre: A Chronicle.* New York: Oxford University Press.

Brown, Steven, Michael J. Martinez, Donald A. Hodges, Peter T. Fox, and Lawrence M. Parsons. 2004. "The Song System of the Human Brain." *Cognitive Brain Research* 20: 363–375.

Brown, Steven, and Lawrence M. Parsons. 2008. "The Neuroscience of Dance." *Scientific American* 299: 78–83.

Budd, Malcolm. 2002. *The Aesthetic Appreciation of Nature: Essays on the Aesthetics of Nature*. New York: Oxford University Press.

Burgess, Geoffrey, and Bruce Haynes. 2004. *The Oboe*. New Haven, CT: Yale University Press.

Buzsáki, G. 2006. *Rhythms of the Brain*. New York: Oxford University Press.

Chapin, Heather, Kelly Jantzen, J. A. Scott Kelso, Fred Steinberg, Edward Large, and Antoni Rodriguez-Fornells. 2010. "Dynamic Emotional and Neural Responses to Music Depend on Performance Expression and Listener Experience," *PLoS ONE* 5 (12): e13812.

Clark, Andy. 1997. *Being There: Putting Brain, Body, and World Together Again*. Cambridge, MA: MIT Press.

Clark, Andy. 1999. "An Embodied Cognitive Science?" *Trends in Cognitive Sciences* 3: 345–351.

Clarke, Eric F. 2005. *Ways of Listening: An Ecological Approach to the Perception of Musical Meaning*. New York: Oxford University Press.

Cook, Nicholas. 1998. *Music: A Very Short Introduction*. New York: Oxford University Press.

Cooke, Deryck. 1959. *The Language of Music*. New York: Oxford University Press.

Copland, Aaron. 1957. *What to Listen for in Music*. New York: McGraw-Hill.

Cross, Ian. 2009. "The Evolutionary Nature of Musical Meaning." *Musicae Scientiae* 13: 179–200.

Cross, Ian. 2010. "Whatever Music Is, It's a Basic Part of Being Human." *Science News* 178: 36.

Cuddy, Lola, and Carole Lunney. 2005. "Expectancies Generated by Melodic Intervals: Perceptual Judgments of Melodic Continuity." *Perception and Psychophysics* 57: 451–462.

Damasio, Antonio R. 1994. *Descartes' Error: Emotion, Reason, and the Human Brain*. New York: G.P. Putnam.

Davies, Stephen. 1994. *Musical Meaning and Expression*. Ithaca, NY: Cornell University Press.

Dewey, John. 1896. "The Reflex Arc Concept in Psychology." *Psychological Review* 3: 357–370.

Dewey, John. 1925. *Experience and Nature*. Chicago: Open Court.

Dewey, John. 1929. *The Quest for Certainty*. New York: Minton, Balch.

Dewey, John. 1934. *Art as Experience*. New York: Capricorn Books.

Dipert, Randall. 1983. "Meyer's Emotion and Meaning in Music: A Sympathetic Critique of Its Central Claims." *Journal of the Michigan Music Theory Society* 6 (18): 3–18.

Dunbar, Robin I. M. 1998. "The Social Brain Hypothesis." *Evolutionary Anthropology* 6: 178–190.

Eliot T. S. 1943. *Four Quartets*. New York: Harcourt Press.

Elster, Jon. 2000. *Ulysses Unbound: Studies in Rationality, Precommitment, and Constraints*. New York: Cambridge University Press.

Foster, Russell G., and Leon Kreitzman. 2004. *Rhythms of Life: The Biological Clocks That Control the Daily Lives of Every Living Thing*. New Haven, CT: Yale University Press.

Gallagher, Shaun. 2005. *How the Body Shapes the Mind*. New York: Oxford University Press.

Gallese, Vittorio. 2007. "Before and Below 'Theory of Mind': Embodied Simulation and the Neural Correlates of Social Cognition." *Philosophical Transactions of the Royal Society of London B: Biological Sciences* 362: 659–699.

Geissman, Thomas. 2002. "Duet-Splitting and the Evolution of Gibbon Songs." *Biological Reviews* 77: 57–76.

Gigerenzer, Gerd. 2000. *Adaptive Thinking: Rationality in the Real World*. New York: Oxford University Press.

Gjerdingen, Robert O. 2007. *Music in the Galant Style*. New York: Oxford University Press.

Goodman, Nelson. 1955. *Fact, Fiction, and Forecast*. Cambridge, MA: Harvard University Press.

Goodman, Nelson. 1968. *Languages of Art: An Approach to a Theory of Symbols*. Indianapolis, IN: Bobbs-Merrill.

Guthrie, William Keith Chambers. 1955. *The Greeks and Their Gods*. Boston: Beacon.
Hanson, Norwood Russell. 1958. *Patterns of Discovery: An Inquiry into the Conceptual Foundations of Science*. Cambridge: Cambridge University Press.
Hanson, Norwood Russell. 1971. *Observation and Explanation: A Guide to Philosophy of Science*. New York: Harper & Row.
Hebb, Donald Olding. 1949. *The Organization of Behavior: A Neuropsychological Theory*. New York: Wiley.
Humphrey, Doris, and Barbara Pollack. [1959] 1991. *The Art of Making Dances*. New York: Grove Press.
Humphrey, N. 1973. "The Illusion of Beauty." *Perception* 2: 429–439.
Huron, David. 2001. "Is Music an Evolutionary Adaptation?" *Annals of the New York Academy of Sciences* 930: 43–61.
Huron, David. 2005. "The Plural Pleasures of Music." In *Music, Music Science Proceedings: Stockholm 28–30 October, 2004*, edited by William Brunson and Johan Sundberg, 1–13. Stockholm: Kungliga Musikhögskolan.
Huron, David. 2006. *Sweet Anticipation: Music and the Psychology of Expectation*. Cambridge, MA: MIT Press.
Huron, David. 2008. "Lost in Music." *Nature* 453: 456–457.
James, Jamie. 1993. *The Music of the Spheres: Music, Science, and the Natural Order of the Universe*. New York: Grove.
James, William. 1890. *The Principles of Psychology*. New York: H. Holt.
Jeannerod, Marc. 1997. *The Cognitive Neuroscience of Action*. Cambridge, MA: Blackwell.
Johnson, Mark. 1987. *Body in the Mind: The Bodily Basis of Meaning, Imagination, and Reason*. Chicago: University of Chicago Press.
Johnson, Mark L., and Steve Larson. 2003. "'Something in the Way She Moves': Metaphors of Musical Motion." *Metaphor and Symbol* 18: 63–84.
Joseph, Charles M. 2011. *Stravinsky's Ballets*. New Haven, CT: Yale University Press.
Juslin, Patrik N. 2001. "A Brunswickian Approach to Emotional Communication in Music Performance." In *The Essential Brunswik: Beginnings, Explications, Applications*, edited by K. R. Hammond and T. R. Stewart, 426–430. New York: Oxford University Press.
Kant, Immanuel. [1787] 1965. *Critique of Pure Reason*. Translated by L. W. Beck. New York: St. Martin's.
Kant, Immanuel. [1792] 1951. *Critique of the Power of Judgment*. Cambridge: Cambridge University Press.
Kivy, Peter. 1991. "Is Music an Art?" *Journal of Philosophy* 88: 544–554.
Kivy, Peter. 2001. *New Essays on Musical Understanding*. Oxford: Clarendon.
Koelsch, Stefan, Tomas Gunter, Angela D. Friederici, and Erich Schrööger. 2000. "Brain Indices of Music Processing: 'Nonmusicians' Are Musical." *Journal of Cognitive Neuroscience* 12: 520–541.
Krumhansl, Carol L. 2002. "Music: A Link between Cognition and Emotion." *Current Directions in Psychological Science* 11: 45–50.
Kruse, Felicia E. 2007. "Vital Rhythm and Temporal Form in Langer and Dewey." *Journal of Speculative Philosophy* 21: 16–27.
Lakoff, George, and Mark Johnson. 1999. *Philosophy in the Flesh: The Embodied Mind and Its Challenge to Western Thought*. New York: Basic Books.
Langer, Susanne K. 1953. *Feeling and Form: A Theory of Art Developed from Philosophy in a New Key*. New York: Scribner.

Langer, Susanne K. 1957. *Problems of Art: Ten Philosophical Lectures*. New York: Scribner.
Lashley, Karl. 1951. "The Problem of Serial Order in Behavior." In *Cerebral Mechanisms in Behavior*, edited by Lloyd A. Jeffress, 110–133. New York: Wiley.
Levi, Isaac. 1967. *Gambling with Truth: An Essay on Induction and the Aims of Science*. Cambridge, MA: MIT Press.
Lévi-Strauss, Claude. 1969. *The Raw and the Cooked*. New York: Harper & Row.
Loewenstein, G., and J. Lerner. 2003. "The Role of Emotion in Decision Making." In *Handbook of Affective Sciences*, edited by Richard J. Davidson, Klaus R. Scherer, and H. Hill Goldsmith, 619–642. New York: Oxford University Press.
Marler, Peter. 2000. "Origins of Music and Speech: Insights from Animals." In *The Origins of Music*, edited by Nils L. Wallin, Björn Merker, and Steven Brown, 31–48. Cambridge, MA: MIT Press.
Mead, George Herbert. 1938. *The Philosophy of the Act*. Chicago: University of Chicago Press.
Merker, Bjorn. 2005. "The Conformal Motive in Birdsong, Music, and Language: An Introduction." *Annals of the New York Academy of Sciences* 1060: 17–28.
Merleau-Ponty, Maurice. 1962. *The Phenomenology of Perception*. NJ: Humanities Press.
Meyer, Leonard B. 1956. *Emotion and Meaning in Music*. Chicago: Chicago University Press.
Meyer, Leonard B. 1967. *Music, the Arts, and Ideas: Patterns and Predictions in Twentieth-Century Culture*. Chicago: University of Chicago Press.
Meyer, Leonard B. 1973. *Explaining Music: Essays and Explanations*. Chicago: University of Chicago Press.
Meyer, Leonard B. 2001. "Music and Emotions: Distinctions and Uncertainties." In *Music and Emotion: Theory and Research*, edited by Patrik N. Juslin and John A. Sloboda, 341–360. New York: Oxford University Press.
Mithen, S. 1996. *The Prehistory of the Mind: A Search for the Cognitive Origins of Art, Religion, and Science*. New York: Thames and Hudson.
Moore, F. L. 1992. "Evolutionary Precedents for Behavioral Actions of Oxytocin and Vasopressin." *Annals of the New York Academy of Sciences* 652: 156–165.
Morgan, Barbara Brooks. [1941] 1980. *Martha Graham: Sixteen Dances in Photographs*. Dobbs Ferry, NY: Morgan & Morgan.
Narmour, Eugene. 1991. "The Top-Down and Bottom-Up Systems of Musical Implication: Building on Meyer's Theory of Emotional Syntax." *Music Perception* 9: 1–26.
Narmour, Eugene. 2008. "My Intellectual Father." *Music Perception* 25: 485–487.
Neuhaus, Christiane, Thomas R. Knösche, and Angela D. Friederici. 2006. "Effects of Musical Expertise and Boundary Markers on Phrase Perception in Music." *Journal of Cognitive Neuroscience* 18: 472–493.
Nettl, Bruno. 1956. *Music in Primitive Culture*. Cambridge, MA: Harvard University Press.
Nettl, Bruno. 2005. *The Study of Ethnomusicology: Thirty-One Issues and Concepts*. New ed. Urbana: University of Illinois Press.
Nettl, Paul. 1948. *The Book of Musical Documents*. New York: Philosophical Library.
Nietzsche, Friedrich Wilhelm. 1927. *Die ewige Wiederkunft; Die fröhliche Wissenschaft; Dichtungen*. Kroner: Leipzig.
Noë, Alva. 2004. *Action in Perception*. Cambridge, MA: MIT Press.
Parrott W. Gerrod, and Jay Schulkin. 1993. "Neuropsychology and the Cognitive Nature of the Emotions." *Cognition and Emotion* 7: 43–59.

Peirce, Charles S. 1868. "Questions Concerning Certain Faculties Claimed for Man." *Journal of Speculative Philosophy* 2: 103–114.

Peirce, Charles S. 1992. *Reasoning and the Logic of Things: The Cambridge Conferences Lectures of 1898*. Edited by Kenneth Laine Ketner. Cambridge, MA: Harvard University Press.

Premack, David, and Ann James Premack. 1983. *The Mind of an Ape*. New York: Norton.

Ramachandran, V. S., and William Hirstein. 1999. "The Science of Art: A Neurological Theory of Aesthetic Experience." *Journal of Consciousness Studies* 6: 15–51.

Repp, Bruno. 2005. "Sensorimotor Synchronization: A Review of the Tapping Literature." *Psychonomic Bulletin and Review* 12: 969–992.

Richardson, Alan. 2010. *The Neural Sublime: Cognitive Theories and Romantic Texts*. Baltimore, MD: Johns Hopkins University Press.

Richter, Curt Paul. 1965. *Biological Clocks in Medicine and Psychiatry*. Springfield, IL: Thomas.

Rosen, Charles. 2010. *Music and Sentiment*. New Haven, CT: Yale University Press.

Rozin, Paul. 1976. "The Evolution of Intelligence and Access to the Cognitive Unconscious." In *Progress in Psychobiology and Physiological Psychology*, vol. 6, edited by E. Stellar and J. M. Sprague, 245–281. New York: Academic Press.

Rozin, Paul. 1998. "Evolution and Development of Brains and Cultures: Some Basic Principles and Interactions." In *Brain and Mind: Evolutionary Perspectives*, edited by Michael S. Gazzaniga and J. S. Altman, vol. 5, 111–123. Strassbourg: Human Frontiers Science Program.

Rozin, Paul, Alexander Rozin, Brian Appel, and Charles Wachtel. 2006. "Documenting and Explaining the Common AAB Pattern in Music and Humor: Establishing and Breaking Expectations." *Emotion* 6: 349–355.

Russo, Frank A., and Lola Cuddy. 1999. "Motor Theory of Melodic Expectancy." Lay language paper presented at the Acoustical Society of America ASA/EAA/DAGA '99 Meeting, Berlin, March 18.

Sachs, Curt. 1937. *World History of Dance*. New York: W. W. Norton.

Scheijen, Sjeng. 2010. *Diaghilev: A Life*. New York: Oxford University Press.

Schmidt, Louis A., and Laurel J. Trainor. 2001. "Frontal Brain Electrical Activity (EEG) Distinguishes Valence and Intensity of Musical Emotions." *Cognition and Emotion* 15: 487–500.

Schulkin, Jay. 2004. *Bodily Sensibility: Intelligent Action*. New York: Oxford University Press.

Schulkin, Jay. 2007. *Effort: A Behavioral Neuroscience Perspective on the Will*. Mahwah, NJ: Erlbaum.

Schulkin, Jay. 2009. *Cognitive Adaptation: A Pragmatist Perspective*. New York: Cambridge University Press.

Schultz Alfred. 1962. "Making Music Together." In *Collected Papers*, vol. 2, 159–178. The Hague: Martinus Nijhoff.

Siegel, Marcia B. 1979. *The Shapes of Change: Images of American Dance*. Boston: Houghton Mifflin.

Simmons, W. Kyle, and Alex Martin. 2012. "Spontaneous Resting-State BOLD Fluctuations Reveal Persistent Domain-Specific Neural Networks." *Social Cognitive and Affective Neuroscience* 7: 467–475.

Sloboda, John A. 1985. *The Musical Mind*. New York: Oxford University Press.

Sloboda, John A. 1991. "Music Structure and Emotional Response: Some Empirical Findings." *Psychology of Music* 19: 110–120.

Sloboda, John A., ed. 2000. *Generative Processes in Music: The Psychology of Performance, Improvisation, and Composition.* New York: Oxford University Press.

Sloboda, John A. 2008. "Leonard Meyer: Embracing Uncertainty." *Music Perception: An Interdisciplinary Journal* 25: 489–491.

Smith, J., and Robert Melara. 1990. "Aesthetic Preference and Syntactic Prototypicality in Music: 'Tis the Gift to Be Simple." *Cognition* 34: 279–298.

Spitzer, Michael. 2004. *Metaphor and Musical Thought.* Chicago: University of Chicago Press.

Steinbeis, Nikolaus, Stefan Koelsch, and John A. Sloboda. 2006. "The Role of Harmonic Expectancy Violations in Musical Emotions: Evidence from Subjective, Physiological and Neural Responses." *Journal of Cognitive Neuroscience* 18: 1380–1393.

Suskin, Steven. 2010. *Show Tunes: The Songs, Shows, and Careers of Broadway's Major Composers.* 4th ed. New York: Oxford University Press.

Taruskin, Richard. 1996. *Stravinsky and the Russian Traditions: A Biography of the Works through Mavra.* Berkeley: University of California Press.

Temperley, David. 2001. *The Cognition of Basic Musical Structures.* Cambridge, MA: MIT Press.

Thaut, Michael H. 2003. "Neural Basis of Rhythmic Timing Networks in the Human Brain." *Annals of the New York Academy of Sciences* 999: 364–373.

Thomas, E. 2001. "Empathy and Consciousness." *Journal of Consciousness Studies* 8: 1–5.

Thompson, William, Lola Cuddy, and Cheryl Plaus. 1997. "Expectancies Generated by Melodic Intervals: Evaluation of Principles of Melodic Implication in a Melody-Completion Task." *Perception and Psychophysics* 59: 1060–1076.

Tomasello, Michael, and Malinda Carpenter. 2007. "Shared Intentionality." *Developmental Science* 12: 121–125.

Trimble, Michael R. 2007. *The Soul in the Brain: The Cerebral Basis of Language, Art, and Belief.* Baltimore, MD: Johns Hopkins University Press.

Varela, Francisco J., Evan Thompson, and Eleanor Rosch. 1991. *The Embodied Mind: Cognitive Science and Human Experience.* Cambridge, MA: MIT Press.

von Frisch, Karl. 1955. *The Dancing Bees: An Account of the Life and Senses of the Bee.* New York: Harcourt, Brace.

Vuust, Peter, and Morten L. Kringelbach. 2010. "The Pleasure of Making Sense of Music." *Interdisciplinary Science Reviews* 35: 166–182.

Vuust, Peter, Leif Ostergaard, Karen Johanne Pallesen, Christopher Bailey, and Andreas Roepstorff. 2009. "Predictive Coding of Music-Brain Responses to Rhythmic Incongruity." *Cortex* 45: 80–92.

Wilson, Margaret. 2002. "Six Views of Embodied Cognition." *Psychonomic Bulletin and Review* 9: 625–636.

Wiora, Walter. 1965. *The Four Ages of Music.* New York: Norton.

Zbikowski, Lawrence M. 2011. "Musical Gesture and Musical Grammar: A Cognitive Approach." In *New Perspectives on Music and Gesture*, edited by Anthony Gritten and Elaine King, 83–98. Burlington, VT: Ashgate.

Zbikowski, Lawrence M. 2012. "Music and Movement: A View from Cognitive Science." *Bewegungen zwischen Hören und Sehen: Denkbewegungen über Bewegungskünste*, edited by Stephanie Schroedter, 151–162. Würzburg: Königshausen & Neumann.

Zeki, S., J. D. G. Watson, and R. S. J. Frackoiak. 1993. "Going beyond the Information Given: The Relation of Illusory Visual Motion to Brain Activity." *Proceedings of the Royal Society B: Biological Sciences* 252: 215–222.

Further Readings

Johnson, Mark. 2007. *The Meaning of the Body: Aesthetics of Human Understanding.* Chicago: University of Chicago Press.

Narmour, Eugene. 1990. *The Analysis and Cognition of Basic Melodic Structures: The Implication-Realization Model.* Chicago: University of Chicago Press.

Schultz, Wolfram. 2007. "Multiple Dopamine Functions at Different Time Courses." *Annual Review of Neuroscience* 30: 59–88.

CHAPTER 3

THE SCIENCE OF VOICE AND THE BODY

MARINA GILMAN

SINCE the beginning of recorded history (and probably well before that) people have been singing. Singing has always been and continues to be an integral part of every music-making culture. The ultimate musical instrument is the human body. All other musical instruments are played by hitting, strumming, plucking, bowing, or blowing into an object that resonates. The objects are made of wood, metal, skins, shells or nuts, animal guts, and many other natural and manufactured materials. In all instances, when played well they have the ability to make sounds that we call music. These sounds can be percussive (drums), melodic (wind instruments), or a combination of both (pianos, xylophones, marimba, etc.), depending on the vibratory characteristics of the instrument.

The human voice is the only musical instrument that has the ability to do it all. It can project over an orchestra. It is percussive and melodic. It can combine melody and text, and some forms of solo singing—known variously as overtone singing, throat singing, or harmonic singing—can even create harmony. And it is 100 percent portable; it goes wherever its owner goes. Yet the voice is so much more than just a musical instrument. It is the source of human sounds, beginning with the baby's first cry. The voice is the primary source of human communication, whether verbal or nonverbal, melodic, or percussive. It is the complex interaction of the vocal mechanism (respiratory system, vocal folds, and the vocal tract) with the rest of the body that is used to produce speech, song (melody, rhythm, and language) as well as so-called vegetative sounds such as coughing, crying, screaming, and so on.

All instruments share a basic set of components. They all must have a sound source such as strings, a mouthpiece, drumhead, or other surface that vibrates the air, which in turn generates sounds. These sounds produce not only pitch but also sound color or timbre. The second element is the resonators—the body, box, or tube of the instrument—needed to reinforce or amplify the sound produced by the source. The final element is a radiating surface, the wood, metal, or material of the instrument or cavity that transfers the sound to the listener (Titze 2008). The size of an instrument generally determines its relative pitch—a piccolo (small flute) is a high-pitched instrument, and a tuba (large brass instrument) is a low-pitched instrument. A substantial difference between instruments and the voice is the respective sizes and interplay of these three elements. The sound source of the voice is the vocal folds (previously referred to as vocal cords). They are located in the neck in a structure called the larynx that is easily identified in males by the prominent structure called the Adam's apple. The vocal folds sit on top of the windpipe or trachea. The vocal folds in an adult male are between 17 mm and 25 mm long. The female vocal folds are between 12.5 mm and 17.5 mm in length. Contrast their small size with a violin string with an average of 32 cm in length or the stings on a piano which can range from 3 to 6 in. for the small high strings to 2.5–3.5 ft. for the longer low strings depending on the size of the piano! The vibrating chamber of the voice, the vocal tract, begins just above the vocal folds and extends only about 14–17 cm to the lips. Comparing length of the vocal tract, at 17 cm, to that of an uncoiled trumpet that is approximately 200 cm, one can begin to appreciate how extraordinary the voice is in its compact size alone.

Our appreciation and understanding of voice and vocal production has been transformed in recent decades by developments in visual imaging and through the new discipline of vocology, the scientific study of voice. For centuries vocal pedagogues have been training world-class singers of all genres. With the advent of formal classical music training, Western vocal pedagogues developed a vocabulary that was to a large extent figurative. After all, how does one tell someone how to breathe; how to color the voice; or how to sing freely with great flexibility for long periods of time? Teaching someone to play an instrument is relatively straightforward. The teacher not only can verbally explain what needs to happen but also can show the student where to place the fingers, or how much pressure to apply to a bow. The vocal mechanism, on the other hand, is entirely inside the body making the task of teaching much more complex. Modeling the sound and providing appropriate images were the only tools available to the voice teacher or vocal coach. It took over one hundred years—beginning with the invention of the laryngeal mirror, the development of a stable light source, the discovery of the optical illusion known as strobe light together with the development of high-speed photography, and the development of high-power microscopes—to appreciate the complexity of vocal fold motion. Innovations in sound technology, such as the development of the spectrogram, have also allowed us to see the sound, the fundamental frequencies, and overtones we are producing.

To fully appreciate the complexities of this instrument of our bodies, it is important to understand how the parts function individually and yet are totally dependent on the whole body.

It was well known that vocal sounds are produced in the larynx, a structure suspended in the neck forming the top of the trachea (windpipe). It consists of two interlocking cartilages: the cricoid and the thyroid cartilages (Figure 3.1). The vocal folds, V-shaped muscles with an overlay of thin vibratory membranous layers, are suspended within the larynx horizontally between the cricoid cartilage in the front and the thyroid cartilage

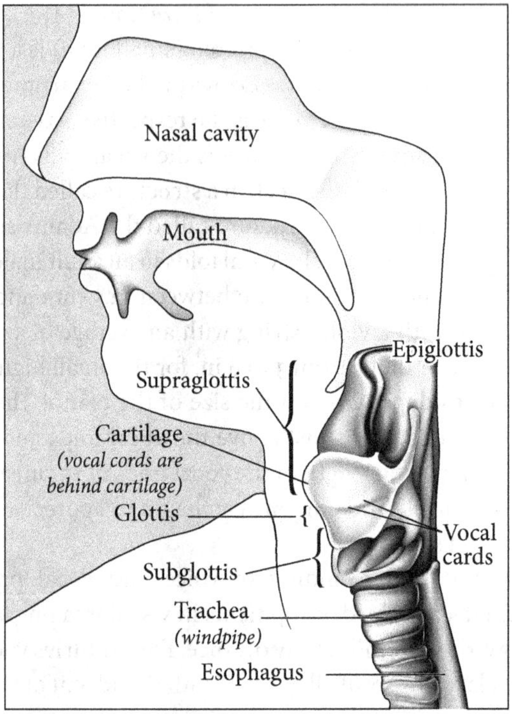

FIGURE 3.1 The larynx, consisting of the cricoid and thyroid cartilages, sits on top of the trachea (windpipe). The vocal folds are suspended between the cricoid cartilage (lower cartilage) and the larger thyroid cartilage, forming an airflow valve at the top of the trachea. The esophagus (the tube leading to the stomach) sits behind the trachea. The glottis is the space between the open vocal folds.

in the back. Small joints in the back allow them to close and open during swallowing and breathing, respectively. The only way anything can get into the lungs is between the vocal folds (Figure 3.2). Directly behind the larynx in the back of the throat is the esophagus, the tube that moves food from the mouth to the stomach.

Above the larynx is a very important small bone, called the hyoid bone. It is the only free-floating bone in the body. This bone serves as the point of attachment to the larynx, and the muscles in the neck, base of tongue, the shoulder girdle, and back of the throat. With all its attachments, the hyoid facilitates the movement of the larynx up and down in the neck essential for swallowing, pitch, and vocal timbre (Rosen and Simpson 2008) (Figure 3.3).

Until it was possible visualize the vocal folds via magnification, the only knowledge scientists had about the voice came from what could be dissected. They had theories about function, but little else. That is, until the advent of the laryngeal mirror.

Guy Babington (1794–1866) and the well-known singing teacher Manuel Garcia (1805–1906) developed the laryngeal mirror or the first laryngoscope at about the same time. However, Manuel Garcia is credited with the discovery, having presented his

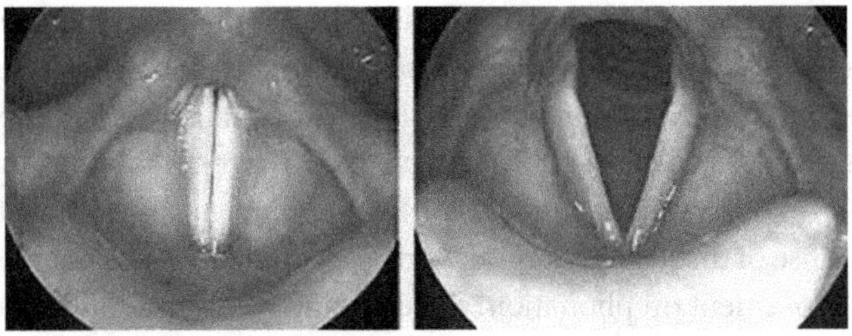

FIGURE 3.2 Seen from the top, the top of the image is the back of the throat and the bottom of the image is the front. Vocal folds are seen here looking down in the throat, in the closed (left) and open (right) positions. In the open position, note the tracheal rings below. Notice the pockets or piriform sinuses on the upper right and left. These form the opening of the esophagus during swallowing.

Emory Voice Center.

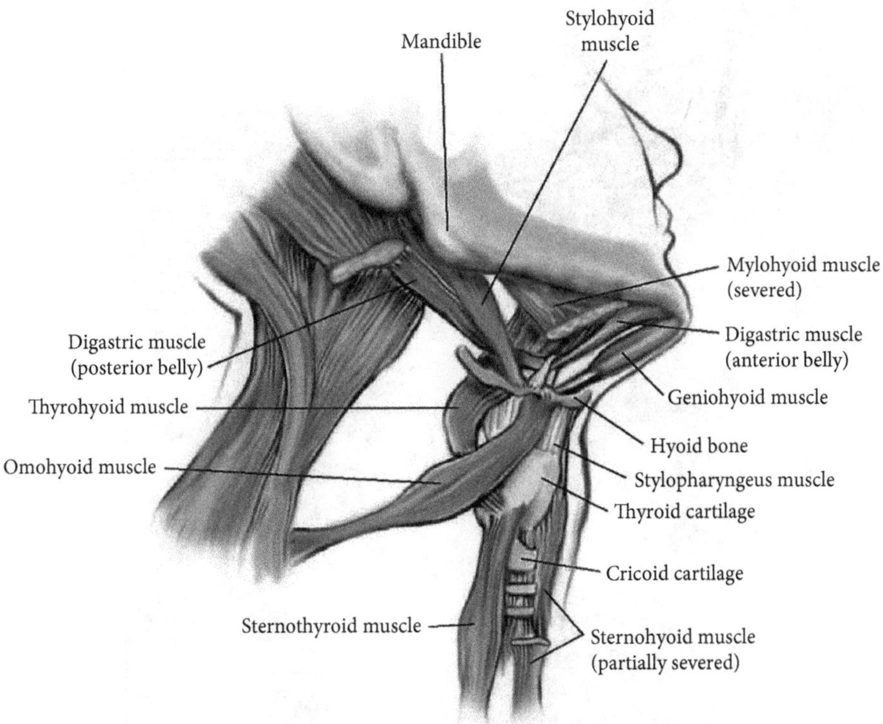

FIGURE 3.3 The hyoid bone is the only free-floating bone in the body. It sits above the larynx, underneath the base of tongue muscles. It attaches to the larynx, first rib and sternum, the base of tongue muscles, and the anterior (front) part of the scapula, just below the clavicle at the shoulder. It is very important for elevating the larynx during swallowing.

http://what-when-how.com/wp-content/uploads/2012/04/tmp26119.jpg.

findings based on visualization of the vocal folds with a laryngeal mirror to the Royal Society of London in 1854. Soon thereafter Morel Mackenzie (1837–1892) electrified the laryngoscope, as the reconfigured instrument was called, based on the work of Johann Nepomuk Czermak (1828–1873). Mackenzie published the first treatise containing illustrations of the larynx based on his observations in 1865. For the first time it was possible to see the vocal folds and get a sense of their shape, size, and gross movement. It was not possible to see the details of their movement on phonation, since "seeing" the vocal folds in action was akin to "seeing" the movement of butterfly wings in flight. In short all one could get was a rough sense of the shape and form through very blurred movement (Figure 3.4) In order to really observe the details of vocal fold motion, the image needed to be slowed down.

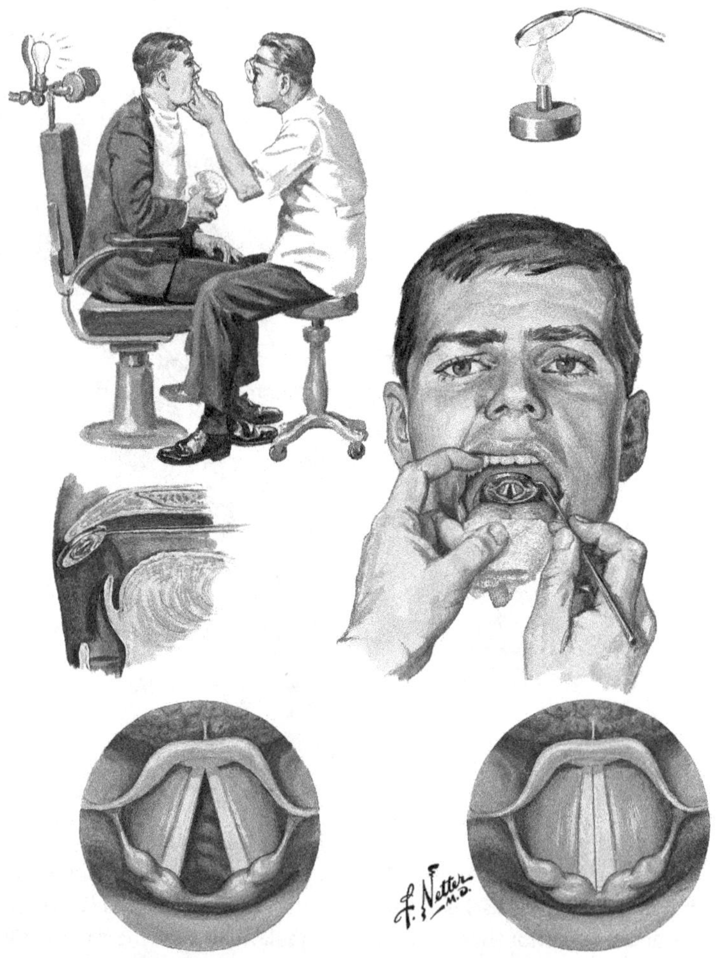

FIGURE 3.4 This is the view of the vocal folds using a laryngeal mirror. It is impossible to see any small anomalies of the vocal fold tissue. When vibrating (making sound) the vocal folds look similar to butterfly wings in flight.

Peter Mark Roget (1779–1869) published an article in 1825 explaining the optical illusion of motion by looking at a number of discrete images seen through slits in a revolving disc. Max Joseph Oertel (1835–1897) was the first person to publish a description of vocal fold movement based on stroboscopic images in 1895 (Mehta, Deliyski, and Hillman 2010). However it was not until 1937 that the scientists at Bell Laboratories, using a high-speed camera attached to a laryngoscope, were able to photograph the movement of the vocal folds (Pressman 1942, 9) (Figure 3.5). For the first time it was possible to see the movement of the intralaryngeal structures, specifically the length, width, and glottal area in motion. It took until the 1950s for electronic stroboscopes to be combined with a microphone placed on the patient's neck to pick up sound waves that the sound and moving images could be coordinated.

Only in 1977 did further developments in fiber optics, combined with stroboscopy and high-speed cinematography, allow laryngeal imaging to finally become a viable tool

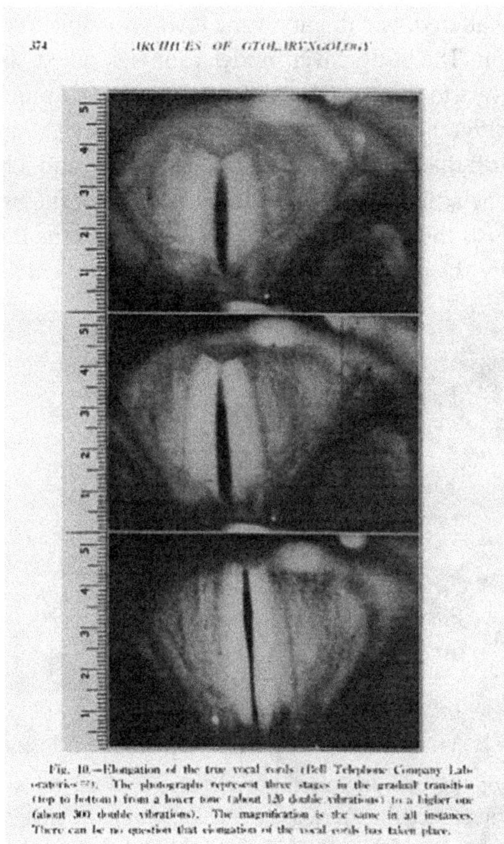

FIGURE 3.5 These images are from the earliest high-speed photographs of vocal folds in motion.
Pressman (1942, 9).

for use by laryngologists (Hirano and Bless 1993). Videostroboscopy as we know it today only became clinically viable with the advent of high-speed imaging in the 1990s.

As a consequence of all of these inventions, we are now able to see the vocal folds move and to identify lesions, scarring, or other subtle anomalies in the mobility and tissue of the vocal folds (Figure 3.6). The major innovations in laryngeal imaging and microbiology have totally changed our understanding of how these small structures work. The vibrating portion of the vocal folds are about 2 mm in length in a newborn, growing to an average of 16 mm in the adult male and 10 mm in the adult female. As we grow and mature the vocal folds also thicken, creating more vibratory mass. Until we were really able to see the microscopic detail of the vocal folds we did not have a clear idea how they vibrated. Early anatomists understood that air from the lungs was necessary to bring them into vibration, but no one really understood how they vibrated. Once it was possible to visualize the vocal folds, various theories began to emerge. In 1977 M. Hirano and his colleagues published their seminal paper describing the structure of the vocal folds accounting for the movement behaviors that can only be visualized with high resolution/high speed cameras (Hirano 1981). Since then it has been refined, but not substantially altered. For the first time it was possible to explain the subtleties of vocal fold vibration. The body-cover model, as it was called, determined that there are three primary layers to the vocal folds. Each layer has a unique histology or structure that allows for differences in stiffness or pliability, as well as contractile properties. These layers consist of: the cover, consisting of epithelium and a basement membrane that divides it from a gelatinous layer called Reinke's space or the superficial lamina propria; the middle layer is the vocal ligament, which includes the intermediate and deep lamina propria; finally, the vocalis muscle is the body of the vocal folds. Each layer of the

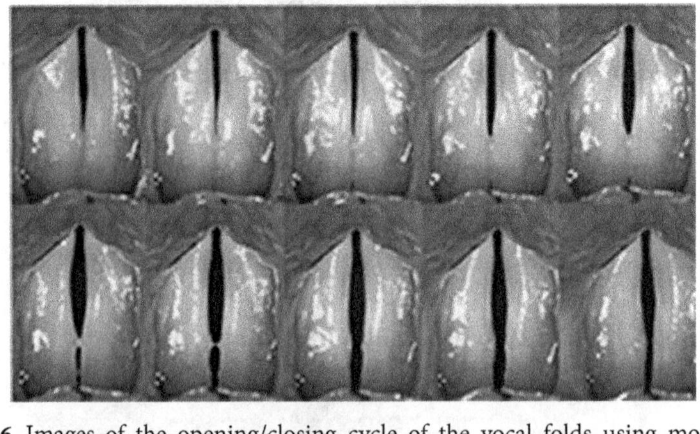

FIGURE 3.6 Images of the opening/closing cycle of the vocal folds using modern videostroboscopic equipment. Notice the increased detail including the small areas of fullness in the anterior (lower) area of the vocal folds, where there is a little stranding of mucous in the lower left frames. The vocal fold edges should be straight, and one should be able to see the point of the V (anterior commissure) at the bottom of the images.

Emory Voice Center.

lamina propria is stiffer than the one above. The cover, the gelatinous superficial layer of the lamina propria, and the epithelium together vibrate over the stiffer ligament and muscle underneath creating sound waves (Rosen and Simpson 2008). In other words, each layer has a different amplitude of vibration.

As a result of these discoveries, previous assumptions about the vibratory properties of the vocal folds needed to be reexamined. For years, scientists assumed that the laws of physics relating to the vibration of strings also applied to vocal fold vibration. The principles stated that the frequency of vibration was determined by the stiffness and mass (length and thickness) of the string. A bass piano string in an upright piano is shorter than a string for the same pitch in a grand piano, but the upright piano string is much thicker, allowing the pitch to be the same when played. When the strings are tuned it is the tension on the strings that adjusts the pitch. Recent research into the vibratory properties of vocal folds has shown that the vocal fold dynamics are much more complex. The vocal fold tissue is only able to vibrate on 3 of its 6 sides (the vocal folds are attached to one another at the anterior or front to form a V and to the cartilage at the posterior or back, and attached to the interior surface of the larynx on the sides, leaving only the anterior and posterior surfaces and one side free to vibrate). So instead of vibration that implies equal displacement of the whole (think of the side-to-side movement of a vibrating string), the vocal folds are in reality deformed when vibrating, since one side along the length of the vocal fold is continuous with the surrounding tissue. In other words, the vocal fold tissue does not move uniformly like a string. So the vocal fold vibration could not be defined by the mass of the vocal fold, because the mass at any point was different from the previous point. Frequency is still defined by vocal fold dimensions, but new equations were necessary for length, thickness, and depth. As a result, laryngologists, vocologists, and voice scientists now have a better understanding of the origin of phonotraumatic lesions—areas of scar tissue or nodules and polyps that develop on the vocal folds at the vibrating edge. No other musical instrument has these capabilities (Titze 2011; Stemple, Glaze, and Klaben 2010).

For as long as people have been living together, singers, actors, and orators, including politicians and preachers, all understood that in order to be heard when speaking to a large group they needed to be able to "project" their voice. Not just yell over the crowd, but find a way in which their voice would carry across the room, cathedral, or field with minimum effort on their part just as opera singers are trained to "project" their voice over the orchestra. In our current age of amplification, it seems that we are losing this ability to project our voices over distances.

How is it possible for people to be able to project their voice so everyone in the theater or at a political rally can hear them without amplification or strain? As discussed, the vocal folds are brought into vibration by the airstream. At rest they are open, but once the brain sends the signal to make sound they begin to close. The airflow from the lungs brings them into vibration, creating sound waves. The resultant sound waves move through the small epilarynx (the area within the larynx but above the vocal folds) up into the mouth, where they are shaped by the tongue, jaw, and lips into the specific sounds such as vowels and consonants, as well as vocal colors, bright, dark, round, and so on. Our ability to get vocally louder and softer comes from the vocal folds vibrating together with more or less impact or collision forces. This ideally comes from increased airflow from the lungs or, if the airflow is inadequate, using the muscles around the larynx to help squeeze them together.

Increased ability to "project" the voice however is a completely different process. The seed of our understanding of the harmonic and resonant properties of the voice can also be traced to the early nineteenth century. Jean-Baptiste Joseph Fourier (1768–1830), the French mathematician and physicist, was able to use trigonometric functions to convert a time-based signal, such as sound, from the original domain to a representation in the frequency domain and back again (Titze 1994). Fourier transform images include the images we see on heart monitors and representations of musical sounds or speech sounds. We could hear the voice, sense that some sounds were more resonant, but how and why was this possible? As with the laryngoscope, it was not until 1951 that the first commercially available machine that could transform auditory signals to visual images or spectrographs was developed by Kay Electric Company (Pieplow 2009) (Figure 3.7).

Our ability to color the voice and project is the result of the interplay of the overtones in the sound. Every complex tone has a basic or fundamental frequency as well as overtones or higher frequencies at the same time. Without going into too much detail, the sound waves bounce around in our mouth. In musical terms, the basic harmonics are the octave, the fifth, the fourth, the major third, and a minor third. This represents an unfiltered spectrum (Boseman 2013). The manner in which these overtones or harmonics interact with each other (i.e., filtered) will either amplify or dampen the sound. The epilarynx, the small space just above the vocal folds but within the larynx, together with the cavities in our head, primarily the mouth and nose, act as filters. So how we change the shape of our mouth, our breath, the position of the jaw, the position

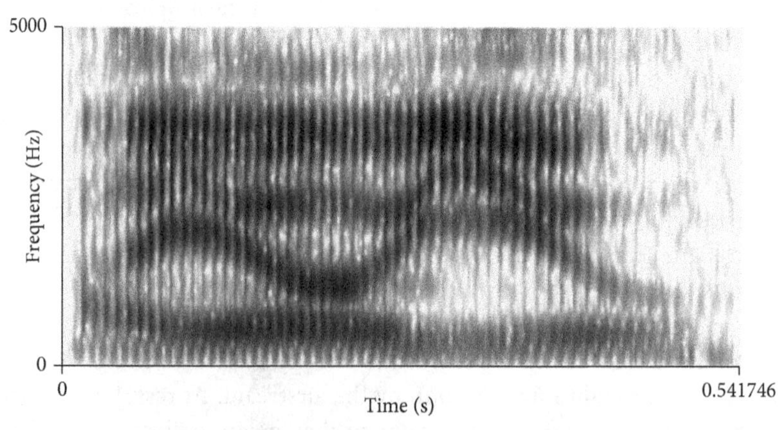

FIGURE 3.7 This is a spectrogram of someone saying, "I owe you." Notice how the middle dark horizontal lines (the 2nd and 3rd formant frequencies) change as the speaker goes from the ah-I in the word "I" to the word "owe," which are both what we call diphthongs (vowels composed of two pure sounds).

of our neck—in short, our posture—determines how well our voice will carry, the color, and overall resonance of our voice. How we shape the oral cavity in turn depends to a large extent on the flow of air, which is dependent on the efficiency of the respiratory system that in turn relies on our overall stability.

Singing and elocution teachers have preached the importance of posture for many centuries. "Stand up straight!" "Chest out!" "Stand with a noble posture." These are all familiar to anyone who studies singing or acting.

What does that mean? We all know that we should stand and sit up straight, and generally use good posture. However, what is posture? Is it standing in an "ideal" alignment that is "mechanically efficient" for the body, defined by dropping an imaginary plumb line aligning the relative positions of the ears, specifically the external auditory meatus, vertically above the midpoint of acromion on the shoulder, with the chest high (not military) through the highest point of the pelvis, through the knee joint, and in front of the ankles (Figure 3.8)? Or perhaps posture is more complex, involving not just body parts stacked up like bricks, but a dynamic interactive system that allows us to move with ease. This idea of posture as dynamic rather than static is very different.

The back specialists at the Cleveland Clinic define good posture in much the same way:

> **Good posture** involves training your body to stand, walk, sit and lie in positions where the least strain is placed on supporting muscles and ligaments during movement or weight-bearing.
>
> **Proper posture** involved keeping bones and joints in the correct alignment so that muscles are being used properly. (Cleveland Clinic 2015)

It is interesting to note that this definition presupposes that we need to train or work at good posture, that it does not come intuitively to us.

The 1947 Report of the Posture Committee for the American Academy of Orthopedic Surgeons defines both good and ideal posture in terms of injury prevention and biomechanical efficiency:

> **Good Posture** is the state of muscular and skeletal balance which protects the supporting structures of the body against injury or progressive deformity irrespective of the attitude in which these structures are working or resting.
>
> **Ideal posture** is an alignment of maximum physiological and biomechanical efficiency, which requires a minimum of stress and strain.

At the mid-twentieth century, Mabel Todd (1880–1956), the founder of Ideokinesis, and Moshe Feldenkrais (1904–1984), developer of the Feldenkrais Method, provided definitions of posture that not only included biomechanical efficiency but also recognized that there is an important gravitational aspect to posture that effects the body's internal interrelationships. They understood that internal biomechanical

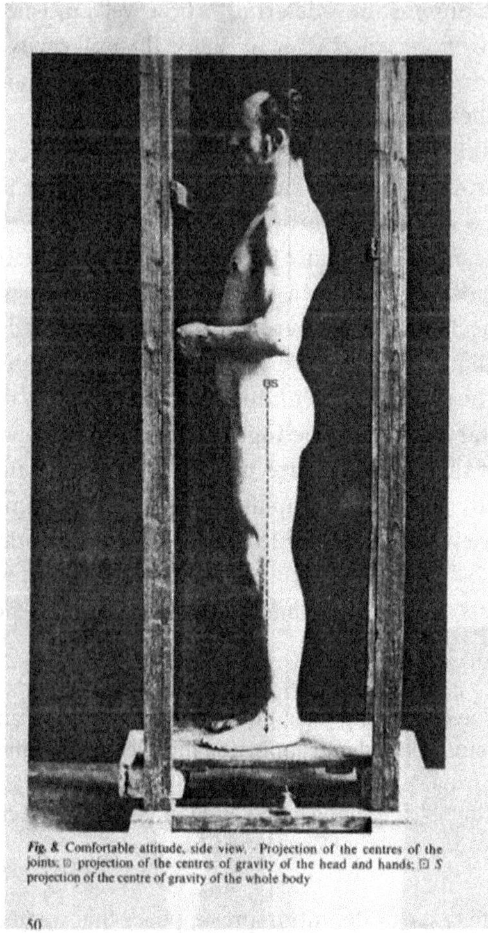

FIGURE 3.8 This image depicts the "ideal posture" based on the plumb line.

Braune and Fischer. 1899.

efficiency and the body's relationship to gravitational forces are related. Todd wrote, "the postural pattern is that of many small parts moving definite distances in space, in a scheme perfectly timed, and with the exact amount of effort necessary to support the individual weights and to cover the time-space-movement" (1937, 22). Feldenkrais defined proper posture in terms of relative movement within the whole body, agreeing with Todd that no segment of the body could move in isolation from the whole. Functional posture then defines the proper posture of the body

> (1) such that it can initiate movement in any direction with the same ease;
> (2) that it can start any movement without a preliminary adjustment;
> (3) that the movement is performed with the minimum of work, i.e., with the maximum of efficiency. (1949, 74)

If we broaden our view of posture to include all the moving parts, it becomes easier to understand how all the elements of voice production depend not only on one another in a linear fashion but also on the whole in a complex interrelationship of all the parts (Gilman 2014).

There has been a significant shift in our understanding of the role of the respiratory system in recent years. It was long assumed that the sole function of the respiratory system was gas exchange: bringing oxygen in and carbon dioxide out. However, there are an increasing number of studies focusing on a postural aspect to respiration. The respiratory system consists of the airway (nose and mouth, pharynx, larynx, and trachea), the lungs, and muscles of respiration. The diaphragm, intercostal muscles of the ribs, and many of the muscles of the back and neck are involved in respiration. Accessory muscles of inspiration include the sternocleidomastoid, the scalenes (small muscles in the neck that attach to the first or second rib), the pectoral muscles, and the serratus muscles attaching to the ribcage. All these muscles contribute to elevation/expansion of the ribs during inhalation. Accessory muscles of exhalation include the rectus abdominus, internal and external oblique muscles, and the transverse abdominus (Pettersen and Eggebo 2010; Kolar et al. 2009; Pettersen 2005; Hoit 1995).

Not defined as part of the respiratory system for gas exchange, but important when thinking about the postural role of the diaphragm, are the ribcage, spine, and shoulder girdle.

The vocal folds are incapable of producing periodic vibration without adequate airflow, which then in turn generates the sound. As discussed, the vocal folds are embedded with the larynx at the top of the trachea and below the nasal and oral passages. From the top down, the only access to the lungs is between the vocal folds and down the trachea. The heart sits between the two lobes of the lungs. The trachea, lungs, heart, and liver are protected by the ribcage. The twelve pairs of ribs attach in the back to the spine. In the front ribs 1–7 attach to the breastbone or sternum by means of costal cartilages. Ribs 8–10 are called false ribs, as they share a costal cartilage attachment with rib 7. Ribs 11 and 12 are the floating ribs, as they do not attach to the sternum at all. This configuration allows the lower portion of the ribcage more mobility.

It is the diaphragm that forms the floor of the chest cavity and the roof of the abdominal cavity. The center of the diaphragm consists of a central tendon. Running out from the central tendon like spokes on a wheel are the muscle fibers. The front fibers attach to the xiphoid process, a small bone at the lower end of the sternum. The fibers on the side attach to ribs 7–12 (the costal cartilages of 7–10 and the inner edge of 10–12). These muscle fibers interdigitate with the transverse abdominis muscles.

The posterior fibers of the diaphragm attach to the lower lumbar vertebrae by means of the bifurcated crura (or tendinous extensions on the lower spine). The quadratus lumborum and the psoas, two strong muscles involved with side bending and lifting of the legs, also attach to the upper lumbar vertebrae (Figure 3.9).

While the diaphragm only attaches to the lower portions of the ribcage as described, the displacement of the upper ribs with inhalation or exhalation has an influence on the ability of the larynx to move freely in the neck. Remember that the larynx is anchored in

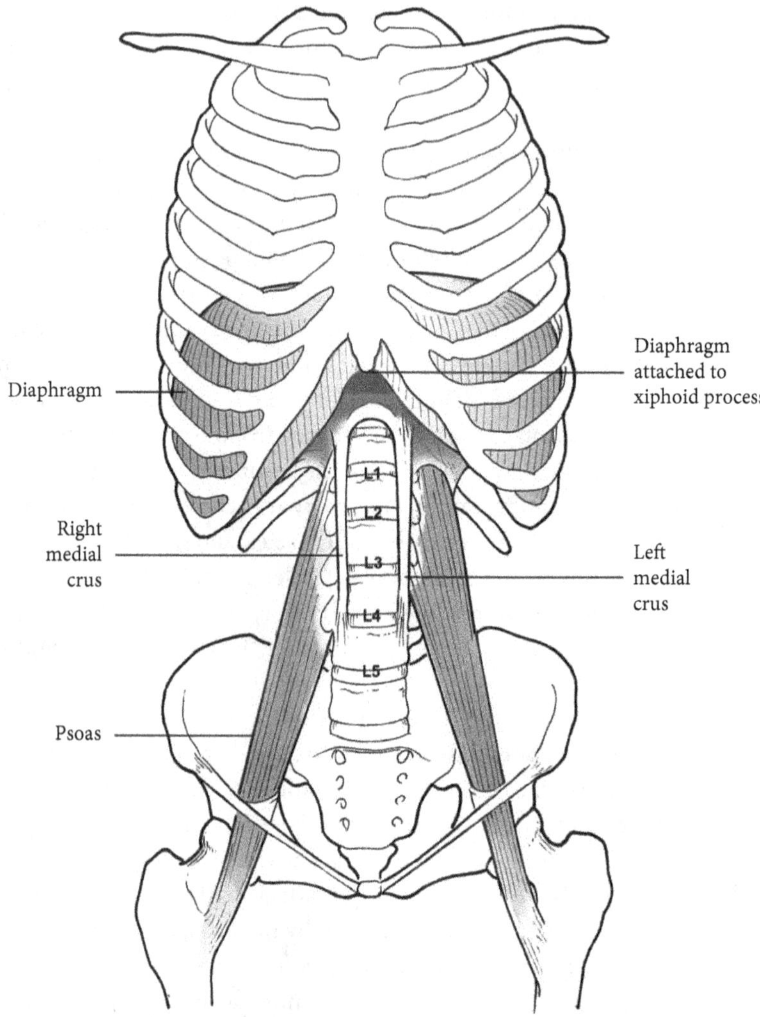

FIGURE 3.9 The diaphragm sits at the base of the ribcage below the lungs. It attaches to the lungs, the base of the breastbone (sternum) in front, and the ribs, and has two long extensions (crura) along the spine. As we inhale, the diaphragm descends, expanding the lungs and ribs compressing the intestines. On the exhale, the chest wall, which includes the intestines and ribs, contracts, pushing the diaphragm up and the air out.

Gilman personal collection.

the neck by the sternothyroid and sternohyoid muscles, both of which attach on the posterior sternum. If the sternum is pulled downward toward the pubic bone, these muscles are elongated, restricting the ability of the larynx to move freely up and down to accommodate the resonance demands of various vocal genres. The omohyoid muscle attaches at the hyoid bone, then travels under the clavicle and inserts on the top of the scapula. If the shoulders are too far forward this muscle may restrict further elevation of the hyoid, and therefore the laryngeal complex, including potentially the base of tongue muscles that also attach to the hyoid. Recent research investigating the relationship of the diaphragm and spinal stability as well as postural deformities of the chest, ribs, and spine in patients with chronic asthma has shown strong relationships between chest, rib, and spine deformities and the position of the diaphragm during quiet breathing, and with upper and lower limb movement (Lunardi et al. 2011; Kolar 2010; Lee et al. 2010; Hodges and Gandevia 2000). Although the patients studied had a serious pathology (i.e., asthma or spinal deformities), these same, though less exaggerated, somatic patterns are often observed in healthy individuals.

On normal inhalation the diaphragm contracts downward, pressing against abdominal contents and expanding the lower ribs. The abdominal contents have a fixed volume. In order to allow for optimal contraction of the diaphragm the abdominal muscles must relax, allowing for forward displacement of the viscera. In the ideal state this means that the abdominal muscles need to relax so the belly expands slightly, creating the extra space. It is the diaphragm contracting downward that facilitates inhalation due to the pressure differential between the outside atmospheric pressure and pressure inside the lungs. The lower ribs elevate with the downward movement of the diaphragm. This makes sense if one imagines that the pivot point of the diaphragm is along the inner edge of the ribs or costal cartilages. If the abdominal muscles are held in a contracted state, the intestines do not have room to move. At that point the ribcage must elevate and the lower ribs widen, facilitating the contraction of the diaphragm. In any discussion of respiration and breathing it is very important to keep in mind that we breathe differently depending on our posture and what we are doing. Breathing for singing is very different from breathing for running, lifting weights, or swimming.

During exhalation the diaphragm is passive. The air is pushed out of the lungs by means of the recoil of the chest wall (ribs and abdominal muscles). For voice use there is also an inward and slightly upward movement of the abdominal muscles, facilitating the control of the airflow.

The fact that the diaphragm and the accessory muscles of respiration attach both to the upper body, ribs, shoulder girdle, and neck, and to the lower extremities, pelvis, and legs, has direct bearing on the role of posture not only for normal breathing but also for voice. Respiration—the uptake of oxygen at the cellular level—is distinguished from breathing—the movement of air in and out of the body. In functional terms these are not separable. The efficiency of breathing and airflow is dependent on the coordination of all the parts, from the feet to the top of the head.

Vocal pedagogy texts used to include at least a page or two on posture. The image was most often that of a barrel-chested male standing chest high, maybe one arm bent, and if the whole body was depicted, then standing with one foot slightly in front of the other. Garcia's "noble posture" is a well-known example.

If on inhalation, the abdominal muscles are not released or unable to be released due to the position of the pelvis, the breathing pattern changes and becomes more about elevation of the ribs. When standing with knees locked, for example, the top of the pelvis is tilted forward, increasing the lumbar curve. In this position the abdominal muscles are not able to release as they are stretched. Standing with the weight primarily on the heels or on the balls of the feet will alter the center of balance in such a manner as to engage the muscles in the upper body by way of a counterbalance. While in this scenario the abdominal muscles might not be as contracted, the muscles of the back and ribs contract to create increased stability. Performing in high heels can be a problem as they destabilize the body. It is very important to practice in the shoes worn, so that the body can learn to accommodate the reduced flexibility of the abdominal muscles and compensate for the limited downward contraction of the diaphragm.

On exhalation there are many variations as well. During optimal exhalation the viscera are pulled up against the diaphragm with the help of the exhalatory muscles and the recoil of the ribcage. If, however, the ribcage is already elevated and the abdominal muscles are already contracted, the only movement option is for the chest (sternum and ribs) to move down. There are several observable patterns when this happens. The sternum slides down toward the pubic bone. This is the pattern we can observe when someone takes a big breath and sighs. The shoulders and chest often are lifted on inhalation and then released down on the exhalation. A subtler pattern appears as a folding of the upper body just below the sternum. The abdominal muscles press out on the exhalation, pulling the sternum downward. This is a pattern that often results with overexuberant down and out breathing, where the emphasis is on the outward movement of the abdominals rather than the balance of inspiratory and exhalatory muscles often encouraged by singing teachers. This pattern can draw the sternum down, limiting laryngeal movement; it draws the head forward, distorting the internal oral space and thus limiting resonance.

Finally, moving up to the head, exaggerated head forward and head back positions can result in increased perceived phonatory effort, even on short moments of phonation at comfortable pitch and loudness (Gilman and Johnson 2017). On laryngeal examination, the oral cavity changes shape depending on the position of the head. When the head is back, the back pharyngeal space widens and the larynx lowers—making this a popular position for classical singers. When the head is forward, the larynx is elevated and the back pharyngeal space narrows. As a result, resonance is altered. Resonance is also compromised when the shape of the vocal tract is altered.

Recently, sonography, functional MRI (magnetic resonance imaging), and radiography have been used to learn more about the subtle ways our bodies shift when using our voices. The movement of specific abdominal muscles during breathing for voicing, as well as small shifts in the length of the airway or change in shape of the vocal tract and cervical spine, can now be observed in real time. A recent series of studies by Miller and colleagues (Miller et al. 2012; Miller et al. 2014) suggest that there is a subtle shift in the cervical vertebrae of the neck during ascending glides, even with the mouth closed. We are increasingly able to look inside the body, observing even subtle changes in the position of vertebrae altering imperceptibly the complex coordination and muscle relationships throughout the body. This ability only increases what voice teachers, trainers, singers, and actors have known for years. Posture is not something apart from voice. How we move, stand, and sit is part of what makes our voices sound and be able to do all the wonderful things we can do vocally—the voice *is* the body, in more ways than we know.

References

Boseman, Kenneth. 2013. *Practical Vocal Acoustics: Pedagogic Applications for Teachers and Singers*. Hillsdale, NY: Pendragon.

Braune W., and O. Fischer. 1899. *On the Centre of Gravity of the Human Body as Related to the Equipment of the German Infantry Soldier*. Reprint ed. Spring Verlag.

Cleveland Clinic. 2015. "Posture for a Healthy Back." Accessed April 9, 2016. https://my.clevelandclinic.org/health/ns_overview/hic_Posture_for_a_Healthy_Back.

Feldenkrais, M. 1949. *Body and Mature Behavior.* Reprint ed. London: Routledge and Kagan Paul Ltd.

Gilman, Marina. 2014. *Body and Voice: Somatic Reeducation*. San Diego, CA: Plural.

Gilman, Marina, and Michael M. Johns. 2017. "The Effect of Head Position and/or Stance on the Self-Perception of Phonatory Effort." *Journal of Voice* 31 (1): 131.e1–131.e4.

Hirano, Minoru. 1981. *Clinical Examination of the Voice*. New York: Springer.

Hirano, Minoru, and Diane M. Bless. 1993. *Videostroboscopic Examination of the Larynx*. San Diego, CA: Singular.

Hodges, Paul W., and Simon C. Gandevia. 2000. "Activation of the Human Diaphragm during a Repetitive Postural Task." *Journal of Physiology* 522: 165–175.

Hoit, Jeannette D. 1995. "Influence of Body Posture on Breathing and Its Implications for the Evaluation and Treatment of Speech and Voice Disorders." *Journal of Voice* 9 (4): 341–347.

Kolar P., J. J. Sanda Neuwirth, V. Suchanek, Z. Svata, J. Volejnik, et al. 2009. "Analysis of Diaphragm Movement during Tidal Breathing and during Its Activation with Breath Holding Using MRI Synchronized with Spirometry." *Physiological Research* 58: 383–392.

Kolar P., J. Sulc, M. Kynci, J. Sanda, J. Neuwirth, A.V. Bokarius, et al. 2010. "Stabilizing Function of the Diaphragm: Dynamic MRI and Synchronized Spirometric Assessment." *Journal of Applied Physiology* 109: 1064–1071.

Lee, Linda-Joy, Angela T. Chang, Micheal W. Coppieters, and Paul W. Hodges. 2010. "Changes in Sitting Posture Induce Multiplanar Changes in Chest Wall Shape and Motion in Breathing." *Respiration Physiology Neurobiology* 170 (3): 236–245.

Lunardi, Adriana Claudia, Marques da Silva, Cibele Cristine Berto, Rodrigues Mendes, Felipe Augusto, Amelia Pasqual Marques, et al. 2011. "Musculoskeletal Dysfunction and Pain in Adults with Asthma." *Journal of Asthma* 48 (1): 105–110.

Mehta, Daryush D., Dimitar D. Deliyski, and Robert E. Hillman. 2010. "Commentary on Why Laryngeal Stroboscopy Really Works: Clarifying Misconceptions Surrounding Talbot's Law and the Persistence of Vision." *Journal of Speech, Language, and Hearing Research* 53: 1263–1267.

Miller Nicola A., Jennifer S. Gregory, Richard M. Aspden, Peter J. Stollery, and Fiona J. Gilbert. 2014. "Using Active Shape Modeling Based on MRI to Study Morphologic and Pitch-Related Functional Changes Affecting Vocal Structures and the Airway." *Journal of Voice* 28: 554–564.

Miller, Nicola A., Jennifer S. Gregory, Scott I. K. Semple, Richard M. Aspden, Peter J. Stollery, and Fiona J. Gilbert. 2012. "The Effects of Humming and Pitch on Craniofacial and Craniocervical Morphology Measured Using MRI." *Journal of Voice* 26 (1): 90–101.

Pettersen, Viggo. 2005. "Muscular Patterns and Activation Levels of Auxillary Muscles and Thorax Movement in Classical Singing." *Folia Phoniatrica et Logopedica* 57: 255–277.

Pettersen Viggo, and Torbjrn M. Eggebo. 2010. "The Movement of the Diaphragm Monitored by Ultrasound Imaging: Preliminary Findings of the Diaphragm Movements in Classical Singing." *Logopedics Phoniatrics Vocology* 35: 105–112.

Pieplow, Nathan. 2009. "A Brief History of Spectrograms." Accessed April 9, 2016. http://earbirding.com/blog/archives/1229.

Posture Committee of the American Academy of Orthopaedic Surgeons, ed. 1947. "Posture and Its Relationship to Orthopaedic Disabilities: A Report of the Posture Committee of the American Academy of Orthopaedic Surgeons." Evanston, IL.

Pressman, Joel J. 1942. "Physiology of the Vocal Cords in Phonation and Respiration." *Archives of Otolaryngology* 35 (3): 355–398.

Rosen, C., and B. Simpson. 2008. "Anatomy and Physiology of the Larynx." In *Operative Techniques in Laryngology*, 3–8. Berlin: Springer.

Stemple, Joseph C., Leslie E. Glaze, and Bernice G. Klaben. 2010. *Clinical Voice Pathology: Theory and Management*. 4th ed. San Diego, CA: Plural.

Titze, Ingo R. 1994. *Principles of Voice Production*. Englewood Cliffs, NJ: Prentice Hall.

Titze, Ingo R. 2008. "The Human Instrument." *Scientific American* 298 (1): 94–101.

Titze, Ingo R. 2011. "Vocal Fold Mass Is Not a Useful Quantity for Describing Fo in Vocalization." *Journal of Speech, Language, and Hearing Research* 54 (2): 520–522.

Todd, M. *The Thinking Body: A Study of the Balancing Forces of Dynamic Man*. Princeton, NJ: Dance Horizons Book; 1937.

CHAPTER 4

THE BODY AS MUSICAL INSTRUMENT

ATAU TANAKA AND MARCO DONNARUMMA

Introduction

Musical instrument performance solicits the human body into interaction with an acoustic, sound-producing object: the instrument. This engages the performer in forms of corporeal interplay not just with the instrument but also with the music being played, the resulting physical sound, and the space in which it is manifest. More than just a manipulation of mechanical aspects of an instrument—pressing keys, closing holes, or exciting strings—this interaction takes on a visceral dimension. Brass instruments, for example, offer acoustic resistance back to the player's lips, helping them with intonation. This interlocking of acoustic and body physiology takes on a phenomenological dimension and can be thought of as a cybernetic human-machine extended system. The idea of an extended system goes beyond any architecture of technology and becomes a set of McLuhan-esque "extensions," where the medium and characteristics of the medium—in this case sound—affects us (McLuhan 1964).

Technologies of whole-body interaction today are widespread beyond music. Game controllers use cameras and sophisticated computer-vision technology to detect user movement for gameplay. Mobile phones are equipped with inertial sensors that detect rotation and orientation in three-dimensional space, allowing tilting motions to scroll through lists and other forms of interaction without touching the screen. These devices, and the sensors in them, have been used by artists and musicians to create new musical instruments (Jensenius and Lyons 2016). But despite the embodied interaction that these digital technologies allow, do they create the same, tight corporeal coupling that we observe with the trumpet player?

Physiological sensors internalize the otherwise exterior nature of movement detection by directly sensing signals from the human body. Rather than external sensors (such as gyroscopes) reporting on the results of bodily gesticulation, biosensors are electrodes placed directly on the body that report on the corporeal activity at the source of body movement. Originally used in analytical contexts in the biomedical field to study muscle development, atrophy, or gait, they are increasingly used in interactive contexts to aid in stroke rehabilitation or even allow prosthetic limb control. The digitization and miniaturization of these technologies have allowed them to come out of the medical laboratory to be used in everyday contexts, including going on stage to be used in musical performances. Could these technologies of physiological sensing afford a connection between performer and digital music system that recalls the visceral acoustic coupling between performer and acoustic instrument? If so, can we imagine configurations of such technologies that would allow the body itself to be thought of as a musical instrument?

This chapter presents the history of embodied interaction in music, leading up to physiological sensing in contemporary experimental musical practice. It draws on theories of the body and philosophies of phenomenology (Merleau-Ponty 1962) in proposing body schemata that might be considered "instrumental." This prompts a definition of "instrument," musical and otherwise. It also demands an examination of our relationship to technology in order to understand what part of the human-machine interface becomes instrumental. Concepts of technological embodiment in cultural studies of the body (Hayles 1999; Haraway 1985) and embodied interaction in the scientific field of human–computer interaction (Dourish 2004) are thus considered to provide a cross-disciplinary view on the issue. The chapter ends by proposing two further notions drawing on musical performance: coadaptation (Mackay 2000), where the system learns the user while the performer learns the system; and configuration (Donnarumma 2012), where the capacity of the instrument and the performer are interlinked and thus mutually affect each other.

Proprioception

Proprioception is the mechanism that allows the body to determine the position of neighboring parts of the body and the effort exerted to perform a physical gesture. Schmidt and Lee (1988) describe how this is made possible by the integration of information from a broad range of sensory receptors located in the muscles, joints, and the inner ear. Proprioception is situated between two other modes of self-perception described by Merleau-Ponty: exteroception and interoception. Exteroception organizes tactile sensitivity to external objects, whereas interoception organizes the sensitivity to the movement of the body's internal organs.

According to Merleau-Ponty, "there is not a perception followed by a movement, for both form a system which varies as a whole" (1962, 111). Perception and movement function together, constituting a delicate balance between intention and performance, between the movement as intended and as it actually occurs. Proprioception can be thought of as a kind of closed-loop motor control mechanism, where the body uses the senses to compare the desired movement to the performed motion, assessing a margin of difference. This information aids in calibrating continuous movement and establishes forms of feedback where action and perception continuously complement one another.

For Merleau-Ponty, proprioception can be both conscious and preconscious. An example of conscious proprioceptive mechanism is the case where one touches the tip of the nose with the eyes closed. In this case, one does not learn the position of the nose through sight, but it is the sense of proprioception that provides this information. On the other hand, preconscious proprioception is demonstrated by a "righting" reflex, an involuntary reaction that the human body produces to correct body orientation when falling or tripping. For instance, when one falls asleep while sitting on a train, the head repeatedly tends to fall on one side and the body moves the neck muscles autonomously to correct the head position. These two modes of proprioception show that "the body and consciousness are not mutually limiting, they can only be parallel" (1962, 124).

Playing musical instruments at first glance would seem to be a clear example of exteroception. However, the masterful playing of an instrument requires more than the tactile sensitivity for manipulating the mechanics of an instrument as object. The parts of the body that are not in direct contact with the instrument, and their position and movement, can be as crucial (if not more) to tone quality and expressive musical phrasing as the body parts that actuate the mechanics of the instrument. Control of the diaphragm is fundamental in wind instrument performance and has a direct effect on the sound quality achieved by the performer, and can also lead to extended technique such as circular breathing. This can be thought of as a case where interoception becomes an important part of musical instrument performance. Elbow position of the bowing arm on a stringed instrument is fundamental not just for producing good tone but also for avoiding physical injuries such as carpal tunnel syndrome. This requires a sense of proprioception for the instrumentalist to be aware of and adjust the location of limb joints in free space. With practice, these techniques for musical instrument performance become increasingly fluid and intuitive for the performer. Can we think of these as conscious processes that, through repetition, enter the preconscious of the musician? If technologies of physiological interaction might allow the body, in the absence of a physical object, to become itself the musical instrument, exteroception may disappear completely, leaving interoception and proprioception as direct conduits to electronic sound production.

Body Schemata

To further understand the superposition of conscious and preconscious factors determining human movement in musical instrument performance, it is worthwhile to look at the notion of body schemata. These are schemes used by the body to govern posture, movement, and the use of physical implements. Body schemata can be thought of as motor control programs. Merleau-Ponty gives an example by observing the case of a blind man's white cane. For Merleau-Ponty, the stick is not an external object to the person who carries it, but rather, for its owner the stick is a physical extension of touch. The stick becomes an additional source of information on the position of the limbs, so with continuous training it becomes integrated into the body schema. It is converted into a sensitive part of the body, or extends the senses, and becomes part of the blind person's proprioceptive sense and motor programs.

Interestingly, Merleau-Ponty (1962, 145–146) also gives the example of musical instrument performers, describing the case of the organist. When rehearsing for a performance with a new organ, the organist, according to Merleau-Ponty, does not commit to memory the objective position of pedals, pulls, and stops. Rather, she incorporates the way in which given articulations of pedals, pulls, and stops let her achieve given musical or emotional values. Her gestures draw "affective vectors" mediating the expressiveness of the organ through her body. The organist does not perform in an objective space, but rather in an affective one. Body schemata constitute "knowledge in the hands," in the words of Merleau-Ponty. This form of corporeal epistemology is the basis of "enactive knowledge" proposed by both the psychologist Jerome Bruner (1968) and the scientist Francisco Varela (Varela, Thompson, and Rosch 1992).

In musical performance, body schemata drive the way the performer physically interacts with the instrument in accord with the musical or affective expression that the instrument affords. The instrument goes beyond the blind person's stick in the sense that it is more than an extension of perception. It also goes beyond the intuitive sense Merleau-Ponty's organist has in moving between different organs of different dimensions. A musical instrument invokes body schemata in ways that extend the human body's expressive potential in the projection of sound. The instrument becomes an extension in the sense of McLuhan, where a medium, or technology, shapes the relationship between us and the world in which we live. Can we, with electronic musical instruments, go beyond the affective dimension Merleau-Ponty describes in the organist? If physiological sensing technologies extend the sonic expressive potential of the performer's body directly without an intervening stick or organ, where does the enactive knowledge reside? In the sound itself?

The Technological Body

Body theory provides one lens, through a cultural analysis of the body. By going beyond Cartesian mind-body dualism to regard the body not just in its physiological manifestation, theories of the body accommodate a plurality of bodies—bodies that can be natural, social, material, or immaterial (Blackman 2008). Part of the diversity of potential bodies is as an entity enmeshed with technology. Haraway (1985) and Hayles (1999) provide interwoven definitions of what it means to be human in the increasingly intimate relations between humans and "machinic" or computational technologies, with

Hayles proposing the notion of the posthuman body. Haraway analyzes the notion of the cyborg—a cybernetic organism part human and part machine—in contemporary science fiction and modern medicine (1991). In her view, the cyborg blurs the humanist models of a unitary gender, species, or human nature by placing humans, living beings, and machines on the same ontological ground. Haraway conceives of living (human, animals, plants) and nonliving beings (machinic and computational technologies) as unbounded entities or cross-species characterized by a blurring of boundaries. From there, it follows that the body is not bounded by a membrane such as skin (the dermis) but that human and nonhuman bodies meld continuously—they are "taken apart and put together" in the realization of hybrid living entities.

Hayles's reading of the posthuman is based on a twofold premise of the meaning of embodiment. On one hand, embodiment signifies the material instantiation of an organism in a given context; on the other, it signifies the information pattern that an organism yields (1999, 2). In this view, both human being and computer code can be considered to be embodied. They are both situated—their existence is meaningful in a specific context, be it a city or a computer, and they yield and share information patterns. For Hayles, human beings and machines share information patterns through their interaction, and in so doing, they constitute each other's modalities of embodiment.

The posthuman discourse posits that technologies can be seen not simply as extending the body, but as hybridizing with it. This can be achieved at a number of different levels: through mediating interfaces, or directly though biomedical technologies. Technologies of human–machine interaction digitally mediate relationships between the body and the surrounding environment. The advancement of biomedical technologies has facilitated the development of physiological interfaces, making this link more direct with the lower level functions of the human body. This has captured the imagination of musicians, from making electronic musical instruments that capture the spatial perturbation of human movement to detecting lower level bodily function as a musical source—perhaps coming closest to the sense of body as instrument.

Embodied Interaction

If the broad deployment of new human interface devices (including the computer mouse) represented the development of computer–human interaction in the 1980s, the development of interfaces like gloves and motion capture systems in the 1990s brought interest in whole body and embodied interaction. The exponential increase in processing speed, and the possibility to render digital media in (or faster than) real time, changed not only the forms of interaction possible with computing machines but also the contexts in which they were used.

Embodied interaction, for Dourish (2004), is one in which computers go beyond metaphorical, representational connections to the physical world, but begin to function in ways that draw on how we as humans experience the everyday world. Embodiment seen from this perspective is not just corporeal but also reflects states of participation in the wider world. This parallels, perhaps, the expansive conception of the body we noted in the previous section. In outlining the challenges of embodied human–computer interaction, Dourish draws on philosophies of phenomenology, of Heidegger and Merleau-Ponty, again questioning Cartesian mind/body dualism. He uses a phenomenological framework to look at examples of ubiquitous, tangible, and social computing, offering a theoretically informed analysis of the otherwise technologically deterministic development of the technologies that now constitute the digital nature of contemporary life.

Dourish evokes Heidegger's term of "Being in the world" as a nondualistic way to think about our interaction with digital technology. He starts with the simple notion of familiarity and the tendency of tangible and social computing to exploit familiar real-world situations and metaphors to set the scene for his definition of embodied interaction. If we inhabit our bodies, which in turn inhabit the world around us, then computing systems, for Dourish, should not just present metaphors for interaction but also become mediums of interaction. He continues, by drawing on Heidegger's notions of *zuhanden* and *vorhanden* ("ready to hand" and "present at hand"), to think about the transparency of computer interfaces, and considers whether they were the object of attention, or a medium facilitating another action. If a computer interface, such as a mouse, becomes the object of our attention in an activity, it is *vorhanden* (present at hand). If, on the other hand, the device itself becomes transparent, and we act through the device as an extension (in the McLuhan-esque sense) of our body in realizing an action, it is *zuhanden* (ready to hand).

The application of Heidegger's concepts to human–machine interaction is potentially fruitful when extended to music and embodied interaction with musical instruments, be they technological or not. Learning a musical instrument (or any new piece on an instrument) typically requires an investment of time and concentration on its mechanics, posture, and fingerings. This can be thought of as *vorhanden*—focus on the object. After mastery of an instrument or with increasing confidence in a particular piece, accomplished performers describe "speaking" through the instrument, where they become just the vehicle through which the music speaks. These situations, we claim, are the transfer of instrument as interface from *vorhanden* to *zuhanden*. Is this the affective sense of Merleau-Ponty's organist? Could this be a useful way to consider the potential of physiological sensing technologies to turn the human body itself into an instrument?

The Body in Electronic Music

Despite the potential of electronic media providing extensions to the human capability for producing music, electronic and computer music, at first glance, have traditionally not implicated the human body in visceral ways. Large recording studios and modular electronic music synthesizers full of racks interconnected by patch cables are not portable in ways that we assume musical instruments need to be. Computer music programming languages, despite recent movements such as "live coding," were not originally conceived for physical performance. We tend to think of these technologies of music production, be they analogue or digital, as disembodied technologies. However, Hayles's notion of sharing information patterns indicates that things may not be so simple. Indeed, the broad assumption that digital tools for music production are disembodied is immediately proven false in uncovering the history of gestural and embodied electronic music.

One of the earliest, and most iconic, instruments in the history of electronic music is gestural. The Theremin (Figure 4.1), invented by the Russian scientist Lev Termen (Leon Theremin) in 1920, used two antennas and electrical circuitry converting electric field perturbations into musical tones to create an electronic musical instrument that was played by waving the two hands. Similar instruments, such as the Ondes Martenot and the Trautonium, were also invented in the early twentieth century (Hopkin 1997). This approach to detecting performer gesture for live electronic performance continued in the 1980s. Michel Waisvisz conceived of the instrument, The Hands (Figure 4.2), and other similar instruments at the Studio for Electro-Instrumental Music (STEIM) in Amsterdam. In 1991 Laetitia Sonami combined early glove-based technologies from first-generation virtual reality and early video gaming with domestic utility and fashion imagery to build a series of instruments called the *lady's glove* (Figure 4.3). Both The Hands and the lady's glove predated consumer electronics equipped with motion capture and used custom sensor electronics to transform performer movements into Musical Instrument Digital Interface (MIDI) data, allowing the performer on stage to articulate and sculpt electronic and digital sound synthesis (Krefeld and Waisvisz 1990).

FIGURE 4.1 Clara Rockmore performing on the Theremin in the 1930s.

FIGURE 4.2 Michel Waisvisz performing on The Hands in the 1980s.

Michel Waisvisz Archive.

FIGURE 4.3 Laetitia Sonami performing on the lady's glove in the 1990s.

This early work has spawned communities of practice in the area of new instrument building, or digital musical instruments (DMIs). The research area that encapsulates these practices is the field of new interfaces for musical expression (NIME), where interactive technologies are used to build electronic music devices that could be performed live, through gestural interaction, in an instrumental manner. However, for the most part, the instruments produced are, like traditional instruments, external objects for the performer to hold and manipulate—instruments through which music is performed.

In 1965, the composer Alvin Lucier met the scientist Edmond Dewan, who was conducting research on electroencephalogram (EEG) signals in relation to aircraft pilots' epileptic fits. Using hospital electrodes on a headband and analog circuitry to amplify alpha waves resulting from relaxed, meditative states, Lucier created a seminal work of biomusic, "Music for Solo Performer" (Figure 4.4). In the piece, the performer sits center stage wearing sensors around the head. The output of the brainwave amplification circuits drives loudspeakers connected to percussion instruments, exciting them and causing them to sound, following the alpha rhythm (8–13 Hz) of the brainwave production of the performer. Around the same time, David Rosenboom, in the early 1970s, also worked with brainwaves, in his case, connecting the signal from the brain into the signal path of an analog modular synthesizer (Rosenboom 1976) (Figure 4.5).

Did these works and experiments turn the human body into a musical instrument? Or did they begin to explore other experiential notions for the performance of music that did not depend on volitional acts of an instrumentalist?

FIGURE 4.4 Alvin Lucier preparing (aided by John Cage) for "Music for Solo Performer" in 1988 at Wesleyan University.

Photo by N. Walz.

FIGURE 4.5 David Rosenboom performing "On Being Invisible" in 1978 in Mexico City.
Archives of D. Rosenboom.

Digital Biosignal Interfaces for Music

The miniaturization of transistor electronics in integrated circuits along with the increasing practicality of analog-digital conversion and availability of digital signal processors such as the Motorola 56000 brought about a step change in physiological interface development in the late 1980s and early 1990s. While the facility of prototyping speed and convenience did not match the ease of present-day platforms like the Arduino, or the flexibility of programmable microcontrollers, advances in electronics did allow engineers and small companies to create custom, digital biosignal interfaces for the arts.

The adoption of MIDI as a communications bus for synthesizers provided a standard for intercommunication between musical devices, and a data format for the representation of performer actions—discrete actions in the form of note on/off messages, and gestures in the form of continuous controller messages. Early digital biosignal interfaces for the arts included the BioMuse, the BodySynth, and the IBVA. The BioMuse was created by Ben Knapp and Hugh Lusted, researchers at Stanford University's Center for Computer Research in Music and Acoustics (CCRMA), and was first shown publicly in 1990. It was a digital-signal processing (DSP)-based biosignal-MIDI interface that read biosignals from the heart (EEG), eyes (electro-oculogram, or EOG), and muscle (electro-myogram, or EMG) to output MIDI based on user definable mappings (Lusted and Knapp 1988). The BodySynth, created in 1991 by Chris Van Raalte and Ed

Severinghaus was an EMG music interface used by performance artists such as Pamela Z and Laurie Anderson. The Interactive Brainwave Visual Analyzer (IBVA, 1991) was created by Masahiro Kahata and was an early low-cost, portable digital EEG device for art. More recently, in order to provide an alternative to the MIDI gestural interfaces described already, Donnarumma has created the open-source instrument XTH Sense (2011). This detects mechanomyogram (MMG) signals through a compression cylinder and microphone, allowing biosignals to interface with computer systems through audio interfaces.

With the exception of the last example, these devices can be contextualized in the era from which they came—early digital technologies, MIDI-based control paradigms in electronic music, first-generation virtual reality, and human–machine interface research. Interest in new interfaces for computing accompanied the early success of the Apple Macintosh and the mouse in 1984. While the computer mouse had been invented in 1964 and had been used in specialist systems in the 1970s, it was its mass appeal and ease of use in a personal computer that captured the public imagination for new, engaging forms of interaction. The late eighties was also the first wave of virtual reality technology, and included peripherals like head-mounted displays and glove-based interfaces. There was an interest in all things virtual, with interfaces allowing the user to navigate these parallel, digital "realities." In this spirit, biosensor musical interfaces became the basis for possible "virtual instruments." In this case, the question was not whether the body itself became instrument, but whether, thanks to sophisticated human interface technology, it allowed the musician to create virtual instruments that were immaterial and purely digital.

The human–machine interface paradigm in this time categorized computer peripherals as input devices or output devices. The former were the keyboards and mouses, and the latter screens, speakers, and printers. Following this paradigm, these musical interfaces were input devices—interfaces meant to capture human action to translate as an input to command a computer system. The conception of MIDI as a protocol for music also paralleled this control paradigm, in a master/slave configuration. Seen from this perspective, new interfaces for music were thought of as controllers with which to dictate digital sound production by way of human action. We explore in what follows how this controller paradigm has evolved and changed in the time since.

Recent Performance Practice

In this section, we describe three works by the authors that parallel this evolution. They respond not just to developments in technology, but demonstrate conceptual shifts in the relation of the performer's body and technology in musical performance practice.

Kagami ("Mirror" in Japanese) was the first concert piece for BioMuse, composed and performed in 1991 by Atau Tanaka (see Tanaka 1993) (Figure 4.6). It used two

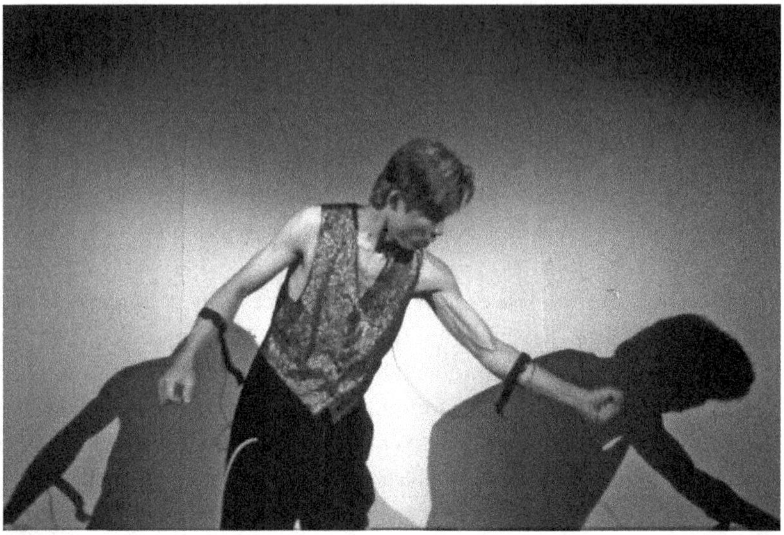

FIGURE 4.6 Atau Tanaka performing *Kagami* on the BioMuse in 1993 at _V2, Rotterdam.
Photo by J. Sprij.

channels of electromyogram (EMG—a stochastic series of electrical pulses resulting from neuron spiking), one on each forearm to track muscle tension of the performer. The BioMuse performed envelope following on the raw EMG signal and translated muscle tension intensity to continuous MIDI controller values. This transformed the performer's body into a MIDI instrument to control digital synthesizers. Instead of the dials, faders, and ribbons that typically generated MIDI continuous control data, the BioMuse captured concentrated, free space arm gesture. The MIDI data reflecting muscle tension of the performer was mapped to different parameters of frequency modulation (FM) and waveshaping vector (WS) sound synthesis on Yamaha and Korg synthesizers. A score, in the form of interactive software, ran on an onstage computer. The software determined the mapping assignment as well as ranges—specifying which synthesis parameter would be controlled by which muscle and how the MIDI control values, ranging from 0 to 127, would be mapped to salient musical values. Oscillator values might be specified as frequency in hertz or in MIDI values for diatonic notes. Modulation and waveshaping indexes would operate on a different set of ranges. In addition to defining controller mappings, the score software set the structure of the piece—the sequence of mappings that would be performed by the musician. Rather than imposing a fixed timeline, each section of the piece was invoked by combinations of muscle gestures. The score would set muscle tension thresholds as triggers. Certain combinations of triggering across the two arms (one followed by another in time, or the two simultaneously) would trip the triggers, advancing the composition to the next section, setting parameters and sequences of melodies which would then be sculpted by the mapped controller data. Building on the idea of sound as mirror for the body, and drawing on the Japanese reference in the title, the sounds of the piece used synthetic approximations of evocative sounds, starting with low throat-singing voices, punctuated by *taiko* drums, leading to rapid melodies of bell tones, finishing with siren-like breaths and *odaiko* percussion.

Ominous (2013) is a sound sculpture generated and manipulated live by a performer. In this piece, created by Marco Donnarumma, music is produced in real time using the mechanomyogram (MMG, acoustic vibrations emitted by muscular tissue). A computer program digitizes the raw signals in the form of audio input and makes it available for live sampling. The piece is based on the metaphor of an invisible object in the player's hands that is made of malleable sonic matter. Similar to a mime, the player models the object in empty space by means of whole-body gestures. A column of red light illuminates the performer's hands. The muscle sounds produced by the contractions of the performer's muscles are amplified, digitally processed, and played back through a circular array of eight subwoofers and eight loudspeakers. The MMG signal of the left bicep flows through a four-stage digital signal processing (DSP) system, whose parameters are driven by synced contractions of the right forearm muscle, with each DSP stage sending its resulting signal to one of the loudspeakers. This creates a multilayered sound where disparate sonic forms can be precisely shaped by coordinating and fine-tuning whole-body gestures that address one or more DSP stages at once. The interplay between instrument and performer relies not only on the MMG sonification but also on strategies of interaction that include extracting expressive features from the muscle sounds, mapping dynamically those features to DSP parameters, and composing the piece sections in real time using neural networks. The MMG sensors on the forearms are analyzed for high-level features, such as abruptness, subtleness, or rhythm of the player's movement. According to these features, the muscle sounds are digitally processed, played back, and spatialized. A neural network compares the stream of muscle sound features with patterns it has learned offline, detects the player's current muscular state, and subsequently loads a new set of mappings and activates or deactivates specific DSP chains, effectively changing the gesture mapping definitions throughout the performance based on the performer's dynamics. Together, the natural muscle sounds and their digital and virtual extensions blend together into an unstable sonic object. As the listeners imagine the object's shape by following the performer's physical gestures molding the red light, a kind of perceptual coupling enables listeners to hear and feel a sculpture that cannot be seen (Figure 4.7).

Myogram (2015) is a recent EMG work by Tanaka and uses a commercially available EMG interface. Two such devices (one on each forearm) each report eight channels of EMG, providing sixteen total channels across the two arms, giving a relatively detailed differentiation of muscle groups around the forearms that are invoked in manual gestures such as wrist rotation, hand flicking, and finger movement. In *Myogram*, the raw EMG data is heard through a process of direct sonification, making musical material out of the corporeal signal where electricity generated by the body provides the sounds heard in the work. The signals from pairs of EMG channels are routed to an individual speaker in an octaphonic sound system. The sensors reporting muscle tension on the ring of muscles on the left arm are heard on four speakers in the corners of the wall, stage left, from the

FIGURE 4.7 Marco Donnarumma performing *Ominous* on the XTH Sense in 2015 at CTM, Berlin.

Photo by S. Kulisch.

front of house to back of the concert hall, from below stage up to the ceiling. Likewise, the eight EMG channels on the right arm are routed to four speakers in the corners of the wall, stage right. By making rotating gestures of the wrists, the muscles in the perimeter of the forearm are solicited in circular sequence. This is heard in the concert hall as spatial sound trajectories of neuron spikes projected in the height and depth of the space, with lateral space divided in the symmetry of the body. Through the composition of the work, these raw physiological signals are subjected to different signal processing treatments. Low pass filters (LPFs) first tune the contour of the sound to de-emphasize the high-frequency transients of the neuron spikes to focus on the stochastic low-frequency fundamental. Ring modulators allow sum and difference frequencies of the EMG data relative to reference frequencies to be heard. Resonators set up resonant filters tuned to specific frequencies, which are excited by the EMG data. Frequency shifters transpose the raw muscle signal data by musical intervals. The performer responds to these different sonic contexts for the sound generated by his own body by gesticulating in space to create physiological signal output that "plays" the space, the filters, and resonators.

Control, Coadaptation, and Configuration

This range of work, from the historical works of the analogue era in the 1960s to today's digital era, demonstrates our changing relationships to technology, notions of the interface, and ultimately evolving visions of the body. The early work of Lucier and Rosenboom sought to track the performer's state and reflect that in the music. By tracking involuntary action, they focused on forms of biofeedback, reflecting the spirit of the sixties era in which these works were conceived. Other works from the time that exemplify this spirit include John Cage's *Variations VII*, performed at Experiments in Art and Technology (E.A.T.)'s seminal 9 Evenings series in 1966. In this piece, Cage created a large-scale work of an environment of multiple sound sources—live telephone lines of the city, radio, and space signal receivers, for performers to "tune in." This list of sound sources includes what Cage described as "Body" to join the sonification of communication waves and urban activity. David Tudor's *Rain Forest* (1968), while not explicitly using body signals, perhaps demonstrates the environmental links imagined at the time between technology and the surrounding world. In *Rain Forest*, sound transformations, of the sort inspired by electronic music, were realized without the use of electronics and instead by transmitting sound sources through resonant physical materials. Both look at forms of indeterminacy as a way of "being in the world."

In contrast, the work of the 1990s reflects the human-interface era. At that time, significant advancements were made in computer science in the elaboration of novel input devices. This zeitgeist influenced the development of DMIs of that era, many of which were conceived of as new musical controllers. Bio-interfaces, including the BioMuse and Bodysynth, were thus conceived of as control inputs to a computer music system, designed to allow a performer to control musical output through corporeal activity. There is a shift in paradigm, from biofeedback in the 1960s to biocontrol in the 1990s. *Kagami*, as described earlier, represents this era. In this regard, the proposition of body-as-instrument in this work plays into a technicity of electronic music, afforded by the MIDI communications protocol, of networking and interoperation of synthesizers. The MIDI specification was conceived on a master/slave model, where the default master controller imagined was a piano keyboard. The MIDI controllers built on other instrumental metaphors, such as woodwind or guitar controllers, were dubbed "alternate controllers." The BioMuse, as physiological MIDI device, proposed the body as MIDI controller, ostensibly with the richness of biosignals opening up new possibilities for synthesizer performance. While the concept body as instrument in this era sought to make these relationships more visceral and organic than the coldness of button presses and slider moves, it remained limited to a unidirectional control metaphor, one of first-order cybernetics (Wiener 1948).

Myogram moves away from control to a form of coadaption between body and sensitive system. The system in its transparency allows the body in its pure state to be heard.

By filling the concert space with sound in abstracted morphogenic dimensions—the left and right walls reflecting the two halves of the body, respectively—the gestures of the performer enter into interaction with the space of the concert hall. This creates a co-adaptation of instrument and performance venue in spatial terms, amplifying the topology of the body as well as creating a direct interaction between the performer's body and the acoustic space shared with the audience. The system becomes less transparent as the piece progresses, taking on resonant characteristics. To play these resonances musically, the performer must find the best gestures to enter into resonance—the "wrong" gesture might create excessive feedback or not excite the system at all, depending on the propagation of sound waves in the specific dimensions of the hall. By playing the space, and the interactive signal processing system, the body as instrument is extended to integrate the acoustic space inhabited by both performer and audience.

In *Ominous*, the continuous changes of the player's physiological state and the way in which the instrument reacts to those changes are both crucial to the performance. It is a relationship of *configuration*, where specific properties of the performer's body and those of the instrument are interlaced, reciprocally affecting one another. The gesture vocabulary, sound processing, time structure, and composition are progressively shaped, live, through the performer's effort in mediating physiological processes and the instrument's reactions to, and influence on, that mediation. The unpredictability of this relationship makes the performance both playful and challenging. The music being played is not digitally generated by the instrument or fully controlled by the performer. Instead, the music results from the amplification and live sampling of the player's bodily sounds. As a performance strategy, it blurs the notion of control by the player over the instrument, establishing a different relationship among them, one in which performer and instrument form a single technological body, articulated in sound and music. Here, an expanded definition of the body parallels the extended notion of the instrument.

Conclusion

This chapter presented the idea of the body as instrument as a notion relying on a multifaceted and cross-disciplinary set of resources, ranging from cultural studies of the body to human–computer interaction, from phenomenology to musical performance. The idea of the human body as musical instrument is shown to be a malleable idea changing across cultural contexts and technological advancements. The understanding of the body as instrument varies according to both the degree of technological intervention and the aesthetic concerns of individual artistic practices. The mechanisms of proprioception, the functioning of body schemata, and the broader understanding of the phenomenological basis of human embodiment, show the human body as inherently open to become, integrate, or combine with an instrument. Physiological technologies for musical performance hold a potential for the body to viscerally interact with machines.

This creates interactions between the performer and system, performer and space, and performer and audience, as forms of musical expression that are embodied, adaptive, and emergent. This enables novel ways of exploring sound and space, linking sonic stimuli and spatial resonances with the physiological processes of a musician in performance.

From feedback to control, coadaptation to configuration, the potential for body as musical instrument has evolved in a half century of practice. These developments are technological and conceptual, and intertwined. The evolution reflects different levels of integration and fusion of body and technology, but the trajectory is not a linear one. The work of the sixties reflects an ideal of integration through feedback where the technology is "ready to hand," or a transparent vehicle between body and environment. The interfaces of the nineties, though an intention to master technology, belie a greater distance between body and technology. The control interfaces are "present at hand," objects focusing attention to carry out a musical task. The body as instrument is technical, cybernetic, and imbued in a musical culture of virtuosity. The recent work extends the boundaries of the body to include the concert space, and imagined editable configurations of body and semiautonomous signal processing. This reflects forms of second-order cybernetics (Glanville 2002), where an organismic view replaces system architecture, and where noise and positive feedback temper entropy. By proposing an expanded body that encompasses technology and space, these last examples respond to the possibility of the posthuman body, proposing hybridized bodies as instruments beyond the extent of the unitary corporeal body.

References

Blackman, Lisa. 2008. *The Body (Key Concepts)*. Oxford: Berg.

Bruner, Jerome. 1968. *Processes of Cognitive Growth: Infancy*. Worcester, MA: Clark University.

Donnarumma, Marco, ed. 2012. "Biotechnological Performance Practice." Special issue, *eContact!* 14 (2). http://econtact.ca/14_2/.

Dourish, Paul. 2004. *Where the Action Is: The Foundations of Embodied Interaction*. Cambridge, MA: MIT Press.

Glanville, Ranulph. 2002. "Second Order Cybernetics." In *Encyclopedia of Life Support Systems*, edited by Georges Feller and Charles Gerday, 59–86. Oxford: EoLSS Publishers.

Haraway, Donna. 1985. "Manifesto for Cyborgs: Science, Technology, and Socialist Feminism in the 1980s." *Socialist Review* 80: 65–108.

Haraway, Donna. 1991. "A Cyborg Manifesto: Science, Technology, and Socialist Feminism in the Late Twentieth Century." In *Simians, Cyborgs and Women: The Reinvention of Nature*, 149–181. New York: Routledge.

Hayles, N. Katherine. 1999. *How We Became Posthuman: Virtual Bodies in Cybernetics, Literature, and Informatics*. Chicago: University of Chicago Press.

Hopkin, Bart. 1997. *Gravichords, Whirlies and Pyrophones*. London: Batsford.

Jensenius, A. R., and M. Lyons, eds. 2016. *A NIME Reader*. Berlin: Springer.

Krefeld, Volker, and Michel Waisvisz. 1990. "The Hand in the Web: An Interview with Michel Waisvisz." *Computer Music Journal* 14: 28–33.

Lusted, Hugh S., and R. Benjamin Knapp. 1988. "Biomuse: Musical Performance Generated by Human Bioelectric Signals." *Journal of the Acoustical Society of America* 84: S179.

Mackay, Wendy E. 2000. "Responding to Cognitive Overload: Coadaptation between Users and Technology." *Intellectica* 30: 177–193.

McLuhan, M. 1964. *Understanding Media: The Extensions of Man*. New York: McGraw-Hill.

Merleau-Ponty, Maurice. 1962. *Phenomenology of Perception*. London: Routledge.

Rosenboom, David. 1976. *Biofeedback and the Arts, Results of Early Experiments*. Vancouver, BC: Aesthetic Research Centre of Canada.

Schmidt, A. Richard, and Tim Lee. 1988. *Motor Control and Learning*. 5th ed. Champaign, IL: Human Kinetics.

Tanaka, Atau. 1993. "Musical Technical Issues in Using Interactive Instrument Technology with Application to the BioMuse." In *Proceedings of the International Computer Music Conference: September 10–15, 1993*, edited by Sadamu Ohteru. San Francisco: International Computer Music Association.

Varela, F., E. Thompson, and E. Rosch. 1992. *The Embodied Mind: Cognitive Science and Human Experience*. Cambridge, MA: MIT Press.

Wiener, Norbert. 1948. *Cybernetics*. Paris: Hermann.

PART II

THE MUSICAL BRAIN, PSYCHE, AND BEYOND

CHAPTER 5

MUSIC CHANGES THE BRAIN

PAUL LENNARD

WHETHER we are listening or playing, music changes our brain. In as little as hours, alterations in the auditory environment around us can modify the way we process sounds for many subsequent hours. From moment to moment our brains adjust how they process the acoustic world. Over time, and particularly during critical periods in development, our brains are determining what sounds and complex auditory patterns are relevant and what should be ignored. While there is strong evidence that the auditory environment helps shape the brain, the question remains as to whether we are all initially "wired" for music in the same way (Levitin 2006; Jourdain 1997). Not only can music remind us of emotions but also it can induce them and lead to observable physiological responses such as changes in heart rate and breathing. Music's direct linkage between mind and body not only holds great promise for treating the ill but also provides a window through which we can elucidate how the brain represents the outside world. Conventional wisdom suggests that language is mediated by the dominant (usually left) cerebral hemisphere and the other side processes music. The reality is far more complex. As we learn to understand music, we seem to shift from a holistic representation of the music to formally treating it as if it were a language. In all of us, musical elements such as pitch, rhythm, melody, musical memory, and imagery are processed in many different areas of both brain hemispheres. However, as we learn the patterns and syntax of music, there seem to be plastic changes in the brain that suggest that the circuitry of language also starts to subserve music. This chapter examines the evidence for how music changes the brain. Neuroplastic changes are mapped in terms of developmental, morphological, and functional parameters. The chapter also looks at shifts in the patterns of lateralization and correlates them with age and musical experience. Finally, this chapter addresses the question of how the relationship between language and music provides insight into how our brain deals with the complex world, both within and around each of us.

Short-Term Neuroplasticity

At every waking instant we are bombarded with an enormous diversity of sensory inputs. We must continually focus on the stimuli that are relevant at that moment. Selective attention provides one major avenue by which we filter the sensory world in order to process needed information. But our brain can also actually be altered by the sensory information impinging on it. This experience-dependent neuroplasticity provides a mechanism of fine-tuning the brain and optimizing appropriate responses. In this section the focus will be on short-term changes. The plastic brain changes produced by exposure to "constrained" music for durations as short as several hours will be examined. Next, experience-dependent brain changes in the human infant during critical periods for both music and language will be described. This experience-dependent fine-tuning represents a form of deafferentation. That is, the brain is learning to focus on auditory stimuli often present, to the relative exclusion of stimuli not normally experienced. The brain also can have an immediate emotional and physiological response that has the potential to provide an important pathway to wellness. Finally, multimodal changes resulting from intensive music training for as little as eight days will set the stage for a subsequent discussion of the long-term effects of musical training and the differences between the brains of musicians and nonmusicians.

What if you suddenly stopped hearing, or had greatly reduced the sound you normally heard at a specific frequency? How quickly would your brain adjust to this new sensory reality? In a fascinating experiment, Christo Pantev and colleagues (Pantev et al. 1999) asked subjects for their three favorite CDs and processed them with a notch filter centered around 1 kHz (0.7–1.3 kHz, 96 dB/oct). Subjects indicated that they could clearly hear the difference in the processed CDs and that they were able to adapt easily and still appreciate the music. Each subject was asked to listen to his or her altered CD for three hours. Before and after each session, Magnetoencephalograpic (MEG) measurements were used to test the subject's responsiveness to a band-passed noised burst centered at 1 kHz. This procedure was repeated three times, usually on consecutive days. The cortical response to the 1 kHz test response was significantly reduced after listening for three hours to music with the 1 kHz frequency dramatically attenuated. By the beginning of testing the next day, baseline sensitivity to 1 kHz was restored. Within a mere three hours, the brain can fine-tune itself by reducing its sensitivity to a sound frequency that is no longer a major constituent of the auditory environment. The authors suggest that this functional deafferentation is likely due to a reduced responsiveness of auditory neurons to the 1 kHz frequency. This proposal is consistent with findings from animal studies. For instance, instead of a functional deafferentation, a restricted cochlear lesion in guinea pigs led within hours to a shift in the characteristic frequency of clusters of cortical cells toward frequencies not impacted by the lesion (Robertson and Irvine 1989). Rapid plastic changes have not only been seen in deafferentation studies but also have been observed in learning paradigms. Once again in guinea pigs, classical conditioning using a tone conditioned stimulus (CS) caused an increased response to the CS frequency in the receptive field of clusters of cortical cells after as little as five conditioned stimulus-unconditioned stimulus (CS-US) pairings (Edeline, Pham, and Weinberger 1993). In part based on the 1999 plasticity study described previously, Pantev and collaborators have explored a promising new technique to help patients with tinnitus that they call tailor-made notched music training (Pantev, Okamoto, and Teismann 2012).

Short-term plastic changes speak to the dynamic abilities of the brain to adapt to an ever-changing sensory world. In essence, based on recent history the brain is adjusting its own sensitivity. But over developmental time, human infants are also faced with the task of shaping input-output response properties of the nervous system to be suited for the culture in which they find themselves (see Trehub's chapter for more on this subject). It is not surprising that by a year of age, the infant brain is profoundly sensitive to language (Kuhl and Rivera-Gaxiola 2008) and music (Hannon and Trainor 2007). While a full discussion of this process of enculturation is beyond the scope of this chapter, an example of this form of developmental plasticity from both music and language acquisition should help highlight the importance of experience-dependent modification of the auditory system. The ability to detect disruptions in isochronous meters (e.g., 2:1, 3:1) characteristic of Western rhythms was compared to the ability to detect disruptions in non-isochronous Balkan rhythms (3:2) in six-month-old (Hannon and Trehub 2005a) and one-year-old (Hannon and Trehub 2005b) North American children. Western and Balkan meters with preserved and disrupted patterns were presented to the infants. Increased look-time was interpreted as a novelty preference for patterns that did not preserve the "expected" pattern. Infants six months of age were able to detect disruption in both Western and Balkan rhythms. By one year of age, infants look times indicated that they were able to detect violations in Western rhythms but no longer distinguished disruptions in Balkan meters. This can be interpreted as a fine-tuning of the auditory system between six months and a year, such that it preferentially responds to the familiar Western rhythmic patterns that it hears every day. A similar situation can be observed during language acquisition. Event-related potentials (ERPs) were used to evaluate the phonetic perception of American infants from monolingual households to English and Spanish consonant contrasts (Rivera-Gaxiola, Silva-Pereyra, and Kuhl 2005). Brain potentials indicated that the seven-month-old infants were able to discriminate both the English and Spanish contrasts, but as a group, the discriminatory ERP responses of the eleven-month-old infants to the nonnative contrasts had disappeared.

So far, the focus has been on the role of auditory experience in shaping future processing of sound. Music, however, can also have immediate and profound effects on the brain and brain-mediated physiological functions. Considerable research has been devoted to examining the areas of the brain that are activated when listening to pleasurable music (Zatorre and Salimpoor 2013). For some, certain specific musical passages give such an intense pleasurable response that it is described as giving the person "chills," "thrills," or "shivers" down the spine (Sloboda 1991). In a fascinating study (Blood and Zatorre 2001), university students were asked to select a classical music passage that reliably elicited intense pleasure or "chills" in them. The students indicated experiencing chills during 77 percent of the trials. When compared to control passages (music selected by a different subject), passages eliciting chills were correlated with increases in heart and respiration rates, as well as changes in cerebral blood flow in brain areas implicated in reward and emotion. The study indicates a rather direct and robust activation of the autonomic nervous system by music. It is likely that the relationship between music and the autonomic nervous system represents a major underlying foundation for the success of music therapies in helping to treat a variety of medical issues. Our choice of music can either excite us or calm us. As an example, for patients diagnosed with mild Alzheimer's disease, when compared to silence, playing the "Spring" concerto of Vivaldi's *The Four Seasons* caused an improved autobiographical recall (Irish et al. 2006). The authors found no differences in arousal (measured by galvanic skin response) or attention, but attributed the improved performance to an observed significant reduction in state anxiety as measured on the State Trait Anxiety Inventory.

While it is clear that passive listening to music can cause neuroplastic changes, does short-term musical training enhance brain changes, and if so, how? In a pair of studies, training versus simply listening with no specific task (passive listening) was examined for melodies (Lappe et al. 2008) and rhythms (Lappe et al. 2011). Nonmusicians received piano lessons for eight sessions over two weeks. In each study, the group that played the piano was compared to a control group that simply listened to the lesson. Groups were tested with MEG for discrimination of violations in expectations of melodic fragments or rhythms based on an increased mismatched negativity (MMN). In both the melody and rhythm study, the group that played the piano showed enhanced MMN when compared to the controls that just listened. The increased MMN of the piano playing group is interpreted as an enhanced plasticity in the auditory cortex and attributed to the involvement of active sensorimotor training as well as passive auditory input.

Musical Training Shapes the Brain

If hours to days of music exposure can result in neuroplastic changes, what is the consequence of extended musical training? In other words, how does the musician's brain differ from the brain of a nonmusician? One approach to answering this question is to employ voxel-based morphometry. A voxel is the smallest resolvable three-dimensional (3-D) space measured by a magnetic resonance imaging (MRI). A voxel-by-voxel morphometric comparison of gray matter volume in various cerebral areas showed marked differences between musicians and nonmusicians (Gaser and Schlaug 2003). The demands of playing an instrument are multimodal, and it is not surprising that a number of cortical areas showed volume increases in the musician. Increases in volume were found in a variety of areas including: Heschl's gyrus in the temporal lobe, which contains the primary auditory cortex (processing sound); primary motor areas (producing movements); somatosensory areas (processing tactile information); and premotor and cerebellar areas (involved in motor planning). Volume increases were also found in the superior parietal cortex, which plays an important role in visual-spatial processing and integrating visual, auditory, and tactile information. As the authors suggest, this multimodal integration, along with the superior parietal area's connection to the premotor cortex, position this area as a prime player in coordinating the translation of musical notation and auditory feedback into the motor plans necessary for playing an instrument. For a more detailed discussion of studies comparing the brains of musicians and nonmusicians the reader is referred to two excellent reviews (Schlaug 2015; Wan and Schlaug 2010).

If musicians have brains that are different from the brains of nonmusicians, are these differences somehow attributable to starting musical training prior to some developmental critical period? It appears that while the majority of changes can most likely be correlated with the length and intensity of training, there does seem to be a critical period that can influence certain structural and functional properties. The corpus callosum (CC) is the major pathway connecting the two cerebral hemispheres. A morphometric MRI study (Schlaug, Jancke, Huang, Staiger, et al. 1995) showed that the midsagittal area of the anterior half of the CC is statistically larger in musicians than in nonmusicians. This in itself is not terribly surprising and may relate to more developed bimanual dexterity. What was notable in the study is that the difference between musicians and nonmusicians was attributable to an increased area only in the musicians who had begun training before the age of seven years old. There was no area difference between nonmusicians and musicians who began training at the age of seven or older. There are other indications that the period around the age of seven represents a critical period for brain changes related to musical training. In a MEG study of use-dependent functional reorganization (or augmentation) of the right auditory cortex, it was found that the strength of activation was significantly greater for piano tones than an equivalent pure tone for musicians but not for nonmusicians (Pantev et al. 1998). When the age that the musicians started training was factored in, a significant linear correlation between age and strength of activation was seen. Enhanced cortical responses were seen for musicians who started before the age of nine, and the earlier they started the stronger the activation. A similar age-dependency was seen in a MEG study comparing the somatosensory cortex representations of the digits of the left hand in string players and nonmusicians (Elbert et al. 1995). Not only was cortical activation stronger in the string players but also the strength of activation was much greater if the musician began playing before the age of twelve. The influence of musical training can be observed as early as ages four to five years. An EEG study (Shahin, Roberts, and Trainor 2004) comparing four- to five-year-old children taking Suzuki music lessons with age-matched nonmusician controls showed the Suzuki pupils had larger auditory evoked potentials (AEPs). In Suzuki pupils, both pure tones and piano tones showed greater P1 responses, which were attributed to greater activation of the primary auditory cortex.

While cross-sectional studies provide an informative "snapshot" of differences before and after musical training, one cannot eliminate the influences of preexisting differences. This is particularly true when using the small sample sizes dictated by practicality in MEG and MRI studies. Longitudinal studies provide an appealing and robust alternative study design. Several longitudinal studies have looked at cortical changes in children over the course of approximately one year of musical training. In a one-year study (Fujioka et al. 2006), MEG response to violin tones and noise bursts were compared in a group of students taking Suzuki music lessons with students receiving no musical training. Children were four to six years of age at the start of the study and were tested four times with an inter-test interval of three to four months. Over the course of the four sessions a change was seen in the auditory evoked field for violin tones but not noise bursts in the Suzuki trained children. The authors suggest that the distinction between violin tones and noise may reflect the development of neuronal networks for the categorization of important sounds. The differential response to violin tones versus noise bursts was not seen in the control group. In a large longitudinal study (Schlaug et al. 2005), a cohort of fifty five- to seven-year-old children about to begin music lessons and a matched control group of twenty-five children were tested using both functional MRI (fMRI) and voxel-based volumetric MRI. Preliminary reports after one year are reported for half the children. A trend toward greater volume of gray matter, but no change in CC size, was seen for the instrumental group. The fMRI results of melodic and rhythmic discrimination showed no changes in the control group, but suggest functional changes in the instrumental group in the temporal parietal junction and in the auditory association areas of both temporal lobes. In another longitudinal study, a group of six-year-old children about to begin weekly half-hour private keyboard lessons were compared to a control group that did not receive formal music training, but participated in a weekly group music class that engaged in singing and playing percussion instruments (Hyde et al. 2009). In this study, structural MRI data using deformation-based morphometry was used to test the children at the onset of the study and after approximately fifteen months. After fifteen months, the keyboard group, when compared to controls, showed increases in a number of areas, including the right primary motor cortex in the area controlling left hand movements, the corpus callosum, and the right primary auditory cortex. The next section discusses further the issue of hemispheric localization of music functions and examines changes in lateralization that seem to be correlated with musical training.

Localization: Music, Language, and Lateralization

It is generally accepted that the dominant cerebral hemisphere is correlated with both speech production and handedness. For most people this represents right-handedness and the localization of major speech centers to the left cerebral hemisphere. While it is most accurate to refer to the dominant and nondominant hemisphere, as a matter of convention and convenience, the literature simply talks about left and right hemispheres with the assumption that the left hemisphere is the dominant hemisphere. The importance of the left temporal gyrus as a major center for speech production has been known since the pioneering works of Pierre Paul Broca in the 1860s and Karl Wernicke in the 1870s. In contrast, behavioral (Bever and Chiarello 1974), lesion (Zatorre 1985), and imaging studies (Hyde, Peretz, and Zatorre 2008) indicate that the superior temporal gyrus (STG) of the nondominant right cerebral hemisphere plays a similar role for processing music. Robert Zatorre and colleagues have proposed that the left auditory cortex is anatomically and functionally fine-tuned for the rapid temporal processing required for speech tasks, while the right auditory cortex is specialized for fine resolution of the spectral information necessary for processing music (Zatorre, Belin, and Penhune 2002; Zatorre and Belin 2001). It should be emphasized that while the left auditory cortex is the "seat" of language processing and the right auditory cortex plays a similar role for music, both left and right auditory cortices are involved in processing speech and music. A dramatic example of this was reported in a study that examined the activation of a single fMRI voxel in the right auditory cortex when subjects were instructed to speak a trisyllabic word, sing the word to a specific melody, and sing the same melody using the syllable "vi" (Schön, Gordon, and Besson 2005). In this particular volume of the right auditory cortex, the activation was seen to be highest for the sung word, intermediate for the sung "vi," and lowest for the spoken word. The authors emphasize that the results do not indicate that the voxel was specific for song, but rather that it participates in both song and spoken word.

I am married to a professional musician with absolute pitch and it is abundantly clear that she and I hear and process music differently. In the simplest of terms, I listen to music holistically, either liking or disliking a piece and never being aware of the structural components that provide the framework of the composition. On the other hand, for my wife music has a specific syntax and set of expectations. It has a grammar quite as precise as the grammar of speech. She not only is acutely aware of intonation but also hears the quality of the notes, the subtlety of the rhythm, and the line of the music in a way that is opaque to me. I hear a favorite piece several times and each time I like it. Each time she hears the piece, she assesses whether the musicians have something new to say. She notices single wrong notes in what to me is the blur of a rapid run during a piano performance. We listen differently. Does that mean our brains have been wired differently, and, if so, in what way? As the neurosciences have begun to focus on the musical brain there has been a growing body of evidence that suggests that while the nonmusician primarily depends on the right auditory cortex, localization of function in musicians is rather dramatically shifted to the left hemisphere. Are the brains of musicians changing in such a way that they are utilizing the neural networks of speech to process music? As early as 1974 there was an indication that musicians may exhibit a shift from holistic to analytic musical processing concomitant with a leftward cerebral hemisphere lateralization (Bever and Chiarello 1974). Auditory information from the ear primarily projects to the contralateral auditory cortex. The experiment examined whether a melodic recognition task would be easier or harder when played monaurally through one ear. The results indicated significant differences between musically naive and musically experienced subjects. Musicians performed best when presented with the stimulus via the right ear/left hemisphere, while the nonmusicians performance on the melodic recognition task was better when the stimulus was presented to the left ear and presumably processed in the right auditory cortex. These behavioral results have been confirmed using voxel-based morphometric MRI measurements (Gaser and Schlaug 2003), which found that professional musicians had greater volume of gray matter in the left Heschl's gyrus than did nonmusicians (with amateur musicians possessing an intermediate volume). This leftward increase in Heschl's gyrus containing Brodmann Area 41 (the primary auditory area) is in direct contrast to the bilateral volumetric changes seen in many of the other areas examined when musicians are compared to nonmusicians.

It is of interest to note another voxel-based morphometric study that compared male symphony orchestra musicians to nonmusicians and found a significant increase of gray matter in the musician's Broca's area of the left cerebral hemisphere (Sluming et al. 2002). Broca's area (Brodmann areas 44 and 45) is involved in both speech production and language comprehension; it is connected to Wernicke's speech area in the STG by the arcuate fasciculus (AF) white matter tract. When the volume of the AF was measured using diffusion tensor imaging, it was found that tract volume was highest in singers, intermediate in instrumentalists, and lowest in nonmusicians (Halwani et al. 2011). When the left and right AF were compared, it was found that the left AF was significantly larger than the right in singers when compared to either instrumentalists or nonmusicians.

Rather than examining leftward brain asymmetries as a function of musical training, another approach has been to look at brain differences in musicians with absolute pitch, which is a relatively rare ability to name individual tone pitches (e.g., a "C" or a "D♯") without an external reference. Absolute pitch is contrasted with relative pitch, which is the ability to identify intervals (e.g., a third or fifth) without being able to name the individual notes, an ability shared by all trained musicians. Notice that this is a matter of naming, typically requiring training at an early age, and thus represents an intersection between music and language training. There are various issues and controversies surrounding absolute pitch (Levitin and Rogers 2005). An analysis of brain differences in absolute pitch musicians versus relative pitch musicians and nonmusicians has focused on the planum temporale (PT). The PT is located just posterior to Heschl's gyrus on the STG and is coincident with Wernicke's speech area and the secondary auditory cortex (Shapleske et al. 1999). The planum temporale's claim to fame is that it is one of the most asymmetrical structures in the human brain that is for the most part morphologically bilaterally symmetric. In fact, by the twenty-ninth week of gestation the left PT is measurably larger the right PT (Wada, Clarke, and Hamm 1975). What is interesting for this discussion is the considerable body of evidence that the left asymmetry of the PT is greater in musicians with absolute pitch than in musicians with relative pitch or control groups. This enhanced asymmetry has been documented by means of positron emission tomography (Zatorre et al. 1998), region-of-interest–based MRI morphometry (Schlaug, Jancke, Huang, and Steinmetz 1995), and voxel-based MRI morphometry (Luders et al. 2004). Using diffusion tensor imaging, it has also been found that fiber tract connectivity within the posterior area of the STG is greater in absolute pitch musicians (Loui et al. 2011). Functionally, pitch naming has been shown to preferentially activate the left PT in subjects with absolute pitch (Wilson et al. 2009). Gottfried Schlaug reports on an experiment designed to evaluate the function of the enhanced leftward asymmetry in musicians with absolute pitch (2004). Subjects were evaluated with fMRI while performing either a phoneme or tone test. In each case the task was to identify whether the phoneme or tone was the same as the one presented two trials before. The absolute pitch musicians exhibited a strong activation of the left temporal lobe for both the phoneme and tone test, while relative pitch musicians and nonmusicians showed a leftward response to phonemes but a rightward or symmetric response to tones.

If the leftward lateralization seen in musicians subserves a shift from a holistic analysis to a more formalized syntactical analysis, one might expect a certain overlap of brain areas processing both language and music. Aniruddh Patel has suggested that linguistic and musical syntax share common cortical processes (2003). This assertion is supported by studies that show musical deficits in people with language disorders. In contrast to controls, children diagnosed with specific language impairment were seen to have an absent or diminished early right anterior negativity (ERAN) ERP when tested on chord progressions for a violation of expectancies (Jentschke, Koelsch, and Friederici 2005). It has also been found that patients with Broca's type aphasia do not exhibit the harmonic priming seen in age- and education-matched controls (Patel 2005). As mentioned earlier in the chapter, both cerebral hemispheres participate to some degree in both language and music processing. Functional overlap of music and language areas not only can exacerbate cross-performance on tasks, but may also hold the key to dramatic improvement of patients with significant language deficits such as nonfluent aphasia. The technique of melodic intonation therapy, which entails singing lyrics (rather than speaking words) along with left-hand finger-tapping of each syllable, is turning out to be a remarkably effective procedure for presumably engaging the right hemisphere and improving speech productions in aphasia patients (Schlaug, Marchina, and Norton 2008).

There is overwhelming evidence that music and musical training can change your brain. But, when we hear music, are we all hearing the same thing? When an orchestra tunes to A4, all the instruments are playing a note with a fundamental frequency of 440 Hz, yet we can discriminate each instrument. Timbre, the unique set of harmonics and their strength, is what gives each instrument its specific character. Even when a note is missing its fundamental frequency, the brain can fill in the fundamental and "re-create" the pitch. Lesion studies have implicated Heschl's gyrus of the right hemisphere as important for determining the missing fundamental (Zatorre 1988). Taking advantage of the brain's ability to fill in a missing fundamental, Peter Schneider and colleagues (Schneider, Sluming, Roberts, Scherg, et al. 2005) have employed a fascinating protocol to examine whether we all hear the same way. Pairs of tones were created such that the second tone had a shift in one direction for the missing fundamental and a shift in the opposite direction for the bottom order harmonic present. Subjects were asked to indicate the dominant direction of pitch change. The authors found that subjects could be divided into f_0 fundamental pitch listeners, who based their determination of pitch change on the direction of change of the fundamental, and f_{sp} spectral pitch listeners, who processed the pair tones based on the directional shift of the harmonics. Musical instrument preference was correlated with whether an individual was categorized as a fundamental or spectral pitch listener (Schneider, Sluming, Roberts, Bleeck, et al. 2005). It was found that fundamental pitch listeners are more likely to play percussive or high-pitched instruments with short sounds and had a larger volume of gray matter in the left lateral Heschl's gyrus. On the other hand, spectral pitch listeners tended to player lower pitched instruments that produced sustained tones and had a dominant right lateral Heschl's gyrus. It is pointed out that this distribution is consistent with the notion proposed by Zatorre and colleagues (1998) that the left temporal lobe is fine-tuned for rapid temporal processing and the right temporal lobe tailored for fine spectral analysis. Of course, one cannot distinguish whether the instrument "changed" the musician or the musician was born with a listening type that made them predisposed to a particular type of instrument.

This chapter has examined how listening to music and music training can change the brain with a specific focus on the auditory cortices. But music also involves memory and often elicits emotional and visceral reactions. Playing an instrument not only involves the auditory system but also must depend on motor planning, motor output, memory, the visual system, and somatosensory feedback. We are all impacted by music. The human brain has been uniquely adapted for processing this important part of our lives, and the study of the neural mechanisms subserving music and music-making can ultimately uncover important general principles of human cognitive processing.

References

Bever, Thomas G., and Robert J. Chiarello. 1974. "Cerebral Dominance in Musicians and Nonmusicians." *Science* 185 (4150): 537–539.

Blood, Anne J., and Robert J. Zatorre. 2001. "Intensely Pleasurable Responses to Music Correlate with Activity in Brain Regions Implicated in Reward and Emotion." *Proceedings of the National Academy of Sciences of the United States of America* 98 (20): 11818–11823.

Edeline, Jean-Marc, Phuc Pham, and Norman M. Weinberger. 1993. "Rapid Development of Learning-Induced Receptive Field Plasticity in the Auditory Cortex." *Behavioral Neuroscience* 107 (4): 539–551.

Elbert, Thomas, Christo Pantev, Christian Wienbruch, Brigitte Rockstroh, and Edward Taub. 1995. "Increased Cortical Representation of the Fingers of the Left Hand in String Players." *Science* 270 (5234): 305–307.

Fujioka, Takako, Bernhard Ross, Ryusuke Kakigi, Christo Pantev, and Laurel J. Trainor. 2006. "One Year of Musical Training Affects Development of Auditory Cortical-Evoked Fields in Young Children." *Brain* 129 (10): 2593–2608.

Gaser, Chritian, and Gottfried Schlaug. 2003. "Brain Structures Differ between Musicians and Non-Musicians." *Journal of Neuroscience* 23 (27): 9240–9245.

Halwani, Gus F., Psyche Loui, Theodor Rüber, and Gottfried Schlaug. 2011. "Effects of Practice and Experience on the Arcuate Fasciculus: Comparing Singers, Instrumentalists, and Non-Musicians." *Frontiers in Psychology* 2: 156.

Hannon, Erin E., and Laurel J. Trainor. 2007. "Music Acquisition: Effects of Enculturation and Formal Training on Development." *Trends in Cognitive Sciences* 11 (11): 466–472.

Hannon, Erin E., and Sandra E. Trehub. 2005a. "Metrical Categories in Infancy and Adulthood." *Psychological Science* 16 (1): 48–55.

Hannon, Erin E., and Sandra E. Trehub. 2005b. "Tuning in to Musical Rhythms: Infants Learn More Readily Than Adults." *Proceedings of the National Academy of Sciences of the United States of America* 102 (35): 12639–12643.

Hyde, Krista L., Jason Lerch, Andrea Norton, Marie Forgeard, Ellen Winner, Alan C. Evans, et al. 2009. "Musical Training Shapes Structural Brain Development." *Journal of Neuroscience* 29 (10): 3019–3025.

Hyde, Krista L., Isabelle Peretz, and Robert J. Zatorre. 2008. "Evidence for the Role of the Right Auditory Cortex in Fine Pitch Resolution." *Neuropsychologia* 46 (2): 632–639.

Irish, Muireann, Conal J. Cunningham, J. Bernard Walsh, Davis Coakley, Brian A. Lawlor, Ian H. Robertson, et al. 2006. "Investigating the Enhancing Effect of Music on Autobiographical Memory in Mild Alzheimer's Disease." *Dementia and Geriatric Cognitive Disorders* 22 (1): 108–120.

Jentschke, Sebastian, Stefan Koelsch, and Angela D. Friederici. 2005. "Investigating the Relationship of Music and Language in Children: Influences of Musical Training and Language Impairment." *Annals of the New York Academy of Sciences* 1060: 231–242.

Jourdain, Robert. 1997. *Music, the Brain, and Ecstasy: How Music Captures our Imagination.* New York: W. Morrow.

Kuhl, Patricia, and Maritza Rivera-Gaxiola. 2008. "Neural Substrates of Language Acquisition." *Annual Review of Neurosciences* 31: 511–534.

Lappe, Claudia, Sibylle C. Herholz, Laurel J. Trainor, and Christo Pantev. 2008. "Cortical Plasticity Induced by Short-Term Unimodal and Multimodal Musical Training." *Journal of Neuroscience* 28 (39): 9632–9639.

Lappe, Claudia, Laurel J. Trainor, Sibylle C. Herholz, Christo Pantev, and Li I. Zhang. 2011. "Cortical Plasticity Induced by Short-Term Multimodal Musical Rhythm Training." *PLoS ONE* 6 (6): e21493.

Levitin, Daniel J. 2006. *This Is Your Brain on Music: The Science of a Human Obsession.* New York: Dutton.

Levitin, Daniel J., and Susan E. Rogers. 2005. "Absolute Pitch: Perception, Coding, and Controversies." *Trends in Cognitive Sciences* 9 (1): 26–33.

Loui, Psyche, H. Charles Li, Anja Hohmann, and Gottfried Schlaug. 2011. "Enhanced Cortical Connectivity in Absolute Pitch Musicians: A Model for Local Hyperconnectivity." *Journal of Cognitive Neuroscience* 23 (4): 1015–1026.

Luders, E., C. Gaser, L. Jancke, and G. Schlaug. 2004. "A Voxel-Based Approach to Gray Matter Asymmetries." *Neuroimage* 22 (2): 656–664.

Pantev, Christo, Hidehiko Okamoto, and Henning Teismann. 2012. "Music-Induced Cortical Plasticity and Lateral Inhibition in the Human Auditory Cortex as Foundations for Tonal Tinnitus Treatment." *Frontiers in Systems Neuroscience* 6: 50.

Pantev, Christo, Robert Oostenveld, Almut Engelien, Bernhard Ross, Larry E. Roberts, and Manfried Hoke. 1998. "Increased Auditory Cortical Representation in Musicians." *Nature* 392 (6678): 811–814.

Pantev, Christo, Andreas Wollbrink, Larry E. Roberts, Almut Engelien, and Bernd Lütkenhöner. 1999. "Short-Term Plasticity of the Human Auditory Cortex." *Brain Research* 842 (1): 192–199.

Patel, Aniruddh D. 2003. "Language, Music, Syntax and the Brain." *Nature Neuroscience* 6 (7): 674–681.

Patel, Aniruddh D. 2005. "The Relationship of Music to the Melody of Speech and to Syntactic Processing Disorders in Aphasia." *Annals of the New York Academy of Sciences* 1060: 59–70.

Rivera-Gaxiola, Maritza, Juan Silva-Pereyra, and Patricia K. Kuhl. 2005. "Brain Potentials to Native and Non-Native Speech Contrasts in 7- and 11-month-old American Infants." *Developmental Science* 8 (2): 162–172.

Robertson, Donald, and Dexter R. Irvine. 1989. "Plasticity of Frequency Organization in Auditory Cortex of Guinea Pigs with Partial Unilateral Deafness." *Journal of Comparative Neurology* 282 (3): 456–471.

Schlaug, Gottfried. 2004. "The Brain of Musicians." In *The Cognitive Neuroscience of Music*, edited by Isabelle Peretz and Robert J. Zattore, 366–381. New York: Oxford University Press.

Schlaug, Gottfried. 2015. "Musicians and Music Making as a Model for the Study of Brain Plasticity." *Progress in Brain Research* 217: 37–55.

Schlaug, Gottfried, Lutz Jäncke, Yanxiong Huang, Jochen F. Staiger, and Helmuth Steinmetz. 1995. "Increased Corpus Callosum Size in Musicians." *Neuropsychologia* 33 (8): 1047–1055.

Schlaug, Gottfried, Lutz Jäncke, Y. Huang, and H. Steinmetz. 1995. "In Vivo Evidence of Structural Brain Asymmetry in Musicians." *Science* 267 (5198): 699–701.

Schlaug, Gottfried, Sarah Marchina, and Andrea Norton. 2008. "From Singing to Speaking: Why Singing May Lead to Recovery of Expressive Language Function in Patients with Broca's Aphasia." *Music Perception* 25 (4): 315–323.

Schlaug, Gottfried, Andrea Norton, Katie Overy, and Ellen Winner. 2005. "Effects of Music Training on the Child's Brain and Cognitive Development." *Annals of the New York Academy of Sciences* 1060: 219–230.

Schneider, Peter, Vanessa Sluming, Neil Roberts, Stefan Bleeck, and André Rupp. 2005. "Structural, Functional, and Perceptual Differences in Heschl's Gyrus and Musical Instrument Preference." *Annals of the New York Academy of Sciences* 1060: 387–394.

Schneider, Peter, Vanessa Sluming, Neil Roberts, Michael Scherg, Rainer Goebel, Hans J. Specht, et al. 2005. "Structural and Functional Asymmetry of Lateral Heschl's Gyrus Reflects Pitch Perception Preference." *Nature Neuroscience* 8 (9): 1241–1247.

Schön, Daniele, Reyna Leigh Gordon, and Mireille Besson. 2005. "Musical and Linguistic Processing in Song Perception." *Annals of the New York Academy of Sciences* 1060: 71–81.

Shahin, Antoine E., Larry J. Roberts, and Laurel J. Trainor. 2004. "Enhancement of Auditory Cortical Development by Musical Experience in Children." *NeuroReport* 15 (12): 1917–1921.

Shapleske, J., S. L. Rossell, P. W. Woodruff, and A. S. David. 1999. "The Planum Temporale: A Systematic, Quantitative Review of Its Structural, Functional and Clinical Significance." *Brain Research Reviews* 29 (1): 26–49.

Sloboda, John A. 1991. "Music Structure and Emotional Response: Some Empirical Findings." *Psychology of Music* 19: 110–120.

Sluming, Vanessa, Thomas Barrick, Matthew Howard, Enis Cezayirli, Andrew Mayes, and Neil Roberts. 2002. "Voxel-Based Morphometry Reveals Increased Gray Matter Density in Broca's Area in Male Symphony Orchestra Musicians." *Neuroimage* 17 (3): 1613–1622.

Wada, J. A., R. Clarke, and A. Hamm. 1975. "Cerebral Hemispheric Asymmetry in Humans: Cortical Speech Zones in 100 Adults and 100 Infant Brains." *Archives of Neurology* 32 (4): 239–246.

Wan, Catherine Y., and Gottfried Schlaug. 2010. "Music Making as a Tool for Promoting Brain Plasticity across the Life Span." *Neuroscientist* 16 (5): 566–577.

Wilson, Sarah J., Dean Lusher, Catherine Y. Wan, Paul Dudgeon, and David C. Reutens. 2009. "The Neurocognitive Components of Pitch Processing: Insights from Absolute Pitch." *Cerebral Cortex* 19 (3): 724–732.

Zatorre, Robert J. 1985. "Discrimination and Recognition of Tonal Melodies after Unilateral Cerebral Excisions." *Neuropsychologia* 23 (1): 31–41.

Zatorre, Robert J. 1988. "Pitch Perception of Complex Tones and Human Temporal-Lobe Function." *Journal of the Acoustical Society of America* 84 (2): 566–572.

Zatorre, Robert J., and Pascal Belin. 2001. "Spectral and Temporal Processing in Human Auditory Cortex." *Cerebral Cortex* 11 (10): 946–953.

Zatorre, Robert J., Pascal Belin, and Virginia B. Penhune. 2002. "Structure and Function of Auditory Cortex: Music and Speech." *Trends in Cognitive Sciences* 6 (1): 37–46.

Zatorre, Robert J., David W. Perry, Christine A. Beckett, Christopher F. Westbury, and Alan C. Evans. 1998. "Functional Anatomy of Musical Processing in Listeners with Absolute Pitch and Relative Pitch." *Proceedings of the National Academy of Sciences of the United States of America* 95 (6): 3172–3177.

Zatorre, Robert J., and Valorie N. Salimpoor. 2013. "From Perception to Pleasure: Music and Its Neural Substrates." *Proceedings of the National Academy of Sciences of the United States of America* 110: 10430–10437.

CHAPTER 6

MUSIC AND PSYCHOANALYSIS

SANDER L. GILMAN

Music as Abstraction

Immanuel Kant, in the *Critique of Judgment*, saw instrumental music as beautiful but ultimately trivial—compared to the other fine arts, it cannot engage the understanding sufficiently, and thus it lacks moral purpose (Parret 1998). Sight provides the rational mind with greater substance. Thus, for him the visual is ranked aesthetically higher than the auditory. By the nineteenth century the exact opposite is also argued. Arthur Schopenhauer, in *The World as Will and Representation*, saw "Absolute Music," music that refuses to mimic the auditory nature of the world, as the highest art exactly because of its refusal to represent the phenomenal world. It is uniquely capable of evoking the metaphysical organization of the Will, the thing in itself, not the phenomenal world, and is therefore how one can best, even if partially, access the Will (Ferrara 1996, 183–199). The auditory is now aesthetically higher than the visual. But this view of sound and aesthetics as the modality to reach the soul or reveal the hidden nature of the body is not uncontested (Edström 2008, 129–130). For Hegel, music has its origin in the essence of what it means to be human, "interjection"—"the 'och' and 'oh' of the heart" (1975, 2:903). It is structured, unlike the unmediated expression of the emotions, but it also moves the "emotions in sympathy with it" (2:894; see also Johnson 1991). This means that music "captures our consciousness which does not find itself opposite an object, and due to this loss of liberty it is carried away by the flowing current of tones" (Edström 2008, 130; see also Biddle 2011; Addis 1999). The psychological power of music rests on its nonrepresentational quality.

In nineteenth-century Vienna these debates influenced how Sigmund Freud and then psychoanalysis shaped a double understanding of listening as cure and music as evidence for the unconscious. The debates about the nature of listening and music were made concrete in Eduard Hanslick's (1825–1904) defense of Johannes Brahms's musical forms against the Wagnerian "Music of the Future," with its claims of representing the visual through the acoustic world. Freud was quite aware of Hanslick's position in these debates, if only because he had read his autobiography (Freud, *Standard Edition* 1955–74 [hereafter cited as *SE*], 8:1). As early as Hanslick's 1854 *Vom Musikalisch-Schönen* (*On the Musically Beautiful*), he stressed that music is music; sound is sound; listening is listening. The creative composer "does not have in mind the representation of a specific content. If he does this, he places himself at a wrong standpoint, more alongside music than within it" (Hanslick [1891] 1986, 35; see also Gay 1976, 257–277). It is independent of any reception by a listener: "Beauty is and remains beauty even if no feelings are aroused and even if it be neither perceived nor thought" (Hanslick, 3). Musical effect is an enigma "neither psychology nor physiology can answer" (33). He stressed that the active listener is quite the opposite of the passive or pathological one: "The most significant factor in the mental process which accompanies the comprehending of a musical work and makes it enjoyable will most frequently be overlooked. . . . Only such music as brings about and rewards this mental pursuing, which could quite properly called a musing (*Nachdenken*) of imagination, will provide fully artistic satisfaction" (64). Music is a thing in itself, and it is reflected in the very nature of the psyche as it is actively captured in the act of listening. Hanslick, in spite of his contention that music is only active in the conscious mind, does not dismiss the therapeutic function of music: "The conception that through music specific feelings such as love, sorrow, anger, and rapture, are aroused in the soul, which heal the body through salutary agitation, has a convincing ring. . . . it is possible for certain emotional states to induce a turn for the better in bodily illnesses. However, it is not possible to produce on demand whatever emotional state one pleases" (53). That would come in the future, when psychoanalysis came to wrestle with the problem of music in a culture of listening as therapy. Indeed, the entire program of psychoanalysis in regard to music can be found in positive or negative terms in this Viennese setting.

These debates about the nature of listening and music as beyond meaning reflect nineteenth-century constructions of the hierarchy of the senses. Once what comes, after Kant, to be called "Absolute Music" is seen as an "empty sign" or a "blank flag," sound increases its aesthetic dominance in a world in which the visual demands more and more attention as an accurate or scientific representation of the world. The medical world spawned by the invention of photography in the 1840s and the discovery of X-rays in the 1890s is a world of the visual, but the world of aesthetics demands music as being beyond the real, a claim echoing in European thought in the Romantic age, as "a 'meta-sign' for their system of philosophy as the *absolute zero* that both participates within the system and organises it from the outside. . . . The problem for the Romantics was that they had two absolutes in their hands—'God' in terms of totality (the world), and 'God' in terms of autonomy (the self). . . . Thus music, as *the* sign of zero, was used to balance the books" (Chua 1999, 168). Lydia Goehr refers to this seeming neutrality of music as a "double-sided autonomy" (1992, 211). "Absolute Music" was not so much music about music as it was "a transcendental sign of absence that enabled the German Idealists and the early Romantics to make instrumental music mean nothing in order that it might mean everything" (Chua 1999, 170). Such claims underpin the rise of the importance of listening within European culture, a central feature of psychoanalysis. Music reveals the inner self, but which self does it reveal?

Darwin and the Music of Human Biology

The self as revealed through speech is at the heart of psychoanalysis as a therapeutic practice. Yet if Freudian psychoanalysis is not only a means of therapy but also a means of understanding normal human development and its multiple pathways, one further facet must be added to the mix. Again, it is from the world of nineteenth-century science, now biology not medicine, but it becomes one of the central pillars of psychoanalysis, and it is the question of evolutionary theory as an explanation for the centrality of music in understanding the self. Here it is not music as an abstraction but rather as a universal human activity that defines the power of music as an aspect of evolutionary theory (Ritvo 1990; for a summary of the recent literature, see Kleinman 2015).

Charles Darwin, in his *Descent of Man* (1871), redefines music's function not as an abstraction but as part of a social pattern dealing with reproduction. He imagines "some early progenitor of man, probably first used his voice in producing true musical cadences, that is in singing, as do some of the gibbon-apes at the present day." Singing "would have been especially exerted during the courtship of the sexes." It is "charming the opposite sex" that drove human beings to not only sing but "the imitation of musical cries by articulate sounds may have given rise to words expressive of various complex emotions" (Darwin [1871] 1902, 2:736–737). Such a functional definition of music is analogous to the notion of cure. Here is an abstraction, but it takes on a new function while retaining its transcendental meaning as a function of the human. It, parallel and within psychoanalysis, has a long reach into contemporary theories of music (Cross 1999, 2007; Huron 2001).

Darwin is responding in part to Herbert Spencer's functional argument in his 1857 essay that music is the means by which emotionally laden language is translated into cultural form. Inherent in expression, the topic of Darwin's much later study, is the relationship between sound and emotion—the scream of horror or the moan of pain. Spencer claims, "there is a consequent physiological relation between feeling and vocal sounds; that all the modifications of voice expressive of feeling are the direct results of this physiological relation." Music is merely the concentration of all "these modifications . . . as it ascends to its higher and higher forms, and becomes music simply in virtue of thus intensifying them" (1857, 312–330, at 329; see also Francis 2014, 145ff). It is the means of expressing within culture the empathy latent in expression, creating a language of feeling, and is thus the highest of all of the arts: "these various modifications of voice become not only a language through which we understand the emotions of others, but also the means of exciting our sympathy with such emotions" (Spencer 1857, 329). Music is also a means of understanding normal psychological development. Both of these voices are heard within psychological theories of music at the end of the nineteenth century.[1]

Freud, Listening, and Music

Sigmund Freud was certainly aware of the debates concerning the hierarchy of the senses of vision and hearing when he applied evolutionary theory to the pathological psyche. At the beginning of his career he is just as certain that he can see his patients manifest visual symptoms that define their neurosis. The senses are put to a specific function in curing the psyche during the late nineteenth century, and as a neurologist Freud shares common presuppositions about the centrality of the visual as evidence.

When Freud went to Paris in 1885 to study with the neurologist Jean-Martin Charcot, he was confronted with a scientist who relied on the photograph to document states of psychopathology, and his view that "seeing is believing" was reified (Hodgkiss 2000, 64). He returned to Vienna desirous of seeing his patients' symptoms manifest themselves. In the case of the pseudonymous "Anna O.," actually Bertha Pappenheim, Freud's older colleague Joseph Breuer had seen a physical symptom of neurosis triggered by hearing music. "She began coughing for the first time when once, as she was sitting at her father's bedside, she heard the sound of dance music coming from a neighbouring house, felt a sudden wish to be there, and was overcome with self-reproaches. Thereafter, throughout the whole length of her illness she reacted to any markedly rhythmical music with a *tussis nervosa*" (*SE* 1:38). But as Freud later recognizes, it is the interaction between Breuer and his patient that made the etiology of the symptom visible. Freud observes the pitfalls of such an approach in his account of his initial response to the case of "Fraulein Elisabeth von R," actually Ilona Weiss, whose case is described in the *Studies on Hysteria* (1895) and whom he treated in the fall of 1892 and July 1893: "I was not able to observe an attack; otherwise I should no doubt have been able to guess from the nature of the movements what it was that had occasioned them" (*SE* 2:171).

By the mid-1890s, as he was formulating the initial theory of psychoanalysis within his rereading of evolutionary theory, Freud had broken with Charcot's understanding of representation, dismissing him in 1893 as "not unduly reflective, not a thinker: he had the nature of an artist—he was, as he himself said, a 'visuel,' a man who sees. Here is what he himself told us about his method of working. He used to look again and again at the things he did not understand, to deepen his impression of them day by day, till suddenly an understanding of them dawned on him" (*SE* 3:10). But, of course, this was a false "seeing," and Freud replaces seeing in his mode of representation (and therapy) with listening. He listens to his patients, hears their slips of the tongue, their dreams, and their reconstruction, always flawed, of their own conscious and even unconscious life. Music now appears in the spoken narratives of his patients as a reflection of the sexual, not the sexual in Darwinian terms as reproductive, but the sexual as the core of psychopathology. In the above-cited case Freud was amazed to listen to his patient:

> To my astonishment the patient produced a whole number of scenes, without hesitation and in chronological order, beginning with her early childhood. They seemed to have in common her having had some injury done to her, against which she had

> not been able to defend herself, and which might have made her fingers jerk. . . . But it was otherwise with one scene from her girlhood which followed. Her bad uncle, who was suffering from rheumatism, had asked her to massage his back and she did not dare to refuse. He was lying in bed at the time, and suddenly threw off the bedclothes, sprang up and tried to catch hold of her and throw her down. Massage, of course, was at an end, and a moment later she had escaped and locked herself in her room. . . . It was only after relating this scene that she came to the one of the day before, after which the sensations and jerking in her fingers had set in as a recurrent mnemic symbol. The uncle with whom she was now living had asked her to play him something. She sat down to the piano and accompanied herself in a song, thinking that her aunt had gone out; but suddenly she appeared in the door. Rosalie jumped up, slammed the lid of the piano and threw the music away. We can guess what the memory was that rose in her mind and what the train of thought was that she was fending off. (*SE* 2:171–172)

Here, even music becomes symbolic of the sexual violation that Freud believed the patient had experienced and is read as symptom. Freud defines music as the symbolic center of the ineffable nature of the creative process itself and therefore as a central means of understanding the complexity of the psyche.

Freud early confronts what creativity is and relates it to psychic processes (Rauchfleische 1990). For if the window to the unconscious (according to Freud) can be dreams, slips of the tongue and hand, jokes, or paranoid fantasies, the nature of creativity is also one of the prime objects that psychoanalysis has used to access the unconscious. And yet it is clear that it is not "creativity" per se that has been the object of study—it is its concrete form in the work of art, the aesthetic object, which is parallel to the dream, the joke, the slip of the tongue, or the fantasy. And yet it is in this arena more than in the others that there is an easy equivalence between the "object" and the "process."

How much more difficult it is to imagine what is "creativity" as an abstraction, especially when separating the idea of "creativity" from the notion of artistic or aesthetic excellence or accomplishment. The reason this must be done, and for Freud it was clearly part of the dilemma, is that taste changes. Freud wrote at the break in an epistemic system, a break which we now call the "modern" and in which we include Freud and psychoanalysis. And yet Freud's "taste" in art, literature, philosophy, and science was clearly of the nineteenth century. He much preferred George Eliot to James Joyce, Leonardo to Salvador Dali, he abjured the cinema and the gramophone, and tolerated the literary work of his modernist friends in Vienna, such as Arthur Schnitzler, because they were friends, not because he enjoyed their writing. Yet all music, he claimed, remained beyond his reach and thus remained proof of the ineffable nature of the creative.

For Freud, as he writes in "The Moses of Michelangelo" (1914), music does not open itself up in ways similar to literature and art, as "when I have been contemplating such things, [I need] to spend a long time before them trying to apprehend them in my own way, i.e. to explain to myself what their effect is due to. Wherever I cannot do this, as for instance with music, I am almost incapable of obtaining any pleasure. Some rationalistic, or perhaps analytic, turn of mind in me rebels against being moved by a thing without knowing why I am thus affected and what it is that affects me" (*SE* 13:210). Freud reframed this in a letter to the novelist (and the first professor of musicology at the Sorbonne, a chair which he held until he retired from it in 1912) Roman Rolland in 1929 that stressed the irrationality of the object as well as his response: "How remote from me are the worlds in which you move! To me mysticism is just as closed a book as music.... And yet it is easier for you than for us to read the human soul!" (E. L. Freud 1970, 389). For Freud music remains the most inaccessible of the arts because it is as ineffable as the mystical, defining the very nature of the creative process. This is a Viennese musical world suspended between Mozart and Schoenberg.

Freud was not totally insensitive to the tradition of music that Hanslick purported to represent in his defense of Brahms. In his letters to his fiancée Martha Bernays, as early as 1886 he evokes missing the beautiful music of Mozart's *Le Nozze di Figaro* (E. L. Freud 1988, 128). The opera reappears in the *Interpretation of Dreams* (1900), when he recounts that in a dream he sang to himself Figaro's cavatina "Se vuol ballare" from *Le Nozze di Figaro* when waiting for the train that was to bring him to his vacation. As is often the case with Freud's accounts of his self-analysis, it is the footnotes or the parentheses that illuminate, and he concludes his account by noting, "It is a little doubtful whether anyone else would have recognized the tune (*SE* 4:206–207). The recounting of the dream does not center on singing Mozart badly, but rather on the power of uncovering hidden meanings, yet that parenthetical comment does provide some insight into Freud's anxiety about music. It is a form over which he does not seem to have command.

Music is inaccessible, yet listening comes to define the psychoanalytic method. Listening is, for Freud, the rejection of being seen, of the visual, the ultimate sense in the hierarchy of the medicine of the world. Freud explains that he places himself out of the field of vision of the patient, because "I cannot put up with being stared at by other people for eight hours a day (or more). Since, while I am listening to the patient, I, too, give myself over to the current of my unconscious thoughts, I do not wish my expressions of face to give the patient material for interpretations or to influence him in what he tells me" (*SE* 12:133). This is Hanslick's active listening as the core of therapy (Barale 2009).[2] Yet it is, of course, active listening that reveals hidden meanings, something that Hanslick's idea of music cannot accept. The odd connections of music, listening, and therapy do indeed come to define the early world of psychoanalysis.

Music and Listening within Psychoanalysis

Music and listening continued their complex relationship within psychoanalysis to an extent unimagined by Freud. In 1916 the American psychoanalyst William A. White located music as an elemental aspect of the unconscious as it is, following Hanslick, the least representational of the arts:

> Psychoanalytic investigation shows the psychic system to be of a dual nature, originally the primary pleasure-principle, and later the recognition of reality. The reality-principle is productive of directive, purposeful thinking, while from the pleasure-principle is developed the sphere of affectivity, expressed in creative social activity, expressions of art, and the gratifications of sex, or in an equivalent of repression. Music, the most subjective of arts, belong[s] to the primary unconscious. (White 1916, 458)

Thus music comes to provide access to a basic substratum of the psyche, and White acknowledges a Darwinian strain, reproduction, as one of its functions. But it comes to be understood as part of the pleasure principle, the id's desire to seek pleasure while avoiding pain that drives forms of repression such as sublimation in forming the ego.

By 1917 psychoanalysis came to acknowledge White's claim (Sterba 1965). Interest had remained in the visual as the expression of the aesthetic, but at least one Prague analyst, Frida (Frieda) Teller, turned to music and its emotional reception as the subject of analysis. She argued that music has a double effect on the listener (Teller 1917). The emotional effect of musical sounds consists in their ability to stimulate the formation of pleasurable erotic phantasies. Music also has a more elementary impact on our mind. Hanslick stressed that this was the nature of an active listening to music; for Teller this emotional surrender to the tones abolishes the flow of intentional processes and leads to an abandonment of the outside world and a submission to a hallucinatory regression in the form of phantasies and memories. There is no question that a Romantic rereading of music as a healing force is at the heart of this argument. For Teller, the experience of music can serve, like dreams and neurotic symptoms, as acceptable expressions of those suppressed emotions that are causal for neurotic symptoms. Such an approach resonated with this generation of psychoanalysts, as Sándor Ferenczi noted in a 1917 letter to Freud (Brabant-Gerö, Falzeder, and Giampieri-Deutsch 1996, 189).

What one needs to note is that while the interest in music had begun to infiltrate psychoanalysis, it was equally true that psychoanalysis had begun to infiltrate musicology. Just as with art history, in the work of important Viennese figures, beginning with Otto Rank and continuing through Ernst Kris and Ernst Gombrich, musicology began to incorporate Freudian approaches. As early as 1913 Robert Lach's work repeatedly refers to the psychological findings and theories of Freud (Lach 1913). He uses the concept of regression to explain musical ornament, which he considers the product of a regression to a more primitive organizational level. Primitive music is more narcissistic, modern music more connected to the elementary drives in the libido, the sexual drive that underlies the unconscious. Lach emphasizes the close relationship between song and reproduction, which was already pointed out by Darwin. Sexual excitement is articulated through language, and he considers the sensual pleasure in the vocal musical sound to be the expression of sexual excitement. According to the Freudian understanding of inherited associations, tone and rhythm excite the strong affects of a primitive past as articulated in Freud's *Totem and Taboo*, which appeared that same year.

In 1910 the Viennese music critic Max Graf, whose son's anxieties formed the basis for Freud's case of "Little Hans" (1909), had already written his study of how sublimation functions as the core of musical creativity (Graf 1910). Graf had been a student of Hanslick but broke radically with him before discovering Freud. His stress lies on the productive nature of musical creativity rather than its neurotic basis. And this tension between the "natural" function of music as an expression of the psyche and the neurotic repression of symptoms that become the wellspring of musical creativity lies at the center of the appropriation of psychoanalysis in musicology.

The tradition of musicological appropriation of psychoanalysis continues today, both in looking at the biographies of musicians and composers as well as the very nature of music itself. Yet such an approach initially had limited traction in mainstream psychoanalysis. A. van der Chijs contended in a lecture in 1922 to the Dutch Psychoanalytic Society, following White and others, that it should be possible to find latent content in music as in a dream (Van der Chijs 1926). He tried to obtain verbal associations to the different parts of musical compositions from an unnamed contemporary composer. For Lach, such a return to more primitive levels is the reason that "unisono" music, choral music in octaves, becomes a symbol for sexuality, especially for homosexual or pseudo-homosexual desire. Music, according to van der Chijs, often symbolizes the identification of the listener with the composer, at times depicting in sound both ejaculation and orgasm. He cites both Freud and Darwin to substantiate his reading.

Within this reading of music as a symbolic language of sexuality is a strong Darwinian tradition that haunts and shapes psychoanalytic discussions of the nature of music. A major contribution was made by the Budapest psychoanalyst Sigmund Pfeifer in a lecture to the 7th International Psychoanalytic Congress in 1922 in Berlin on music-psychological problems, and then expanded in further lectures in Budapest (Pfeifer 1923). Pfeifer starts his treatise with the statement that music has biological and reproductive roots, as Darwin had argued. But Pfeifer expands the very notion of music to make it continuous with other forms of sound production (and listening) in the animal kingdom. He saw the first such connection in the croaking of the frogs at mating time. The essential substance by which the sound is produced is air, and the experience of the lungs filling with air is connected with the drives in the earliest pregenital period of development and helps shape one's sense of self. In humans, the very passage of air through the throat in speaking reflects the tension felt by the libido, that part of the unconscious where the drives are present. Music is a means of escape from reality through rhythm that through the process of psychic economy provides pleasure through compulsive repetition, thereby releasing unconscious fantasies untrammeled by the limitations of language (Spitzer 2002, 363).[3]

In 1939, the émigré Richard Sterba, a member of the first graduating class of the Vienna Institute in 1924 and one of the most articulate commentators on psychoanalytic approaches to music, spoke on this topic at the Chicago Psychoanalytic Society. He stressed that music was inherently different from other forms of representation, especially the visual, as Freud had centered his claims about psychic processes on the dream-work in which the visual symbols of daily experience translated deep psychic struggles into pictures that are both common and coded. Music cannot undertake such a process as "in every work of art conscious and unconscious instinctual wishes are expressed and represented in the form of an image of the outside world. . . . But to apply this method of investigation to music is impossible, since in music the conscious and unconscious emotions are not expressed in the form of images of the outside world. Music is what we may call an objectless art" (1946, 37). Sterba focuses on the cultural function of such "Absolute Music," limiting his analysis to the claims of Western music, as a symbolic register that translates unconscious processes into accessible yet coded musical experiences. This view comes to dominate discussions of the process of both music production and comprehension.

Music in psychoanalysis remains "Absolute Music," in contrast to the other "concrete" visual and literary arts, which, with the rise of modernism and its reception of psychoanalysis, claims to be more and more abstract. One of Freud's earliest Viennese collaborators, Hanns Sachs, emphasized in the early 1940s:

> music, the art-form which comes nearest to being the pure embodiment of beauty, is as enigmatic as beauty itself. We can learn something, however, by comparing it with poetry; we can conceive music as the highest form that poetry attains when it has shed the last "*Erdenrest, zu tragen peinlich*" (rest of earthiness, repugnant to bear). [From the conclusion to Goethe's *Faust II*, when Faust's earthly remains ascend, having escaped Mephistopheles's clutches.] Music acts on the acoustic sense (only some less purified forms like march and dance-music are concerned with the crude kinaesthetic side) without even the association of visual imagery such as words bring with them. Hearing is a purely passive way of sensory reception; all the activity therefore is left to the mind with the least possible interference. The material used by music, the sounds, are selected with an exclusive regard to their own intrinsic quality as high or low, long or short etc. or their own interaction among themselves: diatonic or atonal, rhythmic integration or disintegration, harmony or discord or whatever the principle may be. They are quite free from any conscious ideational content. They have no other message to convey than their own effect on the hearer's ear and mind. (1941, 114)

Likewise, one of the few American-born psychoanalysts of the time, the Boston-based Isador Coriat, at the 44th meeting of the American Psychoanalytic Association in 1942, observed that "the aesthetic effect of music is the result of three factors: compulsive repetition, pleasure in economy and the force of attraction exerted by the unconscious" (Coriat 1945, 407). He goes on to claim that

> music is one of the highest, if not the highest form of beauty that has revealed itself to mankind; its rhythm and melody have no tendency to imitate reality, even in a strictly stylized form. . . . Music produces an artistic illusion, its rhythmic repetition

> leads to the highest narcissistic gratification, thereby releasing fantasies like the associative freedom of a day dream. It is this day dreaming which forms the nuclear basis of poetry and music alike; both are ego-centric and both lead to the highest form of enjoyment, an intensity of aesthetic attitude which is not to be found in any of the other arts, such as painting or sculpture. (408)

For him, music is pure libido symbolism; it lacks objective content because the libidinal aspect of music has not reached the object level of development. It is thus an objectless art and therefore "Absolute Music." It is the only mental creation in which these libidinal processes can be found in pure culture. Coriat importantly makes an analogy between musical composition and the dream-work. Thus, the unformed libido is given form.

In 1951 Heinrich Racker noted, "[it is claimed that] music does not represent in itself the objects of the external world, as is the case with the majority of the other arts; and it is just these objects with their mutual relationships which comprise the central theme of psycho-analysis." Racker attributes this both to "the infrequent application of psychoanalysis to musicology," as well as the fact that "there must exist a resistance of an affective nature, that behind this lack of preoccupation with the subject there is a rejection of the subject" (1951). Music cannot be so easily rejected, as it is a core component of culture. The Argentinian psychoanalysts A. de Pichon-Rivière and L. G. de Alvarez de Toledo contend that "music, like any other artistic creation, has form and content. Its form has harmony and time. Harmony is the objective form of the sound and time corresponds to life. The form is like a symbol; it is a means of communication that goes beyond, yet is less than, its true meaning. Therefore, music as a symbol is an æsthetic concept behind which we hide unconscious fantasies" (1955, 183). The tension between the notion of music as beyond meaning and music as holding inherent symbolic meaning shapes much of the concerns of mid-twentieth-century psychoanalysis.

By the 1950s music is one of the loci classici of psychoanalytic interpretation. One of Freudian psychoanalysis's major figures, the Viennese émigré Theodor Reik (1888–1969) devotes an entire book to the topic. For him "human speech denotes the material reality, music is the language of psychic reality" (1953, 8). It is the difference between spoken and sung speech that defines the psychological structure of the musical, as "the tune expresses . . . the immediate quality of experience. It is an emotional expression much more adequate than words" (249). Music is a ghostly presence within the psyche: "When I am singing a melody that haunts me, I am expressing emotions. It has the same meaning as when I am laughing, crying, sighing or sobbing" (250). Music is, for Reik, primitive, infantile, narcissistic (objectless), and preverbal in nature. As the philosopher Susanne K. Langer said, "the tonal structures we call 'music' bear a close logical similarity to the forms of human feelings . . . music is a tonal analogy of emotive life" (1953, 27). Since they are "felt" rather than "seen," emotions become, within the model of abstract music, its analogy.

Subsequent psychoanalytic debates centering on these core concepts evolved between the 1900s and the 1950s with nuanced differences, but reflecting the Freudian understanding of music now as an essential access point to the study (and treatment) of the psyche (Nagel 2013).

Therefore, it is of little surprise that music therapy, which emerged prior to World War I and began its professionalization in the 1950s under the influence of psychoanalytic models of healing, accepts a set of psychoanalytic definitions of music, even while distancing itself from psychoanalytic practice. While modern music therapy tends to note its origin in the Enlightenment (Gouk 2015), it is only at the end of the nineteenth century that the first recorded systematic experiment in music therapy as a treatment of psychological disorders is undertaken (Leonard Corning's [1855–1923] use of music to alter dream states during psychotherapy). Diverse approaches, such as the Orff-Schulwerk (Carl Orff), Eurhythmics (Émile Jaques-Dalcroze), and the work of Zoltán Kodaly in the 1930s as well as the Nordoff-Robbins approach in the 1950s, assume a special space for music as a means of psychic healing. The pioneering psychiatrist and music therapist Ira Altschuler commented in the early 1950s: "One should keep in mind that music has always been an important factor in the instinctual, emotional, intellectual and spiritual life of people, and as such, from time immemorial, exercised a sort of therapeutic influence. Music . . . lends itself as a therapy because it meets with little or no intellectual resistance and does not need to appeal to logic to initiate action" (Altschuler, 1953, 26). The return to the debates about the hierarchy of the senses in therapy continued even into this realm. Music is not rational; therefore, it must be the royal road to therapy.

Defining music and listening come to be keys to claims about psychoanalytic treatment as well as evidence for a universal theory of the psyche. Recently, a special issue of a professional journal of psychoanalysis was devoted to such a broad understanding of music, as "relevant to individuals, in different ways, in all cultures. People can listen to evocative sound and experience a variety of feelings simultaneously. This experience can enhance ego identity, evoke memory, assist working through unconscious issues, and also inform group dynamics" (Nagel and Bradshaw 2013, 97). Can we argue as ethnomusicologists have argued, in a manner clearly indebted to Freud's theory of the dream-work, that music seems to be a universal human quality but its expression is culturally bound? Can you have music without form, and is not form culturally defined? Is not the very act of listening as much cultural as it is biological? All of the debates about the functional or ineffable nature of music reflect cultural presuppositions about its nature and the listener's relationship to it. Yet we tend to respond to such cultural claims as natural categories. "Unlike a biological category, a musical one is not a natural form . . . but a tradition, from which the singular entity may emancipate itself" (Dahlhaus 1973, 840). But this is not to say that music and listening are unproductive to cure or indeed to insight into complex psychic formations such as creativity as well as the psyche itself. Cultural formations are as real in their effect as the biology that underlies them and can have equally powerful resonance.

Notes

1. By the twenty-first century the Darwinian turn evolved into a discourse, among other things, about music and the counterculture. Sex, drugs, and rock 'n' roll are seen as being linked by evolutionary theory. Steven J. Mithen (2005) returns to Darwin's notion of courtship rituals as the origin and prime mover of music. Rob Brooks (2011) turns the old adage about why young people join rock bands into a further argument about evolutionary theory and reproduction. As Janice Joplin said,

   > All my life I just wanted to be a beatnik. Meet all the heavies, get stoned, get laid, have a good time. That's all I ever wanted. Except I knew I had a good voice and I could always get a couple of beers off of it. All of a sudden someone threw me in this rock 'n' roll band. They threw these musicians at me, man, and the sound was coming from behind. The bass was charging me. And I decided then and there that that was it. I never wanted to do anything else. It was better than it had been with any man, you know. Maybe that's the trouble. (Cooke 2014, 391)

   Other recent biological theories postulate an organic relationship between human sexuality and hormones: "Music suppressed testosterone in the male. . . . [L]istening to music affects the testosterone level in females and . . . sex differences exist in the effect of listening to music on testosterone level" (Fukui 2001, 448). Thus we have another hypothesis for the evolutionary function and origin of music as a means for controlling the male's aggressive libido and instituting "civilization." Such readings turn every sound into music: "music throughout the world, and even in animal sounds (monogamous birds), music and certain sounds function to ease tension or strain, strengthen social bonds, bring pleasure or ecstasy, and bring cohesiveness among people. Music is used empirically and universally in various social situations where stress, tension, and anxiety exist, such as ceremonies, funerals, war, and even banquets" (451). And for humans it creates "civilization."
2. I am bracketing in this discussion Freud's interest in and knowledge of psychoacoustics in the form of Hermann Helmholtz's work, which he read and critiqued as much too biological in its formulation. Contemporaries such as Wilhelm Wundt saw, like Freud, a distinction "between the psychophysiological problem of consonance or dissonance and the psychological-cultural problem of harmony—a distinction he thought Helmholtz had not made sufficiently clear" (Barale 2009, 42).
3. Such views extended well beyond the impact of evolutionary theory on psychoanalysis. The confusion between structured sound and music in evolutionary approaches to music is a powerful one. It has been argued recently that humpback whales produce "smash hits," because the sounds they produce seem to be organized in repeated structures and themes, and frequently resemble a sonata or at least an ABA form (Spitzer 2002, 363).

References

Addis, Laird. 1999. *Of Mind and Music*. Ithaca, NY: Cornell University Press.

Altschuler, Ira M. 1953. "The Past, Present and Future of Musical Therapy." In *Music Therapy*, edited by Edward Poldolsky, 24–36. New York: Philosophical Library.

Barale, Francesco. 2009. "At the Origins of Psychoanalysis: Freud, Lipps and the Issue of Sound and Music." *Italian Psychoanalytic Annual* 3: 37–54.

Biddle, Ian. 2011. *Music, Masculinity and the Claims of History: The Austro-German Tradition from Hegel to Freud*. Farnham: Ashgate Publishing.

Brabant-Gerö, Eva, Ernst Falzeder, and Patrizia Giampieri-Deutsch, eds. 1996. *The Correspondence of Sigmund Freud and Sándor Ferenczi*. Translated by Peter Hoffer. Cambridge, MA: Harvard University Press.

Brooks, Rob. 2011. *Sex, Genes and Rock 'n' Roll: How Evolution Has Shaped the Modern World*. Cardiff: University of New South Wales Press.

Chua, Daniel K. L. 1999. *Absolute Music and the Construction of Meaning*. New York: Cambridge University Press.

Cooke, John Byrne. 2014. *On the Road with Janis Joplin*. New York: Berkley.

Coriat, Isador. 1945. "Some Aspects of a Psychoanalytic Interpretation of Music." *Psychoanalytic Review* 32: 408–418.

Cross, Ian. 1999. "Is Music the Most Important Thing We Ever Did? Music, Development and Evolution." In *Music, Mind and Science*, edited by Suk Won Yi, 10–39. Seoul: Seoul National University Press.

Cross, Ian. 2007. "Music and Cognitive Evolution." In *Oxford Handbook of Evolutionary Psychology*, edited by R. Dunbar and L. Barrett, 649–667. Oxford: Oxford University Press.

Dahlhaus, Carl. 1973. "Zur Problematik der musikalischen Gattungen im 19. Jahrhundert." In *Gattungen der Musik in Einzeldarstellungen: Gedenkschrift Leo Schrade*, edited by Wulf Arlt, 840–895. Bern: Francke.

Darwin, Charles. 1902. *The Descent of Man and Selection in Relation to Sex*. Vol. 2. New York: American Home Library.

Edström, Karl-Olof. 2008. *A Different Story: Aesthetics and the History of Western Music*. Hillsdale, NY: Pendragon.

Ferrara, Lawrence. 1996. "Schopenhauer on Music as the Embodiment of Will." In *Schopenhauer, Philosophy and the Arts*, edited by Dale Jacquette, 129–130. Cambridge: Cambridge University Press.

Francis, Mark. 2014. *Herbert Spencer and the Invention of Modern Life*. New York: Routledge.

Freud, Ernst L., ed. 1970. *Letters of Sigmund Freud, 1873–1939*. London: Hogarth.

Freud, Ernst L., ed. 1988. *Brautbriefe: Briefe an Martha Bernays aus den Jahren 1882–1886*. Frankfurt am Main: Fischer.

Freud, Sigmund. 1955–1974. *The Standard Edition of the Complete Psychological Works of Sigmund Freud*. Edited by James Strachey, Anna Freud, and Angela Richards. 24 vols. London: Hogarth Press.

Fukui, Hajime. 2001. "Music and Testosterone: A New Hypothesis for the Origin and Function of Music." *Annals of the New York Academy of Sciences* 930 (1): 448–451.

Gay, Peter. 1976. "For Beckmesser: Eduard Hanslick, Victim and Prophet." In *Freud, Jews, and Other Germans: Masters and Victims in Modernist Culture*, 257–277. New York: Oxford University Press.

Goehr, Lydia. 1992. *The Imaginary Museum of Musical Works: An Essay in the Philosophy of Music*. New York: Oxford University Press.

Gouk, Penelope. 2015. "An Enlightenment Proposal for Music Therapy: Richard Brocklesby on Music, Spirit, and the Passions." In *Music, Neurology, and Neuroscience: Evolution, the Musical Brain, Medical Conditions, and Therapies*, edited by Eckart Altenmüller, Stanley Finger, and François Boller, 159–186. Amsterdam: Elsevier.

Graf, Max. 1910. *Die innere Werkstatt des Musikers*. Stuttgart: F. Enke.

Hanslick, Eduard. 1986. *On the Musically Beautiful: A Contribution Towards the Revision of the Aesthetics of Music.* Translated by Geoffrey Payzant. Indianapolis, IN: Hackett.

Hegel, G. W. F. 1975. *Aesthetics: Lectures on Fine Art.* Vol. 2. Translated by T. M. Know. Oxford: Clarendon.

Hodgkiss, Andrew. 2000. *From Lesion to Metaphor: Chronic Pain in British, French and German Medical Writings, 1800–1914.* Amsterdam: Rodopi.

Huron, David. 2001. "Is Music an Evolutionary Adaptation?" *Annals of the New York Academy of Sciences* 930 (1): 43–61.

Johnson, Julian. 1991. "Music in Hegel's Aesthetics: A Re-Evaluation." *British Journal of Aesthetics* 31: 152–162.

Kleinman, Kim. 2015. "Darwin and Spencer on the Origin of Music: Is Music the Food of Love?" In *Music, Neurology, and Neuroscience: Evolution, the Musical Brain, Medical Conditions, and Therapie*, edited by Stanley Finger and François Boller Eckart Altenmüller, 3–26. Amsterdam: Elsevier.

Lach, Robert. 1913. *Studien zur Entwicklungsgeschichte der ornamentalen Melopoeie.* Leipzig: C.F. Kahnt Nachf.

Langer, Susanne K. 1953. *Feeling and Form: A Theory of Art.* New York: Charles Scribner's Sons.

Mithen, Steven. 2005. *The Singing Neanderthals: The Origins of Music, Language Mind and Body.* London: Weidenfeld & Nicolson.

Nagel, Julie Jaffee. 2013. *Melodies of the Mind: Connections between Psychoanalysis and Music.* New York: Routledge.

Nagel, Julie Jaffee, and Samuel Bradshaw. 2013. "Overture." *International Journal of Applied Psychoanalytic Studies* 10 (2): 97–99.

Parret, H. 1998. "Kant on Music and the Hierarchy of the Arts." *Journal of Aesthetics and Art Criticism* 56: 251–264.

Pfeifer, Sigmund. 1923. "Musikpsychologische Probleme." *Imago* 9: 453–462.

Pichon-Rivière, A. A., and L. G. de Alvarez de Toledo. 1955. "Music and Musical Instruments. Part I." *Revista de Psicoanálisis* 23: 183–200.

Racker, Heinrich. 1951. "Contribution to Psychoanalysis of Music." *American Imago* 8: 129–163.

Rauchfleische, U. 1990. "Psychoanalytische Betrachtungen zur musikalischen Kreativität." *Psyche* 44: 1113–1140.

Reik, Theodor. 1953. *The Haunting Melody: Psychoanalytic Experiences in Life and Music.* New York: Farrar, Straus and Young.

Ritvo, Lucille B. 1990. *Darwin's Influence on Freud: A Tale of Two Sciences.* New Haven, CT: Yale University Press.

Sachs, Hanns. 1941. "Beauty, Life and Death." *American Imago* 1: 81–133.

Spencer, Herbert. 1857. "On the Origin and Function of Music." In *Essays on Education and Kindred Subjects*, 312–330. London: J.M. Dent & Sons.

Spitzer, Manfred. 2002. *Musik im Kopf: Hören, musizieren, verstehen und erleben im neuronalen Netzwerk.* Stuttgart: Schattauer.

Sterba, Richard. 1946. "Toward the Problem of the Musical Process." *Psychoanalytic Review* 33: 37–43.

Sterba, Richard. 1965. "Psychoanalysis and Music." *American Imago* 22: 96–111.

Teller, Frida (Freida). 1917. "Musikgenuss und Phantasie." *Imago* 5: 8–15.

Van der Chijs, A. 1926. "Über das Unisono in der Komposition. Beitrag zur Psychoanalyse der Musik." *Imago* 12: 23–31.

White, William A. 1916. "Individuality and Introversion." *Psychoanalytic Review* 3: 455–464.

Further Readings

Thaut, Michael H. 2015. "Music as Therapy in Early History." In *Music, Neurology, and Neuroscience: Evolution, the Musical Brain, Medical Conditions, and Therapies*, edited by Eckart Altenmüller, Stanley Finger, and François Boller, 143–158. Amsterdam: Elsevier.

CHAPTER 7

MUSIC SOCIOLOGY MEETS NEUROSCIENCE

MIA NAKAMURA

"The power of music"—or more precisely, music's effects on humans and society—has been a controversial theme in recent discussions of music and social issues, such as conflict transformation, social inclusion, disaster recovery, regional revitalization, and community empowerment (Rice 2014; Bergh and Sloboda 2010; Urbain 2008). Although most scholars acknowledge that music has some influence on our feelings, thoughts, and behaviors, exactly how musical experience can be a resource for improving quality of life still remains unclear. This chapter opens the black box of the power of music and illuminates the mechanism of musical effects through interdisciplinary considerations of sociology and neuroscience.

In the twenty-first century, music has become a hot topic of cognitive science (Pearce and Rohrmeier 2012; Peretz 2006). Particularly, neuroscientific researchers are enthusiastic about examining brain responses to musical stimuli. However, music is not just of sound composition but also of human phylogeny and cultural history (Stige 2002, 81–84). If this is the case, interdisciplinary studies of sociology and neuroscience should be encouraged. Nevertheless, such collaborations have rarely been conducted. Music ethnographers are rather indifferent toward new findings in neuroscience because outcomes in laboratory settings seem distant from real-world experiences, whereas scientific researchers think of ethnographic understandings of music as uninformative because they are too concerned about cultural specificities.

An emerging sociological view of music and a recent development in cognitive theory point to an intersection between these two academic realms. While traditional sociology regards music as a medium connecting people, recent sociologists informed by the actor-network theory consider that music functions as "mediation" (Born 2011; DeNora 2003, 2000; Hennion 2015, originally published as *La passion musicale* in 1993). In other words, music is a reconciling medium between two different parties and at the same time the medium of communication itself, in which its meaning is mediated through communication. Some ethnomusicologists who have bridged the gap between ethnographical accounts and cognitive understanding hold a similar view (Monson 2008; Becker 2004, 2001). On the other hand, the latest cognitive theories regarding musical emotion consider that music is not a one-way stimulation that works on the brain, but rather a bodily activation that affects neural networks contingent on our internal conditions and external environments (McGuiness and Overy 2011; Juslin et al. 2010; Molnar-Szakacs and Overy 2006). Researchers are concerned not only with physiological reactions in the brain but also with musical experiences and episodic memories. Here, an intersection between sociology and neuroscience emerges.

The aim of this chapter is to deepen the understanding of sociological findings with neuroscientific insight. The chapter consists of five sections. The first three sections provide an overview of intersections between sociology and cognitive science. Recent developments in these fields are reviewed in a scope relevant to this study. The last two sections consider interdisciplinary implications of musical effects. Studies on musical activities in Japan from sociological perspectives—including those of sexual minorities and the 2011 East Japan Great Earthquake (Nakamura 2013, 2014a, 2014b, 2017, 2018)—have shown certain patterns and conditions of musical effects for empowerment and social transformation. These ethnographic findings are reanalyzed from neuroscientific perspectives. Moreover, sociological implications for a future neuroscientific study, particularly the effects of musical *retelling*, are discussed.

Recent Developments in Sociology

Although the study of music has a long-standing tradition, the power of music has remained mystical. Starting with Plato's famous remarks in *The Republic*, scholars have believed that music intrinsically holds magical power, investigating musical contents, such as motives, rhythms, harmonies, and structures. Musicologists have assumed that music affects us because it embodies special entities that appeal to us (Small 1998). Though interested in its social contexts, they have scarcely tested the relationships between musical contents and social contexts empirically (DeNora 2003). On the other hand, sociologists who emphasize empirical methods have studied the social effects of music. Assuming that music is simply a communication medium, however, they rarely paid attention to the musical contents. Music sociology, defined broadly, has been practiced conventionally either "by music scholars with an interest in society" or "by sociologists with an interest in music" (Shepherd and Devine 2015, 10). Neither musicologists nor sociologists have had an effective tool to reveal the secret of music's power.

The recent sociological concept of music as mediation presents a challenge for integrating these two views in a sophisticated manner. To put it simply, mediation is the idea that what conveys a message is itself a message. More specifically, it refers to "the all-encompassing process of simultaneously producing, changing, and transforming both the artistic object and its audience at any given point in time" (Acord and DeNora

2008, 226). For instance, Antonin Hennion (2015) discusses how we have become able to appreciate the music of the eighteenth-century composer J. S. Bach. Examining the recovery and adapting process of Bach's music in the nineteenth century, Hennion demonstrated that we ourselves have found our ways to appreciate it through social interactions. He argues that production and consumption of music are not conflicting, as generally assumed, but people who consume music actually produce the way they consume, while they transform themselves to become able to consume it (247–260).

Inspired by Hennion's work, Tia DeNora developed the concept of "affordance" (2000, 39–41). Although James Gibson's original emphasis was "how objects may provide opportunities for perception and/or action" (Acord and DeNora 2008, 227), DeNora focuses attention on how users attach meanings to objects. For example, the past memory of a loved one who used to listen to a particular piece of music may shape the meaning of the music; further, in aerobics, instructors edit and appropriate musical pieces according to their physical activities (DeNora 2000). In these cases, the original purpose of music is no longer relevant. As each musical piece has a certain quality that affords its use, we appropriate it consciously or unconsciously in the range of its applicability to our needs. Importantly, music does not determine its effect by itself; rather, humans produce, negotiate, and transform the use of music through social interactions.

These studies also emphasize that musical experiences are linked to one's past experiences. Traditionally, ethnomusicology presupposes that the meaning of music is constructed culturally, but in the era of globalization and technology, personal experience is more important in forming the meaning of music. The scholarly focus has shifted from cultural identity to bodily experience, which has, consequently, made a connection between sociology and music therapy (DeNora 2013a, 2013b, 2014). Although some music therapists still naively believe in the magical power of music, recent trends are more concerned with musical interactions and bodily engagement (Stige et al. 2010; Pavlicevic and Ansdell 2004; Stige 2002). The concern has clearly shifted from naively using music to how to use music.

In summary, sociology has found a way of discussing the power of music. It is now conceived of more in human terms and may preferably be paraphrased as "the power of humans that use music." This mediation-oriented and human-centered view of music suggests a link to neuroscientific thinking.

Recent Developments in Neuroscience

Cognitive scientists may be the most enthusiastic group of people to reveal the mystery of music's power. Conventionally, the focus of this field has drawn on musical perceptions and cognition regarding Western classical music, but its scope has expanded considerably (Thompson 2009). For the purpose of this study, approaches in the field are classified into traditional psychology, neuroscientific research, and systematic considerations of neuroscientific findings.

The traditional psychological approach tests examinees' perceptions of acoustic stimulation with questions and answers. This approach is widely known, but it should be noted that this sort of study often confuses the acoustic effects of music with its cognitive effects (North and Hargreaves 2006); it also confuses "intramusical" meanings with "extramusical" meanings (Patel 2008, 306). Therefore, research outcomes are often relevant only to similar conditions, presenting few implications for the embodied mechanism.

The second is a neuroscientific approach, which examines physiological responses to acoustic stimulation inside the body using brain-imaging techniques (Pearce and Rohrmeier 2012; Peretz 2006). What is referred to as the "brain science of music" in mass media falls into this category. The first psychological approach requires language cues, which can be biased (Adorno 1976, 4), but this approach directly measures a physiological change. The research has revealed, for instance, the part of the brain that mainly functions in musical listening and the ways neurons react when we engage in musical activities (Thompson 2009, 151–181; see Lennard's chapter for more on this subject). Although this approach is innovative and certainly brings valuable insights, each study can only tell a part of the musical experience because the research method confines a musical experience within a scheme of experimental efficiency.

The third approach is theoretical, focusing on the embodied communication system, which is based on the findings of recent neuroscientific studies mentioned earlier. This approach proposes systematic views of musical reactions, taking seriously the fact that musical experience involves not only auditory perception but also other embodied activities (Bharucha, Curtis, and Paroo 2006).

While this approach is more concerned with real-world experiences, it paradoxically challenges the familiar view of music in two senses. For the first paradox, the new view contradicts the conventional premise of music studies that musical effects arise from musical contents and the performance context. The neuroscientist Antonio Damasio ([1994] 2005) discusses our emotional response in two evolutional phases: an innate response, which is rather simple and automatic, and a learned response, which is a combination of the innate response with past learning that encompasses intellectual activities. Although emotional responses are always automatic, most of them are not innate but learned. Learned responses work instantly because our perceptual systems are developed through our past experiences. In other words, past learning influences our instant responses. To extend this view, current musical experience is shaped not only by acoustic stimuli and ongoing activities but also by past experiences. This fact reminds us that the way of perceiving music depends on musical training, which develops auditory sensitivity (Reybrouck and Brattico 2015).

For the second paradox, the new view undermines the uniqueness of our listening experiences, regarding them as something related to other forms of bodily experience. Another neuroscientist, Vittorio Gallese (2003), demonstrated that an emotional response is basically contagious, persuasively arguing that imitation, empathy, and mind-reading actually share a common embodied mechanism, although they are usually understood as different kinds of activities. He writes:

> The discovery of mirror neurons has changed our views on the neural mechanisms at the basis of action understanding. The observation of an action leads to the activation of the same neural networks active during its actual execution: action observation causes in the observer the automatic simulated re-enactment of the same action. It was proposed that this mechanism could be at the basis of an implicit form of action understanding. (522)

"Mirror neurons," or perhaps more preferably "the mirror neuron system," refers to "groups of neurons in the frontal and parietal regions of the brain that are activated not only when individuals perform intentional actions themselves, but also when they perceive another individual performing that action" (McGuiness and Overy 2011, 248). Bodily responses are now conceived in the interpersonal scheme.

In a discussion of music, Molnar-Szakacs and Overy (2006) point out that "humans may comprehend all communicative signals, whether visual or auditory, linguistic or musical, in terms of their understanding of the motor action behind that signal, and furthermore, in terms of the intention behind that motor action," emphasizing the possibility that "the mirror neuron system is involved not only in the intersubjective representations of actions but also in emotion-representations that allow us to feel connected with other agents" (238). Further, McGuiness and Overy (2011) examine the mechanism of emotional contagion through musical experience. Here, they cite DeNora's sociological account at the beginning, showing the link to sociological insight (245). Their attention to the empathic aspect of music also parallels the sociocultural view of Felicity Laurence (2008) that musical effect is achieved mainly on an emphatic basis; and their focus on the role of music as communion rather than communication has an affinity for the ethnographic work of Thomas Turino (2008).

This new view inevitably prompts us to reconsider aesthetic meaning, which has long been conceived of according to a Cartesian mind-body dichotomy. Although aesthetics had been a matter of the aesthetic object and the mind, meaning-making activity is now understood in a more interpersonal term, and the mind is a part of the body. Jessica Lindblom (2015) lists the following four functions of the body in aesthetic meaning making:

> - The body functions as a social resonance mechanism.
> - The body functions as a means and end in communication and social interaction.
> - Bodily actions and gestures function as a helping hand in shaping, expressing and sharing thoughts.
> - The body functions as a representational device. (11)

The body is viewed as a medium that connects one person to another, transforms a person's thoughts, and changes its own function and meaning through its very working. This line of study, though still in a developing stage, seems promising for revealing music's role in society.

Embodied Communication Models of Musical Emotions

A series of neuroscientific discoveries has led to new theoretical views regarding the embodied communication of musical emotion. Juslin and colleagues (2010), for example, hypothesize seven different psychological mechanisms through which music may induce emotions in listeners: (1) brain stem reflex, (2) rhythmic entrainment, (3) evaluative conditioning (conditioned reflex), (4) contagion, (5) visual imagery, (6) episodic memory, and (7) musical expectancy. The authors argue that when we receive a musical stimulus, our brain processes it, filtering step by step (631). For instance, when something is heard, the lower brain areas ask whether it features extreme acoustic characteristics. If the answer is yes, the brain stem reflex responds to the stimulation. If the answer is no, then the limbic system in turn questions whether the stimulus has been encountered before. If the answer is no, the response is made via evaluative conditioning. However, if the answer is yes, the next question comes up. In this way, a response to a musical stimulus is determined according to the filtering system.

Similarly, Terasawa and colleagues (2013) propose a roadmap of musical emotion, showing the degree of social-personal activities in the vertical axis (1, 1:1, 1:N, N:N) and the degree of unconscious-conscious activities in the horizontal axis (physiological response, musical participation, and contribution to making music). For instance, simple reactions such as changes in heartbeat are placed at the lower left; complex reactions such as group improvisation are at the higher right. They also illustrate embodied models of emotional systems, showing the relationships between the internal embodied mechanism and the external environment. Relatedly, Aniruddh D. Patel (2008) compares music with language in both philosophical and neuroscientific terms.

Although these models still remain speculative in some respects, there are certain aspects that have become evident. First and foremost, musical response occurs with three variables: stimulus, internal condition, and external environment. With music, a stimulus obviously refers to a series of aural stimuli conveyed by music. Internal condition refers to the current bodily condition (both physical and mental), as well as perceptual sensitivity developed through past experience and musical training. External environment includes the environmental conditions surrounding the given stimulus, as well as environmental information accompanied with the stimulus, including visual and linguistic information. Our musical responses are defined by these three variables. A similar remark has been made from a socio-psychological point of view (Hargreaves et al. 2005).

Second, embodied response differs according to one's degree of involvement with sound. For instance, a brain stem reflex to a sudden louder sound and a conditioned reflex to a familiar tune occur regardless of the perceiver's interest in the sound. Rhythmic entrainment and recalling an episodic memory are also induced in casual listening, but whole-body coordination and deep reflection are brought about only by focused listening. Further, musical expectancy with reflective thought and syntactic/analytic understanding of music only occurs in intellectual listening. In addition, embodied response to music differs between listening and performing. In other words, the degree of involvement (e.g., unconscious response, passive listening, active listening, participation, and absorbed participation) results in different effects.

Third, as the discussion of mirror neurons highlights, embodied workings are largely on a simulation basis. Except for brain stem reflex, our ways of responding to musical stimuli involve a certain degree of bodily coordination.

Fourth, in reality, since the musical stimulus induces responses in a multilayered fashion (Dolar 2006), its workings of reactive parameters depend greatly on one's perceptual development. Conditioned reflex and episodic memory are obviously related to past experience. Brain stem reflex (e.g., response to an extremely loud sound) and motor coordination (e.g., rhythmic entrainment) are more innate, their workings thereby being less varied from person to person. In reality, however, brain stem reflex and motor coordination work along with other reactive parameters. For example, a response to a rhythmic, familiar melody may be activated by a combination of motor coordination, conditioned reflex, and episodic memory, where the response inevitably relies on one's past experience.

To sum up, sociology and neuroscience intersect when they take mediation-based and human-centered perspectives. In particular, the interdisciplinary consideration in this section has pointed out four key aspects related to musical response and embodied work. Once these points are understood, sociological findings and neuroscientific explanations will generate synergy in advancing our understanding of the social and embodied mechanism concerning musical effects, in other words, the power of music.

Neuroscientific Reconsideration of Sociological Findings

The remaining sections explore how one can explain musical effects in the real world using the interdisciplinary perspectives of sociology and neuroscience. This section, in particular, reinterprets ethnographic findings using neuroscientific insight; conversely, the next section suggests ethnographic implications for future neuroscientific studies.

Examples in this section are taken from musical activities after the East Japan Great Earthquake (Nakamura 2014a, 2017, 2018). One of the largest earthquakes in history hit the Tohoku region of Japan (the northeast part of the main island) on March 11, 2011. Together with the subsequent gigantic tsunami, it brought catastrophic damage to the region, taking the lives of approximately 20,000 people. Even worse, it triggered a nuclear meltdown at the Fukushima Daiichi Power Plant complex, resulting in the evacuation of 14,000 residents.

Although music was not a primary concern in the post-quake chaos, once recovery efforts began, musical topics received considerable attention. News media reported musical activities among evacuees, repeatedly featuring famous musicians' visits to the disaster areas. Charity concerts were held all over the world, and numerous recovery songs were created. At this time, the phrase "the power of music" (*ongaku no chikara*, in Japanese) was used frequently, which implies the musical effects of empowerment and/or behavioral transformation. Although some usages were simply clichés, there were certainly cases in which this phrase rang true.

A common testimonial concerning the power of music was that afflicted people were moved deeply by musical performances, resulting in empowerment. For instance, a local school chorus group offered a concert at a temporary evacuation center on March 19, 2011—about a week after the disaster. The students chose the piece "Tomorrow Is Sure to Come" (*Asu to iu Hi ga*) for this particular occasion. The song written by a local composer consists of plain melodies and familiar harmony progressions with a catchy refrain: "Tomorrow is sure to come; believe in our future happiness." When the students sang wholeheartedly, some evacuees began to shiver and burst into tears. They also said that they were empowered by the singing. This episode was broadcast nationwide on news shows via Japan's public TV station, NHK, which in turn inspired many viewers outside the disaster area to take actions to help the victims.

Another example is a "Recovery Concert" (*fukko konsāto*) at a temple in Sendai City on March 26—only two weeks after the earthquake, presented by a local professional orchestra, Sendai Philharmonic Orchestra. When a small ensemble played somewhat familiar and soothing musical pieces such as Bach's "Air," the audience was deeply moved; many people cried. The orchestra immediately founded the "Center for Recovery through the Power of Music, Tohoku" with other local musicians, and consequently delivered approximately five hundred concerts at various venues in the disaster area within five years.

Considering the same music would not move the same audience so dramatically on ordinary occasions, these were certainly peculiar phenomena. Three variables concerning musical response discussed earlier (i.e., stimulus, internal condition, and external environment) give a clue to the reason. Right after the earthquake, the internal conditions of evacuees and their external environment were unusual and irregular. People were emotionally unstable from losing loved ones, houses, and almost everything else. They were also tense, having no private space in the evacuee center. Furthermore, the electricity was completely out for days after the earthquake; even after it came back, the sound environment was rather eerie in the evacuee center. Under these circumstances, any stimulus caused strong reactions. Just talking to someone could be problematic.

After the earthquake, people were so sensitive that they hardly wanted to listen to anything. Even a song could have hurt them. An interviewee said, "the recovery songs composed after the earthquake would not work well because they emphasized words. Whereas the well-known melody becomes a vessel to release our feelings, the verbal meaning does not sink inside." In this respect, the familiar and soothing tune of Bach's

"Air" must have met the evacuees' needs. Though containing lyrics, "Tomorrow Is Sure to Come" also worked well, having an equivocal meaning unrelated to the disaster. At the sincere and genuine performance of the local students, the entrainment effect of music let the evacuees relax physically. The unknown but somewhat familiar and nostalgic tune permitted reflections on past occasions, providing a time and space to forget their difficult reality.

On the other hand, there were cases in which music failed to empower the people. It was reported that some musicians, who were overconfident in their own music, upset the people in the disaster area; most of the recovery songs—created with good intentions—did virtually nothing to comfort victims. For instance, Inawashirokos, a famous rock group from Fukushima, wrote a song called "I Love You & I Need You, Fukushima." It was intended to encourage Fukushima residents with the straightforward words against the stigma of the nuclear accident. However, an interviewee said that his father hated the song. A local TV station continuously broadcasted a video of the song, in which ordinary people from places other than Fukushima joined Inawashirokos in singing the lyric, supposedly as a gesture of encouragement. In reality, their weak voices, often barely heard over that of Inawashirokos, gave the impression of superficiality. Evidently, the song did not meet the needs of Fukushima residents. The father's conditional reflex to the song has been fixed as disgust.

Meanwhile, there was a recovery song that was well received in the disaster area and throughout the country. "Flowers Will Bloom" (*Hana wa Saku*) was originally a theme song for a series of NHK television programs on the Tohoku earthquake in 2012; almost instantly, it became popular and continued to be sung across the country. The NHK commissioned a famous playwright and composer born in the disaster area to write the song. It was introduced with a music video in which famous musicians, entertainers, and athletes with a strong connection to the disaster areas took turns singing, just as "I love You & I Need You, Fukushima" was performed. In NHK's case, however, all characters sang their parts in their own singing style and with deep feeling.

The song was widely sung by afflicted people. In 2013, a choral group was formed in Sendai by older people who had lost their houses in the tsunami; this song was the group's focus. As the organizer of the group explained, " 'Flowers Will Bloom' is a song for reflecting on one's life and passing on wishes to the next generation. Some people even think that the lyrics were written through the eyes of the dead. Because many of the older people feel sorry for having survived when promising young people lost their lives, the lyrics perhaps genuinely touch their hearts." The lyric of the song is written from diverse perspectives and affords survivors a chance to express their inner feelings.

These examples suggest that in certain conditions music brought positive effects after the earthquake. For example, soothing and somewhat familiar music—with no words or equivocal text—performed by people sympathetic toward the victims worked well. Clearly, empathic music works better than encouraging music, and a certain equivocality better suits the needs of victims. In terms of sociological concerns, it is thus evident that the effect of music varies according to what kind of music is played, by whom it is performed, and how it is contextualized.

To add neuroscientific insight, another finding is that the reactive parameters of musical response were shared broadly among people for some time after the earthquake. The internal condition and external environment were fairly homogeneous, as afflicted people were emotionally fragile from being placed in a comfortless and inconvenient environment. Thus, the parameters of musical response, except for perceptual sensitivity based on past experience, were shared. The chances that music would affect a large number of people at once were high, which confirms why the power of music was featured and a certain kind of music was effective. The communal effect of music also elevated a cooperative spirit for recovery; therefore, music's effect was empowering.

As time passed, however, music's role changed. Its homogeneous effect diminished; conversely, the importance of musical participation increased. As in the case of the "Flowers Will Bloom" chorus, music became a vessel for bonding community members through musical retelling. The next section further addresses this retelling aspect of music from an interdisciplinary perspective.

Sociological Implications for the Future of Neuroscience

Traditional music scholarship regards musical performance as representing a musical piece. Unless the piece is modified compositionally, its meaning is considered to be basically the same (Madrid 2009). Sociologically, however, performance is an opportunity in which music allows people to do something with it while it itself does something. Therefore, performance is not a mere act of representation; even reperformance has its unique role, producing renewed meanings. Here, the issue of retelling comes into play.

Retelling is a key concept for empowerment in feminist and postcolonial studies as well as narrative and trauma therapy (e.g., Scott 1992; Gilroy 1993; White 1993; Herman 1997). Identity is often thought of as something given and fixed—even innate—particularly in the context of race and gender. However, although genetic factors should not be entirely dismissed, identity is a reflective construction of one's past with a perceived social context (Elliott 2007). Ultimately, identity is a narrative of oneself. Therefore, to be able to feel "that life is a story in the telling, in the retelling, and that one can have some control over that" is a way of recovering from a trauma (Brison 2002, 115). Retelling oneself with re-envisioned relationships gives a clue to transforming perceived difficulty into something manageable.

Retelling works not only linguistically but also artistically. Artistic retelling is often called "memory work" (Onyx and Small 2001). The geographer Karen Till (2008) discovered memory work in participatory art activities associated with reimagining the history of an apartheid district in South Africa. She referred to it as "the difficult process of working through the losses and traumas resulting from (revisiting) past violence and injustice, and of imagining more socially just futures" (110). Musical retelling is also a powerful means of relating in a new way to the world. Particularly, the development of shared perceptual sensitivities through musical retelling may be a key to empowerment.

Examples here are taken from musical activities for sexual minorities in Tokyo, who have aimed for community empowerment and social transformation of the boundary between minority and majority (Nakamura 2013, 2014b). Particularly, an annual music festival for sexual minorities and their allies called "Prelude" is discussed. In Japan, sexual minorities began to attain significant public visibility in the 1990s. Alongside their liberation movements, various community activities were launched in Tokyo's metropolitan area during this period. Prelude began in 2005, and the activities of musical groups for sexual minorities have since become vibrant. In the 6th Prelude in 2010, for instance, three chorus groups, four wind ensembles, one jazz band, and one string orchestra showed up on the stage of a large public hall; they played pop and classical music between occasional comic stage performances. Although the repertoires and levels of performance varied, the players' enthusiasm for their performance, dedication of the volunteer staff, and receptive attitude of the audience yielded a unique musical space in the public hall.

Prelude may appear as just another music festival, but ethnographic work revealed that there were many creative ingenuities empowering the participants, based on trial and error since the 1990s. First of all, the participants have an opportunity to retell musical works, superimposing their own sentiments on the musical texts performatively. For instance, the first chorus group sang "Yell," a choral version of a sentimental song by a young popular singing group called Ikimonogakari. The song expresses the lonely feelings typical of adolescents and quietly cheers them for living through difficulties. Their next song was "Climb Every Mountain" from *The Sound of Music*, which is, needless to say, a solemn song of encouragement, advising metaphorically and resolutely to live with high spirits. While they sang for the adolescent persona and the characters in *The Sound of Music*, they had a chance to express themselves with the songs. The adolescent persona in "Yell" may overlap their own adolescent experience and fear of coming-out; "Climb Every Mountain" may be an encouragement for their fellows. Although no one spoke of any political agenda or shouted for minority liberation, the context of the concert—the persons who sang the pieces were not adolescents but adult gay men—suggested to the audience that the personas of the musical works could be them.

This kind of deliberate musical selection is fairly consistent throughout Prelude. For example, a brass band gave an excellent performance of Tchaikovsky's *Slavic March*. Since the composer had experienced homosexual suffering, their superb performance of this piece exhibited a special meaning at the concert. The ending of Prelude every year is a communal performance with members from all participating groups. In the case of the 2010 festival, they played Elgar's *Pomp and Circumstance, No. 1*. The conductor told the audience prior to the performance that he had traveled to the UK previously, where he listened to this piece in the BBC Proms and was inspired to conduct it. Subtly but clearly, the symbolic scene of the nationalistic event was appropriated for the unity of people at Prelude. Music at Prelude may be appropriated in a unique framework, where the participants retell the music for their needs, which in turn the audience appreciates.

Rehearsing for the festival is also characterized as musical retelling. While they rehearse to perform musical pieces better, their rehearsal may be a bit unusual. For example, when wanting to tell a romantic allegory to improve a melodic or harmonic expression in performance, the conductor would use a story in which homosexual characters love each other. This kind of use of homosexual metaphors are fairly common in music groups in Prelude so that rehearsing music together is a way of crystallizing their unique experiences into musical works. As participants noted, making music with people considered to be fellows becomes a lasting memory, particularly for sexual minorities who usually feel isolated.

Even though comedy occasionally takes place onstage, the concert maintains a certain prestigious quality, in part because of its excellent stage progression, led by a member who is a professional stage manager. This formality is important because it makes the concert "public." Small (1998) argued that a symphonic concert embodies the public sphere in the sense of Jürgen Habermas's monolithic bourgeois space. In Prelude, however, the public hall is transformed into an alternative sphere of the public through their musical performance (Fraser 1991). Sociologically speaking, Prelude is thus a collective artwork created by participants' programming, rehearsing, and staging, which presents a set of new social values.

Neuroscientifically, on the other hand, what is presumably true is that when we play music together with a particular experience, we develop a shared perceptual sensitivity to the music that allows us to believe that we are connected to each other. Of additional importance here is that through musical retelling this connection is made not only with other participants but also with the history of music, in other words, with those who have been involved in that particular music in the past as a composer, a performer, and a listener. We can find a connection with the history embedded in the music through the very music itself (Stige 2002, 105–106; Gilroy 1993). With the embodied workings of the music, musical retelling allows us to fuse the present self with the present other, and the present self with the past other, through shared embodied workings to the music. Intersubjectivity, which has traditionally been viewed in terms of philosophy, may now be explained as a bodily (i.e., neuroscientific) phenomenon. The performative experience of music could thus change our relationships to the world.

Conclusion

The present chapter has explained part of the process in which musical stimuli exert power, based on the interdisciplinary exploration of sociology and neuroscience. The first three sections searched for common ground of these two disciplines. Four key aspects for interdisciplinary collaboration are presented: (1) musical response occurs with three variables: stimulus, internal condition, and external environment; (2) embodied response differs according to one's degree of involvement with sound; (3) embodied workings pertaining to musical responses are largely on a simulation basis; and (4) its workings of reactive parameters depend greatly on one's perceptual development.

The last two sections demonstrate the interdisciplinary considerations of specific ethnographic examples. Discussions about musical activities after the earthquake suggest that the sensitive use of music may become an effective tool right after a disaster because a victim's internal condition and external environment are maintained rather homogeneously. As recovery proceeds, the role of music changes from something to listen to, to something to participate in.

Subsequently, neuroscientifically informed discussions of Prelude have generated the idea that musical retelling involves processes in which we develop a shared perceptual sensitivity of the music that allows us to believe that we are connected to each other—to other participants as well as individuals involved with the music in the past. Even though the motivations of the retelling may be different from those of the preceding telling, they share the way of telling: that is, the embodied workings of the music.

The musical experience is performative and temporal; however, musical retelling provides us with a shared embodied mechanism that ensures both mutuality and diversity, which presumably becomes a resource for our coexistence. The audience's empathic approval of the performance further crystallizes it as accessible musical memories. This retelling aspect of music may be a future theme to explore in more detail through the interdisciplinary efforts of sociology and neuroscience. Since musical performance is, by nature, always retelling, this theme is worth pursuing further.

Acknowledgment

Part of this study was supported by the Japan Society for Promotion of Science (JSPS) KAKENHI Grant Number 25580023.

References

Acord, Sophia Krzys, and Tia DeNora. 2008. "Culture and the Arts: From Art Worlds to Arts-in-Action." *Annals of the American Academy of Political and Social Science* 619 (1): 223–237.

Adorno, Theodor W. 1976. *Introduction to the Sociology of Music*. Translated by E. B. Ashton. New York: Continuum (originally published as *Einleitung in die Musiksoziologie*).

Becker, Judith. 2001. "Anthropological Perspectives on Music and Emotion." In *Music and Emotion: Theory and Research*, edited by Patrik N. Juslin and John A. Sloboda, 135–160. New York: Oxford University Press.

Becker, Judith. 2004. *Deep Listeners: Music, Emotion, and Trancing*. Bloomington: Indiana University Press.

Bergh, Arild, and John A. Sloboda. 2010. "Music and Art in Conflict Transformation: A Review." *Music and Arts in Action* 2 (2): 2–17.

Born, Georgina. 2011. "Music and the Materialization of Identities." *Journal of Material Culture* 16 (4): 376–388.

Bharucha, Jamshed, Meagan Curtis, and Kaivon Paroo. 2006. "Varieties of Musical Experience." *Cognition* 100: 131–172.

Brison, Susan J. 2002. *Aftermath: Violence and the Remaking of a Self*. Princeton, NJ: Princeton University Press.

Damasio, Antonio. [1994] 2005. *Descartes' Error: Emotion, Reason, and the Human Brain*. New York: Putnam. Reprint, New York: Penguin Books.

DeNora, Tia. 2000. *Music in Everyday Life*. New York: Cambridge University Press.

DeNora, Tia. 2003. *After Adorno: Rethinking Music Sociology*. New York: Cambridge University Press.

DeNora, Tia. 2013a. "'Time after Time': A Quali-T Method for Assessing Music's Impact on Well-Being." *International Journal of Qualitative Studies on Health and Well-Being* 8: 1–13.

DeNora, Tia. 2013b. *Music Asylums: Wellbeing through Music in Everyday Life*. Farnham: Ashgate.

DeNora, Tia. 2014. *Making Sense of Reality: Culture and Perception in Everyday Life*. London: Sage.

Dolar, Mladen. 2006. *A Voice and Nothing More*. Cambridge, MA: MIT Press.

Elliott, Anthony. 2007. *Concepts of the Self*. 2nd ed. Cambridge: Polity.

Fraser, Nancy. 1991. "Rethinking the Public Sphere: A Contribution to the Critique of Actually Existing Democracy." In *Habermas and the Public Sphere*, edited by Craig Calhoun, 109–142. Cambridge, MA: MIT Press.

Gallese, Vittorio. 2003. "The Manifold Nature of Interpersonal Relations: The Quest for a Common Mechanism." *Philosophical Transactions of the Royal Society B* 358: 517–528.

Gilroy, Paul. 1993. *The Black Atlantic: Modernity and Double Consciousness*. Cambridge, MA: Harvard University Press.

Hargreaves, David J., Raymond MacDonald, and Dorothy Miell. 2005. "How Do People Communicate Using Music?" In *Musical Communication*, edited by Dorothy Miell, Raymond MacDonald, and David J. Hargreaves, 1–26. New York: Oxford University Press.

Hennion, Antoine. 2015. *The Passion for Music: A Sociology of Mediation*. Farnham: Ashgate. Originally published as *La passion musicale* (Paris: Editions Métailié, 1993).

Herman, Judith Lewis. 1997. *Trauma and Recovery: The Aftermath of Violence—From Domestic Abuse to Political Terror*. New York: BasicBooks.

Juslin, Patrik N., Simon Liljeström, Daniel Västfjäll, and Lars-Olov Lundqvist. 2010. "How Does Music Evoke Emotions? Exploring the Underlying Mechanisms." In *Handbook of Music and Emotion: Theory, Research, and Applications*, edited by Patrik N. Juslin and John Sloboda, 605–642. New York: Oxford University Press.

Laurence, Felicity. 2008. "Music and Empathy." In *Music and Conflict Transformation: Harmonies and Dissonances in Geopolitics*, edited by Olivier Urbain, 13–25. London: I.B. Tauris in association with the Toda Institute for Global Peace and Policy Research.

Lindblom, Jessica. 2015. "Meaning-Making as a Socially Distributed and Embodied Practice." In *Aesthetics and the Embodied Mind: Beyond Art Theory and the Cartesian Mind-Body Dichotomy*, edited by Alfonsina Scarinzi, 3–19. London: Springer.

Madrid, Alejandro L. 2009. "Why Music and Performance Studies? Why Now? An Introduction to the Special Issue." *Transcultural Music Review* 13. Accessed April 11, 2016. http://www.sibetrans.com/trans/trans13/art01eng.htm.

McGuiness, Andie, and Katie Overy. 2011. "Music, Consciousness and the Brain: Music as Shared Experience of an Embodied Present." In *Music and Consciousness: Philosophical,*

Psychological and Cultural Perspectives, edited by David Clarke and Erik Clarke, 245–262. New York: Oxford University Press.

Molnar-Szakacs, Istvan, and Katie Overy. 2006. "Music and Mirror Neurons: From Motion to 'e' Motion." *Social Cognitive and Affective Neuroscience* 1 (3): 235–241.

Monson, Ingrid. 2008. "Hearing, Seeing, and Perceptual Agency." *Critical Inquiry* 34 (S2): S36–S58.

Nakamura, Mia. 2013. *Ongaku wo hiraku* [Possibility of Music: Art, Care, and Culture]. Tokyo: Suiseisha.

Nakamura, Mia. 2014b. "Higashinihon-daishinsai wo meguru 'ongaku no chikara' no shoso: Mirai no bunkaseisaku to ato manegimento no tame no kenkyu 1" [The 'Power of Music' after the Great East Japan Earthquake: A Study for Future Cultural Policy and Arts Management 1]. *Geijutsukougaku Kenkyu* 21: 13–29.

Nakamura, Mia. 2014a. "Retelling, Memory-Work, and Metanarrative: Two Musical-Artistic Mediations for Sexual Minorities and Majorities in Tokyo." *Music and Arts in Action* 4 (2): 3–23.

Nakamura, Mia. 2017. "'Bunka-ryoku' towa nanika?: Higashinihon-daishinsai-go no 'ongaku no chikara' ni kansuru gakusai-kenkyu ga sisasuru koto" [What Is "Cultural Power"?: Suggestions from the Interdisciplinary Study of "The Power of Music" after the Great East Japan Earthquake]. *Bunkaseisaku Kenkyu* 10: 30–48.

Nakamura, Mia. 2018. "The 2011 Japan Earthquake and Music: Recovery Concerts, Recovery Songs, and 'the Power of Music.'" *Music and Minorities* (tentative title), edited by Ursula Hemetek, Inna Naroditskaya, and Yoshitaka Terada. Osaka: The National Museum of Ethnology, forthcoming.

North, Adrian C., and David J. Hargreaves. 2006. "Music in Business Environments." In *Music and Manipulation: On the Social Uses and Social Control of Music*, edited by Steven Brown and Ulrik Volgsten, 103–125. New York: Berghahn Books.

Onyx, Jenny, and Jennie Small. 2001. "Memory-Work: The Method." *Qualitative Inquiry* 7 (6): 773–786.

Patel, Aniruddh D. 2008. *Music, Language, and the Brain*. Oxford: Oxford University Press.

Pavlicevic, Mercédès, and Gary Ansdell. 2004. *Community Music Therapy*. Philadelphia, PA: J. Kingsley Publishers.

Pearce, Marcus, and Martin Rohrmeier. 2012. "Music Cognition and the Cognitive Sciences." *Topics in Cognitive Science* 4 (4): 468–484.

Peretz, Isabelle. 2006. "The Nature of Music from a Biological Perspective." *Cognition* 100 (1): 1–32.

Rice, Timothy. 2014. "Ethnomusicology in Times of Trouble." *Yearbook for Traditional Music* 46: 191–209.

Reybrouck, Mark, and Elvira Brattico. 2015. "Neuroplasticity beyond Sounds: Neural Adaptations Following Long-Term Musical Aesthetic Experiences." *Brain Sciences* 5 (1): 69–91.

Scott, Joan W. 1992. "Experience." In *Feminists Theorize the Political*, edited by Judith Butler and Joan W. Scott, 22–40. New York: Routledge.

Shepherd, John, and Kyle Devine, eds. 2015. *The Routledge Reader on the Sociology of Music*. New York: Routledge.

Small, Christopher. 1998. *Musicking: The Meanings of Performing and Listening*. Hanover: University Press of New England.

Stige, Brynjulf. 2002. *Culture-Centered Music Therapy*. Gilsum, NH: Barcelona Publishers.

Stige, Brynjulf, Gary Ansdell, Cochavit Elefant, and Mercédès Pavlicevic, eds. 2010. *Where Music Helps: Community Music Therapy in Action and Reflection*. Farnham: Ashgate.

Terasawa, Hiroko, Reiko Hoshi-Shiba, Takuro Shibayama, Hidefumi Ohmura, Kiyoshi Furukawa, Shoji Makino, et al. 2013. "Shintai kino no togo ni yoru ongaku jodo komyunikeshon moderu" [A Network Model for the Embodied Communication of Musical Emotions]. *Cognitive Studies* 20 (1): 112–129.

Thompson, William Forde. 2009. *Music, Thought, and Feeling: Understanding the Psychology of Music*. Oxford: Oxford University Press.

Till, Karen E. 2008. "Artistic and Activist Memory-Work: Approaching Place-Based Practice." *Memory Studies* 1 (1): 99–113.

Turino, Thomas. 2008. *Music as Social Life: The Politics of Participation*. Chicago: University of Chicago Press.

Urbain, Olivier, ed. 2008. *Music and Conflict Transformation: Harmonies and Dissonances in Geopolitics*. London: I. B. Tauris in association with the Toda Institute for Global Peace and Policy Research.

White, Michael. 1993. "Deconstruction and Therapy." In *Therapeutic Conversations*. Edited by Stephen Gilligan and Reese Price, 22–61. New York: W.W. Norton.

PART III

EMBODIED MIND, EMBODIED RHYTHM

CHAPTER 8

SOUND-MOTION BONDING IN BODY AND MIND

ROLF INGE GODØY

Introduction

There are countless links between sound and body motion in music-related contexts: for one thing, music is (at least traditionally) made by various body motions such as hitting, stroking, shaking, plucking, bowing, or blowing on instruments, or by motion of the vocal apparatus in singing. And people often move when listening to music, such as in dancing, walking, gesticulating, nodding heads, or swaying whole bodies. Also, people may tend to use motion-related metaphors (such as fast, slow, agitated, calm, etc.) when speaking about music. These and similar cases of links between sound and body motion in music-related contexts reflect the basic tenet of this chapter, namely that sound and body motion are indeed so intertwined that we may be hard put to separate these two when we dig deeper into perceptually salient features of musical experience.

Focusing on what we may call *sound-motion bonding in body and mind*, the aim of this chapter is first of all to present a coherent theoretical framework for understanding this bonding, followed by various findings from relevant research on this bonding in music as well as some ideas on how these insights on sound-motion bonding may be further developed and put to use in various contexts.

Motor Theory

The strong links between sound and body motion in our experiences of music may be seen as instances of the so-called *motor theory* of perception. This theory claims that we tend to spontaneously (albeit variably so) associate whatever sensory input we have, be that of sound, vision, or motion, with some mental image of body motion. In other words, this theory claims that there is an incessant mental simulation of body motion going on in perception, so that, for instance, when we hear sound, we mentally simulate the sound-producing motion or some other sound-related motion that we believe belongs to what we are hearing: when we hear ferocious drumming, we may mentally simulate energetic mallet-hand-arm motion, and when we hear soft string ensemble music, we may mentally simulate slow, protracted bowing motion, and, also, when we hear loud, pounding electronic dance music, we may simulate vivid whole body swaying motion.

Versions of the motor theory of perception appeared in linguistics in the 1960s with the suggestion that language acquisition and language comprehension are not only a matter of becoming familiar with the acoustic features of spoken language but also just as much a matter of becoming familiar with the sound-generating motion of the vocal apparatus, the so-called phonological gestures, implying that sound-generating body motion also serve as mental schemas for making sense out of what we are hearing (Liberman and Mattingly 1985). Initially met with quite a lot of resistance, the *motor theory* has in recent decades become increasingly well supported by noninvasive brain observation methods (i.e., by so-called functional MRI observations of the spontaneous activation of motor areas of the brain when listening to speech or music, activations that resemble the actual generation of the sound by the listening person (see, e.g., Bangert and Altenmüller 2003; Haueisen and Knösche 2001). In fact, research from the last couple of decades seems to suggests that links between sound and body motion, besides being intertwined on a behavioral level, also are "hard-wired," that is, they are found on a neurophysiological level, hence part of our basic biological constitution (Kohler et al. 2002).

The motor theory seems to apply to other domains of human perception as well (Galantucci, Fowler, and Turvey 2006), including visual perception, so that when we, for instance, see an ornate pattern such as in a wrought iron gate, we may mentally trace the twists and turns of this pattern with our hands (Berthoz 1997). Furthermore, it has been suggested that this kind of mental simulation extends not only to social cognition and to the ability of having empathy with others (Wilson and Knoblich 2005) but also to more abstract thought such as various mental schemas and concepts (Gallese and Lakoff 2005). In short, it has been suggested that the motor theory perspective is applicable to most human cognition, leading to the idea of "motor ontology," an idea suggesting that most crucial aspects of our existence are all related to sensations of body motion (Gallese and Metzinger 2003).

The role of motion in perception is furthermore linked with that of imitation, so that simulating body motion associated with what we are perceiving may function as a source of learning, something particularly evident in so-called oral musical traditions. The general capacity for imitation is also manifest as vocal imitation, for instance in onomatopoetic expressions that imitate various environmental sounds, and eminently so in so-called scat singing, which imitates musical instruments, and beatboxing, which can imitate almost any sound (Proctor et al. 2013).

One essential feature of such imitative behavior, what I have earlier called "motormimetic cognition" (Godøy 2001, 2003), is that of approximate recognition and also approximate rendering. For instance, I do not speak Russian or Chinese, yet I believe I am capable of distinguishing between these two languages because their respective repertoires of sounds and their associated phonological gestures are so distinct that even an approximate familiarity with these repertoires is sufficient for me to distinguish between these two languages. I would also dare to try to make sounds that are very approximately Russian versus very approximately Chinese. This kind of approximate knowledge seems to be at work in perception and category formation in general, making Steven Harnad suggest the term "approximationism" as an attribute of categorical perception (Harnad 1987), that is, the tendency of human (and animal) perception to put variants of what we perceive (e.g., different kinds of chairs from baroque to modern) into one single container (e.g., a piece of furniture to sit on).

Another way of denoting this phenomenon of approximate imitation is that of variable resolution in motormimetic cognition, meaning variable degree of detail in our imitations, with a continuum extending from rather coarse, nonexpert type of imitation to fine-grained, expert type of imitation. This is evident in observations of listeners with little or no musical training who may still be able to render basic sound-producing motion in listening, as may be seen in cases of so-called air instrument performance, however the amount of detail seems to vary with the level of musical expertise (Godøy, Haga, and Jensenius 2006). Also, we may learn more about how people perceive music-related features by so-called sound tracing, where listeners are asked to make spontaneous body motion to sound excerpts, either on a digital tablet or by arm or whole body motion in three-dimensional space, with the motion recorded using so-called motion capture technology (Nymoen et al. 2013). These sound-tracing studies tended to reveal that most people have quite clear associations between salient musical features and body motion, in particular those of pitch and loudness, but less so for timbral features and articulation detail features. In Figure 8.1, we can see one example of such sound tracing on a digital tablet. Clearly, there are differences, yet also clear similarities, between several of the tracings.

What these and similar findings on music-related body motion attest to is that most people may indeed make spontaneous links between sound and motion, yet that there are expertise-dependent variations in detail (i.e., in the resolution of these links). Another element here is that music often is composite in the sense of having several features in parallel, for instance with a foreground melody and a background accompaniment, with this accompaniment in turn having roles of chords, bass line, rhythmical patterns, and so on, which may result in different listeners focusing on different elements in the one and the same musical excerpt. For this reason we could also speak of multiple motion affordances of musical sound—that we may have several different motion patterns in parallel, yet that usually there might be some kind of synchrony relationship between the different motion patterns (Godøy 2010a).

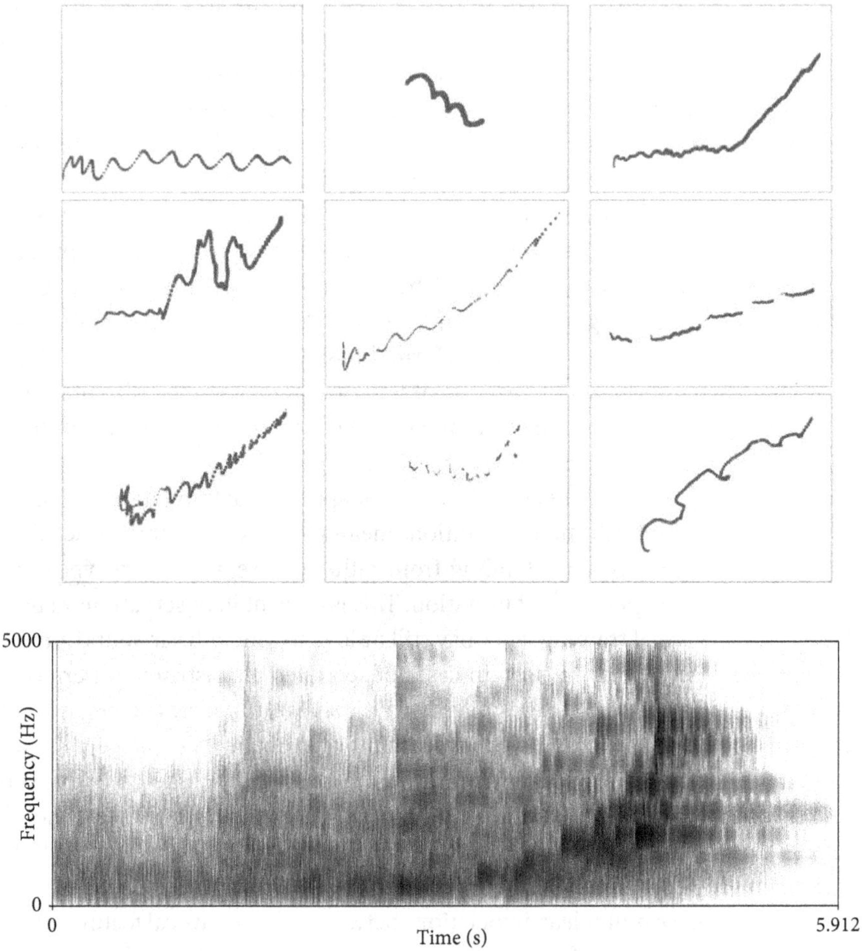

FIGURE 8.1 Spontaneous sound-tracing on a digital tablet by nine different listeners, some with little or no musical training, of a 6-second excerpt (represented as a spectrogram at the bottom) from the beginning of A. Scriabin's 5th piano sonata. In spite of obvious differences, several tracings reflect well the "rumbling" sound in the beginning and the following upward surge of the sound in this excerpt.

Music-Related Motion

As a result of several years of international research cooperation and discussions, there seemed to emerge a consensus that music-related body motion could for a start be classified according to function (i.e., according to its role in musical contexts) and then be progressively differentiated according to various other attributes. The two main classes of functions may be summarized as follows (see Godøy and Leman 2010 for details):

- *Sound-producing motion.* This includes all kinds of body motion directly or more indirectly involved in the production of sound on a musical instrument or by the human vocal apparatus. Basically, this is about energy transfer from the human body to the instrument (or within the human body in the case of singing), hence what we call *excitatory motion.* But sound production also includes motion that modifies the sound, what we call *modulatory motion* (e.g., vibrato with the left hand on a string instrument that is bowed with the right hand, or the opening-closing of a mute on a brass instrument while continuously blowing). Furthermore, we have what may be called *ancillary motion*, meaning motion that in itself is not strictly sound-producing, such as changing posture or making arm, shoulder, and torso motion while playing with the fingers on a keyboard in order to avoid strain injury or in order to help shape the expression of the music. Lastly, we also include here what we call *communicative motion*, motion that communicates with fellow musicians in the ensemble or motion aiming to make theatrical impressions on the audience.
- *Sound-accompanying motion.* This includes all kinds of body motion that occurs in parallel with musical sound, and will typically often (but not always) have some degree of synchrony with the sound events of the music such as in dancing, walking, gesticulating, or other kinds of body motion to the music. Sound-accompanying motion may also variably mimic sound-producing body motion (e.g., some rudiments of drumming with the arms while swaying the whole body to the overall pulse of the music). Lastly, in some cases, the sound-accompanying motion may seem to be rather unrelated to the sound events of the music, yet serve to add a new layer of features to the combined sound-motion experience, as may be seen in some cases of modern ballet or use of music in films.

Although such function differentiations may be useful in our research on sound-motion bonding, it should be noted that the boundaries between categories here may often be gradual and also that an instance of motion may be multifunctional: the raising of the arm, shoulder, and even whole torso at an upbeat location in the music may be part of a sound-producing motion (i.e., preparing an accented chord on the following downbeat), an upbeat cue to fellow musicians, and a theatrical show of effort for the audience.

Also, when listening, it may in some cases not be so clear what sound feature is actively made by sound-production and what is the "passive" response of the instrument: a continuous stroking motion across a washboard or maracas may produce a similar rapid iterative sound as by a shaking, back-and-forth tremolo-type of motion as in a drum roll. This means that there may be differences between the patterns of the sound-producing motion and the patterns of the resultant perceptual features of the sound, or between the causality and what we with Pierre Schaeffer shall call the "morphology" of the sound (Schaeffer 1966; see also Chion 1983). However, the overall energy envelope of a music-related motion may in many cases be similar to both the sound-producing and the sound-accompanying kind of motion. This means that we have generic types of energy envelopes that cut across variant sound and motion instantiations. This can be seen in the famous barber scene in Chaplin's *Dictator* movie, where the barber (Chaplin) makes shaving motions to the sound of Brahms's *Hungarian Rhapsody no. 5*, sounds that are the result of various instrumental motions (Godøy 2010a).

Furthermore, we need to specify what timescale of music-related motion we have in mind, as we may of course focus on different levels of detail as well as on more global or macroscopic features. In our work on music-related motion we have found it useful to differentiate between three main timescales, and, notably so, we have found that these three timescales are equally valid for motion and sound:

- *Micro*: continuous sound and motion, typically concerned with sustained sound or rapid and consistent fluctuations in the sound (e.g., rapid vibrato or tremolo).
- *Meso*: chunks typically in the 0.5- to 5-second duration range, what we shall call *sound-motion objects*, sufficient in duration for the recognition of most salient stylistic features of music (Gjerdingen and Perrott 2008), but also typical for the duration of human motion units in general (Schleidt and Kien 1997) as well as for human attention spans and meaning unit durations (Pöppel 1997).
- *Macro*: the timescale of whole sections and movements, consisting of several meso timescale chunks in succession, resulting in narrative or dramaturgical effects.

The macro timescale leaves us with cumulative, overall memory images of whole works or even whole concerts and other extended musical events. With available methods, it is indeed possible to give informative feature summaries for extended works and even collections of works (e.g., by so-called quantity of motion, an index of the overall activity level, and correspondingly for the sound, various overall index of event density). However, the perceptually most salient features of both sound and motion are to be found at the meso timescale, and this is the timescale that we focus on in the reminder of this chapter.

Motion Features

The main challenge of doing research on sound and on motion is that they are both ephemeral: sound and motion may be saliently present for us here and now, yet gone in the next moment, so we need tools for capturing, storing, processing, analyzing, and representing these ephemeral phenomena in perceptually meaningful ways. Luckily, we now have methods and technologies that allow us to study sound and motion in historically unprecedented ways, methods that enable us to "freeze" sound and motion and examine details at our leisure. Yet we also need to consider knowledge aims (i.e., be more precise about what we are looking for, in our research, and hence also in our choice of methods). Methods may vary along a continuum from qualitative to quantitative, and different methods may be quite legitimate, depending on what we want to explore. First we have:

- *Phenomenological*, more "arm-chair" type methods. These are important because they provide us with a clear conceptual overview of sound and motion features that we are interested in studying, as well as basic issues of memory and imagery.
- *Observation and annotation* methods. These may be practiced both "on the fly" in actual performances and with sound-video recordings. Many important insights are indeed available to the "naked eye" and/or scrutiny "by ear" (i.e., not requiring any use of technologies beyond simple recording and playback, only requiring that an observer have good analytic capabilities).

However, the ephemeral nature of both sound and motion has spurred a quest for more solid means for representations, manifest already at the end of the nineteenth century in ingenious photography methods by pioneers such as Eadweard Muybridge and Etienne-Jules Marey, and later in different motion-tracing methods by music-related pioneers such as Gustav Becking and Alexander Truslit (see Schneider 2010, for an overview). With presently readily available methods and technologies such as the following, we are now in a position to systematically collect motion and sound data:

- *Video-based methods*, including various video processing tools (Jensenius 2013), with video having the great advantage of being the least obtrusive technology in most cases.
- *Sensor-based methods*, including various devices such as inertial sensors (accelerometers and gyroscopes), pressure sensors, proximity sensors, and so on, also including various sensor-equipped traditional musical instruments.
- *Infrared motion capture*, for the moment probably the most accurate and precise method for recording body motion, however with the disadvantage of somewhat cumbersome recording situations.

Methods for the further processing of these various kinds of motion data have been in rapid development, but nevertheless, there still remain substantial challenges here. One major question is actually to find out what features we should look for in the data, and how to represent these features by various means (e.g., graphs, animations). In our past and present research we have tried to focus on the following features:

- *Trajectories* represented by motion curves in time and space (still pictures and animations), providing detail information on both sound-producing and sound-accompanying body motion.
- *Derivatives* of motion curves, such as velocity, acceleration, and jerk, derivatives that provide us with important information on perceptually salient motion features (e.g., affective states reflected in abruptness vs. calmness of motion).
- *Amplitude* and so-called *quantity of motion* reflecting perceptually salient information about the overall energy level of the motion (e.g., "restrained" vs. "extrovert" types of motion).

- *Postures* of the sound-producing effectors (fingers, hands, arms, shoulder, torso, etc.) and/or sound-accompanying motion, reflecting salient orientation points in the music (more on this later).
- *Various motion patterns* related to specific stylistic-aesthetic features, for example, to various rhythmical, textural, melodic, expressive, gait-related, and other patterns.

Another challenge is to try to develop some kind of classificatory scheme for music-related motion in order to enable feature-based navigation in large collections of music-related motion. From our research, and inspired by Schaeffer's work on sonic objects (1966, [1967] 1998; see also Godøy 2006; Chion 1983), we believe we can pick out some basic modes of motion, what we with Schaeffer call a basic typology, consisting of three main dynamic envelopes, each also with distinct biomechanical and motor control features:

- *Sustained*: protracted motion with continuous effort, such as in bowing or blowing.
- *Impulsive*: an abrupt, fast, and brief type of motion, such as in hitting.
- *Iterative*: a rapid back-and-forth or up-and-down motion, such as in a tremolo.

In addition to these three dynamic envelopes, Schaeffer also suggested three basic pitch-related envelopes, something we here may associate with pitch-related motion:

- *Stable pitched*, implying sustained motion as in a protracted tone.
- *Variable pitched*, implying pitch-changing motion as in a glissando.
- *Non-pitched*, that is, noise-dominated, implying various kinds of sound-producing motion.

These dynamic and pitch-related envelopes may then be combined in a 3×3 matrix of basic music-related motion, as demonstrated with examples by Schaeffer ([1967] 1998). However, this matrix is just a beginning, and more complex motion textures may be made by superposition and concatenation of such elements, and by differentiating patterns within the internal fabric of sonic objects, something that was developed in the earlier mentioned morphology of sonic objects in Schaeffer's theory.

Sound Features

As pointed out by Schaeffer on the background of the challenges of the *musique concrète* in the 1950s and 1960s, Western music theory had concepts for only a limited number of sound features, principally those that could be represented by the symbols of the Western common practice notation (i.e., for pitch and duration categories, for indications of tempo, choice of instrument [but these were not always specified], dynamics [also not always specified], and some rudimentary metaphors for expression). In the terminology of Schaeffer, Western music theory was focused on the abstract by its scheme of common practice notation, and Schaeffer saw one of the main challenges for music theory as that of developing tools for handling the nonabstract, what he called the concrete in musical sound, in particular the vast number of timbral, pitch-related, and duration-related nuances that are clearly at work in music, both non-Western and Western.

The main idea of Schaeffer's music theory was to take subjective impressions of musical sound as the point of departure, asking seemingly simple or naive questions as to what we are hearing. This kind of "Socratic" questioning was focused on fragments of sound at the meso timescale, on what became known as sonic objects, and which Schaeffer argued was the most important timescale of music. The crucial element in our context of sound-motion bonding is that this focus on sonic objects also can be extended to motion objects at the meso timescale (Godøy 2006).

In the decades since the publication of Schaeffer's monumental *Traité des objets musicaux* (1966), there have been significant developments in the fields of musical acoustics, psychoacoustics, and music perception. However, there are still the substantial challenges of distinguishing, naming, and exploring a number of perceptually salient and nonabstract features of musical sound. For this reason, we believe it makes sense to extend the basic top-down feature differentiation strategy of Schaeffer to our present research on sound and motion features, enabling us to fulfill what Schaeffer saw as a long-term project, namely to establish correlations between subjective features and the signal (i.e., the acoustic substrate of the sound).

With the above mentioned typology of sound objects as a point of departure, the ensuing differentiation within the morphology of sonic objects follows a similar strategy of (1) discriminating/naming some feature based on a subjective impression, (2) differentiating relative values of the feature, and (3) correlating this with signal features. As an example, consider the morphological feature called grain, meaning a rapid fluctuation in the sound, be that in pitch (e.g., a fast vibrato or a trill), in loudness (a tremolo), or in timbre (spectral flux): our music theory work should then start with designating the feature dimension (pitch, loudness, timbre), and then go on to the relative values (amplitude, rate, regularity, etc., of the fluctuations), and finally correlate this with the acoustic signal as well (based on some clever time domain and frequency domain signal representations that can highlight such features). In this kind of sound analysis work, we may now use various software tools that can help us bridge the gaps between the subjective notions of sound features and the acoustic basis for these features (e.g., *MIRtoolbox, Timbretoolbox, SonicVisualiser*, etc.).

A common strategy in all these studies of sound features is that of thinking of these features as shapes, because shapes are inherently holistic and are inherently capable of representing that which unfolds in time and space (i.e., are inherently concrete, to use Schaeffer's term, capable of going beyond the restrictions of more abstract, symbolic representations). The basic paradigm of Schaeffer's music theory may indeed be understood as that of shape differentiation, and this is what we now can extend to body motion as well. Based on a close reading of Schaeffer's theories, and on elements of the motor theory presented earlier, it seems a good idea to merge concepts from Schaeffer's pioneering sonic object research with more recent motion research (Godøy 2006). But first of all, thinking about musical sound features as shapes, we can list the following:

- *Time domain envelopes* as included in the earlier mentioned typology.
- *Spectral (frequency domain) features* are clear cases of shape cognition and may be represented by various spectrograms and also by body postures (e.g., vocal apparatus shapes for various formants) and on instruments (open-close of mutes, change in position of bowing or plucking, etc.).
- *Chords and harmonic progressions* may likewise be cases of shape cognition, both as spectral shapes and as effector shapes on the instrument (e.g., fingers on the keyboard).
- *Fluctuations and transients* within the sounds (e.g., various kinds of so-called grain, as different kinds of repeated shape patterns, reflecting the rate, amplitude, and evolution of such fluctuations over time).
- *Melodic patterns*, again both as audible sound shapes and as sound-producing motion shapes on an instrument.
- *Textural and rhythmical patterns*, both as sound shapes and as body motion shapes (e.g., the down-up-up motion of a waltz triple meter).
- *Temporal, expressive, gait-related*, and other patterns may likewise be represented as shapes by using various available analytic tools (e.g., extracting tempo curves from performance data).

Shape Cognition

Thinking features of sound and motion as holistically perceived entities is then the basis for what we call shape cognition here. The main point of shape cognition is that shapes can represent temporally extended, nonabstract entities such as the various envelopes found in musical sound (e.g., the dynamic envelopes, the spectral envelopes, the pitch-related envelopes, etc., as presented earlier, as well as body motion features such as posture shapes and motion trajectory shapes). Notably so, shape cognition also eminently applies to more secondary or derived features of sound and motion, that is, to all signal processing of different sound features such as the extraction of spectral centroid, spectral flux, and other perceptually salient sound features, as well as various derivatives of motion data, for example, velocity, acceleration, and jerk, features that are quite important for the perceived overall affective features of the motion.

This means that shape cognition is a very general tool for conceptualizing and differentiating perceptually salient features of sound and motion in music and also for studying links between sound and motion in music. Furthermore, shape cognition has the great advantage of bypassing the restrictions of Western common practice symbolic notation and in principle making any feature of sound and motion accessible for systematic study. In this sense, shape cognition is universally applicable and, it could be claimed, also a very basic feature of our perceptual-cognitive capabilities.

One of the most extensive recent research efforts on such fundamental human shape cognition is no doubt that of so-called morphodynamical theory (Godøy 1997; Petitot 1985; Thom 1983). The gist of morphodynamical theory is that human perception, understanding, and reasoning are based on ordering sensory input as shapes, or, in the words of René Thom, the leading figure of this theory, "the first objective is to characterize a phenomenon as shape, as a 'spatial' shape. To understand means first of all to geometrize" (1983, 6).

Although we may appreciate shape cognition as a viable universal conceptual tool for handling perceptually salient features in different modalities, and for linking different modalities, it may be argued that we also need subjective sensations of sound (of motion, of touch, etc.), in other words, that we, regardless the great analytical powers of shape cognition, also need to bring shapes alive. Put differently, shapes need to be enacted, put into motion, as suggested by the earlier mentioned motor theory of perception.

With this in mind, we now have a look at what could be called *sound-motion objects,* where we try to combine perceptual features of sound and motion fragments at the meso timescale into truly multimodal entities, and in particular look at how these sound-motion objects emerge from our continuous streams of sensory and mental activity.

Thinking Sound-Motion Objects

Sound-motion objects are entities where sensations of sound and motion (including the various components of motion, i.e., the proprioceptive, haptic, effort-related, etc.) are fused into multimodal chunks in our experience. Sound-motion objects are typically in the very approximate meso timescale duration range of 0.5. to 5 seconds, and are perceived as somehow meaningful units in terms of musical expression (i.e., with regard to genre, style, overall "sound," sense of motion, affect, etc.).

In plain language, we are hard put to distinguish between the components in, for example, a rhythmical pattern: tapping a polyrhythmic 3 against 4 pattern with my left and right hands on the table in front of me. Is the rhythmic pattern an auditory phenomenon or a motion phenomenon (in turn including proprioception, haptics, effort, etc.)? From our research, we would answer, "it's both!" We have with this polyrhythmic pattern a sound-motion object that is multimodal in the true sense of the word; however, we may shift our attention focus from one modality to another. Also, although temporally extended, it seems reasonable to suggest that the sequential unfolding of a sound-motion object is perceived as a totality, as something kept in our short-term memory, and where the different stages of this sequential unfolding as a totality contribute to its distinct overall perceptual appearance.

Attempts to explain the emergence of sound-motion objects from the continuous stream of music extends back to early phenomenological philosophy and early gestalt theory toward the end of the nineteenth century. In phenomenology, a leading idea of Edmund Husserl's was that segmenting continuous streams into units was necessary for perception, because without this agglutination into somehow more solid and holistically perceived entities, there would only be an amorphous mass of sensations (1991; see also Godøy 2011, 2010b, 1997). In a similar vein, the crucial issue for early gestalt theory was to try to explain the emergence of distinct entities in perception on the basis of temporally or spatially distributed signal features (be that in sound, vision, or motion) and predispositions of our perceptual apparatus (Smith 1988).

In Schaeffer's theory, the emergence of sonic objects was seen as primarily determined by qualitative discontinuities in so-called found objects, that is, in fragments of sound from the environment or from musical instruments, fragments that had their "natural" beginnings and end points (e.g., discontinuities between sound and silence). But sonic objects could also be made by "artificial" or more arbitrary cuttings from recorded sound, with the result that the cut object would acquire a new status as a sonic object on its own (Godøy 2006; Chion 1983; Schaeffer 1966). Also, psychoacoustic research has given a number of insights on the emergence of sound objects based on a combination of low-level, gestalt-like signal features (Bregman 1990), and more recent research on auditory objects seem to suggest the activation of various multimodal elements in addition to the purely auditory elements (Bizley and Cohen 2013; Griffiths and Warren 2004). In view of sound-motion bonding, such suggestions of multimodal elements are of course very interesting, as they seem to point out the mutual influence of sound and motion on each other.

We may find related ideas on the emergence of meso timescale chunks in motion-related research (i.e., on what we here may call motion objects). For a start, research in human motor control seems to support the ideas of goal-directed and hierarchical organization of human motion (Grafton and Hamilton 2007), something also manifest as posture-oriented motion (Rosenbaum et al. 2007), and also in so-called intermittency in motor control (Loram et al. 2014). We have in our own work toward a general model of sound-producing body motion in music adopted the idea of posture-oriented motion in what we call key-postures, meaning the shape and position of the effectors (fingers, hands, etc.) on instruments (or of the vocal apparatus) at what we call goal-points, which are typically downbeat points or other accented points in time. Furthermore, in this model we designate the upbeat motion toward such key-postures as prefixes, and the ensuing motion after these key-postures as suffixes, with musical performance then consisting of a series of such key-postures interleaved with prefixes and suffixes, as may be seen in Figure 8.2 (for details, see Godøy 2013).

Also, there are a number of constraints on human motion and perception, in particular the so-called psychological refractory period, meaning that human motion has to be planned in advance, hence leading to the notion of action gestalts (Klapp and Jagacinski 2011). This need to preprogram motion, and also to make anticipatory motion of the effectors (e.g., the hand needs to be placed in a particular position before the key

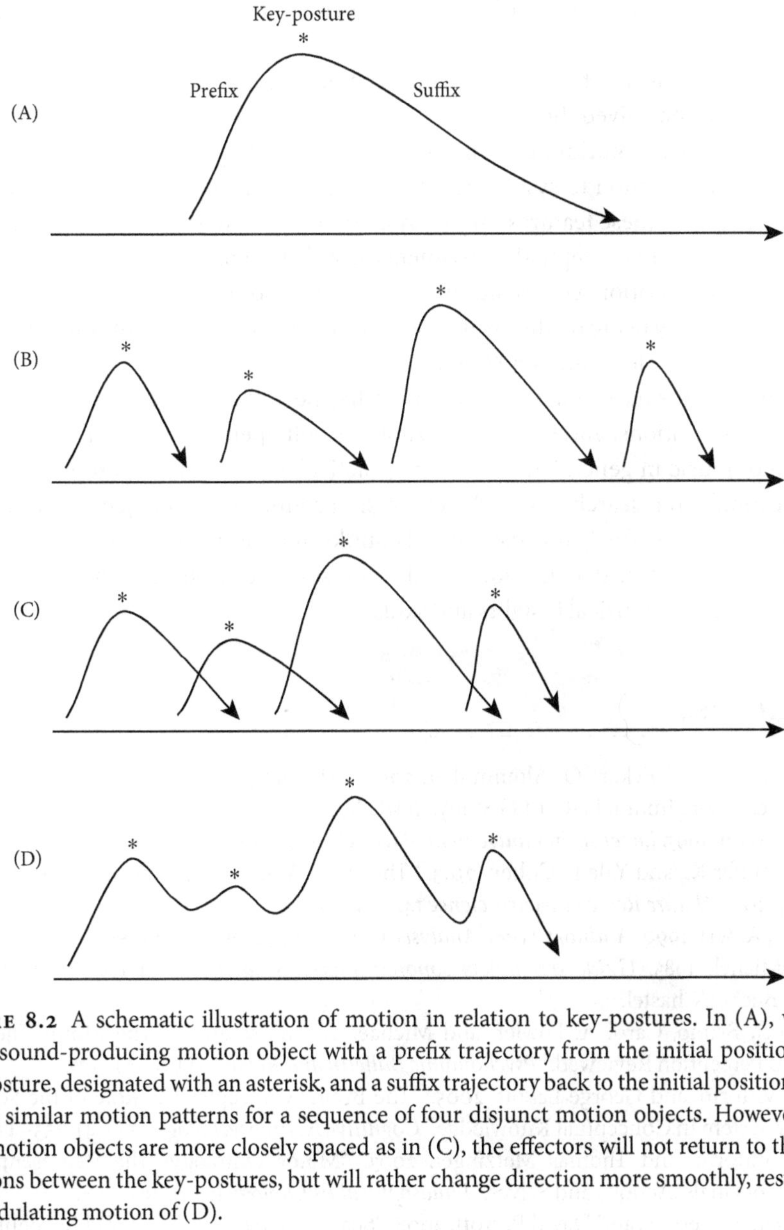

FIGURE 8.2 A schematic illustration of motion in relation to key-postures. In (A), we see a single sound-producing motion object with a prefix trajectory from the initial position to the key-posture, designated with an asterisk, and a suffix trajectory back to the initial position. In (B), we see similar motion patterns for a sequence of four disjunct motion objects. However, when such motion objects are more closely spaced as in (C), the effectors will not return to the initial positions between the key-postures, but will rather change direction more smoothly, resulting in the undulating motion of (D).

depression by a finger), leads to so-called coarticulation (Godøy 2014; see Kim's chapter for more on this subject). We also have changes in grouping of motion by so-called phase transitions (Haken, Kelso, and Bunz 1985), meaning that with an increase in the rate of events, events may tend to fuse into more superordinate events, and conversely, with a decrease in the rate of events, an initially coherent event may become split into separate events. In sum, we should try to understand the emergence of sound-motion objects on the basis of a number of constraints on human motion that also shape human perception.

This means that we have key-posture-centered sound-motion objects that are conceived and perceived holistically in music. Furthermore, these sound-motion objects have a rich, multidimensional set of salient perceptual features, features that we now are in a position to systematically explore by our methods for sound and motion analysis. Also, these features may be systematically tested by so-called analysis-by-synthesis methods in perceptual experiments as well as in practical artistic applications of musical improvisation, composition, and various kinds of dance and multimedia artistic expression, by using readily available technologies for sound synthesis and motion animation (see, e.g., Jensenius and Godøy 2013).

In summary, the basic tenet of this chapter has been that music is a fusion of sound and motion sensations, and that recognizing this will open up for an enhanced understanding of music in general and for some practical applications and tools in both artistic creation and research. Using the expression sound-motion objects here serves to indicate that we may freely translate sound chunks into motion chunks, and conversely, that we may freely translate motion chunks into sound chunks, and that doing so is a way to enhance our musical insights and skills.

References

Bangert, Marc, and Eckart O. Altenmüller. 2003. "Mapping Perception to Action in Piano Practice: A Longitudinal DC-EEG Study." *BMC Neuroscience* 4: 26.

Berthoz, Alain. 1997. *Le sense du mouvement*. Paris: Odile Jacob.

Bizley, Jennifer K., and Yale E. Cohen. 2013. "The What, Where and How of Auditory-Object Perception." *Nature Reviews Neuroscience* 14: 693–707.

Bregman, Albert. 1990. *Auditory Scene Analysis*. Cambridge, MA: MIT Press.

Chion, Michel. 1983. *Guide des objets sonores: Pierre Schaeffer et la recherché musicale*. Paris: Buchet/Chastel.

Galantucci, Bruno, Carol A. Fowler, and Michael T. Turvey. 2006. "The Motor Theory of Speech Perception Reviewed." *Psychonomic Bulletin and Review* 13 (3): 361–377.

Gallese, Vittorio, and George Lakoff. 2005. "The Brain's Concepts: The Role of the Sensory-Motor System in Conceptual Knowledge." *Cognitive Neuropsychology* 22 (3/4): 455–479.

Gallese, Vittorio, and Thomas Metzinger. 2003. "Motor Ontology: The Representational Reality of Goals, Actions and Selves." *Philosophical Psychology* 16 (3): 365–388.

Gjerdingen, Robert O., and David Perrott. 2008. "Scanning the Dial: The Rapid Recognition of Music Genres." *Journal of New Music Research* 37 (2): 93–100.

Godøy, Rolf Inge. 1997. *Formalization and Epistemology*. Oslo: Scandinavian University Press.

Godøy, Rolf Inge. 2001. "Imagined Action, Excitation, and Resonance." In *Musical Imagery*, edited by Rolf Inge Godøy and Harald Jorgensen, 239–252. Lisse: Swets and Zeitlinger.

Godøy, Rolf Inge. 2003. "Motor-Mimetic Music Cognition." *Leonardo* 36 (4): 317–319.

Godøy, Rolf Inge. 2006. "Gestural-Sonorous Objects: Embodied Extensions of Schaeffer's Conceptual Apparatus." *Organised Sound* 11 (2): 149–157.

Godøy, Rolf Inge. 2010a. "Gestural Affordances of Musical Sound." In *Musical Gestures: Sound, Movement, and Meaning*, edited by Rolf Inge Godøy and Marc Leman, 103–125. New York: Routledge.

Godøy, Rolf Inge. 2010b. "Thinking Now-Points in Music-Related Movement." In *Concepts, Experiments, and Fieldwork: Studies in Systematic Musicology and Ethnomusicology*, edited by Rolf Bader, Christiane Neuhaus, and Ulrich Morgenstern, 245–260. Frankfurt am Main: Peter Lang.

Godøy, Rolf Inge. 2011. "Sound-Action Awareness in Music." In *Music and Consciousness*, edited by David Clarke and Eric Clarke, 231–243. Oxford: Oxford University Press.

Godøy, Rolf Inge. 2013. "Quantal Elements in Musical Experience." In *Sound—Perception—Performance: Current Research in Systematic Musicology*, edited by Rolf Bader, 1:113–128. Berlin: Springer.

Godøy, Rolf Inge. 2014. "Understanding Coarticulation in Musical Experience." In *Sound, Music, and Motion: Lecture Notes in Computer Science*, edited by Mitsuko Aramaki, Olivier Derrien, Richard Kronland-Martinet, and Sølvi Ystad, 535–547. Berlin: Springer.

Godøy, Rolf Inge, Egil Haga, and Alexander Refsum Jensenius. 2006. "Playing 'Air Instruments': Mimicry of Sound-Producing Gestures by Novices and Experts." In *GW2005, LNAI3881*, edited by Sylvie Gibet, Nicolas Courty, and Jean-Francois Kamp, 256–267. Berlin: Springer.

Godøy, Rolf Inge, and Marc Leman, eds. 2010. *Musical Gestures: Sound, Movement, and Meaning*. New York: Routledge.

Grafton, Scott T., and Antonia F. de C. Hamilton. 2007. "Evidence for a Distributed Hierarchy of Action Representation in the Brain." *Human Movement Science* 26: 590–616.

Griffiths, Timothy D., and Jason D. Warren. 2004. "What Is an Auditory Object?" *Nature Reviews Neuroscience* 5 (11): 887–892.

Haken, Hermann, Scott Kelso, and Herbert Bunz. 1985. "A Theoretical Model of Phase Transitions in Human Hand Movements." *Biological Cybernetics* 51 (5): 347–356.

Harnad, Stevan, ed. 1987. *Categorical Perception*. Cambridge: Cambridge University Press.

Haueisen, Jens, and Thomas R. Knösche. 2001. "Involuntary Motor Activity in Pianists Evoked by Music Perception." *Journal of Cognitive Neuroscience* 13 (6): 786–792.

Husserl, Edmund. 1991. *On the Phenomenology of the Consciousness of Internal Time, 1893–1917*. Translated by John Barnett Brough. Dordrecht: Kluwer Academic.

Jensenius, Alexander Refsum. 2013. "Some Video Abstraction Techniques for Displaying Body Movement in Analysis and Performance." *Leonardo: Journal of the International Society for the Arts, Sciences and Technology* 46 (1): 53–60.

Jensenius, A. R., and R. I. Godøy. 2013. "Sonifying the Shape of Human Body Motion Using Motiongrams." *Empirical Musicology Review* 8: 73–83.

Klapp, Stuart T., and Richard J. Jagacinski. 2011. "Gestalt Principles in the Control of Motor Action." *Psychological Bulletin* 137 (3): 443–462.

Kohler, Evelyne, Christian Keysers, M. Alessandra Umiltà, Leonardo Fogassi, Vittorio Gallese, and Giacomo Rizzolatti. 2002. "Hearing Sounds, Understanding Actions: Action Representation in Mirror Neurons." *Science* 297: 846–848.

Liberman, Alvin M., and Ignatius G. Mattingly. 1985. "The Motor Theory of Speech Perception Revised." *Cognition* 21: 1–36.

Loram, Ian David, Cornelius van de Kamp, Martin Lakie, Henrik Gollee, and Peter J. Gawthrop. 2014. "Does the Motor System Need Intermittent Control?" *Exercise and Sport Science Review* 42 (3): 117–125.

Nymoen, Kristian, Rolf Inge Godøy, Alexander Refsum Jensenius, and Jim Torresen. 2013. "Analyzing Correspondence between Sound Objects and Body Motion." *ACM Transactions on Applied Perception* 10 (2).

Petitot, Jean. 1985. *Morphogenese du Sens I.* Paris: Presses Universitaires de France.

Poppel, Ernst. 1997. "A Hierarchical Model of Time Perception." *Trends in Cognitive Science* 1 (2): 56–61.

Proctor, Michael, Erik Bresch, Dani Byrd, Krishna Nayak, and Shrikanth Narayanan. 2013. "Paralinguistic Mechanisms of Production in Human 'Beatboxing': A Real-Time Magnetic Resonance Imaging Study." *Journal of the Acoustical Society of America* 133 (2): 1043–1054.

Rosenbaum, David A., Rajal G. Cohen, Steven A. Jax, Daniel J. Weiss, and Robrecht van der Wei. 2007. "The Problem of Serial Order in Behavior: Lashley's Legacy." *Human Movement Science* 26 (4): 525–554.

Schaeffer, Pierre. 1966. *Traite des objets musicaux.* Paris: Editions du Seuil.

Schaeffer, Pierre. [1967] 1998. *Solfege de l'objet sonore.* With sound examples by Guy Reibel and Beatriz Ferreyra. Paris: INA/GRM.

Schleidt, Margaret, and Jenny Kien. 1997. "Segmentation in Behavior and What It Can Tell Us about Brain Function." *Human Nature* 8 (1): 77–111.

Schneider, Albrecht. 2010. "Music and Gestures: A Historical Introduction and Survey of Earlier Research." In *Musical Gestures: Sound, Movement, and Meaning*, edited by Rolf Inge Godøy and Marc Leman, 69–100. New York: Routledge.

Smith, Barry, ed. 1988. *Foundations of Gestalt Theory.* Munich: Philosophia.

Thom, René. 1983. *Paraboles et catastrophes.* Paris: Flammarion.

Wilson, Margaret, and Günther Knoblich. 2005. "The Case for Motor Involvement in Perceiving Conspecifics." *Psychological Bulletin* 131 (3): 460–473.

CHAPTER 9

MUSIC, BACCHUS, AND FREEDOM

HEDY LAW

WHAT are the relationships among music, the body, and freedom? In *Philosophy in the Flesh: The Embodied Mind and Its Challenge to Western Thought* (1999), George Lakoff and Mark Johnson claim that the concept of freedom of action, as with most abstract concepts, is embodied, and is understood metaphorically as freedom of motion (305). This idea of embodied freedom introduces a new dimension to Isaiah Berlin's two concepts of liberty, namely negative freedom (absence of interference) and a positive one (one as one's own master) (2007, 169, 178). Prompted by the bicentennial celebrations of the Fall of the Bastille and the collapse of East Germany in 1989, musicians and music scholars have produced a series of articles on music and freedom. Leonard Bernstein notably conducted in 1989 a performance of Beethoven's Ninth Symphony that celebrated the fall of the Berlin Wall, which modified Beethoven's "Ode to Joy (*Freude*)" to "Ode to Freedom (*Freiheit*)" (Rehding 2005). Freedom, Daniel Chua observes, is an idea from without that can be imposed on Beethoven's music (2005, 14). Chua went on to publish two more articles on Beethoven and freedom (2009, 2014). Yet none of this group of work on Beethoven and freedom addresses the body explicitly. In light of the embodied idea of freedom, one wonders in what ways music in the French Enlightenment, much of which incorporates dancers, constructed Berlin's two concepts of liberty and the embodied concept of freedom.

According to the historian Robert Isherwood, Jean le Rond d'Alembert was the first major writer of the Enlightenment to write about musical liberty in his *De la liberté de la musique* (1759). The spectacles examined by d'Alembert, such as French ballets and operas featuring dancers and singers, may provide the suitable repertories for examining the freedom of motion (Isherwood 1989, 110). In *De la liberté de la musique*, D'Alembert critiques the then-subsided culture war pitting Italian against French music (generally known as the *Querelle des bouffons*, 1752–1754) by advocating for a bipartisan approach. Refusing to side with one national style, either Italian or French, he believes that the partisan dispute hurt musical progress and encourages composers to exercise their artistic freedom by trying all styles in their compositions (Isherwood 1989, 105). D'Alembert justifies the discourse of freedom of music as an extension to that of free press and free trade, implying that the freedom of action is a potentially dangerous concept precisely because it presupposes the embodied concept of the freedom of motion: "all freedoms hold together and are equally dangerous. Freedom of music presupposes freedom to feel, which leads to freedom to think. Freedom to think produces freedom to act. And freedom to act is the destruction of the States" ([1759] 1973, 3:397). Suggestive as this claim was in the French Enlightenment, d'Alembert leaves many threads underdeveloped, not least the critical connection between the freedom of using different musical styles and the destabilizing potential of this type of freedom. As with many men of letters, he also does not illustrate his idea of musical freedom with specific musical examples explicitly, but he nevertheless considers Rameau's *Platée* to be one of the masterworks of French music that incorporates French and Italian styles (388). Thus it is fair to say that d'Alembert's essay offers the promise of understanding the relationship between music, the body, and freedom.

Bacchus, a god revered in Rameau's *Platée*, is an appropriate choice for the investigation of music, the body, and freedom for three reasons. First, the Romans associated Bacchus with Father Liber (*Liber Pater*) and, by extension, the idea of liberty and the process of liberation. Second, Bacchus, the Roman god of wine and revelry and the Roman version of the Greek god Dionysus, draws explicit attention to the body: an inscription of the third century AD described Dionysus as the most epiphanic god, one that is perceived by one or more senses (Seaford 2006, 39). In myth, people possessed by Dionysus eat raw flesh of hunted animals and dismember live animals with their bare hands. Third, Nietzsche connects Dionysus to music in his first book *Die Geburt der Tragödie* (*The Birth of Tragedy*) (1872, rev. 1886). There he makes a case for the productive synthesis of two opposite forces, Apollonian and Dionysian, that gave rise to the ancient Greek tragedy. The Apollonian force honors individuality; it draws and respects boundaries, emphasizing self-control and moderation. By contrast, the Dionysian force destroys individuality; it celebrates transgression or dissolution of boundaries, highlighting excess. Music, especially the orgiastic choral music and dance, owes much to the Dionysiac drive, whereas the plastic arts such as painting and sculpture embody the Apollonian force (Nietzsche 1999, 14). Nietzsche's essay, the first edition of which was dedicated to Richard Wagner, rediscovers the Dionysian force in Beethoven's symphonic music and argues that Wagner's music drama was the first attempt to synthesize the symphonic Dionysian principle and the epic Apollonian speech. Yet if, according to Nietzsche, Beethoven's music played such a momentous role in the cultural history of Dionysus, then how could one account for the convention of Bacchus in pre-nineteenth-century German music?

This chapter answers this question by placing seventeenth- and eighteenth-century French music about Bacchus within the context of three texts: the ancient Greek play Euripides's *Bacchae*, the section "Bacchus and Pentheus" in *Metamorphoses* by the Roman poet Ovid, and Nietzsche's essay "Die Dionysian Weltanschauung" (The Dionysiac World View) (1870), written two years before the completion of his

Birth of Tragedy. Studies of Bacchus should begin with the differences between myth and ritual. Dionysiac cults were practiced in Greco-Roman antiquity. Depictions of Dionysus in myth, poetry, play, or vase-paintings portray the darker sides of Dionysus that have little relevance to the actual practice of Dionysiac rites. By examining Bacchus as represented in myth, and not in rites, I demonstrate that Bacchus justifies the embodied activities of drinking, dancing, and singing as examples of personal freedom. On the one hand, Bacchus offered composers a reason to feature dancing and singing as natural—that is, unmarked and unproblematic—expressions of revelers. Handel's *Alexander's Feast* (1736), a setting of Dryden's ode for St. Cecilia's Day, for example, connects music with revelry: "Bacchus, ever fair and young, / Drinking joys did first ordain. / Bacchus's blessings are a treasure." In Mozart's *Die Entführung aus dem Serail* (1782) K. 384, Pedrillo, servant of the hero Belmonte, tempts the gatekeeper Osmin into drinking wine. Together they sing "Vivat Bacchus," a drinking song celebrating Bacchus that lowers Osmin's guard. On the other hand, some composers such as Rameau and Salieri denaturalize singing and dancing in their musical settings of Bacchus. By investigating settings of Bacchus in select seventeenth- and eighteenth-century French music, this chapter argues that Bacchus connects music, the body, and freedom by representing deviance and defiance. His power to both liberate and destroy fosters artistic creativity in music, marking new origins as well as epoch-defining regeneration.

Bacchus and Deviance

Far from natural, dancing and singing in both Euripides's *Bacchae* and the part "Bacchus and Pentheus" in Ovid's *Metamorphosis* indicate deviant behaviors. In the mountain of Cithaeron, Dionysus's female followers Bacchae sing and dance, "crying in joy to the Bacchic god," "swinging the thyrsus high," and the bacchant "skip[ping] with quick feet as she runs" (Euripides 2008, 46, 47, 49). Dionysus stresses the significance of dancing in the Bacchic rites: "Every one of the barbarians dances in these rituals" (57). A messenger of Pentheus, who had doubts about the power of Dionysus, witnessed the Bacchic rite in mount Cithaeron and considered the otherwise ordinary activities—raising the thyrsi, running, dancing, and singing—unnatural. Their bodily movements were so contagious that even animals mimicked them. "All the mountain and the animals joined them in their Bacchic worship—there was nothing that did not move with the running" (64).

Bacchic dances rejuvenated the old King Cadmus and the prophet Teiresias (Ovid 2004, 111). Their positive experience participating in these supposedly "barbaric" dances unsettles the binary opposition of the civil and the barbaric. Pentheus, Cadmus's grandson and king of Thebes, disapproves of his grandfather Cadmus's participation in these dances: "I am shocked, father, to see you old men divorced from good sense" (Euripides 2008, 51). Pentheus's Apollonian reaction failed to acknowledge any positive aspects of the Bacchic rites. His reaction indicates oppression that interfered with his grandfather's and Teiresias's personal freedom. Did they behave inappropriately because they participated in and enjoyed Bacchic dance? The tension between Pentheus and Cadmus/Teiresias points to the limits of personal freedom. Pentheus indicates not only a lack of imagination but also his oppressive behaviors grounded in ageism and bias against corporeal difference. To remove Pentheus, as Agave did, is to eliminate the possibility of him depriving other individuals of personal freedom (Berlin 2007, 173).

Pentheus's death stems from his doubt about Bacchus's legitimacy as god. Bacchus was an offspring of Cadmus's daughter Semele, a mortal, and the god Zeus. How could Bacchus be a god if his mother was a mortal? Half god and half human, he, as a liminal being, occupied the ambiguous territory between male and female, child and adult, and naiveté and maturity. In Ovid's account, Bacchus was androgynous. He appeared "a little boy formed like a little girl" (Ovid 2004, 114). Yet this child behaved like an adult. Unlike a child, Bacchus spoke as if he were half drunk. Unlike most children, who exerted little control or understanding over their surroundings, Bacchus demonstrated advanced metacognitive skills. He could behave like a child when he saw fit. When the sailors deliberately sailed away from Bacchus's destination Naxos, Bacchus feigned crying and accused the male sailor for tricking a boy like him (116).

Bacchus called attention to the denaturalized body because he had the power to transform the body and the mind. To demonstrate his wrath, Bacchus transformed the nineteen sailors who turned the ship away from Naxos into fish. They became limbless and grew tails. They moved like dancers, "leaping wildly, sucking in water and spraying it into the air" (117). To the eyewitness Acoetes, these bodily transformations proved that Bacchus was a god. Bacchus could transform the mind as well as the body. Euripides's *Bacchae* invited readers to experience alongside Pentheus and Agave by entering and getting out of the altered state of consciousness. The body and the mind became interlinked. For Pentheus, cross-dressing was a necessary step in the process of mental transformation. Once Pentheus cross-dressed as a maenad, he immediately recognized that his perception changed. He thought he saw two suns and two Thebes, and he saw Dionysus as a bull with horns on his head. He did recognize his own mother before his death, but unfortunately Agave, "frothing foam and whirling her twisted eyes," failed to recognize her son (Euripides 2008, 76). Agave regained her consciousness after she returned to Thebes. She found the sky "brighter than before and more translucent" and she noticed that something changed in her mind (80).

The Bacchae's behaviors indicate transcendence through singing and dancing. Nietzsche denaturalizes running and singing in his essay "The Dionysiac World View." The Bacchae run and sing because they have forgotten how to walk and speak. A component of enchantment in these activities enables Nietzsche to make a universalized claim: Bacchus transforms all human beings, not just the Bacchae, through the activities of singing and dancing. Human beings feel different and sing as if they were transformed by magic. "Just as the animals now talk and the earth gives milk and honey, something supernatural now sounds out from within man" (1999, 120). Sounding out from within, the body becomes an instrument of Bacchus. It is through the embodied activities of singing and dancing that human beings feel Bacchus's impact. Rather than simply appreciating external objects such as images and statues, as an Apollonian artist would, the Dionysian singers and dancers internalize art and become art. Since the Dionysiac singers and dancers experience similar ecstasy as gods do, singing and dancing transcend humans into supernatural beings.

The process of transcendence does not necessarily mean that a Dionysiac singer or dancer regresses to an ecstatic state like Agave or the Bacchae, unable to think logically or control their own behavior. Quite the contrary, he or she is deviant by possessing the metacognitive ability to play with intoxication. The crux of Dionysiac transcendence is not about alternating clear-mindedness and intoxication in the way Agave and Pentheus experienced, but about managing the coexistence of these two antithetical states (120–121). This transcendence manifests in multiple forms in music. Michel Pignolet de Montéclair's cantate *Ariane et Bacchus* (1728), for example, features Bacchus's power of transformation. "Bacchus changed the most frightening pain/into durable and charming pleasures" (1728, 28). In Jules Massenet's *Bacchus* (1909), Bacchus promises to use wine to set mortals free from aging and from death. Richard Strauss's intermezzo-turned-opera *Ariadne auf Naxos* (1911–1912, rev. 1916), drawing ideas from French baroque comédie-ballet, Mozart's Piano Sonata K.331, and commedia dell'arte, highlights transformation as a transcending process by showing how one rises above one's present self while retaining one's essence (Gilliam 2014, 127–151).

Liberation and Destruction

The deviance of Bacchus is linked to his ability to free others from stricture. On a personal level, Teiresias noticed that drinking put an end to sorrow. It brought "forgetfulness of their daily round of cares" (Euripides 2008, 52). On a psychological level, Dionysus showed no fear when Pentheus threatened to imprison him: "The god will free me whenever I wish" (58). Indeed, Bacchus freed the imprisoned Acoutes, priest of the Bacchic cult. Negative freedom manifests also in the material form. The doors "flew open" and "chains fell from his shoulders" (Ovid 2004, 118). Sometimes negative freedom brought forth destruction. Bacchus sent the lord of thunder Bromius to set the tomb of his mother Semele on fire and smashed "[Pentheus's] palace to the ground" (Euripides 2008, 62).

Bacchus's destructive side was demonstrated fully in the Bacchic rite in Mount Cithaeron, a place nicknamed the "mountain of joy" (66). The messenger and his followers tried in vain to take Agave out of the mountain. Their attempt infuriated the Bacchae, who, trying without success to kill these intruders, turned against the cows with inconceivable bodily strength: they ripped the young and the mature cows apart apparently with bare hands. Afterward the Bacchae ran and destroyed the lowland plains that produced fine crops for the Thebans. They seized children from the houses and carried them over their shoulders. The Bacchae demonstrated superhuman resistance to physical attacks. The villagers attacked them with spears, but no blood came out of the wounds. Their unusual physical strength and resilience convinced the chorus that Dionysus was a legitimate god: "Dionysus is inferior to none of the gods" (66). The destructive power of Bacchus continued to spread after the death of Pentheus. Bacchus punished Cadmus not only because he had no male heir but also because he refused to accept the murderer Agave as his daughter. On top of that, since Agave went back to Thebes after she killed Pentheus, she introduced terror into the city of Thebes. Cadmus realized how Bacchus destroyed his family, his city, and his kingship and thus believed Bacchus to be a powerful god (81).

Importantly, Dionysus connected negative freedom to creativity. Dionysus asked Pentheus, "Did I not say—or did you not hear—that someone would free me?" To which Pentheus answered with disbelief, "you are for ever mentioning new, unexpected things." Dionysus explained away his observation by connecting alcohol with creativity (62). Nietzsche, however, offers a more in-depth explanation for the connection between Dionysus and creativity. To him, Dionysus has access to and command over the chaotic subconscious of an individual, a notion that counters Descartes's idea that all thought is conscious (Lakoff and Johnson 1999, 392). Thus Dionysus can cause the artist to "bring a new world into being at each creative moment" while retaining the familiar of the old world (Nietzsche 1999, 122).

Because of Bacchus's ability to free men from stricture, the Romans associated Father Liber with Bacchus. Ovid makes clear the connection between Liber and revelers. "Liber has come! The fields reverberate / with the ululations of the revelers" (2004, 111). In the *Encyclopédie*, Chevalier Jaucourt refers the mythological figure of Liber to the adjective "free" (*libre*). "Liber" is the surname of Bacchus "either because he procured liberty in the city of Boeotia, as described in Thucydides's *The Peloponnesian War*, or rather because being the god of wine, he delivers the mind from any concern, and makes one speak freely" (1765, 9: s.v. "Liber"; on Boeotia and freedom, see Thucydides 2009, 1.113). The connection between the etymology of "liber" and carefree nature is unclear. At least Jaucourt emphasizes Bacchus's nonviolent proclivity rather than his destructive side. In the prologue of Voltaire's unperformed tragédie-lyrique *Samson* (1733), Bacchus invites Hercules to speak to him with liberty: "We are the children of the thunder. / Our name, once feared, will not perish on earth. / But let's speak with liberty: / Of all the laurels that encircle your head, / tell me which is the conquest / that the great heart of Alcides was the most revered" (1877–1885, 3:546).

As Robert Wokler observes, the Romans separated *libertas* from *licentias* and thus liberty is different from licentiousness (2012, 158). According to Louis de Cahusac's *La danse ancienne et moderne* (1754), the Bacchanals, priests and priestesses of Bacchus, performed their dances celebrating the intoxicating Bacchic festival. There the properties of convulsions and fury were transformed into expressions of "gaiety, of pleasure, and of voluptuousness." The ancient Greeks imitated the Bacchanals and gave rise to lascivious dances that kept the elements of liberty and joy. The dances of Bacchus feature "the most unbridled license" (Cahusac [1754] 2004, 76–77).

Numerous examples in seventeenth- and eighteenth-century French music connect Bacchus to revelry. Because Bacchus introduces singing and dancing, Bacchus was featured in seventeenth-century French drinking songs called *chansons pour boire*, a genre often coupled with *chansons pour danser*. These short, syllabic, and strophic pieces for one voice and lute accompaniment, compiled in collections from about 1627 to 1670, preceded thousands of *airs à boire* circulated in print or manuscript in over 250 collections called *Airs sérieux et à boire* between 1674 and 1745. One of the *airs à boire* published in *Recueil d'airs sérieux et à boire* by Jean-Baptiste de Bousset (1695) illustrates how Bacchus numbs the pain of unrequited love. "I do not live only for love," the singer consoles himself. "Without him [Bacchus] my despair would have deprived me of the day" (1695, 11–12). In the fourth entrée of *Les festes de l'Amour et de Bacchus*, Silvestre advises Florestan to leave Amour and to follow Bacchus instead. Revelry may liberate him from sorrow. "Come taste life with us; / Leave a fickle beauty / As she has left you: / Benefit from her perfidy, / Come enjoy liberty" (Lully 1672, 36). The one-act ballet *Érigone* (1758), reused as the second act of Jean-Joseph Cassanéa de Mondonville's ballet-héroïque *Les fêtes de Paphos* (1758), glorifies Bacchus: "Victory flies to your voice; / how she follows your steps from sunset to sunrise: / Bacchus, how the universe adores you; / It is pleasure that dispenses your laws" (La Bruère 1748, 13). Bacchus was found infrequently in French music in the second half of the eighteenth century. One exception was the one-act ballet-héroïque *Bacchus et la Minéïde* (Choisy 1773) about the conflicts between Amour and Bacchus. Whereas Amour helps one feel suffering, Bacchus banishes suffering. "Reign, charming Amour, enjoy your victory, / shackle our moments by Mirth and Games; / my liberty made my glory, / my defeat is still more charming to my eyes" (19).

In French spectacles, Bacchus was associated uncritically with wine. In the *Vers pour le ballet du Roy, représentant les Bacchanales* (1623), Bacchus repeats the refrain "my glory is to drink" (Bordier 1623, 3–4). In *Ballet de la nuit* (1653), the character Bacchus wears a costume decorated with vine leaves and a headdress that resembles a wine-bottle basket (Burden and Thorps 2009, 6, 162). In the mascarade *Le mariage de la Grosse Cathos*, a drinking song is followed by a drunken dance (*Air des ivrognes*). This song-and-dance sequence indicates that the dance instantiates the drunkenness described in the song and provides contextual details for musical embodiment. Frequent disruptions of the flow of music in the dance suggest that their drunken state impairs the dancers' sense of balance. Some dancers imitated the unsteady bodily movements of the drunks by means of pantomime. In Lully's *La revente des habits* (1655), the dancers strike a delicate balance between a state of drunkenness and their sensitivity to music. For the fifth entry the poet Issac de Benserade describes six half-drunk thieves who leave the tavern dancing without losing their rhythmic sense: "Being drunk they leave and dance together, staggering often without falling nor losing the beat of the music, so well they have their ear attuned to the sound of the violins in the midst of the wine" (Harris-Warrick and Marsh 1994, 54–55).[1] Yet some dances depict the drunks' impaired sensitivity to music. One ballet de cour, François de Chancy's *Ballet des festes de Bacchus* (1651), for example, illustrates a drunk who struggles to maintain musical sensitivity. "To catch the cadence is a difficult task, / I am losing in this inquiry and my pain and my steps" (10).

Settings of Bacchus in ballets and operas offer fresh communicative opportunities through the expressive body in pantomime. In Lully's *Les festes de l'Amour et de Bacchus*

(1672), the Muse of Polyhymnia, one of the nine ancient Greek muses who rules over hymns and pantomime, "has found the invention of introducing to the theater some characters who express through actions and dances what other express through words" (8). Pantomime was brought to the fore in the hands of a number of early eighteenth-century dancers and choreographers. The dancer Marie Sallé performed in London a *ballet en action, Bacchus and Ariadne*, in 1734. It includes a mimed love scene between Bacchus and Ariadne. The *Mercure de France* considered Sallé's performance as Ariadne too subtle to be described fully in words. Her performance of a woman abandoned by her lover brought forth a wide range of sentiments—sadness, despair, rage, and dejection—by means of her dance steps, bearings, and gestures (*Mercure de France*, April 1734, 772; see also McCleave 2013, 164). The rise of action ballet paralleled the midcentury reform in acting style. D'Alembert complains in *De la liberté de la musique* about the mechanical acting style that indicates actors as marionettes rather than human beings ([1759] 1973, 3:409). The popularity of Sallé's pantomime and d'Alembert's complaint indicate the emergence of a pan-European naturalistic acting style and action ballet.

Commedia dell'arte contributed to the liberation from a mechanical acting style. In the dance titled "Trionfo di Bacco" of the second part of Gregorio Lambranzi's *Neue und curieuse theatralische Tanz-Schul* (1716), two dancers bring Bacchus onstage. Lambranzi's Bacchus is a plump god who, a wine glass in hand, sits atop the thin wood staves carried on the shoulders of dancers thinner than him. It is unclear whether he is floating in the air or he weighs as much as he looks. In any case, the themes of risk and improbability are transferred to the subsequent dance. The description of this dance reads: "These two persons knock their sticks above their heads, under their legs and behind their backs. They carry on a few times with variations and transport Bacchus back again" (Lambranzi [1716] 1975, XV). As with Marie Sallé's rendition of Bacchus, this dance with wooden sticks illustrates liberation by demonstrating the variety of bodily movements that defy precise verbal descriptions. Physical combat gives context to stylistic contrasts in the music. The first entrée of the prologue in *Les festes de l'Amour et de Bacchus* begins with combat between Melpomene, Muse of Tragedy, and Euterpe, Muse of Harmony. When they come onstage and flank the Muse of Hymns and Pantomime Polyhymnia, their entrance is accompanied by "two symphonies of opposite [characters], of which one is very strong and the other extremely sweet, and those form a combat in order to beg Polyhymnia to embellish the divertissements they want to prepare" (Lully 1672, 9).

Musical Deviance and Political Defiance

Bacchus symbolizes liberation and highlights the tension between pleasures and glory in French opera. The prologue of Lully's tragédie-lyrique *Thésée* (1675) complains that

Louis XIV loves glory more than pleasures. It ends with Bacchus, alongside Mars, Venus, and Ceres, who praise peace and abundance (Cowart 2008, 127). In Campra's opéra-ballet *Les Muses* (1703), Bacchus and Ceres, while celebrating the river of forgetfulness, challenge the Muses' glorification of gods and heroes. Likewise, André Cardinal Destouches's *Le carnival et la Folie* of 1704 features Folly, who asks Bacchus to bring forth the sensual pleasures of singing and dancing in place of heroism and glory (Cowart 2008, 192, 194).

The deviant Bacchus provides suggestive materials for political defiance during the late reign of Louis XIV. In June 1697, a party hosted by the marquis de Livry, who belonged to a political clique known as "the cabal of the Dauphin" that resisted the moralizing disposition of Louis XIV, performed a "dialogue" for Bacchus, Comus, and Silenus composed by André Compra. This dialogue, which was a type of dramatic work, appealed to Philippe II d'Orléans, Louis XIV's nephew and the future regent of France, one of the core members of this cabal (Fader 2005, 396). This dialogue for Bacchus set the stage for a more obviously politically dissident opera. In 1703 the amateur composer d'Orléans set music to the tragédie en musique *Penthée*, written by his librettist Charles-Auguste, marquis de La Fare. Since Bacchus justifies pleasures as opposed to the theme of glory, the relationship between Pentheus and Bacchus could provide suggestive material for a political power struggle. Indeed, d'Orléans, Fader surmises, used this opera to present opposition against Louis XIV. Unlike the operas in the second half of the seventeenth century that include Bacchus as a minor character, *Penthée* celebrates Bacchus as "the greatest and the sweetest of all the gods" (La Fare 1740, 219, 241). The librettist La Fare intensified the conflict between Pentheus and Bacchus by turning them into love rivals. Bacchus wants to destroy his rival not just because of Pentheus's disrespect for him but also because they fight for the same woman, Erigone. There is no direct connection between the plot and specific political events at court, Fader emphasizes, but there are suggestive metaphoric parallels between Bacchus and d'Orléans. D'Orléans shared in Bacchus's defiance with his libertine lifestyle, his contempt of politicized religion, his deliberately irreligious attitude, and his disregard of his royal duties and codes of proper behaviors. Since this opera was reported to have been performed in front of Louis XIV, it provides historical evidence of d'Orléans's antimonarchical self-fashioning. Tellingly, D'Alembert notices the politically unsettling potential of this type of symbolic defiance in *De la liberté de la musique* and makes a case for censoring music: "Thus let us preserve the Opéra the way it is, if we want to preserve the kingdom. Let us put an obstacle to the freedom of singing, if we do not want that the freedom of speaking follows it afterwards" ([1759] 1973, 3:397).

In *Penthée* Bacchus represents novelty, which could be understood as a kind of positive freedom. Bacchus is a "new god" who brings forth "new fêtes," "new cults," and "new delights" (La Fare 1740, 242, 259, 266). Based on the extant libretto, it is unclear how this claim of novelty manifests in dance, but the librettist nonetheless calls attention to Agave's uncommon physical strength. In act 3, scene 3, Agave tells her father Cadmus that she has killed a young and proud lion and praises Bacchus for offering her extreme force (217). There are, however, traces of novelty in the music. Bacchic frenzy manifests in Italianate compositional techniques. D'Orléans used a text-painting technique that highlights words such as "sweet" (*doux*). The violent scene in which Autonoë (the character who represents Agave) kills Pentheus is set in the unusual B♭ minor and is filled with dissonant chords and contrapuntal writing. Uses of Italian style in this opera had broader political implications. Since d'Orléans adopted a musical language against Lullian musical convention, this opera embodies pro-Italian and anti-French ideals and has the tendency to be musically deviant and politically defiant (Fader 2014).

Discharge and the Extraordinary

Bacchus's alternative reality incites wonder and even fear. As with the Apollonian Pentheus, his messenger described the holy bacchae in Mount Cithaeron as "strange things, things that go beyond wonder" (Euripides 2008, 63). Euripides's *Bacchae* describes these strange things in detail. A stranger, presumably Dionysus himself, pulled down the top branch of a fir tree by hand and bent it to the ground. He placed Pentheus on the branches and let go of the bent branches. Pentheus was carried to the top of the tree. Trapped in this ad hoc jail above ground, Pentheus became a target for the maenads. A voice from the sky, presumably Dionysus's voice, commands the maenads to take vengeance on Pentheus. Euripides and Ovid further estrange the maenads' bodily movements. The inhuman maenads adopted animalistic properties and act aggressively as if they were predators. They "shot forward with all the speed of a wood pigeon" (75), tore the fir tree out of the ground, attacked Pentheus, and tore his shoulder out. Agave's failure to recognize her own son is conflated with her lack of compassion. She mistook her son for a wild animal (a lion in Euripides's version or a boar in Ovid's version). Her act of infanticide was furthered by the heartless way she treated her son's deceased body. She removed her son's head from his trunk (Ovid 2004, 119) and used it to crown her thyrsus (Euripides 2008, 76) as if she were a trophy hunter. The strange things that go beyond wonder evoke not only fear but also disgust.

Yet Dionysus was also a source of artistic creativity. In *Bacchae* the chorus sang self-reflexively: "it is Dionysus, not Thebes but Dionysus, who has power over me" (74). In *Ballet des fetes de Bacchus*, Apollo and the nine muses pay homage to Bacchus (Cowart 2008, 34). Jean François Marmontel wrote in his *Poétique françoise* (1763) that tragedy was a "hymn in honor of Bacchus" (1763, 122). Rameau's *Platée*, first performed at the wedding of the Dauphin in February 1745, followed *Penthée* by hailing Bacchus as the source of liberty and artistic originality.

The prologue of *Platée*, entitled "The Birth of Comedy," explains why and how Thespis invented comedy. In the opening scene of the Prologue, Thespis's ariette celebrates Bacchus, who allows him to speak the truth and to laugh at the expense of mortals. "Charming Bacchus, god of liberty, / father of sincerity, / you allow us to laugh / at the expense of the mortals" (Autreau 1745, 8). Nietzsche explains why truth-speaking is Dionysian, for the Dionysian artist "spoke the truth in its ecstatic intoxication" (1999,

129). Nietzsche goes on to argue that the Dionysian comedy employs artistic means to "discharge disgust at absurdity." The key is to allow the actor to play with intoxication during performance in order to give rise to a "new form of art" (130–131). The discharge of intoxication manifests in various ways in *Platée*. On a local level, the ariette sung by Thespis in the opening scene includes lengthy melisma on the word "laugh" (*rire*). Discharge of Dionysiac intoxication also operates on a structural level. The purpose of the plot is to have Bacchus liberate Thespis so that he invents a new spectacle that cures Juno's jealousy. Bacchus provides a frame in the prologue for the ensuing three-act comedy. The prologue, Nietzsche would say, "discharges" a new spectacle. On the stylistic level, discharge manifests in the three-act comedy by highlighting what can be called the "topos of the extraordinary" in music and dance.

The three-act comedy, the "new spectacle" promoted in the prologue, is about finding a cure for the jealousy of Juno, Jupiter's wife. This cure is to have Juno realize that Jupiter would not fall in love with Platée, an unattractive nymph that lives in a marsh. Hence *Platée* is about using the nymph's unattractive appearance as a negative strategy in order to reunite Jupiter and Juno. Small wonder that the comedy repeatedly draws attention to Platée's appearance. She is "laughable" (*ridicule*). Unaware of her less-than-desirable appearance, she hopes to get married. She is called sarcastically a "charming object" and a "beauty" (act 1, scene 2). When the king of Greece, Kithaeron, rejects her, she and her followers demand a reason. Rameau employs mimetic music to denaturalize their human voices. They repeat the diphthong /waI/ shared by the words "why" (*pourquoi*) and what (*quoi*) to the point that they sound like croaking frogs (act 1, scenes 3 and 4; see Thomas 2006, 10–11). Jupiter does pay homage to the fake bride Platée, his "new conquest" (act 2, scene 3), but he does so by transforming himself in front of Platée into a creature with four feet and subsequently into an owl. Thus Platée's unattractive appearance blurs the boundary separating human and animal. As with protagonists in American literature, main characters in French operas almost never have disabilities. Yet Platée the nymph is made the protagonist of the opera precisely because she is unattractive. Like representations of people with disabilities in literature, she is featured as a spectacle, a hypervisualized text that elicits comments about her appearance (see Blank's chapter for more on this subject). Her appearance destabilizes the notions of grace and femininity and raises question about beauty, not as an aesthetic category but as a social one (Garland-Thomson 1997, 15).

Platée points out the limits of the human and opens up the animalistic stratum. To Rameau's mind, her unattractive appearance offers him a reason to experiment with the novel and the extraordinary in music and dance. To mark the fake wedding between Jupiter and Platée in act 2, scene 4, an "extraordinary symphony" accompanies the entrance of Folly. This symphony is already used in the Prologue, but its effect as "extraordinary" is unveiled in this scene, where alternations of slow and quick passages both adopt and deviate from the conventional French overture. This symphony of twenty-two measures consists of as many as six tempo markings representing five thematic groups. Excess of thematic materials, abrupt thematic changes, and lack of transitional materials between diverse thematic groups make this passage incoherent and unpredictable. It is extraordinary in two senses. On the one hand, it could be interpreted in the complimentary way as rare and exceptional. On the other hand, it could be interpreted as ridiculous, shocking, bizarre, and extravagant. In the face of this spectrum of interpretations of the extraordinary, Momus refrains from aesthetic judgment and calls it a "new harmony" (*une nouvelle harmonie*).[2]

Folly's entrance is followed by several dances performed by followers of Momus. Some dancers wear pompoms and perform a happy dance; some dress as philosophers and perform a sad *sicilienne*. The pairing of deviance and novelty complicates the self/other axis by adding old/new dimensions. Folly calls Platée the "new Juno" for Jupiter (act 2, scene 5). In act 3, scene 3, a series of dances are performed in the noblest manner for Platée before her pending wedding with Jupiter. These dances are concluded by a tedious chaconne of 231 measures that ends on a diminished seventh chord on D♯ rather than a cadence in A major. This tonally unfinished chaconne is interrupted by yet another "extraordinary symphony." Rameau uses the D♯ as a pivot pitch of E♭ that takes A major efficiently to the remote key of C minor, which announces the entrance of the blindfolded Momus. Presumably refusing to see the unsightly Platée, the blindfolded Momus celebrates Platée's wedding on behalf of Eros and brings with him the perverse wedding gifts of tears, sweet pain, wails, and weeping. In act 3, scene 5 a prelude in "the new character" announces the entrance of three of Momus's followers disguised as the traditional three Graces who dance a supposedly serious loure in a comical way. This burlesqued loure excites Platée, who finds the traditional Graces uninteresting. Folly takes full credit for these new artistic effects. "Admire the effects produced by my immense genius. / I know how to make them sometimes lively, sometimes innocent, naïve, / Always delivering them to charming excess of [artistic] effects" (Rameau 1754, 50).

The body of Platée is decidedly "good to think" (Breckenridge and Vogler 2001, 353), for her unappealing appearance repeatedly bears new styles of music and dance. The negative freedom in the prologue is transformed in the last act into positive freedom, valuing self-expression and creativity. *Platée* does not just point out the opposition between Platée and unmarked, normative beauty, or use Platée as an instrument that reasserts power, it also refers self-reflexively to the opera as novel. Scholars have noted the transgressive, burlesque, or parodic elements in *Platée* (Green 2012; Thomas 2006; Kintzler 1999; Anderson 1983). The point is that *Platée* brings out a side of Bacchus that has the capacity to create something—may it be a dance, an entrance, or a comedy—anew.

Orgy and Regeneration

While most musical settings of Bacchus in the seventeenth and the early eighteenth centuries feature the innocuous side of Bacchus, French writers emphasized his destructive side around the French Revolution. Euripides makes it clear in *Bacchae* that

Dionysus, son of Zeus, is by turns a most terrifying and a most gentle god to mortals (2008, 69). In André Chénier's *Bucoliques* (1794), Bacchus shows his anger in a place near the borders of Gange, where his festivity was despised. Bacchus exercised his power of transformation. The skins of deer worn by the bacchantes were turned into shields, their thyrsi into swords, their wine branches into snakes. There one celebrated the orgies of Bacchus. He destroyed the Indians and returned to Thebes joyfully. The French Revolution recognized the parallel development of destruction and regeneration. Joseph de Maistre, for example, commends in *Considérations sur la France* (1796) the birth of Bacchus as "marvelous" but complains about the anti-Christian aspect of the French Revolution. The temples were either closed or open only for "noisy deliberations and the bacchanals of the unbridled people" (1797, 87, 90).

The positive reception of Bacchus's wrath coincided with settings of Bacchus in operas and ballets in the 1780s and 1790s. In Salieri's *Les Danaïdes*, for example, the Danaids worship Bacchus before they set out to murder their husbands. "Glory, Evan, Eros, Bacchus, oh powerful God! / Terrible and Beneficent! / It is you that the Bacchante calls! / [Mount] Cithaeron trembles at his furious cries; / Iron sparkles in his hand, / death follows the flash of his eyes" (1784, 47). Act 2 of the opéra *Aspasie* (1789) concludes with a bacchanal festival (Darlow 2012, 279). Little is known about Jean-Baptiste Rochefort's one-act ballet-héroïque *Bacchus et Arianne*, performed in 1791 at the Académie Royale de musique. But the fact that it was performed at all during this early phase of the French Revolution attests to the renewed interest in Bacchus.

Conclusion

Bacchus is at once endearing and dangerous. He symbolizes negative freedom by freeing revelers from worries and releasing people from confinement. He symbolizes positive freedom by encouraging candor and fostering artistic creativity. He demonstrates the embodied concept of freedom of action by emancipating people from stricture. He invokes wonder and fear. In *Platée*, Bacchus gives rise to the topos of the extraordinary. This last point invites us to revisit current discourse of the extraordinary in disabilities studies. The concept of the extraordinary body attempts to change time-honored stigmatizing language by reframing the disabled body as corporeal difference, not as a lack (Garland-Thomson 1997, 9). Disability is a matter of representation as much as a physical condition. Embodied exceptionality could be taken as a marvel as much as an anomaly. For this reason, the topos of the extraordinary in musical repertories featuring Bacchus intersects with the theme of the extraordinary in disability studies. Yet in addition to signaling disability as embodied exceptionality, the extraordinary in seventeenth- and eighteenth-century French music on Bacchus also exemplifies the embodied idea of freedom.

Notes

1. The description of this entry is published in Issac de Benserade, *Le ballet de la revente des habits du ballet, comédie, dansé avant le Roy, première partie*, 1655, 4.
2. *Dictionnaire de l'Académie française*, 4th ed. (Paris: Brunet, 1762), s.v. "Extraordinaire."

References

Anderson, Nicholas. 1983. "Rameau's 'Platée': Burlesque or Grotesque?" *Early Music* 11: 505–508.

Autreau, Jaques. 1745. *Platée, ballet bouffon en trois actes, précédé d'un prologue; représenté devant le roi, en son château de Versailles, le mercredi 31 mars 1745*. Paris: Ballard.

Berlin, Isaiah. 2007. *Liberty*. Edited by Henry Hardy. Oxford: Oxford University Press.

Bordier, René. 1623. *Vers pour le ballet du roy représentant les Bacchanales dansé par sa majesté au mois de février 1623*. Paris: Jean Sara.

Bousset, Jean-Baptiste de. 1695. *Recueil d'airs sérieux et à boire par Monsieur Bousset: Pour les mois de Janvier, Février, et Mars 1695*. Paris: Christophe Ballard.

Breckenridge, Carol Appadurai, and Candace A. Vogler. 2001. "The Critical Limits of Embodiment: Disability's Criticism." *Public Culture* 13 (3): 349–357.

Burden, Michael, and Jennifer Thorps, eds. 2009. *Ballet de la Nuit: Rothschild B1/16/6*. Hillsdale, NY: Pendragon.

Cahusac, Louis de. [1754] 2004. *La danse ancienne et moderne ou traité historique de la danse*. Edited by Nathalie Lecomte, Laura Naudeix, and Jean-Noël Laurenti. Paris: Éditions Desjonquères.

Chancy, François de. 1651. *Ballet du roy, des festes de Bacchus, dansé au Palais royal, le 2. & le 4. Jour de May 1651*. Paris: Robert Ballard.

Chua, Daniel K. L. 2005. "The Promise of Nothing: The Dialectic of Freedom in Adorno's Beethoven." *Beethoven Forum* 12 (1): 13–35.

Chua, Daniel K. L. 2009. "Beethoven's Other Humanism." *Journal of the American Musicological Society* 62 (3): 571–645.

Chua, Daniel K. L. 2014. "Beethoven Going Blank." *Journal of Musicology* 31 (3): 299–325.

Cowart, Georgia J. 2008. *The Triumph of Pleasure: Louis XIV and the Politics of Spectacle*. Chicago: University of Chicago Press.

D'Alembert, Jean le Rond. [1759] 1973. "De la liberté de la musique." In *La querelle des bouffons; texte des pamphlets*. 3 vols. 3:2201–2281. Edited by Denise Launay. Genève: Minkoff.

Darlow, Mark. 2012. *Staging the French Revolution: Cultural Politics and the Paris Opéra, 1789–1794*. New York: Oxford University Press.

Euripides. 2008. *Bacchae and Other Plays*. Translated by James Morwood. New York: Oxford University Press.

Fader, Don. 2005. "The 'Cabale du Dauphin,' Campra, and Italian Comedy: The Courtly Politics of French Music Patronage around 1700." *Music and Letters* 86 (3): 380–413.

Fader, Don. 2014. "*Le Régent en Bacchus*? French Noble Self-Construction, Operatic Allegory, and Philippe d'Orléans's *Penthée* (1703)." Unpublished paper presented at the Annual Conference of the American Musicological Society.

Garland-Thomson, Rosemarie. 1997. *Extraordinary Bodies: Figuring Physical Disability in American Culture and Literature*. New York: Columbia University Press.

Gilliam, Bryan. 2014. *Rounding Wagner's Mountain: Richard Strauss and Modern German Opera*. New York: Cambridge University Press.

Green, Robert A. 2012. "Aristophanes, Rameau and *Platée*." *Cambridge Opera Journal* 23: 1–26.

Harris-Warrick, Rebecca, and Carol G. Marsh. 1994. *Musical Theatre at the Court of Louis XIV: Le Mariage de la Grosse Cathos*. New York: Cambridge University Press.

Isherwood, Robert. 1989. "The Conciliatory Partisan of Musical Liberty: Jean le Rond d'Alembert, 1717–1783." In *French Musical Thought, 1600–1800*, edited by Georgia J. Cowart, 95–120. Ann Arbor, MI: University Microfilms International.

Jaucourt, Louis de. 1765. "Liber [Mythologie]." *Encyclopédie ou dictionnaire raisonné des arts et des métiers*, edited by Denis Diderot and Jean le Rond d'Alembert. Paris: Les Sociétés Typographiques, 9:460.

Kintzler, Catherine. 1999. "Platée: La comédie, vérité de la tragédie." *L'Avant-Scène Opéra* 189: 44–49.

La Bruère, Charles-Antoine Le Clerc de. 1748. *Érigone, ballet, représenté devant le Roy, sur le théâtre des petits Appartemens à Versailles*. n.p.: n.p.

La Fare, Charles-Auguste, marquis de. 1740. "Penthée." In *Poésies de Monsieur le marquis de La Farre* [*sic*]. Nouvelle édition, considérablement augmentée, 217–276. Amsterdam: J.F. Bernard.

Lakoff, George, and Mark Johnson. 1999. *Philosophy in the Flesh: The Embodied Mind and Its Challenge to Western Thought*. New York: Basic Books.

Lambranzi, Gregorio. [1716] 1975. *Neue und curieuse theatralische Tantz-Schul*. Teil 1 und 2. Edited by Kurt Petermann. Translated by Michael Talbot. Leipzig: Peters.

Lully, Jean-Baptiste. 1672. *Les festes de l'Amour et de Bacchus, pastorale représentée par l'Académie royale de musique*. Paris: n.p.

Maistre, Joseph de. 1797. *Considérations sur la France*. 2nd ed. London: Mars.

Marmontel, Jean François. 1763. *Poétique françoise*. Paris: Lesclapart.

McCleave, Sarah. 2013. *Dance in Handel's London Operas*. Rochester: University of Rochester Press.

Mercure de France. April 1734. Paris: Pissot.

Montéclair, Michel Pignolet de. 1728. "Ariane et Bachus." In *Cantates à une et à deux voix avec simphonies*, 18–32. Paris: Boivin Marchand.

Nietzsche, Friedrich Wilhelm. 1999. *The Birth of Tragedy and Other Writings*. Edited by Raymond Reuss and Ronald Speirs. Translated by Ronald Speirs. New York: Cambridge University Press.

Ovid. 2004. *Metamorphoses*. Translated by Charles Martin. New York: Norton.

Rameau, Jean-Philippe. 1754. *Platée: Comédie-ballet*. Paris: n.p.

Rehding, Alexander. 2005. "'Ode to Freedom': Bernstein's Ninth at the Berlin Wall." *Beethoven Forum* 12 (1): 36–49.

Roy, Pierre-Charles. 1773. *Bacchus et la Minéïde, ballet héroïque en un acte*. Paris: Pierre-Robert-Christophe Ballard.

Salieri, Antonio. 1784. *Les Danaïdes, tragédie-lyrique, en cinq actes, représentée pour la première fois, sur le théâtre de l'Académie royale de musique, le lundi 19 Avril 1784*. Paris: Lormel.

Seaford, Richard. 2006. *Dionysos*. London: Routledge.

Thomas, Downing A. 2006. "Rameau's 'Platée' Returns: A Case of Double Identity in the 'Querelle des Bouffons.'" *Cambridge Opera Journal* 18: 1–19.
Thucydides. 2009. *The Peloponnesian War*. Translated by Martin Hammond. Oxford: Oxford University Press.
Voltaire. 1877–1885. *Œuvres complètes de Voltaire*. 3 vols. Edited by Louis Moland Paris: Garnier.
Wokler, Robert. 2012. *Rousseau, the Age of Enlightenment and Their Legacies*. Edited by Bryan Garsten. Princeton, NJ: Princeton University Press.

CHAPTER 10

ENTRAINMENT AND EMBODIMENT IN MUSICAL PERFORMANCE

EUGENE MONTAGUE

ACCORDING to the ethnomusicologist Martin Clayton, entrainment "refers to the process by which independent rhythmical systems interact with each other" (2012, 49). This intentionally broad definition gives a useful practical basis for a range of scholarly work, from field studies, to empirical experiment, to score analysis. It suggests links between music and other disciplines, such as biology and psychology, where the notion of entrainment is already deeply embedded. Moreover, it connects to theoretical concepts of time, rhythmic regularity, and intentional action, concepts that are integrated within the notion of what it is to exist as a human. Such connections lend the concept of entrainment considerable potential for the study of embodiment. This potential, placed in the context of musical embodied experience, is the subject of this chapter.

This investigation of relationships between entrainment and musical embodiment in music begins with a discussion of definitions of entrainment and identifies three distinct threads within the scholarly literature on this topic. Each of these threads can be seen to entail complementary perspectives on the nature, function, and operation of musical embodiment. Taken together, they suggest that the study of entrainment in music offers considerable potential to advance research on the concept of the body in music. In a final brief case study, this chapter shows how intertwining strands of entrainment imply different features of musical embodiment in the performance of a keyboard prelude by J. S. Bach.

The concept of entrainment has assumed increasing importance in Anglo-American music scholarship over the past twenty years. The term has become a mainstay for discussions of the complexities of musical rhythm considered as sound, as well as for work on the interactions between humans and the pulses, rhythms, and durational repetitions available within music. Clayton's definition, quoted earlier, provides a useful broad characterization of the concept that underpins most of this work. However, deeper investigation into the "systems" and "interactions" involved in entrainment brings to light some important distinctions between the idea of entrainment that operates across a number of disciplines. This is not a new observation: in 2011 the ethnomusicologist Judith Becker observed: "The term rhythmic entrainment is sometimes used in two different senses, depending on one's disciplinary allegiances" (2011, 49). This chapter goes one step beyond Becker to argue for three distinct, if overlapping, meanings of entrainment in fields as diverse as biology, psychology, and music. In line with the overall focus of this volume, it shows that these three meanings imply distinct perspectives on the notion of embodiment in musical practice: indeed, each entails a specific understanding of what a musical body may be. In order to suggest some of the ways in which these types of entrainment operate within the contexts of playing music, the chapter ends with an analytic case study of a prelude for keyboard by J. S. Bach. This work, with its typical motivic concentration and motoric rhythms, generates a tight relationship between bodily gesture and sonic pattern. It is thus a useful venue for study of the interaction between entrainment and embodiment in performance.

Origins of Entrainment

The concept of entrainment is generally traced back to the discoveries of the Dutch scientist Christiaan Huygens in the second half of the seventeenth century. Huygens was working on navigational devices when he discovered that two pendulums suspended from the same piece of wood tended to synchronize their oscillations over time. Further experiments showed that a similar process occurred in cases where there were more than two pendulums, including those suspended from a board and those placed on one. In all cases, the vital factor in producing the synchronization was the ability of the wooden table or board to transmit the oscillations of the pendula, causing each to affect the other (Pikovsky, Rosenblum, and Kurths 2001, 2–4).

Despite the age of the concept, the term "entrainment" itself does not appear to have achieved common usage until the middle of the last century. It does not appear in regular academic use until the 1960s, where it first emerges in the context of work in biology and physiology. These disciplines tend to deal with the interactions of rhythmical systems over an extended period and many populations, leading to a particular meaning of entrainment, which this chapter terms "long-term" entrainment.

"Long-Term" Entrainment in Biology, Physiology, and Related Sciences

The notion of entrainment as the long-term interaction between two large-scale biological systems emerges through a number of works in chronobiology and physiology beginning in the 1960s and continuing to the present day. It is worthwhile to discuss the topics of some of these studies briefly in order to obtain a sense of their scope. Shepherd Roberts (1965) studied the entrainment of photoreceptors in cockroaches to environmental light cycles, discovering to what degree obstructing these receptors led to a loss of entrainment. In 1966, M. L. Goff studied the effect of light-dark cycles of between sixteen and forty-eight hours on the behavior of rats (Goff and Finger 1966). And Lickey (1969) studied the entrainment between light cycles and a single neuron of the sea hare. There are clear commonalities between these studies that speak to their mutual reliance on long-term entrainment. They are listed as follows:

- All these studies concern the interaction between oscillators over long durations of time of at least several hours.
- All involve the presence of a dominant, external oscillator. This is usually termed the *Zeitgeber*, which the *Oxford English Dictionary* defines as "a rhythmically occurring event, esp. in the environment, which acts as a cue in the regulation of certain biological rhythms in an organism."[1] It is this event, or more precisely, series of events, to which the individual rat, cockroach, or neuron entrains.
- Both the *Zeitgeber* and the dependent oscillations operate by definition unconsciously, at the level of an automatic system.
- All measure entrainment not through close observation of one particular instance but through statistical studies of many, even thousands, of near-synchronicities between the oscillating systems. In these terms therefore, entrainment is a matter of statistical tendency over a sustained period of time.

All four of these characteristics show that long-term entrainment involves thoroughgoing and large-scale interaction between systems. It is quite distinct therefore from the notion of entrainment as an immediately observable synchronization, which is perhaps where musical discussions of entrainment tend to start. However, long-term entrainment does indeed find employment in music studies, including what may be the first clear use of the concept in music-related scholarship.

"Entrainment Music" and the Entrained Body

The first scholar in Anglo-American music studies to introduce entrainment into the vocabulary of music was the music therapist Mark Rider. In his dissertation (1988) and in a number of articles, Rider argues that entrainment involves a particular relationship between sound and the internal workings of the body. Using the long-term concept of entrainment as outlined in biology and physiology, Rider suggests "entrainment music" may have a healing effect on the body in direct terms, bypassing consciousness and operating on the level of individual cells. Rider describes entrainment music as a particular genre of music, improvisatory in character and involving select instruments. It aims at setting up synchronization between the rhythmic impulses of the music and the movement of cells in the subject.

Rider's entrainment music is rooted in the same approach as that used in the biological studies discussed above. Like the studies of entrainment among certain animal populations, the effects of entrainment music occur over relatively long durations: it will take many sessions of entrainment therapy to bring about a cure. Similarly, the effect of entrainment can be felt over large populations: indeed, this is its utility. In addition, the importance of the unconscious character of entrainment is apparent in Rider's argument. Christiaan Huygens, observing two clocks gradually achieve synchronicity from his sickbed in 1665, wrote of "la sympathie des horloges" (the sympathy of the clocks). This phrase brilliantly captures the tension between an observer's acknowledgment of the absence of willpower—clocks cannot intend anything—and the apparently inevitable anthropomorphism—yet the process involves sympathy. In a similar way, long-term entrainment implies a capacity to go beyond and behind conscious impulse, to draw out the healing properties of a body's biological structure. There is no immediate volition involved in this concept of entrainment, yet there is nevertheless a strong appeal to an underlying force or power. This approach to entrainment endures in other, largely therapeutic, approaches in music scholarship. One quite recent example occurs in the controversial literature on binaural beats and their efficacy in treating cognitive disorders (Turow and Lane 2011).

"Short-Term" Entrainment and the Psychology of Attention

The long-term model of entrainment emphasizes the force exerted by the *Zeitgeber* mechanism and the consequent passivity of the subordinate oscillators, which are forced to entrain with the former's phase and period. As a contrast, in what this chapter calls the "short-term" model of entrainment, developed primarily through work in psychology and cognition, the activity of the receiver or listener plays a much greater role.

This latter understanding of entrainment finds its origin in the work of the psychologist Mari Riess Jones (see Jones 2009 for a brief overview of her career). Jones set out to explain the mechanisms that control human attention to the external world and found a large part of her answer in the concept of entrainment. For Jones, human attention is essentially rhythmic and periodic: it interacts with the external world through perceiving and locking into similar rhythmic patterns found in its environment. Entrainment thus becomes a fundamental property of human interaction with the outside world.

> We assume that an attender initially entrains to some relatively regular, prominently marked time period within an event. Perhaps in an utterance, it is a person's

> characteristic articulation rate; in music it may be the beat period or measure span. In any case, this period functions as an anchor or referent time level for the perceiver. . . . This referent period provides (a) a temporal reference frame, "locking the attender to the event," and (b) an event-determined time scale that "calibrates time spans." (Jones and Boltz 1989, 470)

Jones's approach has three central characteristics, (1) it grounds the activity of entrainment in the listener and, given her interest in the immediacy of attention, (2) it emphasizes the short-term importance of locking into attentional rhythms. Indeed, in her earlier work she seems to prefer the term "synchronization" to "entrainment" perhaps in order to avoid the associations of the latter term with long-term durations (Jones 1976). Additionally, given her focus on the cognitive function of attention, her approach to entrainment (3) implies that it interacts with the human potential for expectancy, and thus may link to volitional behavior; an implication mostly absent from the long-term approach described earlier. Developing this third implication, Jones has extended her initial interest in entrainment into deep studies of music as a human activity, studies that investigate questions of attention, expectancy, and rhythmic behavior. Her work has had a foundational influence on the theorization of rhythm and meter in music scholarship. Many scholars have followed her lead in exploring musical rhythm from the perspective of short-term entrainment.

Robert Gjerdingen, for example, studies oscillatory processes of neurons as an example of entrainment, drawing on and developing Jones's emphasis on the function of entrainment in perception (1993). However, the music theorist who has developed Jones's work in the most thorough fashion is undoubtedly Justin London. Working from the concept of entrainment as a fundamental feature of human perception, London proposes, "meter is a specifically musical instance of a more general perceptual facility of temporal attunement or entrainment" (2004, 24). Meter in this sense arises from our capacity to lock into entrainment with music.

This notion of short-term entrainment is primarily a psychological one. London argues the role of musical meter as a type of entrainment is filtered through our capacity for physical movement. "Metric entrainment . . . may be the direct synchronization of our movements to external rhythms" (12). The further question of *what* movements may occur appears to be entirely open, in principle at least, and so the details of what the body does, or how it is, seem relatively unaffected by short-term entrainment. A body may use the entrainment as potential for movement, London suggests, but it is not clear that it has to do so. Rather, if entrainment provides a listener a perception of the potential for movement, then this is enough.

Short-term entrainment implies a listening body, but this body is by and large neutral as regards other aspects of embodiment.[2] In long-term entrainment, the body is understood as the passive receptor of an oscillatory rhythm, which may be targeted to specific bodily parts. In short-term entrainment, the body actively identifies oscillatory rhythms in the environment, seeking out opportunities for entrainment, but without reference to specific physical features. In other words, as an entrained body, I may move my fingers, toes, nose, ears, hips, or all, or none, without altering what is understood as the characteristic quality of short-term entrainment.

"Physical" Entrainment and Embodied Movement

In recent studies of movement to music, the use of techniques such as video and motion capture has allowed for a more precise study of gesture within the context of a natural musical event (MacRitchie, Buck, and Bailey 2013; Nusseck and Wanderley 2009; Palmer et al. 2009). These studies, taken together, suggest a nascent approach to entrainment that is distinct from the others discussed earlier. Beginning from the central observation that a certain movement or set of movements entrain to one or more musical events, this approach emphasizes the local, physical, and (perhaps) low-level nature of this entrainment. In these cases, entrainment implies an (unconscious) elicitation of a particular, synchronous movement. Thus, this approach joins together certain aspects of both long-term and short-term entrainment in a common foundation of physical movement. This third type, which I call "physical" entrainment, suggests a complex relationship between body and sound.

This emphasis on the physical detail of entrainment is founded in part on an extension of the systematic concept of entrainment to include physical actions in general. Thus, Clayton points out that synchronized oscillators are central to many basic physical acts, even before there is any involvement with music or external sound: "Still within the individual but on a somewhat larger scale, many common actions involve the synchronisation of movements between different body parts—for example walking" (2012, 50). On this reading, fundamental habits such as walking, talking, and tapping involve entrainment within the physical systems of the body.

If entrainment is already present in physical movement, then it assumes a richer significance when the event of musical performance is also taken into consideration. Several scholars have focused attention on rhythmic details of the physical performance of music, as well as its reception. For example, the work of Floris Van Vugt has investigated the specific rhythms of scale patterns played by student pianists, showing how finger actions play particular roles in forming temporal patterns (Van Vugt, Jabusch, and Altenmüller 2012). Edith Van Dyck's work on dance music has demonstrated that changes in the sound of the bass drum have a consistent relationship with changes in dancers' head movements (Van Dyck et al. 2013). And work by Caroline Palmer and Marcelo Wanderley has illuminated the role of specific rotation movements by clarinetists in performance (Palmer et al. 2009; Wanderley et al. 2005). These rotations entrain to music on a regular basis, forming an important part of the phenomenology and syntax of musical performance. On a theoretical note, Marc Leman has suggested an amendment to Clayton's discussion of entrainment, arguing that entrainment to music is not merely a psychological feature, but includes spatiotemporal aspects in terms of the place and the time in which the entrained body moves (2012).

All these studies suggest that there is the potential for aspects of musical sound to have a direct effect, through entrainment, on particular parts and movements of the body. Such an effect may depend on either or both of the two types of entrainment presented above. However, the third, physical entrainment, is different both in kind and in the potential it offers for research into embodiment, for the following reasons:

- Such body-sound entrainment will be highly contextual, dependent on personal engagement with music in addition to the spatiotemporal qualities mentioned by Leman.
- The connection between sound and body can be considered as a primary foundation, without the necessity to posit a cognitive, psychological process.
- The direct relationship between sound and body movement means that this entrainment is not a conscious, deliberative movement.

In the specific context of the case study at the end of this chapter, some of the implications of physical entrainment for music and the body are outlined. First, however, it will be useful to set out the ways in which these three types of entrainment imply different conceptions of musical embodiment.

The Entrained Body

Each of the three concepts of entrainment I have outlined has implications for the role of the body in music, and research into musical bodies. Indeed, each may be said to imply a particular concept of musical embodiment.

1. **Long-term entrainment and body schemas:** This notion of entrainment, rooted as it is in biology, finds its significance for music in the study of repeated physical actions made in the process of learning a particular instrument or piece. These actions are often regulated by a *Zeitgeber* such as a metronome: this sets a standard for the development of a particular technique and defines a practical ability that may be shared among a group of instrumentalists. The practice of these techniques, which include scales, arpeggios, chords, and so on, may perhaps be an individual affair, but the common nature of the exercises means that they are shared in effect. And therefore it is by no means difficult to imagine a group of instrumentalists together achieving entrainment with a particular metronome mark. Such an achievement will, in fact, develop a specific bodily ability, much like athletic training in a gym. This suggests that research on this type of musical embodiment could focus on the ways in which musical disciplines develop particular movements that come to define bodies.

   The embodiment created through such long-term entrainment thus includes innervated physical movements. As the entrained physical movements that typify instrumental training sessions become absorbed within the sound, their status as movements typically will fade from the consciousness of the player. This fading is, of course, a benefit: it would be impossible to consciously control all aspects of playing a complex piece (Noë 2009). So, for almost all performing musicians, many of the movements that the body makes to produce sound are no longer a conscious focus of attention. Thus, players may no longer consider performance as a series of physical gestures, preferring instead to understand the music as a flow of sounds.

This experience is not confined to musicians: the same can be said of the movements of many others who use physical actions to achieve specific goals external to their bodies, such as athletes. In his work on consciousness and the body, Shaun Gallagher uses the term "body schema" to describe such integration between action and mind developed through repetitive habits. For Gallagher, a body schema is "a system of sensory-motor capacities that function without awareness or the necessity of perceptual monitoring (2005, 24). Typically, body schemas do not need conscious attention: they underlie interaction between the environment and the body, and are experienced in terms of one's engagement with the world. In this sense, to speak of the body of a musical performer is, at least in part, to speak of this collection of innervated movements and capacities for movement: to speak, that is, of a body schema. Studying the operation of long-term entrainment is one way to come to terms with the development of such body schemas by musicians.

2. **Short-term entrainment and the listening body:** The psychological approach of short-term entrainment depends on someone perceiving an external oscillator. As noted above, such perception is psychological: it does not require any specific bodily activity. This lack of specificity is demonstrated in the traditional use of experiments based on tapping to explore relationships between human perception and musical organization. The bodily activity of tapping is one that offers very little resistance or counterpoint to an external pulse. That, of course, is taken to be one of its benefits for such studies: it appears as a "normalized" activity that represents the best way to gain direct access to the relationship between mind and rhythm. And yet even at the origins of such experiments the details of physicality interrupt this access, thus in the classic studies of Lewis Stevens in the late nineteenth century, the author notes his disappointment at the failure of some of his subjects to overcome tiredness while attempting to complete the required duration of tapping activity (1886). Clearly, the specific context of the physical activity is as an unwelcome distraction: the real interest lies in the mental process of which this activity is, apparently, a product. This attitude illustrates well the particular role of the body in discussions of short-term, psychological entrainment. Bodily movement is held as vital insomuch as it provides evidence of psychological processes, but due to the priority accorded to these processes, the particularity of bodily actions tend to be devalued. The body of this type of entrainment is therefore largely a matter of potentiality rather than actuality. It comes into being as the vehicle of an active, attentive listening that perceives musical rhythms and the potential of entraining to them through unspecified physical actions. It is therefore, at least in part, a virtual, imaginary body.

3. **Physical entrainment and the multifaceted body:** The particularity of movement, absent from short-term entrainment, is the defining characteristic of physical entrainment. In relationship with the sound, certain bodily actions become immediately and directly entrained. Since these actions are, by themselves, evidence of entrainment already taking place within the musculature of the body, we can and should regard such activities as examples of entrainment at more than one level: there must be more than two entrained systems involved in such activities. The body that emerges from this notion of entrainment is a decentralized one, consisting of two or more systems that may, or may not, coordinate. When they do, which is whenever entrainment of physical actions to music occurs, there is a nesting of entrained systems, forming what may be hierarchical structures. This picture of embodiment is complex and multifaceted, suggesting that to account for entrainment in this way may take a number of disparate perspectives.

There is much more that might be said about the characteristics of these three types of entrainment, and the ways in which they may imply musical embodiment. This chapter abjures further theoretical discussion, however, in order to examine some of the ways in which these forms of entrainment interact in an analysis of a single piece of music and to discuss some of the implications these interactions may have for embodiment.

Case Study: The Performer's Body as Entrained and Entraining

The following analysis examines the interactions between the previously outlined three forms of entrainment, the characteristic embodiment each brings forth, and their musical consequences. This takes place in the context of a case study of the Prelude in C minor BWV 847 by Johann Sebastian Bach. Figure 10.1 shows the opening measure of this second prelude from Book 1 of the *Well-Tempered Keyboard*.

FIGURE 10.1 The opening measure of J. S. Bach's Prelude in C minor, BWV 847.

Long-Term Entrainment in the Prelude in C Minor, BWV 847

As in many preludes of this era, there is a constant succession of sixteenth notes. These are grouped in fours, and within each of these groups, the last three notes require the player to make a repeating pattern of movement in which the neighboring fingers of both the right hand and left simultaneously create a shape that descends a single step and then returns. These shapes continue for most of the piece, so that the experience of playing becomes defined, in part, by the continuous making of this finger shape at regular intervals. Figure 10.2 shows a graphic representation of this shape and generates a formulaic statement of the shape in terms of finger action.

As the graph and formula in Figure 10.2 display, the repetition of these finger movements allows them to be felt and heard as a single gesture, with attendant specific rhythmic and durational characteristics. This gesture continues throughout much of the course of the prelude.

In terms of the long-term entrainment discussed previously, the process of practicing, using the score as a reference point, creates entrainment between fingers and the demands of the notation. While the score itself is, of course, silent—a mere set of instructions—the keyboard player will treat the score as an oscillatory reference point for her practice, turning it into a *Zeitgeber* in functional terms. Over the durational experience of practicing this piece, the constant playing of the finger gesture will allow it to entrain with the demands of the score. Clearly, not every instance of the gesture will entrain exactly to the ideal of evenness notated in the score. But there will be a general tendency, which could be expressed in terms of statistics, for this pattern to conform to the oscillation suggested in the score.

As with the other examples of long-term entrainment discussed above, this long-term process of entraining is not individual: it will be shared across the population of pianists practicing this piece, much as a population of neurons exposed to a certain *Zeitgeber* will eventually entrain to it. To be sure, the oscillation of sixteenth notes represented in the score is of much shorter duration than the circadian-type rhythms discussed earlier. However, the process of bodily adjustment to a regulator rhythm over an extended period of time is a similar type of learning. Through sustained practice, the body attunes to the performance of the sixteenth-note figure, so that its performance becomes a habit

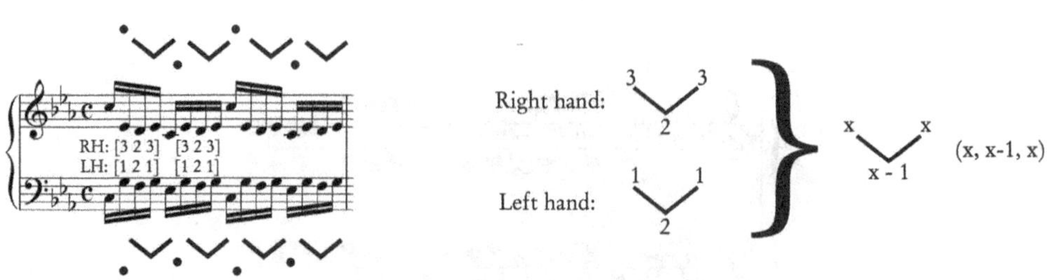

FIGURE 10.2 The basic gestural shape for the player of BWV 847.

in the phenomenological sense: a characteristic capacity of a body that is learned to the point of innervation and no longer demands conscious control (Noë 2009; Talero 2006). The musical body implied by this prelude is, in part, one that has internalized this particular schema. In a general sense, then, this prelude as a piece of music sets up a particular definition of musical embodiment through long-term, background entrainment. A body—any body—that practices this prelude, and successfully realized the score, will incorporate the gestural and rhythmic schemas shown in Figure 10.2 as defining elements of its constitution.

Short-Term Entrainment in the Prelude in C Minor, BWV 847

Within the particular long-term entrainment involved in playing this prelude comes a short-term potential for the creation of metrical entrainment. This specific short-term sense of synchronization emerges against the regular background of long-term finger gesture. In terms of Figure 10.2, the metric entrainment of this piece is generated by the first sixteenth-note of each four-note group; the dots of Figure 10.2 as opposed to the graphic lines.

Figure 10.3 shows the first sixteenth-note of each grouping of four. The patterns of these notes establish a quadruple meter through what is roughly the first half of the piece. This pattern creates several different levels of possible entrainments, and these are shown on the figure with the curved projective arrows used by Christopher Hasty (1997) in his theory of meter as an essential element of rhythm. For example, at the quarter-note durational level the distinction between the notes of gesture A, and the notes alternately above and below it form one level of potential entrainment, as shown by the shortest projective arrows in Figure 10.3. The up-down-up-down contour at the half-note durational level forms a further potential meter: this is shown by the longer arrows on the second level. The changes in pitch pattern at the whole-note level, and particularly the succession created by the C4–B4 major seventh followed by the C4–C5 octave present another level of entrainment, shown by the longest curved lines on Figure 10.3. Since the first level could nest within the second, and the second within the third, it

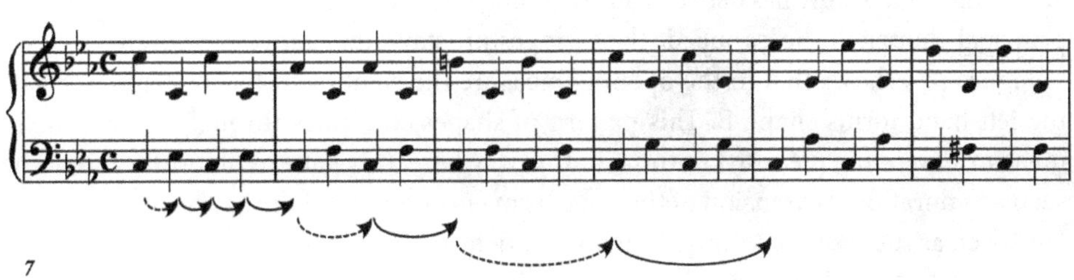

FIGURE 10.3 Three patterns that may generate short-term entrainment in BWV 847.

is possible to shift perception from one level to another during the course of playing Figure 10.3, though one can only attend to one level at any one moment.

To describe these as examples of entrainment raises the question: what is being entrained? In the long-term entrainment that developed the basic gesture as a habit, the answer was clear: the fingers themselves. And it was argued earlier that this entrainment helps define a particular form of musical embodiment. In the case of short-term entrainment, as discussed earlier, the entrainment does not specifically define a musical body. It has, to be sure, bodily correlates. For example, the outlines of the contour in both right and left hands give a rough type of support to the sense of short-term entrainment. More precise and important, however, is the aural distinction between consonance and dissonance in the intervals formed between the notes of Figure 10.3. Thus, the dissonant quality of the major seventh in measure 3 facilitates a strong sense of arrival on the consonant octave downbeat of measure 4, suggesting a phrase ending. There is no specific bodily correlate to this arrival, other than the experience of hearing the octave as a consonant resolution.

The body implied by this short-term entrainment is, therefore, primarily a listening body rather than an active one. Again, as with the long-term entrainment and the definition of a body schema, the activity of playing and listening to this music has potential to define a particular body. In this case, however, this body is defined through what it may perceive: it is a body defined through perception rather than action.

Physical Entrainment in the Prelude in C Minor BWV 847

In tandem with the other two forms of entrainment, the details of specific physical movements form a third and vital rhythmic element in relation to this piece. This form of entrainment reflects the details of physical engagement with the sound over the duration of the music; as such, to track its course through the music would require a lengthy analysis. In this chapter, therefore, attention is confined to one salient point concerning the interaction of hand shape with short-term, metric entrainment in the prelude.

Figure 10.4 shows the four different shapes the hand assumes in playing this prelude, all the while maintaining the basic gesture, as shown by the V-shaped lines in the example. The shapes are A, in which the basic gesture lies between a descending interval, and mirror A, in which the gesture lies between an ascending interval. B and mirror B follow the same principal, though the notes outside the basic gesture stay either above or below it.

Figure 10.5 shows that in the opening measure the right hand forms shape A, while the left hand forms shape B. This pattern of shapes continues through the first three measures: were the piece to continue in this way here, this pattern of motives could be seen as a durational extension of the long-term entrainment of the background gesture. However, as the music continues, measure 4 defeats this potential through a change in

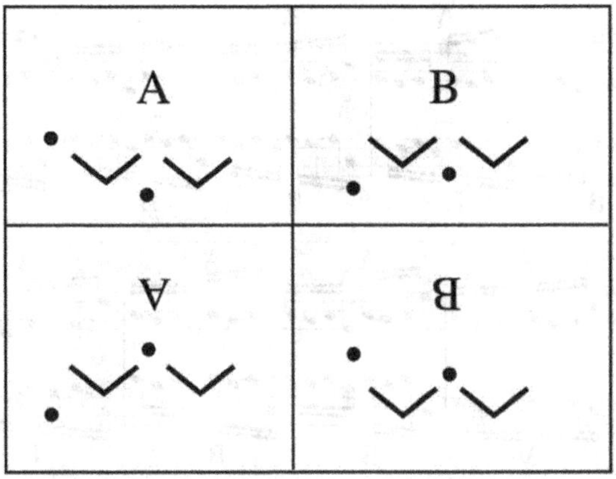

FIGURE 10.4 The four different varieties of hand shape in the motives of BWV 847.

the left-hand shape from B to mirror A. This new beginning coincides with the short-term return to the C octave in measure 4, which marked out a sense of return for the listening body, as discussed previously.

This change of shape in measure 4 is a marked and specific physical change, creating an important alteration in bodily movement. The specific requirement for the left hand to play G3 above the ongoing ground defines an active body that contrasts with the listening body that hears the returning octave. This body counterpoints the listening one, responding to the heard arrival point of the octave C with a specific physical movement that goes against the grain of the listening body. This Barthesian moment is defined by a concatenation of three different aspects of musical embodiment within the experience of the player. The long-term entrainment with the prelude's characteristic gesture combines with the short-term entrainment created by the experiences of consonance and dissonance. Together with this, the rhythm of the physical shapes of performance provides a further level of entrainment in performance. Through this experience, the musical qualities of measure 4 emerge as the direct products of the interaction between the respective types of entrainment, each of which supports different forms of musical embodiment through the sound.

The experience of these different forms of entrainment becomes a musical feature of this prelude, and physical entrainment, defined through the details of the hand shapes that execute this music, emerges as an important factor at several further points in the prelude. The body that plays and hears this prelude is one that is defined by the body schema of the basic gesture; by its attention to the short-term entrainment of meter, together with its physical correlates; and by its occasional conflicts with this meter in terms of the rhythm of hand shapes. In this sense, there is not just one body that plays and entrains to the rhythm of this music, but, as the musical contexts and moments change, there are potentially many different forms of embodiment that emerge throughout the course of the prelude.

FIGURE 10.5 The succession of hand shapes through the first page of BWV 847.

Conclusion

This chapter has introduced and outlined three distinct, though surely related, concepts of entrainment. These concepts bring with them particular implications for musical embodiment, implications that suggest the relationship between music and body is complex and multifaceted. By way of a short discussion of embodiment in the experience of playing and listening to Bach's Prelude in C minor, this chapter suggests some of the ways in which scholarly approaches that include entrainment may be useful in studying the multiple ways in which the human body creates and uses music.

Notes

1. OED Online, "Zeitgeber, n.," accessed June 13, 2016, http://www.oed.com/view/Entry/232755?redirectedFrom=zeitgeber.
2. In this neutrality it adheres to the classical tradition of measuring rhythmic effects in psychological terms through experiments that involve standard or default movements such as tapping. In such studies, tapping is taken to stand for any type of movement: it becomes the default way to demonstrate the human psychological capacity to perceive rhythmic patterns.

References

Becker, Judith. 2011. "Rhythmic Entrainment and Evolution." In *Music, Science, and the Rhythmic Brain: Cultural and Clinical Implications*, edited by Jonathan Berger and Gabe Turow, 49–72. New York: Routledge.

Clayton, Martin. 2012. "What Is Entrainment? Definition and Applications in Musical Research." *Empirical Musicology Review* 7: 49–56.

Gallagher, Shaun. 2005. *How the Body Shapes the Mind*. Oxford: Clarendon Press.

Gjerdingen, Robert O. 1993. "'Smooth' Rhythms as Probes of Entrainment." *Music Perception* 10: 503–508.

Goff, M. L., and Frank W. Finger. 1966. "Activity Rhythms and Adiurnal Light-Dark Control." *Science* 154: 1346–1348.

Hasty, Christopher. 1997. *Meter as Rhythm*. New York: Oxford University Press.

Jones, Mari Riess. 1976. "Time, Our Lost Dimension: Toward a New Theory of Perception, Attention, and Memory." *Psychological Review* 83: 323–355.

Jones, Mari Riess. 2009. "My Life and Times." *Psychomusicology: Music, Mind and Brain* 20: 89.

Jones, Mari Riess, and Marilyn Boltz. 1989. "Dynamic Attending and Responses to Time." *Psychological Review* 96: 459.

Leman, Marc. 2012. "Musical Entrainment Subsumes Bodily Gestures: Its Definition Needs a Spatiotemporal Dimension." *Empirical Musicology Review* 7: 63–67.

Lickey, Marvin E. 1969. "Seasonal Modulation and Non-24-Hour Entrainment of a Circadian Rhythm in a Single Neuron." *Journal of Comparative and Physiological Psychology* 68: 9–17.

London, Justin. 2004. *Hearing in Time: Psychological Aspects of Musical Meter*. New York: Oxford University Press.

MacRitchie, Jennifer, Bryony Buck, and Nicholas J. Bailey. 2013. "Inferring Musical Structure through Bodily Gestures." *Musicae Scientiae* 17: 86–108.

Noë, Alva. 2009. *Out of Our Heads: Why You Are Not Your Brain, and Other Lessons from the Biology of Consciousness*. New York: Hill and Wang.

Nusseck, Manfred, and Marcelo Wanderley. 2009. "Music and Motion: How Music-Related Ancillary Body Movements Contribute to the Experience of Music." *Music Perception* 26: 335–354.

Palmer, Caroline, Erik Koopmans, Christine Carter, Janeen D. Loehr, and Marcelo Wanderley. 2009. "Synchronization of Motion and Timing in Clarinet Performance." In *Proceedings of the Second International Symposium on Performance Science*, edited by Aaron Williamson, Sharman Pretty, and Ralph Buck, 159–164. Utrecht: European Association of Conservatoires.

Pikovsky, Arkady, Michael Rosenblum, and Jürgen Kurths. 2001. *Synchronization: A Universal Concept in Nonlinear Sciences. Vol. 12 of Cambridge Nonlinear Science Series*, edited by Boris Chirikov, Frank Moss, Predrag Cvitanovic, and Harry Swinney. New York: Cambridge University Press.

Rider, Mark S. 1988. "Effect of Cell-Specific, Music-Mediated Mental Imagery on Secretory Immunoglobulin A (Siga)." PhD diss., University of North Texas.

Roberts, Shepherd K. 1965. "Photoreception and Entrainment of Cockroach Activity Rhythms." *Science* 146: 958–959.

Stevens, Lewis T. 1886. "On the Time-Sense." *Mind: A Quarterly Review of Psychology and Philosophy* 11: 393–404.

Talero, Maria. 2006. "Merleau-Ponty and the Bodily Subject of Learning." *International Philosophical Quarterly* 46: 191–203.

Turow, Gabe, and James D. Lane. 2011. "Binaural Beat Stimulation: Altering Vigilance and Mood States." In *Music, Science, and the Rhythmic Brain: Cultural and Clinical Implications*, edited by Jonathan Berger and Gabe Turow, 122–136. New York: Routledge.

Van Dyck, Edith, Dirk Moelants, Michiel Demey, Alexander Deweppe, Pieter Coussement, and Marc Leman. 2013. "The Impact of the Bass Drum on Human Dance Movement." *Music Perception* 30: 349–359.

Van Vugt, Floris Tijmen, Hans-Christian Jabusch, and Eckart Altenmüller. 2012. "Fingers Phrase Music Differently: Trial-to-Trial Variability in Piano Scale Playing and Auditory Perception Reveal Motor Chunking." *Frontiers in Psychology* 3: 495.

Wanderley, Marcelo M., Bradley W. Vines, Neil Middleton, Cory McKay, and Wesley Hatch. 2005. "The Musical Significance of Clarinetists' Ancillary Gestures: An Exploration of the Field." *Journal of New Music Research* 34: 97–113.

CHAPTER 11

RHYTHM AND THE PERFORMER'S BODY

DANIEL B. STEVENS

Tell me, where is fancy bred?

Or in the heart, or in the head?

—Shakespeare, *The Merchant of Venice*, III.ii.

Introduction

Near the end of his distinguished career as a pianist and composer, Robert Helps (1928–2001) performed several of Leopold Godowsky's paraphrase-studies of Frédéric Chopin's études.[1] During one such performance, Helps paired the second of Chopin's *Trois nouvelle études* with Godowsky's first study of that étude, delivering renditions of both pieces that display the pianist's technical and expressive virtuosity. Table 11.1 provides URLs to recordings made at this fall 1997 recital of both pieces.

The wide fluctuations of tempo and dynamics of Helps's performances given in Table 11.1 are animated by "The Performance Worm," a visual representation of musical expressivity developed by Simon Dixon, Werner Goebl, and Gerhard Widmer (2002). These particular Performance Worms were created with the gracious assistance of Werner Goebl using beat-level IOIs (inter-onset intervals).[2] The Performance Worm tracks the expressive elements of tempo (in beats-per-minute) and dynamics (in decibels) within a two-dimensional grid. The Worm's path fades into the distance showing the expressive trajectory over time, and its head displays the current measure number. While some of the tempo and dynamic fluctuations in Helps's performances follow the expressive markings, motivic features, and phrase groups found on the score, others (as I will show) seem to owe themselves entirely to the creative contributions of

Table 11.1 URLs to Performances

Étude	URL
Chopin	https://vimeo.com/171653232
Godowsky	https://vimeo.com/171658051

the performer. The Performance Worms make visible the expressive range of Helps's musical imagination.

But where is such musical fancy bred? While modern technologies have advanced our ability to measure and represent expressive fluctuations, explaining their causes remains no less a challenge today than it was for Chopin's contemporaries. Mendelssohn described Chopin's playing as bringing about "miracles" (Eigeldinger 1986: 267), while Liszt spoke of his "ravishing touch" as "divinely beautiful" (273). Chopin's pupil Karol Mikuli explained the composer's temporal fluctuations in evocative gestural terms, noting that "in keeping time Chopin was inexorable. . . . Even in his much maligned tempo rubato, the hand responsible for the accompaniment would keep strict time while the other hand, singing the melody, would free the essence of the musical thought from all rhythmic fetters, either by lingering hesitantly or by eagerly anticipating the movement with a certain impatient vehemence akin to passionate speech" (Eigeldinger 1986, 276). Another pupil, Georges Mathias, suggests a linkage between Chopin's pianism and his compositions: "he played as he composed" (277). As Berlioz put it, "his playing is shot through with a thousand nuances of movement of which he alone holds the secret, impossible to convey by instructions" (272). Over a century later, the secrets of nuanced musical expression seem equally impenetrable. In their survey of computational, visual models of expressive timing, Goebl and Widmer (2009) question "even if research into expression of movement, motor control and biomechanics advances, how would it ever be possible to measure, quantify and finally model the performer's 'mind and body in the "heroic struggle" to express and communicate'?"

Historically, music research has bypassed this problem by focusing on the musical work and treating performances as the realization of an abstract concept formed in the mind before and during a performance. In this view, the performer's body is an instrument for realizing that abstract idea and presenting it to an audience. This chapter explores an alternative view, in which musical ideas and expression are developed through embodied cognition in the act of performance. Specifically, it addresses some ways in which physical gestures cause, rather than result from, the rhythmic nuances that contribute to expressive performance. In sum, I argue that the concept of embodied rhythm involves not only the physical gestures required to produce notated rhythms but also the generative nature of the rhythmic activities of the body in performance. The body can be the source, not just an instrument, of rhythmic activity.

Embodied Rhythm/Rhythmic Expression: Context and Method

Rhythm can be defined as the temporal organization of musical works: the relative durations of pitches and harmonies, metric structure, phrase rhythm, formal design and proportions, fluctuations of tempo (e.g., rubato), and the timing of expressive aspects such as the addition of vibrato or dynamic changes. The concept of "embodied rhythm" situates these rhythmic phenomena within the broader field of embodied cognition, and within the domain of musical experiences in which rhythms are mediated, experienced, generated, and interpreted through bodily gestures. Despite the body's obvious role in shaping musical time through the embodied actions of performers, these bodily gestures are tacitly marginalized or entirely excluded in many theories of musical rhythm: London (2004), Berger (2002), Caplin (2002), Krebs (1999), Epstein (1995), Yeston (1976), and Cooper and Meyer (1960) treat the body as incidental to their conception of rhythm, if they address it at all.

As the essays in this collection show, the body has gained prominence in studies of musical expression and interpretation. Numerous empirical studies have measured micro-timing differences between performances of the same piece (Repp 1995; Ohriner 2012, 2014; Benadon and Zanette 2015), raising intriguing questions about the role of the body in producing expressive variations. Others (Rothstein 2005; Hudson 1994) have examined expressive effects like rubato and sought to delineate their principles and historical practice. Pierce (2007) has explored how developing a repertoire of bodily gestures (e.g., swinging arms, tapping the air) can help energize and characterize rhythmic expression in performance, while Le Guin (2006) and Cumming (1997) approach the body as a site for developing meaning and musical subjectivity.[3] Early work on musical embodiment (Clynes 1982) sought to provide a neurophysiological basis for understanding musical figurations as "gestures," expressive shapes that could be universally recognized and interpreted.

Rather than treating the body and its physical movements as a metaphor for musical gestures (Larson 2006), I argue here, building on Cook (2012 and 2013), that notated music scripts complex, multilayered physical gestures whose coordination has an immediate, direct impact on musical timing in act of performance. If "musical gestures are musical acts" (Cox 2006, 2016), then physical gestures can also be thought of as a kind of music, whose internal structures and movements can be read alongside the notated musical relationships found in the score. The relationship between musical and physical gestures need not be linear or one-to-one, with performers' movements mapping indexically onto the musical shapes indicated by the composer (Hatten 2004). Performers do not only project musical structures encoded in the score, and the performances they produce cannot be explained strictly by way of structuralist methodologies. Responding to Mitch Ohriner's claim that "performers create structure in much the same way that readers create poems" (Ohriner 2012, paragraph 38), Nicholas Cook suggests that one solution to the structuralist problem "may be to think of performers as more broadly creating meaning within the structural affordances of compositions, meaning that cannot itself be encompassed within a purely structural description" (Cook 2013, 68). While both of these perspectives distinguish and positively valence the contribution of the performers as distinct from pure musical structure, neither goes so far as to recognize that performers operate according to the structural characteristics of their bodies and instruments. By acknowledging that both compositions and bodies operate according to unique sets of structural limitations and possibilities, I seek to explore how compositions can be generated within the structural affordances of performers' bodies. In this view, performers' gestures contribute something as vitally important, impactful, sophisticated, and structured as the musical materials provided by the composer.

This stance attempts to provide a new answer to a question that is rarely asked: what is expressed in an "expressive" performance? Given the work-centric focus of many analyses, musical expression is often presumed to refer to a particular performer's articulation of musical structures and gestures. Because words like "express" and "expression" are so intricately woven into discourse about performance, it can be difficult to see tempo fluctuations such as those in Helps's performances as distinct from the rhythmic relationships and expressive markings on the score. While there exists a relationship between notated rhythms and timing in Helps's performances, I explore the extent, quality, and meaning of this relationship to argue that Helps's inflection of rhythmic relationships does more than express musical structure. The process of articulating structure transfigures the notated content, adding a separate layer of musical action through embodied gestures. Further, I elucidate the structure of these embodied gestures so as to show that the temporal fluctuations in performance offer a unique mode of musical commentary on the score provided by the composer. To anticipate my conclusion, what is expressed in performance includes both notated sound relationships and the bodily mechanics and their related associations (e.g., emotional, gestural, dramatic).

To accomplish these goals, I explore the contributions of embodied gestures to rhythmic inflection from the perspective of a performer. I attempt to bring analysis and performance into dialogue by narrating the performance in terms of the physical activities involved in its production. While I focus exclusively on gestures that directly relate to the production of sound, I contend that some gestures perceived as ancillary may produce expressive rhythmic fluctuations in performance (Wanderly and Vines 2006) and communicate expressive content to an audience.[4] The analyses that follow are written from my perspective as a pianist and are significantly enriched by the formal training I received on this instrument while pursuing two professional performance degrees. Like an analyst who imagines and narrates the musical structure of a work, I endeavor to narrate the gestural content of Helps's performance in a way that I as a performer could (and do) emulate through my own performance of these works. While adopting a first-person approach may involve speculation, it is informed by the four years (1996–2000)

I spent studying as a private student of Helps as an undergraduate music major at the University of South Florida, and by later research in the Helps archives at the university's library.[5] While I focus on another pianist's performance, I take the expressive contour of this performance as an invitation both to imagine how I might coordinate my own gestures in performance to create similar musical effects and to explore the ways in which the embodiment of musical rhythm creates unique musical meanings beyond the abstract relationships found on the score, and may even generate new compositional possibilities. While the reflections that follow are speculative and qualitative, my intention is to highlight the complex dynamics of performers' physical gestures in order to help nonspecialists better appreciate the relationship between notated rhythm, musical gestures, and temporal fluctuations and to raise questions that may point specialists to consider new ways of measuring and interpreting expressive timing in performance.

Chopin, Whiteside, and the Gracious Rhythm of the Body

Helps's performances provide an excellent case study to examine the generative, second-order nature of embodied rhythm in two respects. The pieces themselves admit a variety of clear similarities and differences. Godowsky's study closely follows the primary melodic and harmonic design of each section in Chopin's ternary-form (A–B–A′) étude. However, Godowsky layers onto this musical structure new polyrhythmic relationships, countermelodies, and textural-registral distributions, creating new musical relationships, expressive opportunities, and performance gestures (Table 11.2). Further, Godowsky's étude thrice repeats the opening A section of the original, creating an expanded A^1–A^2–A^3–A^4–B–A^5 design. (One measure in the Godowsky corresponds to two in Chopin's original.) Chopin's étude provides the basic materials for two types of expressive performance: that of the performer (Helps), and that of the composer (Godowsky), whose paraphrase study provides a musical commentary on—and a compositional realization of—the expressive potential of Chopin's original. If Chopin "played as he composed," Godowsky composed as he played, for his paraphrase invites performers to explore a wide range of temporal fluctuations over Chopin's original étude, both within sections and across the piece as a whole. At every point of contact, the roles of Chopin, Godowsky, and Helps as pianists and composers are blurred. Helps's brilliant handling of both études, particularly his attention to the long line, exhibits the intuitions he brings to the piano as a composer, and his consecutive performance of the Chopin and Godowsky études allows us to consider the connection between Helps's embodiment and expression of their materials.

Studying these particular performances admits a second advantage due to the unique pedagogical pedigree of the performer. Robert Helps studied exclusively with the pianist Abby Whiteside (1881–1956) and is widely regarded as one of the finest exemplars

Table 11.2 Godowsky étude, polyrhythmic ratios, section durations, and registers

Section	Measures	Polyrhythmic Ratios	Duration_ *(in seconds)*	Register and position of the original melody
A^1	1–8 (8)	3:4	40	(starts on) G♯ 4 in the top voice
A^2	9–16 (8)	3:2:5	35	G♯ 5 in the top voice
A^3	17–24 (8)	9:6:4:3	45	G♯ 4 in an inner voice
A^4	25–32 (8)	6:8 & 9:8	40	G♯ 6 in the top voice
B	33–44 (12)	3:2:4	61	variable
A^5	45–52 (8)	2:3	65	G♯ 4 in an inner voice
Coda	53–56 (4)	2:3	39	variable

of her method. The synthesis of bodily and musical creativity admired by Chopin's contemporaries and enshrined in his compositions forms the basis of Whiteside's approach to the piano. Therefore, examining Whiteside's approach to piano playing provides a foundation for understanding some of the dynamic qualities of Helps's embodied performance.

Abby Whiteside developed her approach to piano playing based on her lifelong, methodical study of Chopin's études. The "Whiteside technique" offers an alternative to the finger-centric technique epitomized by the études of Czerny and Hanon.[6] Whiteside, like Chopin, embraced the unique qualities, strengths, and weaknesses of each finger while shifting the source and center of physical movement from the hand to the torso. This technical shift led Whiteside to develop her idea of a "fundamental rhythm," which refers to the coordinated rhythms of the body, organized around the movement of the torso, which carry the long line and emotional contour of the piece across a given span. Whiteside taught that all gestures originate in the torso and upper arm, and work from center to periphery as energy is transferred from the upper body through the shoulder, upper arm, forearm, wrist, and fingers, with each segment serving as a fulcrum for transferring energy. "Without a gracious and sensitive rhythm in the body," Whiteside says, "no one can project [a piece's] inner meaning and beauty. . . . This rhythm is the output of physical activity" (Whiteside 1997 [1955], 58). For her, a carefully developed bodily rhythm was necessary to musical expression, and without it, "faults in interpretation will remain no matter how much he has learned in intellectual concepts" (58).

Whiteside's method establishes this fundamental rhythm in her students through a practice method she called "outlining."[7] Using this method, the pianist begins by learning to play the sounds on the first downbeats of two- or four-measure groups (with odd-measure groups as required) and connecting these aural points of arrival and departure with the deeper movements of the torso across a span of time. For example, to learn Chopin's Étude in G♭ major, op. 10, no. 5 ("Black Key"), Whiteside would instruct the pianist to begin by playing through the piece at (or near) performance tempo and playing only the pitches on the downbeats of every other measure. During this first stage, the pianist focuses on the movement of the torso and upper arm as means of initiating other movements, shaping the long line, and coordinating gestures with sonic points of departure and arrival (i.e., the notes being played while outlining). Once the outline of the piece is learned, the pianist adds notes at the measure, half-measure, and beat levels, coordinating additional levels of gesture: the in/out movement of the upper arm and forearm, the rotation of the forearm, and finally the fingers to fill in the beats and direct energy into the keys. This intentional coordinating of gestures worked out through the process of outlining produces what Whiteside called a "blended action," in which each gesture flows from the fundamental rhythm (Whiteside 1997 [1955], 123).

The ingredients of any blended action involve distinct actions of the body involved at each level of the playing mechanism. These actions can be characterized by the fulcrum, the "steady, resistant force" (Whiteside 1997 [1955], 38) that allows more peripheral actions to receive and transfer energy from the higher part of the mechanism (the torso). These actions can also be defined by the quality and range of motion. Table 11.3 summarizes each of these actions, starting with the center, the torso, and moving to the periphery.

Table 11.3 Characteristics of components of the playing mechanism

Component	Fulcrum	Motion	Quality of action	Quality of ease/ release	Musical spans
Torso	Bench (and floor)	Movement around center of balance	Shift forward and around center/ upright	Shift backward and toward center/upright	Full phrases; large measure groups; gestures can be coordinated to shape the long line (both temporally and expressively)
Upper arm	Torso	In/out (shaping phrase and creating gestural subgroups); left/right (following position of hands at keyboard)	Powered by muscles in the chest and back, the upper arm initiates movement away from the keyboard (**"out"**) while tracking left and right along the keyboard.	In the reciprocal movement toward the keyboard (**"in"**), the upper arm is released and is drawn by gravity to return to its starting position near (or in line) with the torso.	One- to four-bar groups; short subphrases

(*continued*)

Table 11.3 Continued

Component	Fulcrum	Motion	Quality of action	Quality of ease/ release	Musical spans
Forearm/ hand	Upper arm	Down/up	Action occurs during the initiating **"down"** gesture (associated with the upper arm's **"out"**); the bicep relaxes, allowing the forearm weight to drop and slightly lowering the wrist	Ease/release occurs during the reciprocating **"up"** gesture (associated with the upper arm's **"in"**); the bicep gently contracts, slightly lifting the forearm and wrist (without interrupting weight transfer into the key).	One- to four-bar groups; short subphrases (in conjunction with the upper arm)
Forearm (rotary)	Upper arm	Pronation (inward rotation) and supination (outward rotation)	Starting from a stable position balanced over the left or right of the palm, the forearm rotates energetically to the opposite side.	When released, the forearm rebounds to the stable position.	Groups of 2–4 notes that alternate around the middle of the hand; trills; stepwise patterns that move across the hand
Fingers	Hand; metacarpo- phalangeal joint	Down (engaged) or suspended (disengaged)	Fingers slightly drop to transfer weight and energy from the arm through the hand and metacar- pophalangeal joint into the keys; once a key is played, the finger relaxes but remains engaged on the key until the point of release.	Fingers release; disengaged support in the hand and joints allows the keys to lift.	Single notes

As the source of movement throughout the playing mechanism, the torso initiates and carries the expressive contour of the long line, sustaining gestural continuity from the beginning of the piece to the first point of repose. The torso is well suited to this task given the large range of motion within which it can move as it shifts slightly away from center balance in any direction. Similarly, subtle movements of the torso can evoke rich expressive connotations and emotional states. The movement of the upper arm (between the shoulder and elbow) serves two purposes. Its left/right movements (parallel to the keyboard) serve the practical purpose of bringing the forearm and hand to the correct position on the keyboard; in extreme cases, the torso can assist by shifting the left or right. The upper arm also serves an important musical purpose: its out/in motion, in which the elbow moves along an elliptical path away from (out) and back toward (in) the keyboard, provides one of the most powerful ways of shaping measure groups, usually two or four bars long. Working in tandem with the out/in gestures of the upper arm, the forearm and wrist move through a down/up cycle. When power is applied through the "out" motion of the upper arm, it is transferred through a slightly lowered forearm and wrist. As the upper arm rebounds and moves in toward the keyboard, the forearm and wrist also relax and respond with a slightly elevated position (up). While it is most common for the "out/down" and "in/up" cycles to work together, it is possible for these gestures to have different periodicities.

The other two components of the playing mechanism—the rotary motion of the forearm and the fingers—are primarily concerned with transferring energy from the forearm, upper arm, and torso into the keys in the most efficient way possible. In musical examples with extended consecutive alternations between the left and right sides of the hand, the forearm can facilitate by using rotary motion (similar to turning a door handle). At the periphery of the playing mechanism, the fingers serve to transfer energy and arm weight into the key. Unlike approaches in which the fingers work like pistons, dutifully playing all the notes, the primary job of the fingers is either to engage, in which the finger is supported at the metacarpophalangeal joint (the large knuckle between the fingers and palm), or to slightly elevate so as to avoid depressing keys. Aside from this slight elevation, the fingers themselves rarely need to lift after playing a key, as the actions of the keys can lift the fingers.

From this brief foray into Whiteside's technique, we may draw five important insights into the function of the pianist's body that have significant implications for how we understand temporal fluctuations in performance. The first is that the bodily gestures do not necessarily follow the internal groupings (phrases or subphrases) or other events at the surface of the piece. Certainly, a pianist's physical gestures may shape themselves to the musical surface, as they presumably would if outlining was used to learn a piece with regular two- or four-bar groupings. Yet highly expressive moments can also be created by using gestures that do not directly follow or project surface-level features. Pianists can coordinate gestures at each level (torso, upper arm, forearm, and so on) in a manner that can follow or resist aspects of musical structure. For instance, an ascending musical pattern might at one point be articulated using an energetic acceleration involving an inward-leaning torso and active arm gestures, and elsewhere be shaped quite differently by relaxing the upper body and arm. That Chopin may himself have placed different parts of the playing mechanism in expressive conflict is suggested by the unusual phrase markings seen most commonly in the Nocturnes (e.g., Op. 37, nos. 1 and 2; Op. 55, no. 1; Op. 62, no. 1), in which the melodic phrase shape is misaligned with harmonic resolution, left-hand figurations and phrase markings, and/or regular hypermetrical groupings.[8]

Second, a blended, well-coordinated technique can influence the expressive temporal relationships in a performance at multiple time-scales (i.e., beat, measure, phrase) and gestural strata (i.e., torso, arm, hand). While one part of the body such as the forearm might energetically push a melodic segment forward, increasing tension, another part may be involved in a pattern of release at a larger time-scale. Alternatively, one component of the mechanism can interrupt or take over a process being guided by another; or, the dominant component can suddenly disengage, leaving it to other higher or lower components to carry the line to its conclusion. At either end of the spectrum of blended actions, we might imagine situations in which each bodily gesture approaches a state of repose or energetic engagement. But at most moments within different pieces of music, pianists are likely to fall somewhere between these extremes. This approach to the body allows pianists and other performers to see pieces as sites for exploring unique configurations of gestural relationships, some of which may be unique to each individual performer.

The process of developing a blended action guided by a fundamental rhythm leads to a third insight: that each part of the body is characterized by different qualities of tension and release, or action and ease, to follow Whiteside. The action of the torso is usually slow, and is initiated by slightly moving the center of balance away from a vertical position: forward, backward, or side-to-side. The movement of the torso in shaping a phrase can itself be a highly sculpted and nuanced activity, and according to Whiteside, the entire mood of the piece finds its source in the gentle, powerful action of the torso. The upper arm functions primarily to transfer the torso's energy and the arm's weight toward the keys. The upper arm primarily moves "in" (toward the piano), and "out" (away from the piano), creating the only horizontal action. While the upper arm's movement is mostly visible through the position of the elbow, its power comes from the weight that is supported by its connection to the torso at the shoulder. The forearm's movement is "up" and "down" according to the slight rise and fall of the wrist. Together, the upper arm and forearm work to create combined gestures of "down/out" and "up/in." A third arm action is the rotary action, which centers on the left-right or right-left rotation of the wrists. The fingers tend to release tension when a key is released, rather than struck. The primary responsibility of fingers and wrist is to transfer energy from body's center (torso, upper arm, forearm) into the key, although they are also capable of acting independently.[9] Given the enormous variety of ways that different elements of a blended action can be emphasized from one moment to another, a wide variety of expressive nuances can be produced using this technique.

Fourth, in the process of developing a blended action, outlining creates gestural spans at every level of the playing action that have a distinctly "passing" quality, the gestural equivalent of a passing tone. In Whiteside's parlance, once a gesture was initiated, the pianist "tucks in" lower level musical figures and physical gestures that occur on the way to the point of arrival. In other words, the starting point for Whiteside's technique is to outline a space that is subsequently filled by lower-level gestures. Usually, gestures at a particular level, once initiated, tend to move through the space they outline in a consistent, smooth, organic, natural trajectory toward their end, that musical time point in which the following gesture commences. The larger, longer, and higher-level gestures of the torso and upper arm tend to subsume lower-level ones in the process of developing a blended action. The longer a higher-level gesture is, the more music it can subsume, and the more powerful it is, the less likely a lower-level gesture can interrupt its natural progression. While it is possible for interruptions to occur, in which a lower-level gesture takes command over the timing and articulation of a particular musical moment against the gestural shaping of a higher-level gesture, such moments are quite exceptional and should be deployed with caution. While such gestural conflicts may charge a performance with moments of high expressive intensity and gestural tension, frequently, needlessly or carelessly allowing lower-level playing to take a role more appropriately executed by the torso and upper arm can negatively impact the performance by breaking the tension of the longer line and bringing a pedantic level of hyperfocus to each surface moment of the music. One need only consider a beginning piano student playing Mozart's Sonata in C major, K. 545 and tediously plunking out one melodic tone after the next to understand the importance of higher-gestures subsuming lower ones. Generally, the greater the distance between gestural levels (fingers to torso being at the farthest remove), the less of an impact a gestural aberration at a lower level can make on the flow of the higher-level gesture. Just as melodic diminutions occur in different varieties and are subject to different qualities of expressive articulation, so to the passing span within an outlined space can be expressively charged or gesturally shaped. Once a gesture has been initiated, it can float, glide, slide, charge, lilt, lunge, press, arc, swoop, or swing its way toward its endpoint. Developing a broad gestural vocabulary through physical movement is central to Alexandra Pierce's teaching (Pierce 2007). Building on Pierce, I propose that highly trained musicians can blend gestures whose outlined spaces at different levels have distinct characteristics, both complementary and oppositional. For instance, the torso can outline a long-ranging span and gently float between onset and terminus, while a lower-level gesture such as the forearm can press through a particular melodic gesture. In short, blended gestures coordinated by a fundamental rhythm enable musicians to create and explore a wide range of expressive possibilities in performance.

Finally, while the blended actions of a musician may be oriented toward a single expressive goal, the blended action itself is not indivisible or uniform. For that matter, the nuances of musical expression often far surpass our ability to articulate them in words, precisely because of the multiple, layered, non-linear relationships between different gestural levels. Being divisible, blended gestures have a complex, yet identifiable structure whose components interact in performance and create an expressive overlay on musical structure. While gestures surely serve a critical role in illuminating and projecting musical structure, musical structures in turn provide opportunities to explore and develop nuanced musical movements in the body. The analysis that follows seeks to elucidate the dance of body and music that occurs in two performances by one of Whiteside's most noted protégés.

A Gracious Rhythm in Action: Helps Performing Chopin

Following Whiteside's dictum, let us move from center to periphery as we consider embodied rhythmic expression in Helps's performance of Chopin's étude. Given that the movements of the torso and upper arm directly shape the longest spans of music, we first consider the overall structure of the A section of Chopin's étude (see Figure 11.1). This section follows a sentential design, with a basic idea (mm. 1–4), its repetition (mm. 5–8), fragmentation and acceleration (mm. 9–12), and cadential material (13–16). Measures 9–12 can themselves be broken into two 2-measure groups, in which

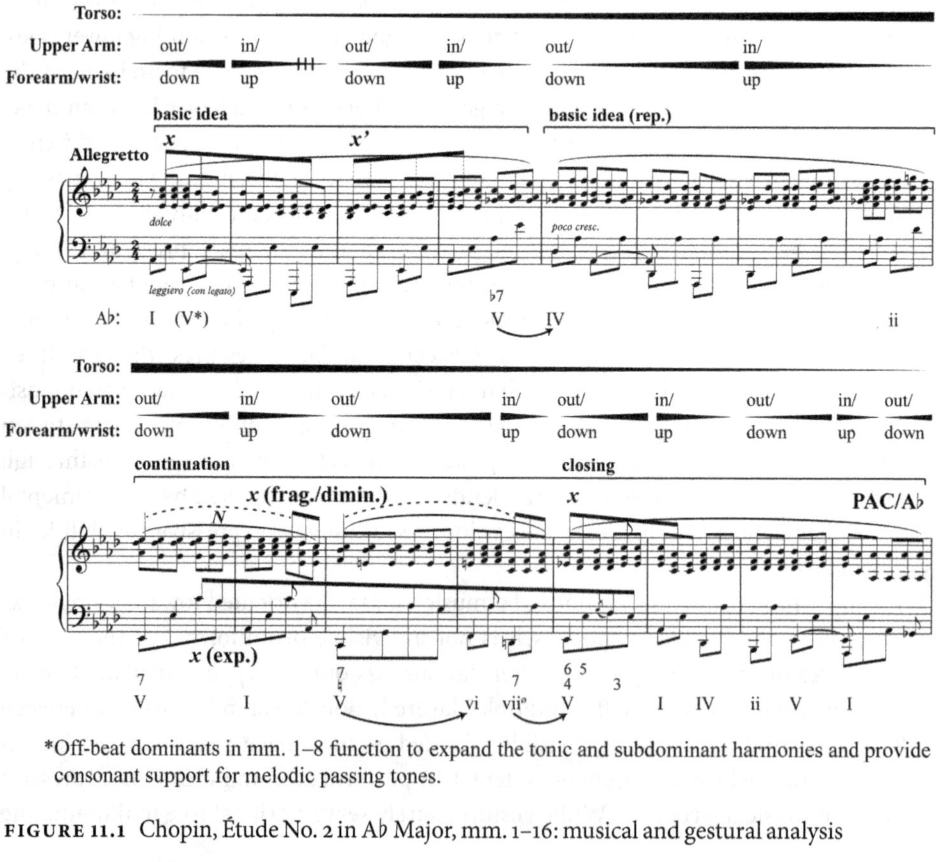

*Off-beat dominants in mm. 1–8 function to expand the tonic and subdominant harmonies and provide consonant support for melodic passing tones.

FIGURE 11.1 Chopin, Étude No. 2 in A♭ Major, mm. 1–16: musical and gestural analysis

a temporally augmented upper-neighbor figuration is followed by an accelerated stepwise third descent in eighth-note triplets that recalls the opening melodic gesture. The fragmented basic idea, shortened measure groups, and faster, dominant-focused harmonic rhythm propel the music through the phrase continuation toward the cadence. And yet, within these same measures (mm. 9–12) are two musical elements that resist the melodic acceleration. In the left hand, there emerges in m. 9 hints of a melodic voice, whose slower eighth-note stepwise descent gently unfolds the same melodic sixth (C4–E♭3) first heard in mm. 1–2 (C5–E♭4), but now on second beats over a six-measure span, working against the four-bar groups in the right hand and resisting its forward drive, and drawing attention back to the opening basic idea itself.

The lines above the staff in Figure 11.1 indicate how I might implement a Whitesidian approach to perform this phrase in a manner that emulates Helps's performance.[10] For the first half of the phrase, the torso initiates movement and energetically builds forward-directed momentum. As this momentum increases (mm. 5–8), the torso seems to override some of the surface-level liberties taken by the forearm-wrist and hand/fingers, whose temporary resistance of forward motion is symbolized by the three vertical lines above m. 2. Put differently, just after the torso initiates movement, more peripheral parts of the gesture seem to have freedom to attend to surface-level details before being swept into the forward momentum of the more central gestures. In Helps's performance, the subtle differences in timing between mm. 1–4 and 5–9 also suggest a different engagement of the arm, which shifts from two-measure cycles to a four-measure gesture. Consequently, the third measure of the basic idea shifts from being a gestural downbeat (m. 3) to upbeat (m. 7). At mm. 7–8, the gestural cycles of torso and arm are at a state of maximum tension, with the torso reaching a state of highest energy while the arm reaches the end of its two-measure rebound.

Helps's performance of mm. 9–10 and 11–12 again features temporal differences. With the torso now in a period of energy release, the arm returns to its two-measure cycles and becomes the dominant gesture. The arm extends its energy expenditure in mm. 11–12 before finally being subsumed by the torso, which draws the arms into a more coordinated, lower-energy approach to the cadence. At the cadence (m. 16), the expansion of the right hand to the A^4 octaves and the descending bass line point the music forward to the middle section in E major. The sectional downbeat in m. 17 seems to be anticipated gesturally by re-engaging the arms in an out/down gesture before downbeat's arrival.

At the periphery of the playing mechanism, the fingers and hands have the dual roles of transferring activity of the torso and arms into the keys and shaping surface-level figurations. The intricate interplay of suspensions and other voice-leading figurations at the surface of Chopin's étude also could represent an embodied action applied to a simpler musical structure. Schenker notes in *The Art of Performance* that trills, suspensions, and arpeggios "belong to those subtle means music uses to break the rigidity of motion and rhythm" (Schenker 2000: 57), and he occasionally describes musical figures as resulting from particular gestures, such as "lingering" or "holding" (Schenker 2000: 53–54, 57). Figure 11.2 shows the derivation of Chopin's musical surface through successive stages of embodied rhythmic play. While Figure 11.2 shows a

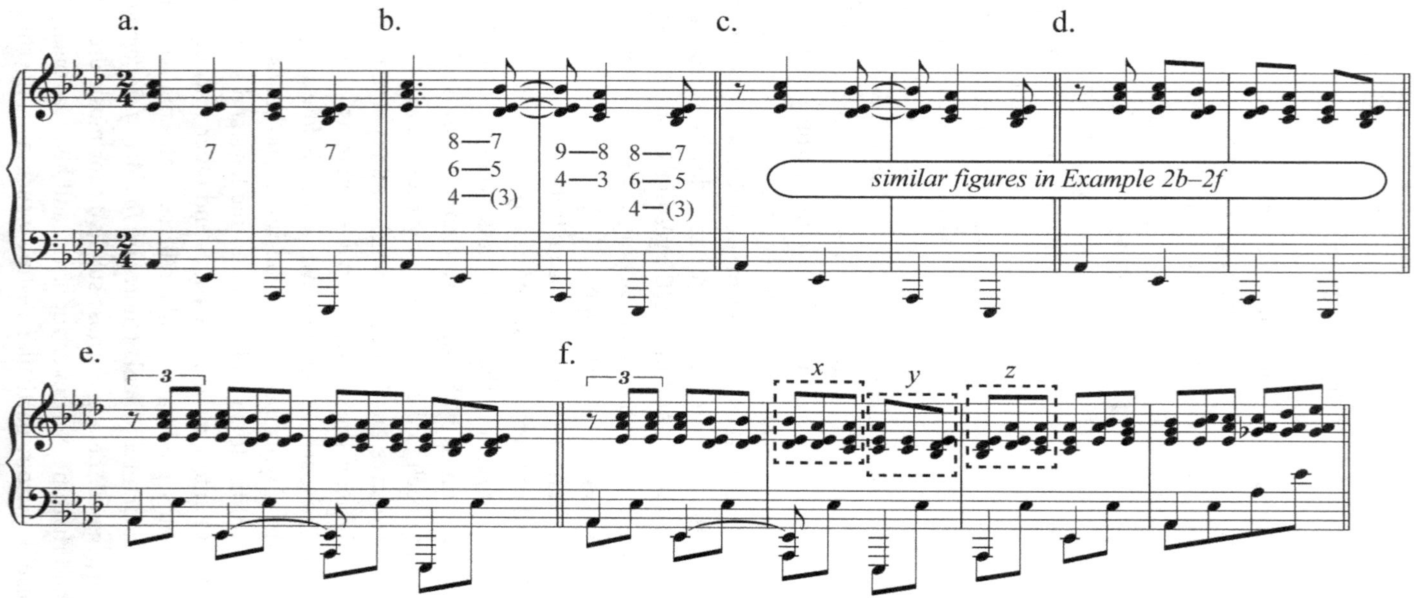

FIGURE 11.2 Chopin Étude, mm. 1–2ff: rhythmic derivation

rhythmic reduction, it also illustrates a process of rhythmic *generation* through successive embodied interactions with the source material in Figure 11.2a. In Figure 11.2b, the performer holds the first chord, delaying its move through the passing V^7 chord and initiating a chain of suspensions. This simple embodied action creates an array of new contrapuntal relationships along with a new composite rhythm (eighth notes), and it displaces the hands, such that the right hand lags behind the left, only to catch up at m. 17. In Figure 11.2c, the right hand is further offset from the left by delaying its entrance, creating a downbeat rest.

The right hand's rhythm is further activated in Figure 11.2d, although the straight eighths do not allow much flexibility to the movement of the individual upper voices, which are still constrained to move together on the off-beats. As if to break this new deadlock, the pianist introduces triplets into the right hand, while the left retains the original eighth-note composite rhythm. The possibility of triplets allows two possible points of dissonance resolution, the second or third triplet eighth of each beat; Figure 11.2e follows mm. 1–2 of Chopin's setting by showing the earlier resolution, projecting a more forward-oriented expressive character. Figure 11.2f expands the scope of reference to the first four measures of Chopin's étude, in which the individual voices in the right hand are allowed to resolve at different locations, opening even more contrapuntal possibilities (not shown) while requiring a new level of attention from the performer, who now tracks the distinct movements of three unsynchronized voices. As Figure 11.2f shows, the staggered resolutions create a general shift from early- to late-resolving voices. In m. 1, all voices resolve on the second triplet eighth; in the enclosed box labeled "*x*," the top voice moves first while the lower two resolve on the third eighth. In box "*y*," all three voices seem to wait until the third eighth note to change. This interplay carries through mm. 3–4, where staggered resolutions and a repeating appoggiatura figuration create a dynamic surface-level tension between forward movement and lingering delay.

Helps's performance responds to these tensions through the flexible timing he employs in mm. 1–4. It is also significant that he does not respond with like flexibility elsewhere when these patterns repeat. At the return of section A, Helps lingers only slightly in mm. 41–44, which repeat the opening four measures, and in mm. 5–8 and 45–48, which transpose these musical figurations to the subdominant, the surface-level activities seem to be subsumed into the large sweep of upper arm and torso. It is not that the gestural impulses disappear entirely; rather, they are present implicitly through the pianist's gestures and through ever more subtle, and at times barely perceptible temporal nuances.

These tensions between surface melodic structure and gestures (both central and peripheral) leave significant room for the pianist to shape time in a way that responds to the musical structure while creating something quite new. Looking closely at Helps's performance, it is immediately striking that the tempo fluctuations do not consistently shape the four- and two-bar phrase groupings identified earlier. Considering only changes in tempo (horizontal movements by the Worm), it is clear to both ear and eye that temporal shaping only loosely conforms to musical patterns bracketed in Figure 11.1. The dynamic interaction between phrase structure, measure groupings, motivic development, and harmonic rhythm sets the stage for pianists to explore an equally dynamic gestural variability as they coordinate elements of the playing mechanism. At the same time, Chopin's étude itself composes out an expressive performance of underlying musical compositions, even one as simple as the chord progression shown in Figure 11.2b. One can easily imagine a pianist playing the music as written in Figure 11.2b using physical gestures associated with pensive delay, anxious acceleration, or a range of other feeling-states. Asked to express these feelings in sound, a creative musician might well add a melody to the progression, or begin altering the progression itself to reveal the musical content of this added expressive layer. In the following section, I propose that Godowsky's paraphrase study does precisely that: Godowsky represents several possible expressive performances of Chopin's étude through the process of layering new rhythms and physical gestures over the original composition.

A Gracious Rhythm in Action: Helps Performing Godowsky (Performing Chopin)

Just as Chopin's étude may be read as a compositional realization of an expressive performance of an underlying progression, Godowsky's paraphrase realizes the expressive potential of performing Chopin's original, a creative process that generates music that goes well beyond that notated in Chopin's score. If Chopin's score provides the script for a range of dramatic performances, Godowsky's étude takes those performances as the basis for new musical exploration. Further still, Helps's performance of Godowsky creates a rich new aesthetic layer that transforms and expands the ways in which we hear and experience each successively deeper layer of musical elaboration. Through its introduction of new gestures and countermelodies, Godowsky's paraphrase étude explores a range of expressive pathways through Chopin's original.

While a full account of all the compositional details of Godowsky's paraphrase and Helps's performance is beyond the scope of this chapter, it will suffice to focus on two aspects: the array of layers and gestures explored in the opening four A sections and the final, reflexive A section that concludes the étude. The fourfold repetition of section A draws attention to the differences between them, and thus to the impact produced by the unique gestures that distinguish them. The repetitions further draw in the listener, who may mentally trace the original melody, even when it gets lost amid the layers of new countermelodies. Through their unique combinations of gestural layers and polyrhythmic relationships, the opening four A sections demonstrate an expansive array of implicit musical possibilities latent in Chopin's original.

Figure 11.3 provides two-measure incipits of the opening four iterations of the étude's A section as well as the new figuration introduced in the fourth iteration

FIGURE 11.3 Godowsky, Étude, excerpts

(Figure 11.3e). Each iteration includes the original melody and follows the same harmonic rhythm and pattern as Chopin's original. From the outset, Godowsky loosens the relationship between harmonic structure and linear embellishment, releasing the inner voices from the suspension and appoggiatura figurations in the Chopin étude to explore a more chromatic inner voice pathway. In the second iteration (Figure 11.3b), the closed right-hand positions of the first are opened, requiring rolled chords in the right hand, with the left hand taking some notes of the inner voices. The five-against-three polyrhythm is made even more difficult to perform given the difficulty of coordinating the rolled chords, with the melody perched on top, with the temporal spaces outlined by the left-hand gestures. The third iteration (Figure 11.3c) allows for the greatest expressive fluctuations, which are considered in greater detail in what follows. Here, the left hand is tasked with articulating the melody, bass, and three-against-two polyrhythm of the original, while the chromatic inner voice from the opening is set free to improvise in the highest voices. In the fourth iteration, the right hand features two rhythmic figurations, both of which employ a rotary gesture. In Figure 11.3d, the rotary action is divided evenly over a duple division of the eighth note into two sixteenths. In Figure 11.3e, the action is split over a triple division, requiring a faster pronating gesture that results in a more elliptical movement of the forearm.

The rhythmic structure and musical texture of Chopin's étude is saturated with new material created with successively smaller note values. This saturation demands a slower tempo so that it is possible for the performer to articulate new musical details, a fact that reinforces Schenker's insight that musical elaborations "break the rigidity" of time. Schenker's insight could be restated in Whitesidian terms by noting the principal that, as the peripheral physical gestures of the performer become more complex, they tend to be coordinated and subsumed by more central gestures over longer temporal spaces. We have already seen this to be the case in Chopin's étude. This temporal prolongation is implicit in the move from Figure 11.2a to 2f; one could conceivable play the first progression rather quickly, but filling in these spaces with suspensions and rhythmic interplay takes significantly more time. Similarly, Helps begins the Godowsky at a significantly slower tempo than the Chopin. And yet, the increased density of material masks the effect of the slower tempo; from the introduction of a new chromatic line in its first measure, there is also a quickening of movement at the musical surface of the Godowsky. The surface rhythmic saturation of the Godowsky also creates a condition in which the central and peripheral gestures may find themselves in conflict. On the one hand, taking a Whiteside approach, pianists would subsume difficult surface-level polyrhythmic elaborations within broader and more centralized gestures of the upper arm and torso. In Helps's performance, despite the increase in rhythmic and melodic activity in mm. 1–8, the temporal fluctuations are less pronounced than those he employs in the Chopin. While fluctuations are present, they seem to be guided by a background gesture in the torso that unfolds over a longer time span than was employed in the Chopin performance. On the other hand, the new melodic segments made possible by the rhythmic saturation invite the peripheral mechanisms (fingers and hands) to take even more time to draw out inner lines or linger on the chromatic turns of the right hand's inner voice. Given this conflict, it is notable that Helps generally employs a more forward-directed approach to timing in the opening A sections, in which surface-level effects employed in the Chopin are largely subsumed by the sweep of the torso and upper arms in the Godowsky. Only section A^3 (mm. 17–24), where the inner chromatic line is set free in the highest voice, does the right hand's shaping exert some momentary control over the longer-range gestural trajectory of the torso. In this section, the left hand's sweeping arpeggiations are absent and the placing of the short values has been inverted in the texture, moving from bottom to top. Here, the left hand takes the 2:3 rhythm of the original while the right hand overlays its elaborate descant.

In contrast with the opening A sections, in which the torso's gestures and the sweeping arpeggiations of the left arm generally override or diminish any dramatic surface-level temporal fluctuations, the temporal articulation of the final A section results from a much more balanced interplay between gestural levels. As Table 11.2 indicates, the duration of the concluding A section is 20–30 seconds longer than any of the opening four A sections. In this section, the guiding fundamental rhythm of the torso enters a stage of overall ease, a gestural contrast to the long-range forward-directed energy that characterized the torso in the opening. The set-up for this final A section begins in earnest at m. 43 (although the deceleration takes hold as early as m. 41), during which all levels of the playing mechanism, from the torso out, seem to exhaust their energetic output and gently release into the cadence that marks the return of section A (m. 45). From this point forward, the torso, once again freed from the role of coordinating expansive left-hand arpeggiations, only gently guides the temporal shaping of this section, allowing the arms, hands, and fingers to take a more active role in shaping time.

The expressive markings on the score throughout the final A section invite a dynamic, nuanced interplay between gestural levels. At the beginning of this section, shown in Figure 11.4, the left hand once again assumes the role of playing both the melody and bass in 2:3 polyrhythm. By recalling the texture from section A^3 and the rhythmic profile of Chopin's original, this section evokes the quality of reminiscence. In contrast with Godowsky's section A^3, the left hand's elaborations are less elaborate and are much like those found in Chopin's original. If the highly ornate rendering of Chopin's original inner voices in the Godowsky composes out an energetic expressive rendering of the original étude, the emptying of these elaborations in the left hand beginning at m. 45 in Godowsky's paraphrase suggests a more relaxed approach. The decrescendo Godowsky assigns the original melody in the left hand in m. 49 suggests an inversion of the torso's energetic shaping of the opening melody toward an arrival at the second measure. Now, the melody approaches the middle point of the opening subphrase with a sustained delicacy. The torso still is a guiding force directing the fundamental rhythm and actions of other gestures, but one floats toward points of arrivals rather than leading to them with energetic arcs. The left hand's pattern of gentle release followed by build is repeated in mm. 47–48.

Beginning in m. 45, the left hand also begins a process of emptying the contrapuntal interplay found in Chopin's original voices (see Figure 11.2), a liquidation that is brought to completion in mm. 49–52 through the placement of rests on the beat. This rendering of Chopin's original highlights the compositional value of a durational symbol that is likely overlooked by many pianists: the downbeat rest at the beginning of Chopin's étude. As we already explored, this rest—the result of an embodied gesture—creates the conditions of possibility that set the rhythmic and contrapuntal process of Chopin's étude in motion. But the possibilities that inhere in that opening rest are

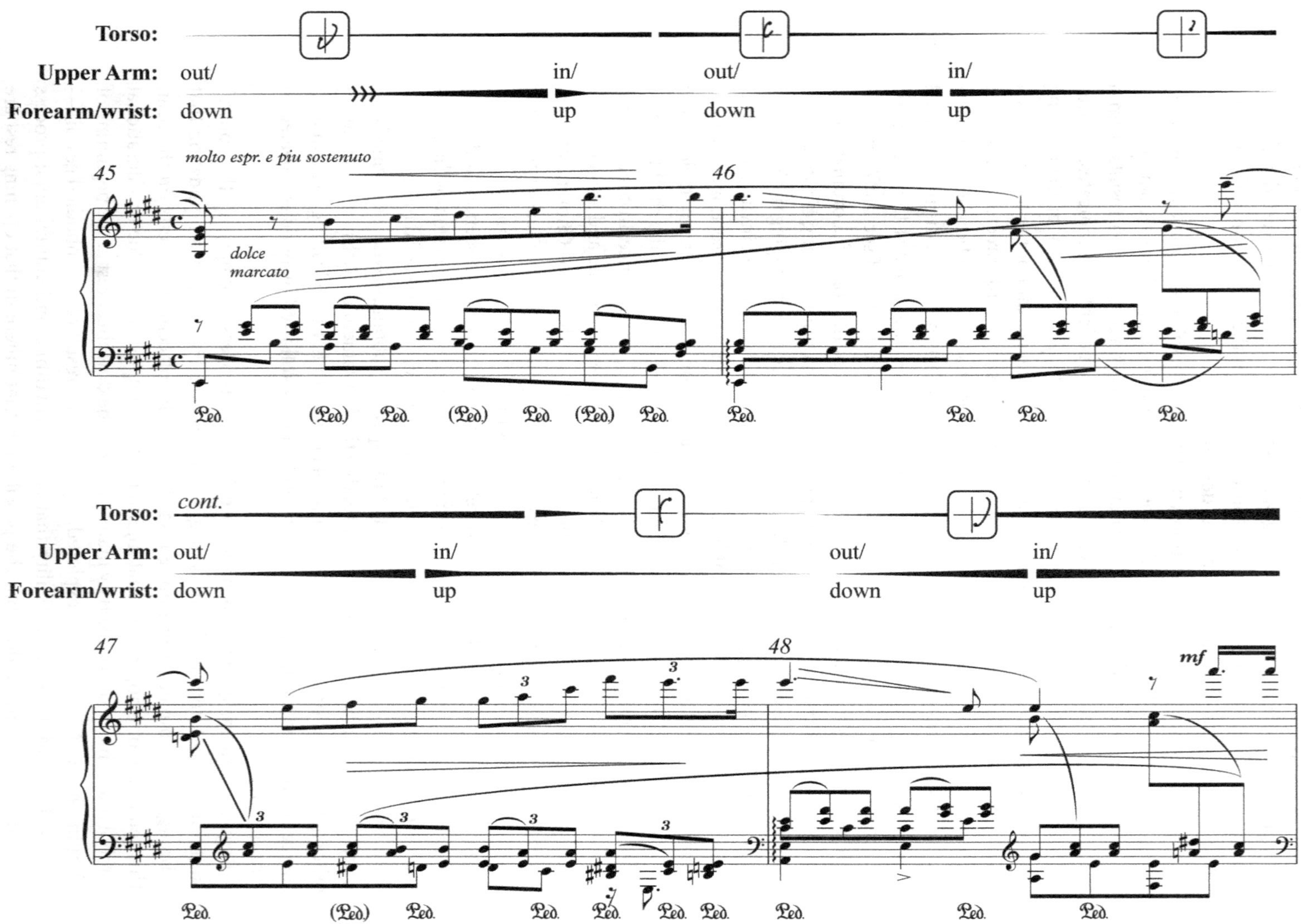

FIGURE 11.4 Godowsky Étude, mm. 45–53

FIGURE 11.4 (CONT.) Godowsky Étude, mm. 45–53

further explored in the final section of Godowsky's study, in which the right hand's descant also starts off beat in mm. 45 and 47 and unfolds in contrary motion with the original tune and bass line.

As the torso relinquishes its forward-directed energy in m. 45, Godowsky indicates that the melody increases in energy by applying a crescendo. The ascending gesture, in contrary motion with the melody and bass in the left hand, and its increasing volume suggest forward-directed motion to the B5. This moment, like many similar ones that follow in this section, places different levels of the playing mechanism in poignant conflict. Just as the torso and left arm and hand are releasing energy, like a musical exhalation, another element, the right arm, hand, and fingers push forward and higher. (The forward arrows in m. 45 indicate that the fingers even play a role in pushing this music suddenly forward.) This oppositional gestural relationship repeats in mm. 47–48 and again in mm. 49–50. Leading into m. 49, the descant reaches to the highest register of the piece, comparable to that achieved in section A^4, and for the remainder of the piece, it works in alignment with the torso, leading to the tender arrival on the sequential repetition in m. 50 and guiding the fundamental rhythm toward the tonic arrival provided by Godowsky's recomposed ending. In mm. 51–52, having exhausted the generative possibilities made possible by the rich gestural interplay, Chopin's original melody itself begins to disappear, allowing Godowsky's countermelody to guide the musical line to the cadence in m. 53. After this cadence, Godowsky provides an entirely new closing, one that again recalls elements of Chopin's original melody, some of the chromatic elements introduced by Godowsky, and a segment of Godowsky's turbulent middle section.

In Figure 11.4, I have included new symbols indicating the movement of the torso around the center/upright (symbolized by the center of the axes). These symbols indicate the manner in which I am inclined to move my torso while playing this passage. Throughout, the torso is kept upright in a posture that, from the hipbones to the head, should feel similar to standing. While each pianist may feel movements forward and backward and left/right of center differently, generally speaking, a forward movement is associated with increased energy, intensity, and forward direction, while a movement backward or toward center can either be used to carry the phrase (energizing and connecting more peripheral gestures) or to release gestural and musical energy. In order not to obscure the starting point of the torso's gestures, I have placed the torso's curve for each gesture after the start of the gesture's line. The movement symbolized should be understood as beginning at the start of that gesture and unfolding over its span (which is variable).

The four interconnected layers (torso, upper arm, forearm/hand, and fingers) represented in this analysis indicate the gestural dynamics at work in Helps's performance and give insight into the ways in which physical gestures influence timing. As the section draws to a close, m. 52 provides one more glimpse into the impact of gestures on timing. This measure begins with three pairs of eighth notes, but Helps stretches the third considerably longer than the others (see 4:32–4:38 on the recording). Given the cadential dominant and pitch repetition that follows on beat four, the final pair of eighth notes on beat three (C♯–B) is taken with a small in/up gesture in the arm, giving those notes a gestural weight similar to the first melodic notes in the measure. A similar gestural pattern repeats on the last beat, with both instances being guided to the arrival on m. 53 by the gentle, continuous gesture in the torso. I invite the reader to experiment emulating the physical gestures associated with this moment and others described above and in Figure 11.4. Both pianists and nonpianists, trained musicians and nonspecialists, can develop a sensitivity to the interaction between physical gestures and musical timing in performance. Indeed, making this embodied knowledge of rhythmic production more accessible has been the goal of this chapter.

Conclusion: Measuring, Representing, and Interpreting Embodied Rhythm

Through the preceding discussion of Whiteside's technical approach to piano playing, and my speculative description of embodied rhythms in Helps's performances, I have attempted to establish and demonstrate both the significant, generative role the body plays in musical performance and the complexity and multilayered quality of the bodily gestures involved. From this viewpoint, I conclude this chapter by addressing some matters related to the ways in which embodied gestures in performance have been represented, measured, and interpreted so as to address and move beyond some of the difficulties scholars have faced when measuring and interpreting performance and listener responses.

The embodied approach to rhythm illustrated in this chapter reveals a significant drawback of the reductionistic tendency to represent fluctuations of tempo as "faster" or "slower." For example, the Performance Worm presents tempo and tempo modulations as points and motion along a single axis. While the animated right and left movements along the x-axis are technologically innovative and reveal remarkable qualities of performances, they nevertheless continue the long-standing use of right and left arrows by performers and theorists (Schenker 2000; Rothstein 2005) to indicate acceleration (right arrow) and deceleration (left arrow) of tempo (see Figure 11.5). These one-dimensional representations of time reify the problematic notion that expressive tempo modifications directly reflect the performer's mental representation of the piece, rather than a more complex process that involves the performer's engagement of the body.

FIGURE 11.5 An example of tempo modification in Beethoven, Trio in D major, Op. 70, no. 1, I, mm. 1–6. From Schenker (2000, 54).

Further, this simplistic representation of tempo modification has served as a foundation for a wide range of analytical observations, prescriptive principles for rubato in performance, and interpretive metaphors (Schenker 2000; Rothstein 2005; and Larson 2012). In particular, Larson's description of musical forces (gravity, magnetism, and inertia) as a conceptual metaphor for analysis could be enriched in its application by exploring the ways these forces are embodied in different combinations simultaneously at different levels of the playing mechanism (Larson 1997–1998, 2012).

The inseparable connection between physical gestures and timing in performance also raises challenging questions about how analysts track the temporal dimension of performances. Given that recordings afford analysts a stable musical object and the ability to compare numerous renderings of the same piece, timing is often measured without regard to performers' bodies. While quantitative studies have raised and addressed numerous important questions about the peculiarities of individual performers and performances, a fuller account that addresses why and how fluctuations of time are generated would seem to demand that more attention be given and new technologies be developed to studying the complex movements of performers along with quantitative measurements of timing. Similarly, studies that track listener responses to timing might well explore a range of embodied responses to performance, including the ways in which physical movements of fingers, hands, wrist, arms, and torso respond to timing fluctuations. While listener-response data produced by measuring gestures generated at the most peripheral part of the listener's body (e.g., tapping while listening) may be experimentally straightforward, such designs seem to be founded on assumptions about rhythmic embodiment that this chapter calls into question. If the rhythmic activity of a performance is directed from the torso and modified through dynamic coordination of bodily gestures at specific instruments, then more work is needed to explore the equally complex ways in which embodied responses to rhythm can be generated and measured in listeners' bodies.

If performances express embodied rhythmic activities and structures, then finding new ways to measure and model the relationship between timing and physical gestures in performance may provide new frameworks for interpreting these events and relating them to structural elements found on the score. Such studies, if performed across a wide variety of instruments and styles, could lead to rich dialogue between music theory and performance and begin to answer the question posed by Widmer and Goebl (2009) regarding the difficulty of measuring, quantifying, and modeling performers' minds and bodies as they communicate expressively in performance. Ironically, one of the most musically satisfying answers to this question seems to lie embedded in compositions themselves. By reading Godowsky's paraphrase study as a realization—a composing out—of the expressive potential of Chopin's étude, we are reminded that musical scores not only provide the script for endlessly complex and differentiated renderings of pieces but also encode embodied, expressive renderings of underlying music. Ultimately, however sophisticated our technologies and tools of measurement may become, the insights we gain have perhaps their greatest payoff as they lead us to experience anew the musical possibilities that inhere within the physical abilities and limitations (Straus 2011) of our own bodies as performers and listeners.

Notes

1. Robert Helps performed the Chopin and Godowsky études on a solo piano recital that also featured works by Helps, Poulenc, Del Tredici, Mendelssohn (arr. Helps), and Ireland (arr. Helps). The recital took place at the University of South Florida School of Music Recital Hall on October 23, 1997. Godowsky's fifty-three studies (Godowsky, n.d.) present extremely virtuosic renditions of many of Chopin's études, often by increasing rhythmic complexity, adding new countermelodies, combining études, inverting textures, and sharing material between the hands.
2. Instead of relying on the algorithms for identifying tempo described by Dixon, Goebl, and Widmer (2002), I manually set the time points for each beat onset using BeatRoot, a beat tracking and visualization program developed by Simon Dixon, accessible at: http://www.eecs.qmul.ac.uk/~simond/beatroot/.
3. For more resources on gesture as a basis for musical meaning, see Hatten (2004), Guck (2005), Leman (2008), Godøy and Leman (2010), and Geeves and Sutton (2014).
4. Wanderley and Vines (2006) explore the role and perception of ancillary gestures in performances on the clarinet. Their work builds on the division of gesture made in Delalande (1988) into (1) effective gestures, those that produce sound; (2) accompanist (or ancillary) gestures, which are a performer's expressive body movements; and (3) figurative gestures, those that are inferred by the audience but may not correspond to a visible gesture by the performer. Delalande's analysis is based on a performance of Bach's *Art of the Fugue* by Glenn Gould, yet despite the enormous differences between performer and repertoire in her study and my own, my appropriation of Whiteside's pedagogical approach to explaining timing in performances of Chopin and Godowsky counters Delalande's description of various movements of the chest and torso as ancillary. For more on the expressive quality of effective and ancillary gestures, see Davidson (1993).
5. As a student of Helps, I regularly observed Helps's piano playing up close. I recall watching Helps work through passages of the Godowsky in his studio and attending the concert at which the recordings used in this chapter were made. Helps's technical approach to the piano was one of ease and efficiency. Helps's playing rarely (if ever) included extraneous or theatrical physical movements of any type. Rather, his physical demeanor at the piano was poised, balanced, and often quite still. While the movements of the fingers and arms were most visible, movements of the torso guiding these gestures were nearly imperceptible to the eye. And yet, from the first lesson on a piece to its final performance, Helps often directed his attention to the torso, often by firmly nudging it while I played mid-phrase, and always with surprisingly effective musical results.
6. Abby Whiteside's writings were originally published in two volumes: *Indispensables of Piano Playing* (published in 1955) and *Mastering the Chopin Études and Other Essays* (published posthumously in 1969). These volumes have been published together in Whiteside 1997. In the 1997 republication, the original volumes receive their own

distinctive sets of page numbers. All citations to the 1997 volume will therefore cite the publication year of the original so that the citation's source is clear.

7. For more information about the process of outlining, see Whiteside's application of the approach to learning Chopin's Étude in C Major, Op. 10, no. 7 (Whiteside 1997 [1969], 49–66).
8. I thank Kevin Korsyn for first pointing out Chopin's unusual phrase markings to me.
9. While Whiteside develops these concepts throughout Whiteside (1997 [1955; 1969]), see in particular pages 30–75 of the 1955 monograph.
10. The multilayered representations of the structure and relationship of embodied gestures in Figures 11.1 and 11.4 (see the General comment above) take their inspiration in part from Judy Lochhead's (2006) descriptive maps of music that cannot be represented in score notation.

References

Benadon, Fernando, and Damián H. Zanette. 2015. "A Corpus Analysis of Rubato in Bach's C Major Prelude, WTC I." *Music Performance Research* 7: 1–26.

Berger, Anna Maria Busse. 2002. "The Evolution of Rhythmic Notation." In *The Cambridge History of Western Music Theory*, edited by Thomas Christensen, 628–56. Cambridge: Cambridge University Press.

Caplin, William E. 2002. "Theories of Musical Rhythm in the Eighteenth and Nineteenth Centuries." In *The Cambridge History of Western Music Theory*, edited by Thomas Christensen, 657–94. Cambridge: Cambridge University Press.

Chopin, Frédéric. 1880. *Trois nouvelles études*. Edited by Karl Klindworth. In *Oeuvres complètes de Frédéric Chopin*, Band 1: *Études*. Berlin: Bote & Bock.

Clynes, Manfred. 1982. *Music, Mind, and Brain: The Neuro-Psychology of Music*. New York: Plenum Press.

Cook, Nicholas. 2012. "Music as Performance." In *The Cultural Study of Music: A Critical Introduction*, 2nd ed., edited by Martin Clayton, Trevor Herbert, and Richard Middleton, 184–194. New York: Routledge. First published in 2003.

Cook, Nicholas. 2013. *Beyond the Score: Music as Performance*. New York: Oxford University Press.

Cooper, Grosvenor W., and Leonard B. Meyer. 1960. *The Rhythmic Structure of Music*. Chicago: University of Chicago Press.

Cox, Arnie. 2006. "Hearing, Feeling, Grasping Gestures." In *Music and Gesture*, edited by Anthony Gritten and Elaine King, 45–60. Burlington: Ashgate.

Cox, Arnie. 2016. *Music and Embodied Cognition: Listening, Moving, Feeling, and Thinking*. Bloomington: Indiana University Press.

Cumming, Naomi. 1997. "The Subjectivities of 'Erbarme Dich.'" *Music Analysis* 16 (1): 5–44.

Davidson, Jane W. 1993. "Visual Perception of Performance Manner in the Movements of Solo Musicians." *Psychology of Music* 21: 103–113.

Delalande, François. 1988. "La gestique de Gould: Élements pour une sémiologie du geste musical." In *Glenn Gould pluriel*, edited by G. Guertin, 85–111. Québec: Louise Courteau.

Dixon, Simon, Werner Goebl, and Gerhard Widmer. 2002. "The Performance Worm: Real Time Visualisation of Expression Based on Langner's Tempo-Loudness Animation." Technical Report. Vienna: Österreichisches Forschungsinstitut für Artificial Intelligence. TR-2002-15, 2002.

Eigeldinger, Jean-Jacques. 1986. *Chopin: Pianist and Teacher as Seen by His Pupils*. New York: Cambridge University Press.

Epstein, David. 1995. *Shaping Time: Music, the Brain, and Performance*. New York: Schirmer Books.

Geeves, Andrew, and John Sutton. 2014. "Embodied Cognition, Perception, and Performance in Music." *Empirical Musicology Review* 9 (3–4): 247–253.

Godowsky, Leopold. n.d. *Studies after Frédéric Chopin*. Berlin: Schlesinger.

Godøy, Rolf Inge, and Marc Leman, eds. 2010. *Musical Gestures: Sound, Movement, and Meaning*. New York: Routledge.

Goebl, Werner, and Gerhard Widmer. 2009. "On the Use of Computational Methods for Expressive Music Performance." In *Modern Methods for Musicology: Prospects, Proposals, and Realities*, edited by Tim Crawford and Lorna Gibson, 93–113. London: Ashgate.

Guck, Marion A. 2005. "Dramatic Progression in Haydn Sonata No. 46 in A♭, *Adagio*." In *Engaging Music: Essays in Music Analysis*, edited by Deborah J. Stein, 180–190. New York: Oxford University Press.

Hatten, Robert. 2004. *Interpreting Musical Gestures, Topics, and Tropes: Mozart, Beethoven, Schubert*. Bloomington: Indiana University Press.

Hudson, Richard. 1994. *Stolen Time: The History of Tempo Rubato*. Oxford: Clarendon Press.

Krebs, Harold. 1999. *Fantasy Pieces: Metrical Dissonance in the Music of Robert Schumann*. New York: Oxford University Press.

Larson, Steve. 1997–1998. "Musical Forces and Melodic Patterns." *Theory and Practice* 22/23: 55–71.

Larson, Steve. 2006. "Musical Gestures and Musical Forces: Evidence from Music-Theoretical Misunderstandings." In *Music and Gesture*, edited by Anthony Gritten and Elaine King, 61–74. Burlington: Ashgate.

Larson, Steve. 2012. *Musical Forces: Motion, Metaphor, and Meaning in Music*. Bloomington: Indiana University Press.

Le Guin, Elisabeth. 2006. *Boccherini's Body: An Essay in Carnal Musicology*. Berkeley: University of California Press.

Leman, Marc. 2008. *Embodied Music Cognition and Mediation Technology*. Cambridge: MIT Press.

Lochhead, Judy. 2006. "'How Does It Work?': Challenges to Analytical Explanation." *Music Theory Spectrum* 28 (2): 233–254.

London, Justin. 2004. *Hearing in Time: Psychological Aspects of Musical Meter*. New York: Oxford University Press.

Ohriner, Mitchell. 2012. "Grouping Hierarchy and Trajectories of Pacing in Performances of Chopin's Mazurkas." *Music Theory Online* 18 (1).

Ohriner, Mitchell. 2014. "Elicited Asynchronies in Recorded Performances of Chopin's Mazurkas." *Empirical Musicology Review* 9 (2).

Pierce, Alexandra. 2007. *Deepening Musical Performance through Movement: The Theory and Practice of Embodied Interpretation*. Bloomington: Indiana University Press.

Repp, Bruno. 1995. "Expressive Timing in Schumann's 'Traumerei': An Analysis of Performances by Graduate Student Pianists." *Journal of the Acoustical Society of America* 98 (5): 2413–2427.

Rothstein, William. 2005. "Like Falling Off a Log: Rubato in Chopin's Prelude in A♭ Major (Op. 28, no. 17)." *Music Theory Online* 11 (1).

Schenker, Heinrich. 2000. *The Art of Performance*. Edited by Heribert Esser. Translated by Irene Schreier Scott. New York: Oxford University Press.

Straus, Joseph N. 2011. *Extraordinary Measures: Disability in Music*. New York: Oxford University Press.

Wanderly, Marcelo, and Bradley W. Vines. 2006. "Origins and Functions of Clarinettists' Ancillary Gestures." In *Music and Gesture*, edited by Anthony Gritten and Elaine King, 165–191. Burlington: Ashgate.

Whiteside, Abby. 1997. *On Piano Playing*. Portland: Amadeus Press. Reprint of *Indispensables of Piano Playing* [1955] and *Masterng the Chopin Études and Other Essays* [1969]. New York: C. Scribner's Sons.

Yeston, Maury. 1976. *The Stratification of Musical Rhythm*. New Haven, CT: Yale University Press.

CHAPTER 12

EMBODIED RHYTHM AND MUSICAL IMPACT OF CORPORAL PUNISHMENT IN TWENTIETH-CENTURY OPERA

SHERSTEN JOHNSON

OPERA as an art form creates a visual-aural locus in which music and bodies tell vivid and extraordinary stories. Operatic vocal music in all of its melodramatic excess emanates from bodies, and bodies function not only as resonating instruments but also as dramatic signifiers. To intensify and foreground this excess, operas often set to music the extremes of human bodily experience—disease, death, and violence—all duly tinged with hues of eroticism. Syphilis, suicide, swords, and lots of sexual innuendo abet the intensity of body-centric dramas (see Blank's and Hutcheon and Hutcheon's chapters for more on this subject).

Not surprisingly then, operatic art forms include a number of works that stage scenes of corporal punishment. These moments of somatic suffering can concentrate attention on the physicality of the body in a way that even Mimi's consumption or Carmen's stabbing cannot. Although punishments in opera range from captivity to execution, this chapter focuses on three exemplary cases of person-on-person violence in the form of flogging. Flogging is one of the cruelest punishments one human being can inflict on the body of another. A symbol of sadistic oppression, the act is fraught with physical suffering and emotional devastation. In terms of staging, the punishing implement—the whip—can be more visually vivid than a gun or knife or an invisible disease. Beyond sheer dramatic impact, the operatic medium reinforces such a portrayal in that the action of whipping is itself peculiarly musical. The repetition of strokes creates a compelling rhythm; the sonic blows, intense in both sharpness and volume, impact the listener, and the visual impression of the arcing strokes seem strangely lyrical as well.

The scenes in question form pivotal moments in the narrative arc of each work by creating a mid-drama moment of intensity; the punishments are devastating but not deadly. One of these scenes, the trial in the 1970 Lloyd Webber and Rice rock opera *Jesus Christ Superstar*, mimetically enacts 39 lashes to an electric guitar ostinato. In another scene from Britten's 1951 opera *Billy Budd* with libretto by Forster and Crozier, the punishment itself is not shown but the music depicts the emotional fallout. Yet another instance occurs in the Adams and Goodman opera *Nixon in China* (1987), in which the Nixons attend a ballet dramatizing the plight of oppressed peasants. The scene sonically and visually crystallizes many layers of emotion: empathy for the brutalized peasant, anger at the oppressor, voyeuristic pleasure, awareness of the propagandist subtext, the ambiguous role of Nixon as oppressor seeing a portrayal of capitalist oppression, and sympathy for upset onlookers.[1]

This chapter analyzes these operatic moments to see how music, action, and text multimodally represent not only the cruel impact of blows but also the emotional impact for victim, punisher, onstage onlookers, and audience. Close readings of the three scenes in juxtaposition draw on theories of embodied cognition including Zbikowski's work on the conceptual modeling of rhythmic passages (2004, 2009) and Cox's "mimetic hypothesis" (2011), which holds that we understand sounds in comparison to sounds we have made ourselves. Observations from literature on corporal punishment provide insight as well, considering how music and drama work together to signify power relationships. Ultimately, this chapter engages the question of not only how music helps to culturally construct the body in these moments but also how this music gets into our bodies as listeners to intensify our experience of the dramas.

Mimetic Impact: *Jesus Christ Superstar*

"To keep you vultures happy I shall flog him."

Offered as a first exploration is a seemingly straightforward enactment of a flogging scene from the rock opera *Jesus Christ Superstar*. As the most "realistic" scene examined in this chapter, it not only has much to offer in terms of music and action analysis but it can also provide a model against which the other analyses can be measured.

Superstar, a rock opera passion play with lyrics by Tim Rice and music by Andrew Lloyd Webber, portrays the last days of Christ's life spanning from right before his palm-bestrewn entrance into Jerusalem to the crucifixion. Purported to be "the Gospel according to Judas," it focuses on the events through Judas's eyes as well as those of other followers of Christ. Initially written by Lloyd Webber and Rice as a concept album in 1970, it was adapted for the stage in 1971. The discussion here concentrates on a further adaptation, the 1973 film version produced by Norman Jewison.[2]

Jewison's production sets the passion story as a play-within-a-play. The film taps into the late 1960s/early 1970s cultural anxiety about authority, war, and government, while capitalizing on the then-current postmodern caché of youth counterculture. The movie, filmed on location in Israel, begins with a troupe of actors arriving in the desert on a bus.

During the overture, they prepare to perform the passion play. The mise-en-scène mixes realistic locations, minimal sets, and a range of historic and current costumes (a blend of period robes and hippy attire) and props (ranging from spears to military tanks). These chronologically juxtaposed elements are not especially disruptive to viewers due to the conceit of the staged action nested within the larger diegesis of the film. The actors, without Jesus (performed by Ted Neely), then leave by bus at the end of the film.

Within the play, as the action advances, political tensions accrue and the crowd turns from adoring to angry. Pilate (performed by Barry Dennen in the film and Broadway premier), in a last-ditch attempt to satisfy the rioting crowd's calls for crucifixion, orders a flogging. The scene, while not as true-to-life as, say, the flogging in Mel Gibson's 2004 film *The Passion of the Christ*, is intended to be generally representative of biblical events, and as such it adds a layer of edgy severity and immediacy to this passion-opera.[3] As the scene begins, the crowd has come to Pilate to demand that Jesus be crucified. Pilate, having had a distressing premonition, argues first with the crowd and then with Jesus, demanding a response to the crowd's accusations, but Jesus refrains from a vigorous defense, perhaps because he is complicit—in Judas's eyes—in his own demise. Eventually Pilate agrees to keep the crowd of "vultures" happy by flogging Jesus.

The crowd is not convinced that a flogging will suffice, and their cries of "crucify him, crucify him" escalate.

As Figure 12.1 shows, the chant alternates two bars of 7/8 with four bars of 2/4. In the 7/8 bars, the chant "Cru-ci-fy him" begins the measure on four eighth notes leaving

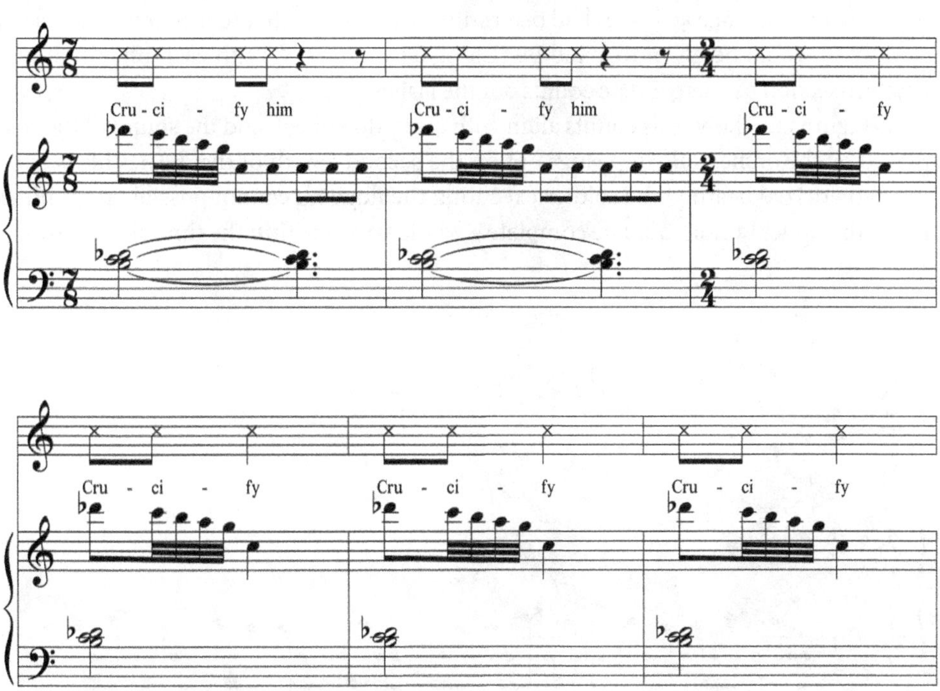

FIGURE 12.1 "Crucify him" chant ("Trial before Pilate," Reh. J).

the remaining three eighths articulated only in the accompaniment. So when successive measures begin, it sounds as if they have entered an eighth early, like there is an eighth note missing. The passage thus performs a simulacrum of escalation with iterations of the chant entering seemingly early. The third through sixth measures articulate the chant—now just "Cru-ci-fy"—in a 2/4 meter. Although these four measures enunciate a more regular meter with no "missing" eighth notes, they also mimic escalating tensions in a new way by repeating the pattern directly so that a downbeat now occurs every four (instead of seven) eighth notes.

As the flogging begins, the tempo abruptly changes from *Adagio* to *Animoso*, from 92 half-note beats per minute to 120 quarter-note beats per minute. Because this new tempo is slightly faster than the preferred range for locating beats, which is held to be in the range of 85 to 100 beats per minute (Zbikowski 2004, 279), the impulse to shift to a higher grouping level—to feel half-note beats rather than quarters—will be perceived by many listeners. The shift mimics the sense of ratcheting up to a higher gear on a bicycle, when pedals suddenly move more slowly but the overall speed is faster. In this situation, it adds to the feeling of escalating intensity.

Pilate does not issue an order for a specific number of strokes, so as the sequence begins and one hears the first count, several questions might come to mind: *How many lashes will be shown? How realistic will they be? How can they perform the flogging without endangering the actor? Will there be thirty-nine, as is traditionally attributed to (although not actually recorded in) the biblical account? Will we have to watch ALL of them?*

The sequence indeed counts out 39 lashes, drawing on the tradition of Mosaic law, which holds that once 40 lashes had been administered the offender becomes blameless, so only 39 would be given to avoid punishing an innocent.[4] With each repetition of a rhythm-section ostinato, Pilate counts out the lashes.

As Figure 12.2 shows, his counts align with every downbeat, and the sound of the lash played by the orchestral whip rings out on the second eighth of that beat. This is truly an aestheticized beating; the soldier executing the flogging cannot possibly respond to Pilate that quickly, nor could he complete a whole stroke within the time allotted by the

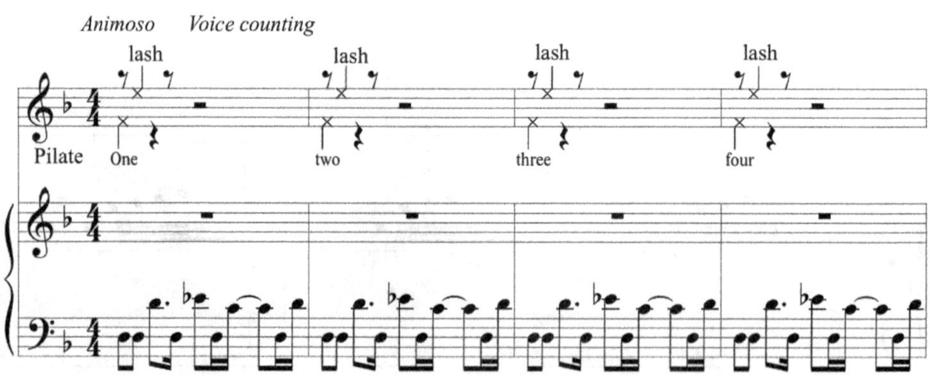

FIGURE 12.2 "39 Lashes" ostinato ("Trial before Pilate," Reh. M–N).

ostinato. The film version furthers the aestheticized visual appeal by focusing on the handsome body of the soldier executing the flogging, showing in slow motion the grace of the uplifted arm and the twirl of the whip. The medium also allowed Jewison to vary the relentless 39 repetitions with a mix of film cuts and camera work that centered at times on Pilate, the executioner, the crowd, and on the anguished face or bloody back of Jesus. Compare this film treatment with that of the Broadway premiere, which featured Pilate being carried about the stage by soldiers, counting out strokes through a stylized megaphone designed to look like a big mouth. The Broadway scene was criticized, according to Nassour and Broderick (1973) for seeming "endless" and "losing some of its effectiveness" (159).

The flogging ostinato motif itself consists of a D Phrygian compound melody that differentiates registral layers with two distinct eighths on the low tonic downbeat followed by shifts between low and high tonics, the upper D elaborated by neighbor notes E♭ and C. Later in the measure, alternating sixteenth- and eighth-note durations syncopate the implicit fourth beat.

Figure 12.3 notates the brass parts in the treble clef of the reduced score, which outline the high tonic melody with its neighbor notes. The brass attacks divide each measure into groups of sixteenth notes in the pattern 4 (resting) + 4 + 3 + 3 + 2, thus adding quickening articulations at the end of the measure. Each measure thus calls for a repetition of itself with a pseudo *accelerando*. Despite this, though, the tempo remains relentlessly the same throughout.

The ostinato repeats with a *Bolero*-like layering of additional entrances that contrast relentless repetition with intensification of instrumentation, increased dynamics, and added dissonance. Successive layers add complexity: the rhythm guitar becomes more active, the brass instruments climb higher and higher, and the electric guitar and electric bass improvise more actively. This layering process, combined with the syncopations, give the ostinato what Jessica Sternfeld (2006) has described as a "driving beat that lends it an air of aggression and strength, a certain forward-rushing, out-of-control feel" (29). Finally, on the 39th stroke, Pilate sobs (almost screams) out the last number, the music stops abruptly, the soldiers release Jesus and he falls on his face, while the executioner breathes heavily.

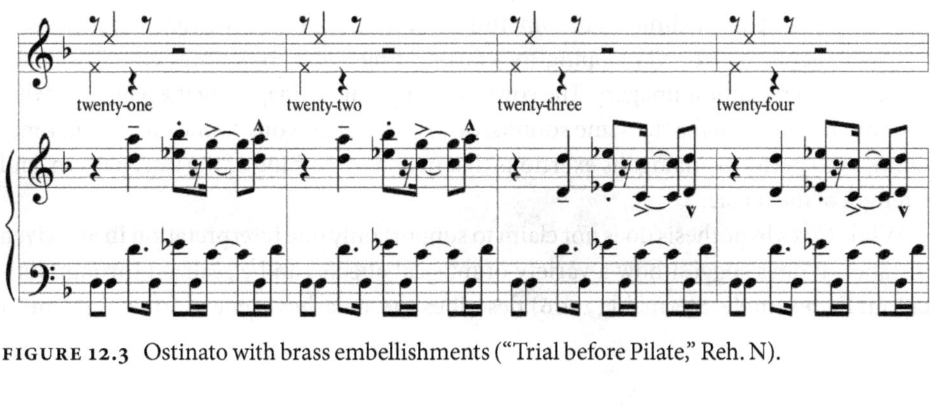

FIGURE 12.3 Ostinato with brass embellishments ("Trial before Pilate," Reh. N).

In writing about conceptual models for rhythmic grooves, Zbikowski (2004) defines a groove as "a large-scale, multi-layered pattern that involves both pitch and rhythmic materials, and whose repetitions form the basis for either a portion or all of a particular tune" (275). He suggests that grooves are heavily embodied and signal how to move. He proposes a model for musical rhythm, shown here:

> Proposition 1: Rhythm concerns regularly occurring musical events.
> Proposition 2: There is differentiation between rhythmic events.
> Proposition 3: Rhythmic events are cyclic.
> Proposition 4: There is a strong sense of embodiment associated with musical rhythm. (276)

He writes, "when musical events summon the concepts of regularity, differentiation, and cyclicity, knowledge about both music and bodily motion is activated. Such events will, in consequence, function as ready targets onto which bodily motions can be mapped" (279).

In *Superstar*, the flogging ostinato is in fact a multilayered pattern that differentiates pitches and regular rhythms that cycle over and over. This ostinato then acts as what Zbikowski (2009, 363) calls a "sonic analog" for repeated motion, particularly appropriate for the beating consisting of counted strokes to which the music is anchored. The compound melody, with its doubly articulated downbeat, acts as a robust mimetic cue for whipping with a low tonic count and stroke and embellished high tonic windup. The regularity of the downbeat is relentless. A mix of rhythmic values syncopates succeeding beats and, as mentioned earlier, compresses articulations into shorter and shorter groupings. One can even imagine these rhythmic compressions mapping iconically on the changing speed of the lash. In short, borrowing from a 1994 *New York Times* headline about violent film music, this riff has "a nice beat, you can torture to it."[5]

In order to unpack the affect of this music a bit more, I turn to the work of Arnie Cox. His "mimetic hypothesis" (2011) holds that (1) we understand sounds in comparison to sounds we have made ourselves, and (2) this process of comparison involves tacit imitation, or mimetic participation, which in turn draws on the prior embodied experience of sound production. Cox bases his hypothesis on the notion that humans understand one another in part via imitation, sometimes overtly—which in the case of musical activity might be realized via tapping toes or nodding—and sometimes covertly, in the form of mimetic motor imagery. The covert activity can be imagining the feeling of what it would be like to make the same sounds, or to make analogous sounds in another medium, or to perform analogous exertions that match something of the rate, pattern, and strength of the music.

While Cox's hypothesis does not claim to support only one interpretation in any given instance, it does suggest how a variety of musical affects can be realized through imitation. For example, Sternfeld (2006) describes the 39 lashes scene as "one of the most terrifying moments in the show, largely because of the music to which it is set" (47). But what musical features might support her reading of "terrifying?" According to the mimetic hypothesis, listeners might ask themselves what it is like to (1) make the same sounds as the musicians and actors; for example, to play the trumpet in increasingly higher registers, or to count out loud until hoarse like Pilate; or (2) make analogous sounds in another medium: for example, to align the body and hold enough tension and attention to make a strong sound like that of the whip precisely on the "and" of beat one 39 times; or (3) perform analogous exertions that match something of the rate, pattern, and strength of the music, for example to repeat exertions with increasing complexity, agitation, and force, like that of the ostinato repetitions. Imagining these actions requires establishing body states that, in this context, can be analogous to those of being frightened, like being frozen with tension or moving with increased agitation.

Although Sternfeld finds this music terrifying, one can imagine ways in which it might be more disturbing: making the lashes less regular on the order of what Adams does in *Nixon* (see discussion later),[6] or say by the addition of a vocal reaction from the victim in the manner of the flogging of Kunta Kinte in the 1970s miniseries *Roots.* Following Cox, covertly mimicking vocal distress in terms of sounds we ourselves have made would tend to establish bodily states of distress.

Listeners focusing on other features of the music might unconsciously ask themselves what it is like to correctly predict the occurrence of a musical event. One of the reasons we like grooves so much is that we can easily predict what will happen via mimetic simulation. As Cox points out, successful anticipation is positively valenced, so it would not be entirely surprising that a listener would feel pleasure from this music. A syncopated, "groovy" ostinato played by instruments we know are used to underscore activities like dancing—guitar, bass, drums, and so on—might invite one to empathize with the smiling, hair-tossing blonde woman in the crowd in the film, leading one perhaps to feel a sort of guilty pleasure. Conversely, one could imagine music that would be more inclined to help us identify with a resigned Jesus, especially as portrayed by Ted Neely with very little reaction. How would music that would help us feel more fully his acquiescence and detachment sound?

As this example demonstrates, the empathy involved in imitation can lead to identification with the condition of a certain character. Even though the 39 lashes ostinato accompanies Judas in his opening aria as well as his last moments, by the time of the trial before Pilate, Judas has already hanged himself. This suicide opens up a space for the listener to direct attention to another character. One might think that person should be Jesus, the victim. Instead, the terrifying intensification and aggressive rushing shares more in common with Pilate's increasing anxiety than with the resigned, silent, slumping body of Jesus. In addition to Pilate's increasing vocal stress and distress,[7] the whip has a mind of its own; it cracks regardless of Pilate's count, almost stepping on it. If the music is terrifying and out of control it also helps unify the drama by reiterating the bewildering question: Why is Jesus fated to be destroyed and why is no one able to stop it?

Closeted Impact: *Billy Budd*

"The shame'll never pass."

Sharing many parallels with *Superstar*, Britten's 1951 opera *Billy Budd*—based on the novella by Herman Melville (published posthumously in 1924)—tells the story of a handsome but naive young seaman who is impressed into service in 1797 aboard a British man-of-war. Billy, well beloved by his fellow sailors, comes to the attention of two powerful men: the sadistic master-at-arms, Claggart, and the caring but necessarily strict ship's captain, Vere. Jealous of Billy's popularity and, one suspects, secretly attracted to him, Claggart becomes bent on destroying him and plots to falsely accuse him of mutiny. Billy, frustrated by his inability to defend himself because he stammers under stress, knocks Claggart down and accidentally kills him. Vere knows the accusation to be false, but to maintain his authority must still sentence Billy to be hanged. Both operas, *Superstar* and *Budd*, feature false accusations, an undeserved (sacrificial) death at the hand of a reluctant authority figure preceded by a pivotal flogging, and allies who are powerless to change the political situation. In the case of *Budd*, it is a different character, the Novice, who receives the punishment. In his innocence, he stands with Billy as a Christ figure. The scourging so demoralizes him, though, that he becomes an accomplice in Claggart's plan to frame Billy. Thus, in contrast to the biblical principle from Deuteronomy mentioned earlier in which the punishment would purify a wrongdoer, the flogging corrupts the Novice.

Much has been written about how Melville deliberately shrouds his narrative in an aura of ambiguity and how Britten follows suit. Rupprecht (2001), for example, writes that Britten's operatic adaptation uses motifs as "keepers of the mysteries and secrets so central to Melville's novel" (75). Billy's inability to speak, Vere's ambivalence, Claggart's questioning of himself, and even the sea's mists serve as markers of uncertainty. In exploring these ambiguities, critics often also point to the behind-closed-doors interview scene in which thirty-four expressive, skillfully orchestrated chords hint at secret meanings while Vere delivers his sentence to Billy in his cabin out of sight and earshot of the audience. But in actuality that is not the only concealed scene in *Budd*. Unlike the flogging in *Superstar*, the Novice's punishment takes place off-stage. This remarkable reworking obscures historical details in that the act of flogging aboard a British man-of-war would of necessity be public, as is Billy's hanging. The open ritual would have been needed in order to make a show for the other impressed men to maintain discipline. On board ship, the captain was king and the whip was his scepter.

This departure from Melville's story was not the original intent, however. An early draft (January, 1949) lists the following stage directions:

> Vere is joined by his other officers (Sailing Master, Ratcliffe and six others), the marines are marched into place below and across the quarterdeck, a gang of seamen

> brings a square grating to which the offender is lashed, the ship's corporal stands ready with the whip, and all the crew stands to watch the scourging. (quoted in Seymour 2004, 146)

But later that year, Britten requested that the flogging take place off-stage and set a lament depicting the emotional results of the punishment. Seymour suggests that Britten deleted this scene because he felt in some way implicated in the sublimation of sexual tension through violence necessitated by the claustrophobic atmosphere on board a ship filled with only men (146). Carpenter (1992) goes a step further suggesting that the flogging was too closely linked in Britten's mind with abuse he endured at school during his youth (289).

These propositions aside, it is important to note that even Melville was complicit to some extent in veiling the flogging:

> When Billy saw the culprit's naked back under the scourge, gridironed with red welts and worse, when he marked the dire expression in the liberated man's face as with his woolen shirt flung over him by the executioner as he rushed forward from the spot to bury himself in the crowd, Billy was horrified. (Melville [1924] 2006, 25)

The passage does not describe a sequence of lashes to which Billy reacts, but rather the event is flattened by the narrative into almost a single instant; not the accumulation of individual strokes but the complete, collected gridiron of welts. Without the phrase "under the scourge," Billy could just as easily have come upon the Novice after the punishment was over, as he does in the opera. Britten's production is not so far off; compare the excerpt from Melville's novella with the 1966 BBC production of the opera,[8] which features lengthy shots of the Novice's "gridironed" back first glimpsed by Billy as the Novice is carried off.

This is not the only one of Britten's operas to have sequestered a flogging. An early scene from *Peter Grimes* drafted in the librettist Montagu Slater's hand was excised in order to focus the libretto more clearly on the oppression of the individual by society. According to Philip Brett (2006), Slater's draft represents "a Grimes who would have severely hampered the opera's chances of success" (43) by revealing himself as a willful sadist rather than a victim. Part of Peter's text reads:

> Your soul is mine.
> Your body is the cat o' nine
> Tail's mincemeat. O! a pretty dish
> Smooth-skinned & young as she could wish.
> Come cat! Up whiplash! Jump my son
> Jump (lash) jump (lash) jump, the dance is on.

But ultimately, if Britten can be accused of closeting the *Budd* flogging, it is because he chose Vere as his protagonist and it is Vere who must shroud the act in the "mist" of ambiguity. Humphrey Carpenter (1992) quotes Britten in a 1960 radio interview, "I must admit that Vere, who has what seems to me the main moral problem of the whole work, round [him] the drama was going to centre" (287). (Although let us also remember that Britten was interested in foregrounding roles for his partner, the tenor Peter Pears (1910–1986). Britten thus decided to let Vere survive into old age and reminisce on the events of the story. Framing the action within Vere's prologue and epilogue filters all of the narrative through Vere's point of view. No wonder the flogging is suppressed. Vere cannot bear to relive the details of it; he feels guilty. He is ambivalent about the need for, and the brutality of, corporal punishment. When we hear the Novice sing plaintively about the shame that will never pass, it is Vere's regret and shame that we hear. Why would the Novice be ashamed when the flogging was not public? Seymour ties this shame back to sexual undercurrents (2004, 147).

Left with the aftermath of the flogging, we hear the Novice's sorrowful dirge even before we see him. During the first notes of the lament the Novice's friend reports to Claggart that the flogging has been done according to king's regulations referencing the officially recognized form of corporal punishment. As in *Superstar*, a certain number of strokes were issued, in this case twenty. This process of indexing the punishment to fit (supposedly) the crime, as in Mosaic law, was meant to set normative boundaries and, as Geltner (2014) writes in his study of flogging, "curb the slippery slope of descent into worse violence" (19). And yet, in another surprising reworking of Melville's story, the indexed number of strokes seems not to fit the crime at all. While the Novice's offense in the novella is a serious one, in the opera he is merely clumsy. Billy rightly sings, "I've never seen [blood] shed for no reason."

The Novice's friend, upon announcing that the beating has been completed, advises Claggart that the boy did not take it well and he cannot walk. Claggart's response to the limping supported sufferer is one of the most chilling moments of the opera. "Let him crawl," he growls in the low extreme of his register after a frightening *molto crescendo* in the brass. What follows is an interchange between the Novice and his friend, punctuated by the chorus of sailors, highlighting the transformation of the men from merely looking on to sharing in the Novice's pain and mourning their loss and lost-ness:

FRIEND: *Come along kid. Novice: I'm done for.*
FRIEND: *The pain'll soon pass. Novice: The shame'll never pass.*
SAILORS: *He's lost forever on the endless sea.*
FRIEND: *The bruises'll heal up kid. Novice: But my heart's broken.*
SAILORS: *Ay, he's heartbroken. We're all broken.*
FRIEND: *I'll look after you. Novice: They've caught me, my home's gone.*
SAILORS: *Ay, his home's gone. They've caught us, they've caught all of us.*
ALL: *We're lost forever on the endless sea. We're all of us lost.*
FRIEND: *Come along kid. Novice: I'm done for.*

Rather than a rhythmic groove that portrays the motion of the lash as in *Superstar*, this lament deploys rhythms to mimic the endless sea and the dissolution of the waves. Britten is a master at writing themes that evoke the sea, but unlike, for example, the "sea phrase" in *Death in Venice*, which depicts the beautiful ocean vista whose waves fall in lazy uncontested thirds, the waves of the lament flow in separate imitative currents tinged with tight seconds that dissolve their energy iconically into dissonant roiling sixteenth-note up- and down-welling eddies. The friend's line "the pain'll soon pass" dissolves into repetitions of "soon pass, soon pass" while the Novice sings, "never pass, never pass." Facilitated by accents and overlapping registers, the currents of their contrary lines along with their rhythms—offset by a sixteenth note—dissolve even the sheer material of language ("soon" and "never") into a mystifying, whirling elided "soonever, soonever."

Figure 12.4 collates their individual lines onto one staff to demonstrate the composite hearing of vortexing currents.

Returning to Cox's mimetic hypothesis, a listener might ask what it is like to make the same sounds as the musicians or analogous sounds, for example, by subvocalizing the Novice's version of the saxophone line with its sobbing rising minor sixth interval and descending half-step. Or what it is like to play the saxophone alone. The saxophone is thought of as an instrument that often performs improvisatory material, unmediated by notation and thus more directly expressive. A listener might also consider the analog exertion schemas of the strings and hear in them bodily states similar to slumping and needing support to walk with slow, irregular steps, and establish a body state isomorphic with earlier experiences of sorrow (slumping posture, slowed motions, halting steps).

Listeners can identify with sorrow via mimetic participation, but it is musically clear that Vere feels it too. Stephen Allen (2006) writes of Vere's "convergence" with the characters he is recalling when one character's text and/or music merges with another. For example, Vere's "I have been lost on the infinite sea" (Prologue) is recapitulated by the Novice

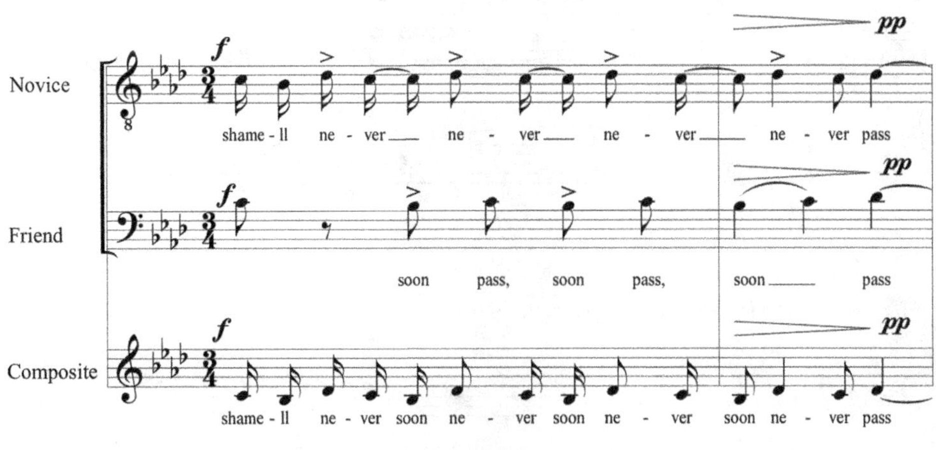

FIGURE 12.4 "The pain'll soon pass/The shame'll never pass" (Reh. 42, mm. 22–23).

after his flogging, and the Novice's "my heart's broken!" anticipates Vere's "my heart's broken, my life's broken" (act 2, scene 2). Allen proposes that shared lyrics and motifs illustrate "Vere's identification with—but simultaneous implication in—the suffering of the Novice" (53).

Figure 12.5 represents these themes. Allen proposes that the "Starry Vere" theme[9]—the sailors' naively awe-filled musical reverence of Vere—shown in Figure 12.5a (with its ascending interval pattern of $2^{nd} + 3^{rd} + 2^{nd} + 2^{nd}$) melodically inverts in the Novice's lament (Figure 12.5b, which descends $2^{nd} - 3^{rd} - 2^{nd} - 3^{rd}$). Both themes have been heard

FIGURE 12.5 Converging themes.

combined and syncopated in the composite cello-clarinet and saxophone counterpoint shown in Figure 12.5c.

I suggest taking Allen's analysis one step further to account for the mutation of the sailors' "Starry Vere" theme. Throughout the passage beginning at Reh. 42, the ascending line moves from the sailors' open pentatonicism, to the more tightly constructed, inverted Novice's lament, and finally to the sailors' "They've caught us, they've caught all of us" at Reh. 43, m.13, which, as Figure 12.5d shows, ascends stepwise (1 + 2 + 1 + 2 halfsteps, etc.). With its tight alternation of half and whole steps, the line also reveals the influence of the sobbing half steps of the Novice and the creepy ascending half step of Claggart's "crawl" motif, so much so that when the men recap the cello-clarinet line all shreds of "Starry" pentatonicism are abandoned for a cramped octatonic scalar line that truly crawls.

In *Budd*, as in *Superstar*, the flogging scene (as well as the whole drama) is filtered through the lens of another character, in this case Vere rather than the Judas figure (who in *Budd* would be Claggart). If Britten were to have more directly "narrated" Billy's story, it might have included a scene like the one the librettists first drafted. Perhaps musical materials could be gathered from the extant opening scene on deck, in which orchestral whip sounds comingle with sounds of mimicked birds and waves, whistles, military fanfares, and the sailors' work rhythms, all as part of the sonic fabric of oppressive naval life. These materials, however, are the same ones that Britten incorporated into the final hanging scene, so he may have made his decision to "closet" the flogging partly based on issues of dramatic pacing, not wanting to steal thunder from the opera's ending.

Propagandist Impact: *Nixon in China*

"Whip her to death!"

The third opera scene under consideration takes the discussion in a new direction; if the mimetic flogging in *Superstar* forms a central point of departure, and the closeted flogging in *Budd* inscribes movement to one attenuated extreme, then the punishment in *Nixon in China* marks motion to an opposite extreme. *Nixon* not only represents the whipping on stage but also treats it with a good deal of operatic excess, and it forms the central topic of fourteen minutes of the three-hour opera.

The 1987 work, with music by John Adams and a libretto by Alice Goodman, was inspired by Richard Nixon's visit to China in February 1972. Although the real-life events were carefully researched and replicated in sets, costumes, and lyrics, in other ways the creators took many liberties, highly fictionalizing much of the drama. My observations are drawn from the 2011 Metropolitan Opera production.[10]

The scene in question, scene 2 of act II, represents an event during the Nixons' time in China in which the visitors observed a performance of the opera-ballet *The Red Detachment of Women*. The scene thus nests a play-within-a-play, while toying with the boundaries of the inner diegesis by involving members of the on-stage audience in the ballet's story. The opera-ballet loosely draws on a real-life group of roughly one hundred peasant women from Hainan province, a tropical island off the coast of southern China, which was at that time a Communist enclave. These women joined the Chinese Workers and Peasants Red Army, and in 1931 they became the first female fighting troop, battling in over fifty skirmishes on the island. As women fighters, they were able to exploit guerrilla tactics, blending into the peasant population to avoid detection.

This story, which glorifies a young peasant girl's heroic overcoming of slavery under a tyrannical overlord and her escape to join the Red Detachment, has formed the basis of many re-enactments through the years. Kristine Harris (2010) documents the adaptations of the story through a continuum of formats including early accounts, film melodrama, comic book, ballet, and various opera versions. Most famous among these were a 1960–1961 feature film directed by Xie Jin and an opera-ballet, whose creation in 1963–1964 was overseen by Jiang Qing (Madame Mao) and Mao's chief diplomat and advisor Zhou Enlai (who arranged Nixon's trip to China). It was this opera-ballet that the Nixons watched during their visit in 1972. At that time, roughly the midpoint of the Cultural Revolution, traditional nonsocialist arts and artifacts were banned and either destroyed or were in deep hiding. This opera-ballet, one of only eight that escaped censorship, was considered a model of the people's art as it fulfilled the ideal purpose of artworks (according to Mao) which was to "serve the interests of the workers, peasants, and soldiers, and conform to proletarian ideology" (Lu 2004, 143).[11]

Rather then trying to reproduce the opera-ballet that the Nixons had watched in some similar form for the opera, John Adams, Alice Goodman, and the producer Peter Sellars staged a very fictionalized version that highlights the political differences of the two countries while refraining from directly reproducing the anti-Western propaganda of the opera-ballet. Adams, rather than mimicking the opera-ballet's music (late nineteenth-century Western art music style peppered with well-known Chinese melodies), composes a pastiche of unrelated themes reflecting the fact that the opera-ballet music was composed by Madame Mao's hand-picked Creation Group committee (writers, musicians, choreographers, stage designers, and principal performers).

In this scene, the curtain rises on a ballet tableau observed by an on-stage audience of the Chinese dignitaries and American visitors. A peasant girl, dressed symbolically in Communist red (and perhaps blood red) is chained to a post. Four dancers surround her and tell her story through movement while Mao's secretaries narrate. The narrative describes floggings that this peasant and others have endured in addition to other acts of unspeakable violence. The dancers mimic the punishment, pantomiming blows. Ironic music in an upbeat tempo and bright, dry strings and brass accompany the tale of the horrors of slavery.

A few moments into the scene, the secretaries sing, "down on all fours our grandfathers swallow abuse as if by choice the humble flesh kisses the lash."

Figure 12.6 reduces the "lash" music to a single staff showing the top line of the Secretaries' chords and an increase in complexity of three kinds of musical accents. The first type of accent, marked above the staff with a C (contour accent), results from the

FIGURE 12.6 "Kisses the lash" (II/ii/41–49).

higher pitches of the melody, which are quite marked since the melody consists of only two different notes, A and C. These accents occur on strong beats at first, but as the passage continues they get out of sync with metrical accents. The second of these accent types is shown above the staff with the symbol ">"—these are the expected metrical accents, which also change with the changing meter. The third type of accent, resulting from cracks of the orchestral whip, appears on the staff with an X note head. An annotation under the staff marks a possible hearing in which a listener might respond to the increase in metrical complexity. It shows the predictable counts of the first three 2/2 measures. In measure 4 the contour accent suggests that the 2/2 meter is continuing, but the downbeat of measure 5 sounds accented enough to cause a reinterpretation of the previous measure as 3/2. The contour accent on beat 2 of the measure suggests a continuation of half-note beats, but the restatement of the eighth-quarter-eighth-quarter syncopated pattern right after that causes a reinterpretation of the meter as 6/4. Another repetition of the syncopated pattern in measure 6 sets a strong expectation for a metrical accent in the middle of the measure, which is when we hear the first crack of the orchestral whip delaying the next syncopated pattern by a quarter note. Measure 7 features three whip sounds on the "ands" of the beats, motivating us to reinterpret the previous 6/4 meter as reverting to 3/2. The metrical play also moves the last "lash" from a weak position to a strong one.

Not unlike *Superstar*, in which the sound of the stroke follows on the half beat after the call of the stroke number, the whip sound here follows the word "lash" a half beat later. Here, though, the pattern is much less predictable. Comparing this version with the original opera-ballet in which the lashes fall on the beat in a regular fashion, Johnson writes, "In this way the original seems less sinister than Adams's version" (2011, 222). Although we have heard a lot of metric dissonance by this point in the opera—it is a hallmark of Adams's style, after all—the direct connection with the sound of the orchestral whip increases the stakes for a listener to be able to predict when it will crack.

The second section of the *Nixon* ballet features the Kissinger character dressed as the henchman of an oppressive overlord. This violation of the ballet's fourth wall emphasizes the perception that capitalists and landlords are similarly abusive. The henchman sings a vulgar song about an earlier rape and beating while menacingly fondling the chained peasant. "When the first cut scored her brown skin I started in, man upon hen!" he sings. The mention of "brown skin" here not only draws attention to the very sheath of the victim's body despite her full suit of clothing but also highlights the displacement of what would have been a brown-skinned Chinese bully by the white Kissinger character, further implicating Americans in the tyranny of the landed class. The music here, with closely positioned chords in regular pounding rhythms is vaguely reminiscent of Stravinsky's "Dance of the Adolescents" from the *Rite of Spring* (1913), alluding to a young woman who dances herself to death and foreshadowing the dancing girl who is to be whipped to death.

The peasant girl temporarily escapes but she is caught again in the third section of the ballet. The Kissinger/bully returns armed with a cane and accompanied by four male dancer accomplices with whips.

Figure 12.7 excerpts a bit of the music of this lengthy section from the piano-vocal score. Kissinger repeats the call "Whip! Whip! Whip!" to A minor chords marked *sforzando, fortississimo*, and *staccatissimo* that jump severely from the high register to low. Meanwhile, the four accomplices strike the stage, thus actualizing the whipping in comparison to the other floggings discussed in this essay in which the executioners pantomime while the crack sound comes from the orchestral instrument. This again is an aestheticized beating. Somehow, in the hands of leaping, acrobatic dancers the whips become artful extensions of already lengthened poses, and their beats are synchronous with the call of "Whip!," again not allowing the executioners time to respond to the command as would occur with a real beating. As soon as Kissinger sings, "whip her to death," the orchestral whip comes in, and either it or the bass drum executes each stroke from there on. The peasant girl reacts accordingly, as if violently struck.

All told, fifty-four lashes strike with great intensity in unexpected intervals. This beating is much more fraught than that of *Superstar*. The victim is more vulnerable, a young woman, a slave, not even partially in control of her environment, and definitely not the son of God. Furthermore, she is the victim of four executioners (five counting Kissinger) and, unlike *Superstar*, there are no "civilized" indexed counts nor is the victim meant to survive. In this case, the audience, like the Pat Nixon character, focuses on the peasant girl and the dancer who plays her, asking (a la Cox) what it would be like to respond to intense and unpredictable blows. Mrs. Nixon becomes afraid for the dancer, as does the audience.[12] She comforts the girl, and sings, "whip her to death, he said, I'd like to give his goddam whip a crack."

FIGURE 12.7 "Whip her to death!" (II/ii/288–91).

These first sections of the ballet thus rehearse a flogging three times, once as told by the secretaries and dancers, once as told by the Kissinger/bully, and a third time violently acted out. Each layer adds complexity and intensity. Fortunately, the peasant girl survives and goes on to become a rifle-toting member of the Red Detachment and the ballet continues for roughly another fifteen minutes. While the flogging in this scene foregrounds the backlash against "all who set small against great" (a line sung by the Kissinger/bully character), it also resonates with the hopes of the downtrodden in the other two scenes discussed in this essay: the flogging of Jesus, who preached that the meek shall inherit the earth, and that of the sailor, whose shipmates almost come to mutiny against the naval officers.

Conclusion

This chapter explored not only the cultural construction of bodies in states of extreme somatic suffering but also how music can intensify our experience of that suffering. But as raw and traumatic as these punishments can be, the three productions create distance from the topic in several ways. First, historical and geographical detachment helps keep these events in the "other" category. Audiences can keep these horrors at arms' length for, after all, those were more uncivilized times.[13] The action is situated in Palestine of biblical times, or on an eighteenth-century British naval vessel at sea, and although the action in *Nixon* takes place in 1972, the ballet foregrounds a "backward" China of the 1930s. Each of these three productions further couches troubling events in the fictional cushions of a play-within-a-play, framed by Jewison's troupe of actors, Britten's Captain Vere's memories, and the Communist propaganda ballet in *Nixon*. Staging and filming decisions further distance viewers from the suffering characters either by slow-motion camera work as in *Superstar*, or the stylized ballet and bawdy humor of *Nixon*, or by leaving the scene out altogether, as in *Budd*.[14]

Not surprisingly, these productions culturally construct punished bodies as "other" in a way that reflects real-life attitudes toward corporal punishment in general. But Geltner argues that we should resist this othering. He writes:

> Taken together, the negotiation of corporal punishment's meaning and its social, cultural, and political uses reveal two things. First, there has seldom been a period, society, or polity that, figuratively and often literally, spared the rod. Secondly and perhaps paradoxically, that corporal punishment is rarely an uncontested measure, including among civilizations long perceived as hungry for or at best indifferent to pain in punishment. (2014, 12)

At some level, however, we as audience members actually *do* resist the sense of detachment. Even with this constructed distance, we feel the impact of Jesus's punishment on Pilate and the crowd, the Novice's sorrow on the sailors, and the peasant's brutalization on the Nixons, and we feel the impact on ourselves as well. These operatic moments help us explore the contested, taboo nature of corporal punishment via embodied rhythm and mimetic participation, and in some small safe way we can de-other flogging by experiencing it through music that leaves no bruises.

Notes

1. For other notable flogging scenes in twentieth-century operas, see also Shostakovich's *Lady Macbeth of the Mtsensk District* (1934) and Gian Carlo Menotti's *The Medium* (1946).
2. The "Trial before Pilate" scene from Jewison's film is available on YouTube: *Jesus Christ Superstar 1973 (Trial before Pilate) HD*, YouTube video, 0:15, posted by detodounpoco82, July 27, 2012, https://www.youtube.com/watch?v=WEzEROSj11Q.
3. For discussions of violence in film, see Grønstad and Henrik (2012), Prince (2003), and Torr (2002).
4. From Deuteronomy 25:3: "but the judge must not impose more than forty lashes. If the guilty party is flogged more than that, your fellow Israelite will be degraded in your eyes."

5. Kenneth M. Chanko, "FILM; It's Got a Nice Beat, You Can Torture to It." *New York Times*, February 20, 1994, http://www.nytimes.com/1994/02/20/movies/film-it-s-got-a-nice-beat-you-can-torture-to-it.html.
6. Also see Allen Kurzweil's story of his real-life flogging performed to the soundtrack of the "39 Lashes," in which he poignantly tells of the affect of the unexpected blows of the belt wielded by a bully who did not follow the rhythmic cues of the ostinato: Kurzweil (2015).
7. Many will agree with Stephenson Humphries-Brooks (2006, 61), who reads this vocal stress more as passion than angst, and in fact reads the whole scene as an incident of stereotypical Hollywood queer sadism.
8. 1964 BBC production available on YouTube: *Britten—Billy Budd—1966 BBCtv Recording Complete*, YouTube video, 2:38:05, posted by Oliver Drake, November 18, 2012, https://www.youtube.com/watch?v=A1lVBfqQyjs.
9. According to the novella, Vere acquired the nickname "Starry Vere" after his kinsman greeted him with that title upon his homecoming. The name comes from a poem by Andrew Marvell, in which it refers to a military leader renowned as a severe disciplinarian (Melville, 18).
10. Metropolitan Opera video available at https://www.metopera.org/Season/On-Demand/.
11. Eng (2009) passionately argues for the original opera-ballet to be considered as a work of art despite its propaganda function.
12. In fact, all of the close unpredictable strikes near the dancer leads Richard Paul Fink (Kissinger in the Metropolitan Opera production) to admit to having to be careful. When asked by Thomas Hampson in an interview during the interval, "is there something extra that you have to find to be among ballet dancers?" He replies, "don't hurt them!"
13. According to Max Frankel, a reporter who accompanied the Nixons to China in 1972, Adams failed to create *enough* distance from actual events despite the fact that "newsreel dramas" can help to overcome the "musty odor that inhabits many opera houses." See "A Witness Sees History Restaged and Rewritten," *New York Times*, February 10, 2011, http://www.nytimes.com/2011/02/13/arts/music/13nixon.html?pagewanted=all&_r=0.
14. See Goehr (2008) for an in-depth philosophical discussion on the topic of violence displacement in operas and other works.

References

Allen, Stephen Arthur. 2006. "*Billy Budd*: Temporary Salvation and the Faustian Pact." *Journal of Musicological Research* 25: 43–73.

Brett, Philip. 2006. "'Grimes Is at His Exercise': Sex, Politics, and Violence in the Librettos of *Peter Grimes*." In *Music and Sexuality in Britten: Selected Essays*. Edited by Philip Brett and George E. Haggerty, 34–53. Berkeley: University of California Press.

Carpenter, Humphrey. 1992. *Benjamin Britten: A Biography*. London: Faber & Faber.

Cox, Arnie. 2011. "Embodying Music: Principles of the Mimetic Hypothesis." *Music Theory Online* 17 (2). http://www.mtosmt.org/issues/mto.11.17.2/mto.11.17.2.cox.html.

Eng, Clare Sher Ling. 2009. "*The Red Detachment of Women* and the Enterprise of Making 'Model' Music during the Chinese Cultural Revolution." *VoiceXchange* 3 (1): 5–37.

Geltner, G. 2014. *Flogging Others: Corporal Punishment and Cultural Identity from Antiquity to the Present*. Amsterdam: Amsterdam University Press.

Goehr, Lydia. 2008. "The Musicality of Violence: On the Art and Politics of Displacement." In *Collected Work: Elective Affinities: Musical Essays on the History of Aesthetic Theory*, 171–203. New York: Columbia University Press.

Grønstad, Asbjørn, and Henrik Gustafsson. 2012. *Ethics and Images of Pain*. New York: Routledge.

Harris, Kristine. 2010. "Re-makes/Re-models: *The Red Detachment of Women* between Stage and Screen." *Opera Quarterly* 26: 316–342.

Humphries-Brooks, Stephenson. 2006. *Cinematic Savior: Hollywood's Making of the American Christ*. Westport, CT: Praeger.

Johnson, Timothy A. 2011. *John Adams's* Nixon in China: *Musical Analysis, Historical and Political Perspectives*. Burlington, VT: Ashgate.

Kurzweil, Allen. 2015. *Whipping Boy: The Forty-Year Search for My Twelve-Year-Old Bully*. New York: Harper Collins.

Lu, Xing. 2004. *Rhetoric of the Chinese Cultural Revolution*. Los Angeles: University of Southern California Press.

Prince, Stephen. 2003. *Classical Film Violence: Designing and Regulating Brutality in Hollywood Cinema, 1930–1968*. New Brunswick, NJ: Rutgers University Press.

Melville, Herman. [1924] 2006. *Billy Budd, Sailor, and Other Stories*. New York: Bantam.

Nassour, Ellis, and Richard Broderick. 1973. *Rock Opera: The Creation of* Jesus Christ Superstar *from Record Album to Broadway Show and Motion Picture*. New York: Hawthorn Books.

Rupprecht, Philip. 2001. *Britten's Musical Language*. Cambridge: Cambridge University Press.

Seymour, Claire. 2004. *The Operas of Benjamin Britten: Expression and Evasion*. Suffolk: The Boydell Press.

Sternfeld, Jessica. 2006. *The Megamusical*. Bloomington: Indiana University Press.

Torr, James D. 2002. *Violence in Film and Television*. San Diego, CA: Greenhaven.

Zbikowski, Lawrence M. 2004. "Modelling the Groove: Conceptual Structure and Popular Music." *Journal of the Royal Music Association* 129 (2): 272–279.

Zbikowski, Lawrence M. 2009. "Music, Language, and Multimodal Metaphor." In *Multimodal Metaphor*, edited by Gitte Kristiansen, Michel Achard, René Dirven, and Francisco J. Ruiz de Mendoza Ibáñez, 359–381. New York: Mouton de Gruyter.

Discography

Adams, John, Alice Goodman, and John McGinn. 1999. *Nixon in China: An Opera in Three Acts*. New York: Hendon Music.

Adams, John, Alice Goodman, and Peter Sellars. 2012. *Nixon in China*. New York: Nonesuch Records.

Britten, Benjamin, E. M. Forster, Eric Crozier, and Herman Melville. 1985. *Billy Budd: Opus 50 an Opera in Two Acts: Revised Version 1961*. London: New York.

Forster, E. M., Eric Crozier, Basil Coleman, Benjamin Britten, and Herman Melville. 2008. *Billy Budd, an Opera*. London: Decca.

Jewison, Norman, Robert Stigwood, Melvyn Bragg, and Andrew Lloyd Webber. 2004. *Jesus Christ Superstar*. 1973. Universal City: Universal.

Lloyd Webber, Andrew, and Tim Rice. 1973. *Jesus Christ Superstar*. New York: Leeds Music Corp.

PART IV

MUSIC AND THE DISABLED AND SEXUAL BODY

CHAPTER 13

MUSIC AND THE EMBODIMENT OF DISABILITY

MICHAEL B. BAKAN

THE ethnomusicologist Martin Stokes has argued that the term "theory," at least as used historically by ethnomusicologists, effectively "implies people . . . lacking theory" (Stokes 2013, 826). He specifically claims that the "people we have historically looked at across the colonial divide" through our research have been denied pathways to agency and voice in the construction of theoretical positions about themselves. "Once this was an easy way of defining ethnomusicology's objects," Stokes posits, but in more recent times we have cultivated "a more sophisticated understanding of our own historical compulsion to define others in terms of what they lack" (826). This new kind of understanding has prompted a paradigm shift toward what Stokes calls a disciplinary condition of "post-theoreticism": "The 'interpretation of others' paradigm, dominant throughout much of the last decades of the twentieth century in anthropology and ethnomusicology, erodes theoreticism by relativizing it. This kind of thinking establishes difference, though hesitates to explain it. This would, after all, be at odds with the relativizing instinct. So 'post-theoreticism' might well be our current theoretical condition" (826).

This chapter takes Stokes's claims regarding post-theoreticism as its key point of departure. It folds readily into emergent perspectives on music-related modes of listening, performing, and embodying premised on what Rosemarie Garland-Thomson and Sander L. Gilman have described as "disabled listening and viewing, that is, moving away from a normative listener/viewer to alternative models, [ones] that could mirror the experiences of disabled individuals but would not be limited to them. This is an alternative to creating 'abled' and 'disabled' performers as complementary or alternative categories." Such a move is significant, for, as Gilman adds, it suggests a disability studies–inspired orientation toward conceptualizing disability not merely as "a question in representation and practice" but also as "an epistemological question,"[1] one that, returning again to Stokes (2013, 826), demands answers that aver "our historical compulsions to define others in terms of what they lack" and, moreover, that does so against the implicit and erroneous premise that such "lack" equates with an inherent incapacity to generate "theory" in the first place. This approach yields an essentially post-theoreticist ethnomusicological consideration of how musics and the human bodies that create and experience them are connected in ways that reflect, embody, and inform one another.

The chapter begins by shifting focus from Stokes's colonial divide to the equally trenchant divide of ableism, which is discrimination against individuals with disabilities and which similarly relies on our historical compulsion to define others—in this case "disabled" others—in terms of what they lack. It then attempts to bridge the ableist divide through an exploration of the relationship of bodily experience and music experience, centering specifically on how that relationship is mediated by two individuals: the autistic[2] self-advocate and political activist Amy Sequenzia, and the professional drummer—and aspiring social dancer—Mitchell Baker (pseudonym). In profoundly different ways, Sequenzia and Baker have had to contend with pervasive challenges rooted in the limiting constraints of their physical bodies, and in each case music has figured prominently in both the make-up of those challenges and their strategies for responding to them.

Two concepts embedded in the opening quoted passages from Stokes inform the post-theoreticism of this chapter: establishing difference and explaining difference. For Stokes, as noted, the kind of thinking that emerges from ethnomusicology's current condition of post-theoreticism "establishes difference, though hesitates to explain it" (2013, 826). This chapter's line of inquiry offers a variation on that theme. It acknowledges the recognition of difference as an inevitable product of music and disability discourse, but it does not aim to establish such difference per se. Concomitantly, it promotes the notion that resistance to explaining difference is often justifiable, especially in instances where evoking personhood takes priority over describing it, as is the case here.

The reasons for taking this approach are both epistemological and political. The triadic formulation of "inability, lack, and loss" (Titchkosky 2007, 8) assigned to people living with disabilities has historically underscored a systemic fetishization of difference—bodily, cognitive, intellectual, and otherwise—usually achieved at the expense of actually listening to and taking seriously what disabled people have to say on their own behalves about who they are, what matters to them, and what they want and need. Explaining difference—or more accurately in this instance, explaining disability *as* difference—has thus had insidious and ultimately compounding effects. The hesitation to perpetuate this practice, along with the desire to forge new paths away from it and toward more productive avenues of inquiry and representation, are thus well founded. One response to such hesitations and desires has been to foster the creation of discursive spaces within which the subjects are the same people who mainly get to define and account for themselves in the text; after all, they are the primary experts at being—and knowing—who they are (see Bakan 2014a, 2014b, 2014c, 2015a, 2015b, 2016, 2018).

The present chapter foregrounds the words and ideas of Amy Sequenzia and Mitchell Baker in this way. The narrative is based principally on transcripts of conversations I had with each of them between 2013 and 2015, this as part of a larger online dialogical ethnography project (see Bakan 2018).

Amy Sequenzia

Amy Sequenzia is in her early thirties. She is autistic. She does not communicate through speech. Rather, she communicates by typing. She types with just a single finger, using a specially equipped iPad and the aid of her full-time associate, Adriana Avila, who supports Amy's elbow while she types. Amy does have ambitions to one day be able to type with multiple fingers and perhaps even without Adriana's manual support, but the physical obstacles posed by her difficulties with motor coordination have thus far been prohibitive.

Amy Sequenzia is a widely published author, having contributed chapters to and/or coedited titles such as *Loud Hands: Autistic People, Speaking* (Bascom 2012) and *Typed Words, Loud Voices* (Sequenzia and Grace 2015). She is also a prolific blogger and one of the most influential and respected spokespersons for autistic self-advocacy and neurodiversity rights.

Rather than focusing on how Amy experiences music as an autistic person, which has been the focus of other publications (e.g., in Bakan 2018), here the orientation is different: Amy's autistic subjectivity is still germane, but the larger interest is in the ways she explains how musical experience mediates her corporeal existence, that is, her life inside her body and her cognitive perceptions thereof. Beyond being autistic, Amy has cerebral palsy, seizure disorder, and other conditions that pervasively affect her physical ontology and functional abilities. Listening and responding to music on multiple sensory levels, and also in relation to multiple realms of bodily physical sensation, has a dramatic impact on the ways in which she hears, sees, and feels the world within and around her. She is extraordinarily eloquent in communicating about such subjects, to the extent that her comments and interpretations stand alone and require little additional framing, explanation, or interpretation.

As mentioned, Amy does not speak. She never really has, nor has she ever wanted to. "I can make noises and sometimes a word comes out, like my name," she explains. "But this is kind of involuntary, probably because of and related to seizure activity. . . . When I was little I had speech therapy and everyone wanted me to speak. I could say a few words but it was nothing like conversing. Then I learned how to type and it is so much easier, I don't miss not speaking, or not being able to."

Amy does not sing either: "I sometimes 'hum' but it is not about specific music. It is like *stimming*," she says, invoking a common abbreviation for certain self-stimulatory behaviors, such as hand flapping and body rocking, in which autistic people frequently engage. Whereas most medical, scientific, and music therapy literature describes stimming negatively in relation to the symptomatology of autism spectrum conditions (American Music Therapy Association 2012; Edelson n.d.), autistic people themselves often counter that stimming is a pleasurable, productive, creative, and even social mode of expression and communication (Bagatell 2010; Sinclair 2012a; Sacks 1995).

Amy may not speak or sing, but she hears very well, and she listens even better. Listening is a profusely multisensory experience for her. "I hear everything!" she exclaims, adding, "I only have trouble understanding when I am too overwhelmed and anxious [to focus], or after some seizures. . . . The way I perceive people and the sound of words, it touches me internally. I feel the words and I feel people. Some people might be too intense, some might calm me down, some are like a beautiful poem."

"The fact that I don't speak makes me listen better, I guess," Amy asserts, but people will often have precisely the opposite impression, assuming on the basis of her outward appearance that she is either oblivious to her surroundings or cannot hear at all. "My body does not necessarily follow my brain commands and that causes false assumptions," she says—in particular false assumptions about a lack of thought, feeling, and sensory experience that belie the extraordinary richness of Amy's inner world.

Music is an integral part of that world, and one that Amy relishes bringing to life through her written (i.e., typed) words: "I just love music. I love live music, I love each instrument's voice, even though I don't know all instruments. I can see colors coming from the orchestra or choir, I can see color in movement. I love when someone sings to me, even someone who cannot sing well because the words dance in front of me. Sometimes I can feel the music entering my blood stream and I even react in a visible way. (My friend [Adriana] says I make some faces and my eyes change.) . . . I just feel good listening to notes, even though I don't know them [the way musicians do]. It is the whole experience, the colors, the movement, the feeling. It embraces me completely. I feel the same when someone reads a poem and does it really well."

That visceral experience of music is deeply valuable to Amy. "I am usually not very aware of my body," she states. "I don't always know when I am in pain, I can't always feel certain types of pain, I rarely know where it hurts. To simply get up from a chair is sometimes hard, as if my body forgets how to move. I don't always know that I need to go to the bathroom. So when I feel the music inside my body, when I feel my blood running with the music, it is an amazing thing."

Amy describes herself as a synesthete, evoking her experiences of music in terms of colors, shapes, textures, and waves of motion. "I hear some notes . . . and they have colors, and the colors are like waves moving above, in front and around the music makers. Sometimes it is a big rainbow dancing, sometimes one color is dominant. . . . [T]he colors and movements make me feel things I cannot just bring up without the music."

The quality of Amy's synesthesia is related to the quality of the musical performances she takes in as well. "To me," she explains "it is like I am immersed in colors, unless someone hits the wrong note. . . . When something disrupts my experience, [like a wrong note,] the colors fade, the feeling subsides. I can go back, but it is an interruption."

The synesthesia experiences Amy described to me during our dialogues had been inspired by a variety of vocal, instrumental, and mixed vocal-instrumental performances, but they had occurred almost exclusively during concerts she had attended. The music "has to be live," she insisted. "I experience the words and sounds of music differently if

I am listening on the radio or [to a] CD. I still enjoy and can see colors, but the feeling is less intense."

As for preferred types of music, her "favorite is classical music because that's when the color, [the] movement experience, is the best." (She also enjoys James Taylor and Elvis Costello, Elvis Presley and the Beatles, and musicals like *Les Miz, Cats*, and *Evita.*) And it is in the context of classical music concerts, too, that Amy experiences music most profoundly as a vehicle for meaningful social experience. Explaining this phenomenon provides her with an opportunity to situate herself relative to music not only as a deep listener, synesthete, and bodily engaged audience member, but as a proudly autistic individual as well.

"The neurotypical [non-autistic] majority believe they are so super awesome, that anyone who can pretend to be, or pass, as neurotypical must be 'high' [functioning] or whatever," Amy opines, adding sarcastically, "the rest of us, we are pitiful and broken." But it "is not what people think it is. We [autistics] are social, we are [just] not conventional. Because our experiences with the world are so different from neurotypicals, because of sensory overload and such, we develop our own 'socialization.' Maybe music gives us the social experience because it is also so private and personal."

When elaborating on this thought, Amy does not attempt to generalize an answer on behalf of autistic people overall, but rather reflects on her personal experiences of the social attributes of musical engagement. She describes the classical concert environment as one that allows for "being social without the societal-imposed traditional or usual ways" of having to interact with other people. Being able to connect while "not feeling overwhelmed by the presence of so many people," she says, "the body and mind are free from the need for regulation and the feelings are more intense. There goes, again, the theory that we are not social. And [with the music] being live, I share the experience with the musician more intensely, even if they don't feel the same way." Amy expands on this last idea: "I think it is because during a concert, the energy of the music comes from the stage, and initially we are just receiving. The audience starts sharing with the musicians, or choir, or singer, but it is also an individual experience. Each one of us shares with them, through music, but not with each other. To me this makes being around all the people easier."

And so, with the effective mediation of music, Amy is able to transform a potentially stressful, even traumatic, social environment into one perfectly attuned to her autistic sensibilities and corporeal realities. Music enables her to joyously embrace her autistic, embodied self, and she insists she would have it no other way, for she is proud of who she is, just as she is, as an autistic woman and in most every other way—though she would happily give up having seizures, which she hates!

I once asked Amy, "If you could wave a magic wand and make your autism 'disappear,' would you?"

She replied, "I would break the wand before anyone could wave it. . . . The way I act, react, understand, and stand in the world is because I am Autistic. Autism cannot be separated from me and is part of every aspect of my life."

Mitchell Baker

Mitchell Baker is in his early fifties. He is, by all appearances, an able-bodied, nondisabled individual. He exercises regularly, adheres to a healthy diet ("most of the time"), and is in good health. He might be better described as hyper-abled than disabled, since he is capable of some rather exceptional feats of physical dexterity, speed, accuracy, and rhythmic acuity, especially when engaged in the activity that largely defines his life as a successful professional musician: drumming.

Yet Mitchell, like Amy, has a complicated relationship with his body, and the complexities of that relationship have much to do with how that body interacts with music in particular. Mitchell has been an accomplished percussionist since early childhood, and much of his livelihood has been earned playing music for dancers, whether as a member of Latin, rock, funk, swing, or "ethnic" bands, or as a percussionist for studio dance classes or ballet and modern dance performances.

But there is also a "dark side" (his phrase) to Mitchell's musical relationship with dance, that being his own dancing. He identifies as a person coping with what he only half-jokingly describes as PTDD: "post-traumatic dance disorder." By his account, the finely tuned rhythmic sensibility that he carries within his body and mind—and which he is able to effortlessly project into the wider world in the form of precisely ordered, hard-grooving percussive sound—"stops at my hips."

Mitchell lives with feelings of shame and frustration regarding his "incomprehensible inability to dance," which he also refers to as his "dance disability." Acute self-consciousness has compelled him to assiduously avoid situations where he might be expected to dance, and on those rare occasions when he has mustered the courage to put himself out on the dance floor, he has suffered some rather astonishing attacks from diverse parties: his teacher in a university West African dance class, the Latina grandmother of a former student of his at a wedding reception where he tried to dance salsa, and even his own wife at a local Bastille Day celebration. Mitchell believes that where others with "dance challenges" like his tend to be given a pass for their ineptitude most of the time, he, as an accomplished dance *musician*, is not. "I should be able to dance. That's what people think, and truth be told, it's what I think, too."

During the same period of time that Mitchell and I were having our dialogues for this chapter, he was also confronting his pervasive dance challenges in a different context. He and his wife, Martina (pseudonym), had at her urging enrolled in couples dance classes at a local Arthur Murray–type studio. Mitchell was gamely attempting to gain at least nominal competence dancing swing, salsa, rumba, waltz, and hustle, plus "a tiny bit of tango and cha-cha as well." His goal: to be able to go to a wedding, a bar mitzvah, or a holiday party with Martina and dance well enough to avoid embarrassment and humiliation.

"Dancing has pretty much always been a struggle for me," Mitchell confessed early on in the first interview.

> Well, check that. I used to feel okay if I could just "jump around" on the dance floor, and when I was a kid I actually thought I danced pretty well, because I was able to "hit" all the rhythmic shots in the music and stuff like that. So up through about high school, I figured I was pretty cool out there [dancing], but then I discovered that people were making fun of me, mocking me behind my back. It was kind of like that episode where Elaine dances at a holiday office party in [the classic television sitcom] *Seinfeld*. She looks really goofy and everyone's ridiculing her, but up to a certain point she's unaware of all that. She thinks she's really happening—then it all falls apart. That was me. [See Suber's chapter for more on this subject.]

It was not until a few years after high school, though, that Mitchell suffered his first outright "dance humiliation." He was a university music student at the time, majoring in percussion performance. As part of the program, his percussion teacher, who had studied Ghanaian drumming as a graduate student, directed a West African drumming ensemble.

"I loved playing in that group," Mitchell recalls.

> And I was really good at it. I always had a knack for complex polyrhythms. When I started getting serious about playing drums in my early teens, I'd listen to Elvin Jones, Tony Williams—all those great jazz drummers—and I'd figure out how to play a lot of the hip polyrhythmic things they were doing. It was so much fun! And then when I went to university and got exposed to the West African music, it was like, "Wow—this is where all those rhythms are coming from." So I really took to it, and I quickly rose in the ranks of the percussion students to a kind of leadership role in the African ensemble.

Bolstered in confidence by his positive experiences with West African music, Mitchell was excited to come across a listing in the schedule of classes for a new West African dance class being offered through the dance department. The teacher was a visiting professor from Nigeria.

"His name was Kwasi, I think," Mitchell recalled.

> On the first day of class, he lined us all up and started showing us some basic steps and motions. He immediately had us following him around the gym, instructing us to imitate everything he did. There were a few different moves, all low to the ground with double steps on each side and a kind of chicken dance–type shoulder motion. They had live drummers in the class—I think there were three of them—and I knew a couple of those guys. Really good players. I recognized a lot of the rhythms from my African drumming class; it seemed like we were doing a piece I had learned called "Agbekor." I was having so much fun, and I thought I was doing really well, too. But then I noticed that Kwasi was giving me these death stares, and shaking his head aggressively at me like I was doing something terribly wrong. He kept frowning, and he had this weird, intimidating expression, like, "What the hell are you doing?!"
>
> I tried to make adjustments. I watched his movements really carefully and tried desperately to get the moves right, but the harder I tried, the more nervous I got and

> the more agitated he looked. So next thing I know, he's shouting at the drummers to stop playing and storming over towards me. He wasn't that big, but he was strong as an ox and he looked *mad*. It was pretty scary.
>
> "You, what's your name, sir?" he barked at me. I told him it was Mitchell.
>
> "Mitchell, I do not understand what is happening with you. You are getting every rhythm correct—I don't know how—but you have *absolutely no flow in your body!!* What is your story?!"
>
> I wasn't sure *what* my story was, but the best I could come up with was that I was a drummer, that I kind of knew this piece from having played it—or something like it—and that this was the first time I had ever tried to do this kind of dancing. But none of that seemed to curry favor or soften his opinion of me.
>
> "I cannot stand to watch you dance, Mr. Mitchell! It is *painful* for me. I will not have it. You said you play the drums?" I nodded yes. "Well, if that is so, go over there with those other men and play the drums. That is what I want you to do for the rest of this course. I will give you an 'A' if you do that, but I never want to see you dance again. Never!"

"And that was that," Mitchell reported. "I drummed in that class for the rest of the semester, I got my 'A,' and Kwasi actually turned out to be a pretty nice guy. He complimented me on my playing on more than a few occasions and was usually friendly and courteous to me. But that public humiliation, his *rage* at me in that first class—I never really got over that. It still stings."

Several years after his West African dance debacle, Mitchell got himself into another precarious situation that did not end well.

> A couple of friends—actually both former percussion students of mine—were getting married. They invited me to their wedding. Both of them were originally from the Miami area—her family was Cuban, his from El Salvador—so it was this huge Latin wedding. Amazing food, amazing music, amazing party, and wow, could those folks *dance*!
>
> So I was just taking everything in, having a blast. Then the bride, Maria (pseudonym), she comes up behind me, grabs my arm, and starts pulling me out onto the dance floor. "No!" I shouted. My reaction was *way* too intense. It was like she was about to throw me in front of a firing squad or something. She seemed genuinely startled. I tried to explain. "It's just, well, it's just that I really can't dance."
>
> "What do you mean you can't dance?!" Maria shoots back at me. "Of course you can dance. Everyone can dance! Come on!" And she grabs me again, but I dig in my heels. She backs off, but just a bit. "OK. You say you can't dance," she says, and I can tell from her tone that this thing is not over yet. "Well, then I will *teach* you to dance. I can teach *anyone* to dance." All my instincts tell me to get out, run away, but then there's this one little part of me, this little hopeful part deep inside, going, "She says she can teach *anyone* to dance. Maybe she *can* teach me." And like an idiot, I decide to listen to that little, stupid voice.

"You're on," I tell Maria. "Give it your best shot, but if it's not happening within about five minutes, you have to promise to give up and just go enjoy yourself. It's your wedding day. Don't waste it on the likes of me." She laughs and takes me by the arm.

> "OK," she says. "So this is what you do. It's easy." She demonstrates the basic steps of the man's part for salsa. Left foot forward, two steps center, right foot back, two steps center:
>
> || Quarter Eighth-Eighth | Quarter Eighth-Eighth |
> | Quarter Eighth-Eighth | Quarter (. . . and so on) ||

It does look easy, and so natural, when she does it, but when I try, no such luck. I'm getting the rhythm right, but it's the old "absolutely no flow in your body" thing. I throw up my hands in defeat. "It's my hips. They're not designed for dancing."

"No! You can do this; you *must* do this!" Maria insists. "You aren't supposed to *move* from your hips. That's your problem. You move from the *knees*. Watch me!" And she shows me again, and this time the light goes on in my brain. I watch her knees. Her hips are *following* her knees, not the other way around. That's it! That's the ticket! I get a little adrenaline rush as I try it that way and discover it feels right, for the first time ever. And I'm convinced it probably looks right, too. I am dancing salsa (well, sort of)! Then, out of the corner of my eye, I see this woman walking over to us. She looks to be in her late sixties, maybe early seventies. She's laughing; I'm not sure at what. Maria, who's been locked in on me up to now, suddenly walks off to greet her. I keep practicing my "moves," growing confidence with each passing forward-step, back-step sequence. "Move from the knees, the knees," I keep reminding myself.

As I continue to practice, I overhear Maria and the older woman exchanging a few sentences in Spanish. The woman says something while pointing at me, and then both of them break out laughing, hysterically. Then they hug and the woman walks off toward the buffet table. Maria comes back over to me. I stop dancing.

"Who was that?' I ask Maria. "Oh, that's my *abuela*, my grandma." "Ah. What were you guys talking about?" "Oh," Maria pauses, awkwardly. "Oh, nothing." "Was it about me?" I ask, unable to tame my morbid curiosity. "Oh, no, we were just"—another awkward pause—"just, joking around." "Come on, you can tell me. I'm a big boy." "You sure?" she says. "Of course," I reply. "Well," Maria says, trying but unable to repress some giggles, "she was saying to me, *Hey, what's that guy doing? Killing cockroaches?!*"

She cracks up, she can't stop laughing. "I'm sorry, I'm sorry!" she blubbers through her laughter. "It's just, it's just, well, it's *funny*!"

"I try to play along, fake-laughing myself to look like a good sport, but inside I'm hurting, really hurting. It wasn't that I thought I had suddenly become a *good* dancer, but I had at least thought that maybe, just maybe, something had clicked—the move from the knees, not the hips thing, I don't know—and that maybe I didn't have to keep being so insufferably *bad*. It just got to me. After that, I went back into my shell, didn't dance again for years."

But eventually he did, for the simple reason that his wife, Martina, loved to dance, and Mitchell was determined to be "a good husband, and being willing to dance with your wife is part of the gig."

There were conditions to that willingness, though.

> Martina wanted us to take dance lessons together, to actually have some dance chops as a couple. I told her to forget it, that it was a lost cause. I told her about the African dance class, the wedding experience, all that stuff. But I also told her I was willing to compromise, at least to some extent. Going out dancing was fine, but it had to be casual, informal—free, not structured. Martina actually liked to dance like that, too—only difference being that she looked beautiful doing it—so we agreed to give it a shot.
>
> On the first couple of occasions, it was okay. As long as I laid low and didn't try anything "fancy," there seemed to be little damage done to my reputation or hers. But then there was this one big dance party at a Bastille Day celebration at this nice French restaurant in town. They had a big outdoor dance floor set up, great band, yummy appetizers and wine—and it was a perfectly gorgeous day. Well, I had a bit too much to drink and I got a bit loosey-goosey. I asked Martina if she wanted to dance. She said sure. She was really getting into it, putting on a bit of a show really. The whole scene got me kind of riled up, I don't know. I guess my well-cultivated inhibitions and instincts of self-preservation were getting drowned out by the wine. Anyhow, I kind of "cut loose," giving in to the impulse to "jump around" the way I used to when I was a kid, you know, before the PTDD and all. And it was fun, a *lot* of fun.
>
> But then I noticed Martina sort of moving away from me, and then she was dancing with her eyes closed, like she was trying to shut me out. And next thing you know, she's shooting me a death stare—like she's channeling Kwasi or something!—and then she's *yelling* at me. "Sit *down*, Mitchell! I don't want to dance with you anymore! I want to dance by *myself!*" Ouch!
>
> But this isn't some Nigerian dance teacher or some grandma at a wedding. It's my *wife*, for [goodness] sake! So I decide I need to stand up for myself—bad call—and things go from bad to worse. There we are, on this beautiful day at this awesome event, arguing on the dance floor about whether I should have the right to dance. It was surreal. I was crushed, and I got all defensive and pouty and acted like a bit of a jerk. And the whole thing really messed with Martina's head as well. For years after that, we *never* danced together, not once, and that and everything swirling around it put a real strain on our marriage. We even ended up in couples' therapy eventually, and while I'm not going to sit here and say that the Bastille Day incident was the main cause of all that, it definitely contributed, and it was definitely something that came up a number of times, from both of us, when recounting our relationship challenges in therapy.

On Valentine's Day of 2015, more than ten years after Mitchell's Bastille Day fiasco, Martina made a bold move. She surprised him with a gift of couples dance lessons at a local studio. It was an introductory "dance package"—four private lessons, plus six group lessons or "dance parties"—for fifty dollars.

"It was really sweet, really touching," Mitchell recalled.

> But it was also scary. I immediately felt pressured by the situation, and I was honest enough to tell her that straight away (I guess the therapy had helped me with some of that communication-type stuff). But she was prepared for my response. She told me she didn't have high expectations at all. She just wanted me—us—to give this a try and to hopefully have some fun with it. At the very least, it would be a nice way for us to spend some quality time together learning something new. And at best, it would give us the tools we needed to, say, dance a number or two at the endless Bar and Bat Mitzvah parties we were invited to in the upcoming months, something we hadn't even had the courage to do at our own daughter's Bat Mitzvah a couple months before.
>
> So I said, "Fine, we'll give it a shot," and the intro package went well enough that we invested in a package with more lessons and parties to go a few steps further.

Some nine months later, the results of the experiment have been mixed. "It's mainly been a lot of fun," says Mitchell.

> We have an excellent—and even more importantly, very *patient*—teacher. She understands my challenges and is creative in finding ways to work with them and help me figure things out. And it turns out I have some real assets, too. I always know where the beat is (you'd be amazed at how many dance students—and even teachers—rarely do!) and my "insider knowledge" of what's going on rhythmically in the music helps me to approach dancing in a, I don't know what you'd call it, a more "structural" way than most folks do. Now, that very ability has also proven to be a *liability*. Lots of times, the teachers demonstrate and have the students do stuff with absolutely *no* regard for the music they're dancing to—I mean, it's like they don't even hear the beat or the rhythm. This drives me nuts, and I can't "separate" from the music that way, so I can get really disoriented and frustrated when that happens. And then I get kind of militant, slavishly adhering to the music's rhythmic structure before I've mastered the steps and moves, which is actually counterproductive to my learning process and, even worse, pisses Martina off and makes it hard for her to learn. Not good! So I'm working on being able to "detach" from the music and just focus on the steps. I'm getting better at that, but it's a work in progress, and it's *important* work, because if I can't learn to do that, it's going to mess things up with Martina, and that's the last thing I want! Sure, I want to dance better, but strengthening our marriage is a *much* higher priority.

A strong upside to Mitchell's dance lessons odyssey has been its value to his musicianship. "Learning to dance to the music I've been playing all these years has been incredibly educational," he says,

> quite enlightening actually in terms of deepening my understanding of how the music actually works, how it works *as dance music*. Like in dancing swing, the basic beat pattern of the dance is actually in three [i.e., triple meter], even though the music is in four. I never knew that, and that knowledge has changed my swing

concept as a drummer dramatically, in a really good way. Seems that all those hemiola "polymeter" things Elvin Jones played were not really "polymetric" at all. They're just consistent with the way swing dance movement lines up with swing as a musical groove. That's not to say that Elvin was playing for dancers—I suspect he rarely did (well, who knows?)—but it might help to explain why, to my ears and body at least, Elvin's *way* of swinging *dances* so powerfully!

And one more example: salsa. Learning to dance salsa, at least at an elementary level (and yes, Maria was right that the movement comes from the knees, not the hips) has taught me that what's going on in the dance steps is all about lining up with the on-beat rhythmic accent patterns played by the *bongocero* [bongó player]; that's true for both the *martillo* rhythm played on the [bongó] drums and the cowbell part in the *mambo* sections [of salsa tunes]. But the *spins* when you dance—well, it's mainly the lady who spins, but anyhow—those seem to all line up with the clave pattern, and in particular, they usually *start* on the *last* beat of clave, so it goes something like this relative to a standard 2-3 clave pattern:

|| Rest - **One** - Two - Rest | **One** - - Two - - **SPIN!** ||

When you know how all that works in the dance, it changes the way you think about the music—and how you play it—and it makes you play *for* rather than *against* the dancers. It makes you a better musician, at least a better dance musician!

Conclusion

Amy Sequenzia's and Mitchell Baker's accounts of the complex relationships they have with their bodies, specifically as mediated through music, reveal alternately related and divergent insights into the nature and nurture of musical embodiment. In both cases, "disability," whether perceived as an externally imposed or self-assigned construct of ontology (or both), factors significantly in the particular ways their bodies work and the particular dynamics of embodied experience music yields.

Listening to music, especially when experiencing it performed "live," provides Amy with access to forms of physical awareness—like the blood running through her veins—that would otherwise be unavailable to her. This provides her with a valued sense of vitality and bodily wholeness. Amy's synesthesia yields further benefits of musical engagement, for when she listens to music (again, mainly live), a vibrant world of shapes and colors and movements is revealed within her imagination, providing pathways to myriad forms of sensory experience that her physical limitations might otherwise preclude. And finally, music serves as a gateway for Amy to experience fulfilling moments of social connection, especially in crowded environments that would otherwise be overwhelming for a person of her particular autistic sensibilities.

Mitchell, like Amy, has gained access to productive ways of tuning into his body with the aid of musical mediation. For him, though, the music–body relationship has been a two-way street, alternately punishing and rewarding, liberating and constraining. Mitchell has always danced within his body while drumming, but he has struggled against his body when pressed to convert that inner rhythmic sense into the outward bodily expression of dancing per se. Exacerbating that struggle have been the liabilities of living in a world where people—including himself—have been unwilling to accept his "dance disability," and have, moreover, been prone to judge him harshly on account of it.

Such lack of sympathy has had serious effects on Mitchell's self-esteem, his capacity to flourish socially, even his marriage. His current course of instruction in social dancing, facilitated by his wife, Martina (whose compassion for his predicament has grown over the years), shows promising signs of moving Mitchell toward having a healthy sense of himself as a dancer, of being accepting of rather than in conflict with the disabling limitations of his own body.

The stories of Amy and Mitchell presented here show that there are many ways to embody music, and that the intersections of music and the body can be highly paradoxical, perhaps especially in contexts where disability is a factor to consider. The brief exploration of music and the embodiment of disability offered in this chapter should inspire new research and new ways of thinking about the many rich intersectionalities of music, bodies, and the ways in which disability is lived and experienced.

Notes

1. These quotations are drawn from Sander L. Gilman's contributions to a series of email exchanges that took place between March 20 and March 22, 2016. More specifically, they represent Gilman's summations of his recent conversations on this topic with the disability studies scholar Rosemarie-Garland Thomson (see also Howe et al. 2016; Lerner and Straus 2006; Straus 2011; Garland-Thomson 2009).
2. Amy Sequenzia, in common with many other people diagnosed with autism spectrum conditions, or ASCs (a preferred designation to the more common "autism spectrum disorders," or ASDs), insists that she is an autistic person, not a "person with autism." Person-first language constructions like "person with autism" have been promoted within many circles of autism-related discourse, but there is strong resistance to their use within the autistic self-advocacy community with which Sequenzia identifies (see Sequenzia 2012; Sinclair 2012b).

References

American Music Therapy Association. 2012. "Music Therapy as a Treatment Modality for Autism Spectrum Disorders." Accessed April 24, 2018. http://www.musictherapy.org/assets/1/7/MT_Autism_2012.pdf.

Bagatell, Nancy. 2010. "From Cure to Community: Transforming Notions of Autism." *Ethos* 38 (1): 33–55.

Bakan, Michael B. 2014a. "Ethnomusicological Perspectives on Autism, Neurodiversity, and Music Therapy." *Voices: A World Forum for Music Therapy* 14 (3).

Bakan, Michael B. 2014b. "The Musicality of Stimming: Promoting Neurodiversity in the Ethnomusicology of Autism." *MUSICultures* 41 (2): 133–161.

Bakan, Michael B. 2014c. "Neurodiversity and the Ethnomusicology of Autism." *College Music Symposium* 54. http://symposium.music.org/index.php?option=com_k2&view=item&id=10673:neurodiversity-and-the-ethnomusicology-of-autism&Itemid=128.

Bakan, Michael B. 2015a. "Being Applied in the Ethnomusicology of Autism." In *The Oxford Handbook of Applied Ethnomusicology*, edited by Svanibor Pettan and Jeff Todd Titon, 278–316. New York: Oxford University Press.

Bakan, Michael B. 2015b. " 'Don't Go Changing to Try and Please Me': Combating Essentialism through Ethnography in the Ethnomusicology of Autism." *Ethnomusicology* 59 (1): 116–144.

Bakan, Michael B. 2016a. "Music, Autism, and Disability Aesthetics." In "Colloquy: On the Disability Aesthetics of Music," convened by Blake Howe and Stephanie Jensen-Moulton. *Journal of the American Musicological Society* 69 (2): 548–553.

Bakan, Michael B. 2016b. "Toward an Ethnographic Model of Disability in the Ethnomusicology of Autism." In *The Oxford Handbook of Music and Disability Studies*, edited by Blake Howe, Stephanie Jensen-Moulton, Neil Lerner, and Joseph N. Straus, 15–36. New York: Oxford University Press.

Bakan, Michael B. 2018. *Speaking for Ourselves: Conversations on Life, Music, and Autism.* New York: Oxford University Press.

Bascom, Julia, ed. 2012. *Loud Hands: Autistic People, Speaking.* Washington, DC: The Autistic Press/The Autistic Self Advocacy Network.

Edelson, Stephen M. n.d. "Self-Stimulatory Behavior." Autism Research Institute. Accessed April 24, 2018. http://www.autism.com/index.php/symptoms_self-stim.

Howe, Blake, Stephanie Jensen-Moulton, Neil Lerner, and Joseph Straus, eds. 2016. *The Oxford Handbook of Music and Disability Studies.* New York: Oxford University Press.

Garland-Thomson, Rosemarie. 2009. *Staring: How We Look.* New York: Oxford University Press.

Lerner, Neil, and Joseph N. Straus, eds. 2006. *Sounding Off: Theorizing Disability in Music.* New York: Routledge.

Sacks, Oliver. 1995. *An Anthropologist on Mars: Seven Paradoxical Tales.* New York: Alfred A. Knopf.

Sequenzia, Amy. 2012. "Non-Speaking, 'Low-Functioning.' " In *Loud Hands: Autistic People, Speaking*, edited by Julia Bascom, 159–161. Washington, DC: The Autistic Press/The Autistic Self Advocacy Network.

Sequenzia, Amy, and Elizabeth J. Grace, eds. 2015. *Typed Words, Loud Voices.* Fort Worth, TX: Autonomous Press.

Sinclair, Jim. 2012a. "Autism Network International: The Development of a Community and Its Culture." In *Loud Hands: Autistic People, Speaking*, edited by Julia Bascom, 22–70. Washington, DC: The Autistic Press/The Autistic Self Advocacy Network.

Sinclair, Jim. 2012b. "Why I Dislike 'Person First' Language." In *Loud Hands: Autistic People, Speaking*, edited by Julia Bascom, 223–224. Washington, DC: The Autistic Press/The Autistic Self Advocacy Network.

Stokes, Martin. 2013. "Afterword: A Worldly Musicology?" In *The Cambridge History of World Music*, edited by Philip V. Bohlman, 826–842. New York: Cambridge University Press.

Straus, Joseph N. 2011. *Extraordinary Measures: Disability in Music*. New York: Oxford University Press.

Titchkosky, Tanya. 2007. *Reading and Writing Disability Differently: The Textured Life of Embodiment*. Toronto: University of Toronto Press.

CHAPTER 14

MUSICAL REMEDIATION OF DISABILITY

BLAKE HOWE

OVER the past twenty years, scholars of disability studies have critiqued the prosthetic function that the disabled body so often plays within narrative discourses: disability usually serves as a point of negative contrast to an able-bodied norm and generates the narrative tension that ultimately serves as a prerequisite for narrative resolution. Arguing that "disability has been used throughout history as a crutch upon which literary narratives lean for their representational power, disruptive potentiality, and analytical insight," David T. Mitchell and Sharon L. Snyder (2000, 49) use the term "narrative prosthesis" to describe this strategy of appropriation. As they explain, the operation of the narrative prosthesis typically unfolds in four steps: (1) "a deviance or marked difference is exposed to a reader"; (2) "a narrative consolidates the need for its own existence by calling for an explanation of the deviation's origins and formative consequences"; (3) "the deviance is brought from the periphery of concerns to the center of the story to come"; and (4) "the remainder of the story rehabilitates or fixes the deviance in some manner" (53). "Cure or kill" is a widely used phrase that describes this fourth and final step (Garland-Thomson 2004, 780): stories appropriate disability to introduce narrative conflict, then resolve that conflict by banishing disability—either by rehabilitating the disability (e.g., Handel's *Orlando* [1733]) or eradicating the disabled character altogether (e.g., Berg's *Wozzeck* [1922]).

In narrow microcosm, Charles Wesley's hymn "O for a Thousand Tongues to Sing" (Wesley and Wesley 1740, 120–123) exemplifies these elements of the narrative treatment of disability. The hymn gathers a cast of unsavory characters ("murtherers, and all ye hellish crew" and "harlots, and publicans, and thieves"), then grants them redemption from their sinfulness: "[Jesus] breaks the power of cancell'd sin, / He sets the prisoner free." Implicitly grouped with murderers, harlots, publicans (tax collectors), thieves, and others in need of redemption are the disabled:

Hear him ye deaf, his praise ye dumb
Your loosen'd tongues employ,
Ye blind, behold your Saviour come,
And leap, ye lame, for joy.

This unpleasant quatrain (from an otherwise cherished hymn, still widely reprinted in Protestant hymnals [Tyson 2006]) prostheticizes physical impairment within the framework of the *religious model of disability* (Hogan 2016; Howe 2016; Wheatley 2010). Human bodies, so this ideology goes, are imperfect corruptions of a godly prototype (the perfect or ideal body of God, or of Adam before the Fall). Disabilities make manifest the human body's accumulation of earthly wickedness, and are thus purged when the soul reaches heaven, redeemed of its sins.

Wesley's hymn draws its quartet of disabled types—the deaf, dumb, blind, and lame—from Isaiah 35:5–6 (later echoed in Matthew 11:5): upon reaching Zion, "the eyes of the blind shall be opened, and the ears of the deaf shall be unstopped. Then shall the lame man leap as an hart, and the tongue of the dumb sing." In Zion (as in so many utopian prophecies), disability is purged in favor of uniform able-bodiedness (and, for Wesley, uniform whiteness: "Cast all your sins into the deep, / And wash the Ethiop white," reads the penultimate stanza). Commemorated in this semiautobiographical hymn, Wesley's own evangelical conversion coincided with his swift recovery from pleurisy, reinforcing his belief in the bond between physical and spiritual health. While recuperating, Wesley was read Matthew 9:2, which describes Jesus healing a paralyzed man by cleansing his spirit: "Son, be of good cheer; thy sins be forgiven thee" (Wesley 1849, 1:89–90). Here, too, disability fulfills a prosthetic function: paralysis is a symptom of sinfulness; forgiven, the disabled man now walks.

As Lennard Davis (2002) writes, "narratives involving disability always yearn for the cure, the neutralizing of the disability" (99). (Wesley's deaf now hear and his blind now see; his "lame" now jump, and his "dumb" now sing.) The ubiquity of the "cure narrative" is but a symptom of the stigma disability this continues to accrue: disability is not a stable, legitimate mode of identity (something to be affirmatively "claimed" [Linton 1998]), but a volatile pathology, requiring rehabilitation before its victims may join the population of the "normate" (Rosemarie Garland-Thomson's neologism for "the veiled subject position of cultural self, the figure outlined by the array of deviant others whose marked bodies shore up the normate's boundaries" [1997, 8]). Investigations of two parallel impulses—the urge to fix broken bodies and to assign meaning to their brokenness—have pursued genres in fiction, film, medical literature, and memoir (Couser 1997, 2009; Quayson 2007; Davis 2002; Keith 2001; Mitchell and Snyder 2000; Hawkins 1999; Frank 1995). More recently, investigations of the cure narrative by musicologists and music theorists have illuminated the ways in which composers also use this archetype—for example, by introducing an errant, "disabling" note and then resolving (or "curing") it by the end of a musical work (Howe 2016; Straus 2011; Grave 2008). This is the trajectory of the standard cure narrative: disability (usually a coded symbol for some destabilizing problem, like sinfulness, malevolence, nonconformity, or alienation) creates tension within a story; that tension resolves when the disability is cured or expelled.

This chapter pursues various formulations of the cure narrative in interrelated traditions of musical healing—examples in which music itself is the agent of rehabilitation. Music has frequently assumed this therapeutic function by invisibly transferring the body from states of distress to states of health in stories and rituals attending to the disabled body (Gouk 2000; Horden 2000; Kümmel 1977). For example, in "O for a Thousand Tongues to Sing," Wesley ties music directly to Jesus's name "'Tis musick in the sinner's ears")—and, by extension, to goal states of fitness and well-being "'Tis life, and health, and peace!"). As J. R. Watson (2010, 29–30) argues, for Wesley, "the word 'music' was constantly associated in his mind with 'the Name of Jesus,' " and "the music on earth [was] a foretaste of heaven." At the conclusion of Isaiah 35 (the biblical source for the hymn's text), the redeemed—former sinners, former murderers, former prisoners, as well as the formerly disabled—enter Zion united in song.

Music's capacity to repair broken bodies is often predicated on its capacity to represent energetic motion (Keller and Rieger 2009; Gur 2008; Larson 2012). Indeed, metaphors of mobility abound in writings about music. Terms like "harmonic progression" and "melodic ascent/descent" have contributed to our understanding that music "moves" through an abstract space, just as bodies move through physical space; similarly, meters proceed incrementally forward in time (beats 1, 2, 3, etc.), and rhythms may embody the physical motions of a dance or a march (see Montague's chapter and Stevens's chapter for more on this subject). Further, as Carl Dahlhaus (1982, 19) writes, "Ever since antiquity . . . it was the concept of motion that provided a connection between music and affection or ethos. The motions of tones sympathetically release those of the soul, a soul that is often described by the simile of a stringed instrument; both musical and psychic motions are subject to the same laws." Dahlhaus here alludes to Aristotle's doctrine of imitation, which states that the character (ethos) of music stimulates that same character within a listener. We are what we hear. Related to this is Boethius's influential notion of the *musica humana*, the musical structures (tunings, proportions, intervals) that govern the human body and soul, aligning in health but misaligning in sickness. This musical body might be consonant or dissonant with the *musica mundana* (the musical order of the cosmos) and the *musica instrumentalis* (audible music, performed on instruments or sung) (Bassler 2016; Howe 2016; Rankin 2005). Music moves; in doing so it agitates the internal motions of the human body and spirit, repairing inoperative bodies. Music also resonates; in doing so it has the power to produce either consonance (health) or dissonance (disease) in the bodies with which it sympathetically vibrates.

If disability (as a narrative prosthetic) obstructs resolution, then music (as an agent of healthful mobility) offers a quick, efficient, and nonsurgical means to negate it. Within many traditions of musical cures, the underlying biological impairment remains hidden under the patient's skin—and because of this, music is free to magnetically work its invisible magic. With energetic melodies instead of scalpels, and resonant harmonies instead of potions, the blind simply blink their eyes open, while the dumb simply open their mouths to speak. (As in traditions of faith healing, musical cures do not work so well on people with external disfigurements. No amount of singing will repair a facial scar or regrow an amputated limb.) Because of the magnetic powers attributed to it, music—sonically modeling a healthful, ideal body—frequently participates in these cure narratives as the medicine of choice, rehabilitating deviancy under a normalizing regimen of active, stimulating pulses and the healthy, robust resonance.

Tarantism

A potent example of music's mobilizing power to cure broken bodies may be found in the anonymously authored introductory essay to William Billings's *The New-England Psalm-Singer* (1770). Concluding a discussion of acoustics, the essay's author (later identified as Dr. Charles Stockbridge) offers his readers this unorthodox bit of medical advice:

> Among the many wonderful Effects, which are said to be produced by Musick, none is more extraordinary than that which it has on those, who have been bit by the *Tarantula*. . . . The *Tarantula*, is a large Spider, about three quarters of an Inch long, and as big as one's little Finger; they are Natives of *Aupulia* [*sic*] in the Kingdom of *Naples* in *Italy*. They are not venomous, but in hot Weather; at which Time, whoever is bit by them after some Time loses both Sense and Motion, and dies if destitute of Help. The most effectual Remedy is *Music*. (9)

Surely most readers would prefer an antivenin, first. But the medicinal advantage of music, according to this text, lies in its capacity to represent movement, stimulating an otherwise paralyzed, motionless body to dance:

> The Musician tries a Variety of Airs, till he hits upon one, that effects the Patient, who upon that begins to move by Degrees; first keeps Time with his Fingers, Arms, and Legs, afterwards is violently agitated in every Part of his Body; and then leaps up, begins to dance, and increases in Activity every Moment. . . . They all require the quickest and briskest Tunes; and are never moved by a slow dull Harmony. (9)

Buoyant music compels the stiff body to rise and dance; it does so via its tempo and rhythm (quick, brisk), plus its harmony (a consonant, resonant major) and the force and purity of the resulting vibrations ("the Nerves of the [Patient's] body are . . . agitated by the Vibrations which are occasioned by the Principle and stronger Notes"). This discussion of spider venom occupies a key position within Billings's tunebook, as a pivot from a scientific description of Pythagorean intervals to Billings's own compositions—signaling, perhaps, that his sacred tunes and anthems will have a similar medicinal value (for the sinful soul, at least, if not the poisoned body).

There are several possible sources for this passage (on account of some pervasive word and phrase borrowings), including the writings of Pére Noël Regnault (1731, 2:267–269), Martin Clare (1735, 295–296), Charles Owen (1742, 153–156), and Richard Mead (1762, 66–75). Together, these and other scientific sources provide a more developed hypothesis on the medical basis for this peculiar musical cure. For example, in a widely copied passage, Regnault describes both the tarantula venom's immobilizing effect and music's mobilizing effect on blood, nerves, and limbs. First, the venom "thickens the Blood, and stops several of its Passages," causing nerves to slacken; this results, ultimately, in "Inactivity, and Defect of Knowledge and Motion." The victim's body is inert—but now sounds music: "the Vibrations of the quick Airs which are played, agitate the Blood and the rest of the animal Spirits," refueling the nerves and fibers that sympathetically vibrate with the music around them. With blood flowing and nerves resounding, the patient (in "successive Motion of the Fingers, Arms, Legs, and Body") begins to dance and sweat, safely dissipating the tarantula's venom. In their speculative explanations, these sources thus trace the trajectory of a musical cure, identifying external signs of physical movement and imagining internal signs of nerve and blood movement to describe the victim's transition from disease to health. Music, in its capacity to stir movement and vibration deep within an organism, choreographs the body into a rehabilitated state.

Remarks on the music therapy of tarantula bites, a southern Italian ritual known as *tarantism*, appear in numerous treatises, particularly in the sixteenth through eighteenth centuries (Daboo 2010, 81–161; Lüdtke 2009, 55–75). The earliest descriptions—a brief first mention by Gulielmus de Marra (1362), followed over a century later by Alessandro di Alessandro's more extensive travelogues from his journey to Apulia—provided the basis for later medical, musical, and magical exegesis (Daboo 2010, 72–76). In his monumental tome *Magnes* (1641), Athanasius Kircher devoted significant space to tarantism, which he considered evidence for the "magnetism" (that is, the invisible attractive forces) of music (Findlen 2004). (Significantly, he was also one of the few authors to provide transcriptions of the musical antidotes themselves, shown in Figure 14.1.)

As Gary Tomlinson (1993) observes, Kircher's conception of tarantism reflects an earlier Renaissance tradition, in which music's curative powers were studied within an "amalgam" of "ancient humoral medicine, Pythagorean and Platonic ideas of music's effects, and a magical worldview" (164). Indeed, this interdisciplinary trio—humoral medicine, mimetic music, and natural magic—underpins much of the discourse on music healing in the Renaissance, which attempted to root its practices in both the rational and irrational, both the scientific and artful. The musician-physician Marsilio Ficino (1433–1499) was an important advocate of this syncretic approach. Responding to a question about why he so frequently mixed the study of music and medicine, Ficino describes Apollo as "inventor of medicine and the king of playing the cithara": he "temperates everything by his strings of sound, that is, by his motions and powers," thus giving "life and health to everyone and removing sickness with its vital rays" (Katz 1987, 90–91). Blending humoral medicine with the doctrine of imitation, Ficino writes that a healthy body consists of the "harmonious circuit of the humours and the movement

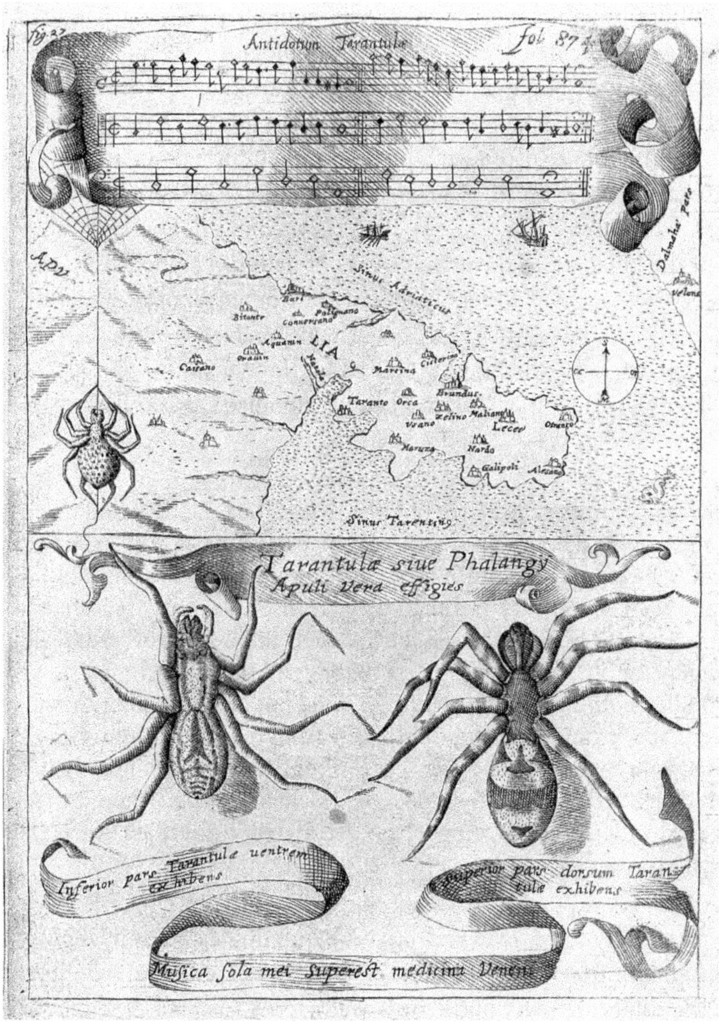

FIGURE 14.1 "Antidotum Tarantulae," Athanasius Kircher, *Magnes* (1641).

of the pulse," and "serious music" may restore this harmony should the body become infected with diseased discord.[1] For Ficino, music may constitute a kind of ideal sonic body: it is both the prototype of health (resonant structure, rhythmic pulse) and the means by which a defect of immobility may be cured. In his influential treatise *De vita libri tres* (1489), Ficino expounds these theories further; as evidence, he briefly mentions tarantism: "That there is indeed in certain sounds a Phoebean and medical power is clear from the fact that in Puglia [Apulia] everyone who is stung by the phalangium [wolf spider, or tarantula] becomes stunned and lies half-dead until each hears a certain sound proper to him. For then he dances along with the sound, works up a sweat, and gets well" (Tomlinson 1998, 389).

Other authors, unconcerned with magic, sought a rational basis for the ritual, whether through medical or sociological methodologies. One interpretive solution that gained traction was the reading of the tarantula as another sort of prosthetic within the performed narrative of the ritual itself: according to this view, the poisoned body was a vessel for a number of other physical, mental, and cultural ailments in need of remediation. One of the most prominent scientific observers of tarantism was Giorgio Baglivi (1668–1707), who studied the topic in *De praxi medica* (1696). After affirming the authenticity of the poison and the effectiveness of its musical medicine in some victims, Baglivi observed that many others (particularly women) had never been bitten in the first place. In a provocative and influential conclusion, he hypothesized that women suffering from melancholy, overheating, boredom, loneliness, or some other emotional crisis might feign a spider bite to partake in the ritualized music-dance therapy (Daboo 2010, 139):

> Whether they be under the Power of Love, or have lost their Fortunes, or meet with any of those Evils that are peculiar to Women, they never leave poring upon the mournful Object, till they run into Despair, or downright Melancholy, or at least are upon the Borders of it. And then they are also under a Disadvantage of living a solitary Sort of Life, like that of Nuns, deny'd the Benefit of free Converse with Men, though never so fair and honest. Besides all this, the Climate is sultry, their Constitution is of the same Nature, their Food hot and very nourishing, and their Life easy. Partly from the former, and partly from the latter of these Causes, 'tis no rare thing with them to turn mopish and melancholy: No wonder then, if they be mightily delighted with Variety of Music and Dancing; whence they feign themselves to be stung by the *Tarantula*, on purpose to enjoy the agreeable Diversion of Music. (Baglivi 1723, 334–335)

As is so frequently the case in narrative treatments of disability, Baglivi's interpretation of tarantism frames the destabilizing states of fixity and blockage (abstracted here from the catatonic body of the poisoned victim) as placeholders for other types of disorder or estrangement. Indeed, the gendered nature of the healing ritual (in which victims were almost always female and healers almost always male) suggests that womanhood, marked as either depravedly overheated or listlessly idle, was itself a disabling condition (see also Hutcheon and Hutcheon's chapter for more on this subject).

In his classic anthropology of tarantism, *La terra del rimorso* (1961), Ernesto de Martino abandoned the medicalization of the dance altogether, focusing instead on its social and cultural function. Through his extensive fieldwork in Salento, De Martino observed the contemporary incarnation of the ritual. His most famous subject was Maria of Nardò, who claimed to have been first bitten by a spider at age twenty, when her lover (shunned by her family) had left her. She then rehabilitated herself through dance and later would repeat the ritual annually. Analyzing these and other similar therapeutic practices, De Martino argued that magic rituals like tarantism were provoked by an existential "crisis of presence" (*crisi della presenza*), rooted in social marginalization. Ritual performance provided its victims a way to reclaim their agency.

Distinguishing from actual spider species in Salento (in reality, seldom poisonous!), De Martino writes of a mythical spider (*taranta*) that provides logic, order, and meaning—in his words, a "horizon"—to a suppressed or oppressed will. Invoking the language of blockage and breakthrough, De Martino writes, "To perform its function as a symbol, the *taranta* must evoke, configure, revive, and release the obscure stimuli of the unconscious which risk submerging consciousness with their ciphered indomitability." Notably, De Martino has relocated the source of the poisoned victim's blockage: it lies not in imbalanced humors or thickened blood or slackened nerves, but rather in a repressed stimulus of the unconscious, which the energies of dance will liberate. He continues,

> As a cultural model, this symbol [the *taranta*] offers a mythical-ritual order for settling these [psychic] conflicts and reintegrating [marginalized] individuals into the group. The symbol of the *taranta* lends a *figura* to the formless, rhythm and melody to menacing silence, and color to the colorless in an assiduous quest of articulated and distinct passions, where a horizonless excitation alternates with a depression that isolates and closes off. This symbol offers a perspective for imagining, hearing, and watching what we lack imagination for and are deaf and blind to, and which nevertheless peremptorily asks to be imagined, heard, and seen. (De Martino 2005, 35–36)

According to De Martino, the ritual of tarantism has the capacity to open previously sealed, inaccessible states of fulfillment. The ritual itself, then, is a performed cure narrative, which frames the impaired body (catatonic from symbolic poison) both as an impediment to social and self-fulfillment and as the necessary antithesis that gives those states of fulfillment their profound meaning. Notably, De Martino relies on other disability metaphors (deafness and blindness) to heighten the symbolism of this rehabilitative trajectory. Indeed, his language here loosely suggests another parsing of Isaiah 35:5–6, with Zion now replaced with secularized ideals of social integration and psychic harmony—as if, perhaps, to say, "then shall the blind see, the deaf hear, and the tarantula-bitten dance."

Mesmerism

In 1825, Franz Schubert (in what is surely the strangest anecdote from his short life) performed music for the hypnotism of an emotionally distressed woman, conducted by the mesmerist and healer Ludwig Schnorr von Carolsfeld (Goldschmidt 1983). As evidenced by the records of Schubert's performance, metaphors of musical motion played an important role in this hypnotic therapy: "It was still very striking to remark the various ways in which the tones and chords could affect a s[omnambulist]. Several, which were very moving, produced the most wonderful motions and contortions of [the patient's] body" (Solomon 2003, 231; Feurzeig 1997, 232).

Advocated by Franz Anton Mesmer in the late eighteenth century, mesmerism tweaked the familiar concepts of the *musica mundana* and *musica humana* to argue that a "universal fluid" (operating according to magnetic or attractive forces) bound the cosmos and the human body together, as well as human bodies to each other (Makari 2015, 298–328). (The title of Mesmer's dissertation, *De planetarum influx in corpus humanum* [On the Influences of Planets on the Human Body], explicitly adopts this Boethian model.) Within a disordered individual, the flow of this fluid was blocked, and a healer may use his own fluid to provoke a violent crisis (spasms, convulsions) within the patient, thus dissolving the impediment. Healings were performed with dramatic—and often erotic—theatricality: mirroring the gender roles of tarantism, a male mesmerist (equipped with numerous props, including a phallic "iron rod" that conducted the universal fluid) guided his hands and instruments over an incapacitated female patient, while audience members watched in amazement as she convulsed her way to health. One notable follower, Armand-Marie-Jacques de Chastenet, Marquis de Puységur, used the principles of mesmerism to hypnotize patients. In their "magnetized" trance, the somnambulists—now completely under the spell of their male hypnotist—could perform extraordinary feats of prophecy (diagnosing their own illnesses and foreseeing their recovery, mindreading the thoughts of others) and virtuosity (heightened intellectual capacity, fluency in new skills). This ritualized cure narrative scripted a reinforcement of male dominance over the female body, while appropriating disability as an eroticized symbol of feminine passivity. Audiences participated as lurid observers, the safe containment of the disabled object granting them social permission to stare.

Because the universal fluid could be "communicated, propagated, and augmented by sound," music often played an important role in mesmeric rituals (Franklin 1785, 21; Kennaway 2012). Indeed, a report on the practice (prepared for the King of France by a group of scholars, including Benjamin Franklin; see also Gallo and Finger 2000) observed that "a piano forté is placed in one corner of the apartment, and different airs are played with various degrees of rapidity; vocal music is sometimes added to the instrumental" (Franklin 1785, 23). The report explains that the piano "diffuses [the fluid] through the air," according to the following magnetic principles:

> That sound, conformably to the theory of M. Mesmer, is also a conductor of the magnetism, and that to communicate the fluid to the piano forté, nothing more is necessary than to approach to it the iron rod [a "conductor of magnetism"]; that the person who plays upon the instrument furnishes also a portion of the fluid, and that the magnetism is transmitted by the sounds to the surrounding patients. . . . It has been observed, that the changing of the key and the time, in the airs played upon the piano forté, had an effect upon the patients; so that a quicker motion agitates them more, and renews the vivacity of their convulsions. (24–26)

The sympathetic vibration of consonant harmonies and the invigorating agitation of rhythmic activity: these two principles are familiar from the above discussion of tarantism, among other healing traditions. In his execution of both, Schubert seems to have been an exemplary musical mesmerist: his performance included a series of pieces in B♭ major (apparently the harmony best tuned to the patient's fluid), including several dances, the tempos of which surely sought the "quicker motion" that might agitate the patient's incapacitated body toward rehabilitation.

Throughout the early nineteenth century, similar tales of healing abound, but without mesmerism's stagecraft (Kramer 2000). In many instances, a male healer plays music that jolts a formerly inert, listless female patient into motion. In his memoir of 1829, a notable nineteenth-century practitioner of hypnosis describes how he used a Jew's harp to "magnetize" the drinking water of his emotionally distressed patient: "she became clearer," "spoke in rhythm," and (after drinking her water) "involuntarily began to sing" (Kerner 1845, 73). An analogous episode is retold by Alan Walker, who describes a musical healing of an elderly woman in a noncommunicative state by a young Franz Liszt: upon hearing his performance, "she appeared to enter a highly charged state, and she vibrated to every chord," her motions resembling an "electrical discharge" (1989, 151–152). More emphatic, in both its gendering of the healing and its depiction of blockage and motion, is an episode from Peter Joseph Schneider's important music-therapy treatise *System einer medizinischen Musik* (1835). With a few singers, a piano, and an Aeolodicon (reed organ or harmonium), Schneider offered musical remediation to a patient diagnosed with "menstrual constipation":

> She sat there silent, silent as a fish, and without motion . . . . Now I began to play my fantasy [on the Aeolodicon] in a key appropriate to the spiritual voice of one sunk deep in melancholy, E♭ minor. Quietly, then swelling into the strongest fortissimo, and from there to a moderate piano; fully at times, at times restrained, sustaining the pure triad and then continuing through the full range of harmonic changes and variations. After the third crescendo . . . the patient deeply drew her breath and violently shuddered. But still she remained mute and otherwise still. I preludized on G♭ major and gave the pitch for the chorus to begin. A crescendo led to their breath, and with the opposite they sang their first words. They had not as yet subsided, and one could hear the leading tenor clearly, discernably and quietly supported by the choir—when suddenly the patient cried out "O! Antonio, Antonio!" Thereupon she cried so much that we had to stop our singing. The unfortunate patient, deeply shaken, had to be brought to bed. Once her room was illuminated we made the pleasing discovery that with the utterance of "O! Antonio, Antonio!" her flow of monthly bleeding had been released; and it had thickly, very thickly covered the entire floor. (Schneider 1835, 2:267–268; translation adapted from Kramer 2000, 345–346)

This strange tale illustrates several important traits of nineteenth-century musical hypnosis, including the gendered hierarchy that grants the male healer dominance over his female subject (here, her menstruation and her emotions), as well as the underlying principles of musical mobility (rhythmic activity) and sympathetic vibration (harmonies provoking motion in a stagnant body).

In light of the list of key characteristics in C. F. D. Schubart's *Ideen zu einer Ästhetik der Tonkunst* (1806), extensively quoted by Schneider throughout his treatise, the therapeutic use of E♭ minor and G♭ major seems especially significant. According to Schubart, Schneider's first key (E♭ minor) reflects the "anxiety of the soul's deepest distress, brooding despair"; from there, Schneider modulates to G♭ major, a key that reflects "triumph over difficulty" and the "free sigh of relief uttered when hurdles are surmounted" (Steblin 2002, 117). From difficulty to triumph, confinement to release, sickness to health, depression to ecstasy, Schneider's harmonic modulations enact the trajectory of the cure narrative. Further, the erotic descriptions of musical thrusting—louder, softer, louder, softer, then climaxing with a sudden outburst ("O! Antonio!")—create a sexual subtext for the patient's vaginal bleeding.

Transcendence

The cure rituals associated with mesmerism also reflect broader Romantic concerns about artistic creativity, in which expression sought freedom from the confines of hermetic corporeality. As John Warne Monroe (2008) writes, "the lucid somnambulist, her mind liberated from bodily constraints, freed to seek enlightenment in the ethereal world of pure essences, gave a powerfully concrete form to a central Romantic ideal" (70). The body, cast as deficient or disabled, impedes the spiritual flight of the subjective imagination. This recurring trope is but a secular reflection of biblical texts like Isaiah 35:5–6 ("the eyes of the blind shall be opened") and an artistic reflection of De Martino's *taranta* (providing "color to the colorless")—but instead of striving past the sinful body to reach Zion, the poet strives beyond corporeal pollutants (trappings of mimetic art) toward a state of pure self-expression. These transcendental metaphors stigmatize the body as inherently weak and frail, and often appropriate disability to concretize this characterization.

Cherished for its ineffability, music offered the Romantic subject one way to bypass (and in doing so, to "cure") corporeality, whether as a conduit for the universal fluid within mesmerism or as an aestheticization of the expressive ideal. Indeed, the erotic subtext of hypnotic healings also finds its parallel in highly suggestive descriptions of the ecstatic nature of creative self-expression. In his autobiographical novel *Hesperus* (1795), Jean Paul (also an occasional practitioner of magnetism) writes of a musician attempting to transparently express his feelings through improvisation, while perspiring sweat and crying tears (among other bodily fluids, presumably): "Nothing exhausts me as much, nothing soothes me more than improvising at the clavichord," he writes. Eroticizing the creative process, while also affirming the limitations that immanence has on creativity, he concludes, "I could improvise myself to death" (Richards 2001, 175). Honoré de Balzac, another devotee of mesmerism, staged the romantic attraction of two fictional lovers, Ursula and Savinien, at the keyboard. Ursula, "threw into [the music] the feelings which were agitating her being," so that "her soul spoke to the young man's soul and wrapped it, as in a cloud, with ideas that were almost visible." Alluding to principles of mesmerism, and to the erotic nature of its practice, Balzac writes, "Genuine feelings have a magnetism of their own, and Ursula was willing to show her soul, as a coquette her dresses to be admired. . . . Ursula was never more honest and candid than at this moment [i.e., during her musical performance], when she was born again into a new life" (1896, 202–203). Just as music might diffuse the numinous "universal fluid" in a healing ritual, here music serves as a conduit for erotic transcendence, as the most pure, unsullied method for the communication of feeling. As the subject enters this electrified state (entering into a kind of spiritual ideal—or "born again into a new life," as Balzac writes), the physical world dissolves. The body no longer impedes self-expression.

Likewise, in the poem "Auflösung" by Johann Mayrhofer (trained in both Catholic theology and Greek antiquity), "embers of rapture" provide the fuel for violent self-immolation:

> From out of all recesses
> of my soul gush pleasing powers
> that envelop me with heavenly singing—
> Sink under, world, and never disturb
> the sweet ethereal choirs. (Howe 2009, 315)

Reflecting a long poetic association between death and orgasm, as well as the erotic connotations of the metaphysical trance, the demise of the poet's body is not bleak, but pleasurable (gushingly so). In its finale, Franz Schubert's setting of this poem (D. 808) abandons a twisting chromatic pathway for euphoric alternations of the tonic (G major) and its subdominant (C major)—suggesting arrival within the transcendent harmonic order of a *musica mundana*, the "sweet ethereal choirs" now unpolluted by corporeal noise. This is a vision of creative ecstasy—but its fulfillment, relying on a narrative prosthetic, achieves meaning only after the disabling constraint of corporeality is burned and annihilated (Howe 2009). Here, fleshliness itself is immobilizing, preventing the transcendental flight of the spirit; but music propels the spirit forward, dissolving earlier confines. Another Schubert setting—"Die junge Nonne" (D. 828)—constructs a similar expressive trajectory, blending fiery immolation and transcendent music: freed from "earthly bonds," the previously immobilized title figure finds guidance in the sound of a bell, directing her spiritual flight heavenward to join the heavenly bridegroom as his bride ("Komm, himmlischer Bräutigam, hole die Braut"). The body, cleansed by immolation, is discarded; the spirit flies on. Notably, Lisa Feurzeig has linked the poem's poet, Jakob Nikolaus Craigher de Jachelutta, to Friedrich Schlegel, who was one of the main figures of the mesmerist circle in which Schubert performed (1997, 241–243). When read in this light, the poem seems to substitute the sexually dominant mesmerist healer with Christ the bridegroom, for whom the nun abandons the physical trappings of corporeality. The body, discarded for all of its sinful and disabling impurities, functions as the narrative prosthetic that shapes the arc of spiritual transcendence: the journey is an aestheticized rehabilitation, given definition by the disabilities it surpasses and conquers. Music—pianos sounding, choirs singing, bells ringing—sparks the metamorphosis from physical fixity to spiritual mobility.

Novalis offers some insight into this secular theology. In a well-known aphorism that recalls Marsilio Ficino's fusion of the physician and musician, Novalis fancifully proposes a demedicalized artfulness in the therapy of illness: "Every illness is a musical problem—its healing a *musical solution*" (Krell 1998, 62). Elsewhere, he muses on the illness's purpose and its social function: "Could not *illness* be a means to a higher synthesis—the more horrific the pain, the higher the pleasure concealed within it," he speculates. "Every illness is perhaps the necessary *commencement* of the more intense conjunction of two creatures—the necessary beginning of love" (50). Implicit here is the punishing ideology of the cure narrative, of course: bodily impairment, illness, and pain are the narrative impediments on which resolutions rely, and they function to allegorize an unrelated struggle or quest—toward synthesis, or love, or pleasure, or some other state of fulfillment. Mesmerists executed this trajectory through healing rituals, in which the incapacitated patient and her healer fused in an erotic transfer of universal fluid. In this scenario, the disabled body allegorized the supremacy of healer over patient, man over woman. The Romantic artist embarked on a similar path, in his flight from the body toward the expressive ecstasy of creativity. In this scenario, the body itself is disabling; it is an index of the corporeal world that prohibits numinousness. In both practices, medical and aesthetic, music enables the cure.

A belated but extremely potent expression of this Romantic trajectory may be found at the end of the nineteenth century, in Richard Strauss's *Tod und Verklärung* (1889). According to the composer, the programmatic work concerns "the dying hours of a man who had striven towards the highest idealistic aims, maybe indeed those of an artist" (Del Mar 1986, 1:77–78). The immobilized subject's ailments are extensive and include "heavy irregular breathing," "horrible agonies," and limbs that "shake with fever." These disorders are musically depicted throughout the work: in the austere opening, the weak heartbeat of the timpani echoes rhythmically jagged "breathing" from the strings; later, interrupting a dreamlike interlude, violent outbursts of angular chromatic melodies gruesomely depict a body consumed by pain. But this suffering is assigned a retrospective purpose by the tone poem's narrative arc. Executing what Rosemarie Garland-Thomson (2004) identifies as the "cultural logic of euthanasia," death "cures" the subject of his body so that he might finally attain the artistic ideals he sought but could not find on earth. As Alexander Ritter's programmatic poem for the work describes with great intensity,

> Coldly and scornfully the world places barrier after barrier in the path of his efforts. Whenever he believes he is near his goal, he is greeted by a thunderous "Halt!" "Turn the barrier into a rung by which to rise ever higher and higher!" Thus he strives, thus he climbs and does not abandon his sacred impulse. What he has sought all this time

> with his heart's deepest longing, he still seeks while bathed in mortal sweat, seeks—but, alas, cannot find it. Even if he conceives of it more clearly, even if it gradually grows upon him, still he can never exhaust its possibilities, can never consummate it in his mind. Then the last blow of death's iron hammer rings out, breaks the earthly body in two and covers his eyes with the night of death. But he hears mightily resounding from heaven that which he sought here longingly: world-redemption, world-transfiguration. (in Strauss 1979, 100–102, adapted from the translation of Stanley Appelbaum)

Suggesting the erotic ecstasy of artistic expressivity, this text is frantic, its short phrases gasping for breath. The subject's creative flight is put to increasingly difficult physical tests—as exemplified by his accumulation of disabilities, forming "barrier after barrier" (*Schrank' auf Schranke*) that require traversal. Impeding self-actualization, these disabilities collectively operate as a narrative prosthetic, stalling resolution while also defining by counterexample its very desirability.

FIGURE 14.2 Richard Strauss, *Tod und Verklärung*, Op. 24, ending measures (in piano reduction by Otto Singer).

Strauss's work ends with a complete statement of a theme previously presented only in clipped, fragmentary form: this is the long-delayed, long-awaited fulfillment of an artistic ideal, now unencumbered by earthly impairment. Outlining a triad, it is a musical signifier of music itself (Figure 14.2). In the final measures, Strauss seems to show us how its melody unfolds from C major (the most fundamental of all keys): first the full theme (G-C-D-E-E-D-C), then just its buried triad (C-E-G), then just its fifth (G-C), then just its tonic (C). The subject has sought for, and now joined through death, this cosmic harmonic order of a *musica mundana*, this mythic ideal of Pythagorean purity: it is here where the artist seeks expression without the confines of corporeality, it is here where bodily difference is shunned in favor of able-bodied uniformity, it is here where (in a cruel, eugenic vision of perfection) the blind shall see and the lame shall leap.

Note

1. On the musicality of the human pulse within late medieval thought, refer to Siraisi (1975).

References

Baglivi, Giorgio. 1723. *The Practice of Physick*. 2nd ed. London.

Balzac, Honoré de. 1896. *Ursula*. Translated by Katharine Prescott Wormeley. Vol. 8 *of La Comédie humaine of Honoré de Balzac*. Boston: Roberts Brothers.

Bassler, Samantha. 2016. "Madness and Music as (Dis)ability in Early Modern England." In *The Oxford Handbook of Music and Disability Studies*, edited by Blake Howe, Stephanie Jensen-Moulton, Neil Lerner, and Joseph N. Straus, 529–538. New York: Oxford University Press.

Billings, William. 1770. *The New-England Psalm-Singer; or, American Chorister*. Boston: Edes and Gill.

Clare, Martin. 1735. *The Motion of Fluids, Natural and Artificial*. London.

Couser, G. Thomas. 1997. *Recovering Bodies: Illness, Disability, and Life Writing*. Madison: University of Wisconsin Press.

Couser, G. Thomas. 2009. *Signifying Bodies: Disability in Contemporary Life Writing*. Ann Arbor: University of Michigan Press.

Daboo, Jerri. 2010. *Ritual, Rapture, and Remorse: A Study of Tarantism and "Pizzica" in Salento*. Oxford: Peter Lang.

Dahlhaus, Carl. 1982. *Esthetics of Music*. Translated by William Austin. Cambridge: Cambridge University Press.

Davis, Lennard. 2002. *Bending over Backwards: Disability, Dismodernism, and Other Difficult Positions*. New York: New York University Press.

Del Mar, Norman. 1986. *Richard Strauss: A Critical Commentary on His Life and Works*. Reprint ed. Ithaca, NY: Cornell University Press.

De Martino, Ernesto. 2005. *The Land of Remorse: A Study of Southern Italian Tarantism*. Translated by Dorothy Louise Zinn. London: Free Association Books.

Feurzeig, Lisa. 1997. "Heroines in Perversity: Marie Schmith, Animal Magnetism, and the Schubert Circle." *19th-Century Music* 21: 223–243.

Findlen, Paula, ed. 2004. *Athanasius Kircher: The Last Man Who Knew Everything*. New York: Routledge.

Frank, Arthur W. 1995. *The Wounded Storyteller: Body, Illness, and Ethics*. Chicago: University of Chicago Press.

Franklin, Benjamin. 1785. *Report of Dr. Benjamin Franklin and Other Commissioners, Charged by the King of France, with the Examination of the Animal Magnetism, as Now Practised at Paris*. London: J. Johnson.

Gallo, David A., and Stanley Finger. 2000. "The Power of a Musical Instrument: Franklin, the Mozarts, Mesmer, and the Glass Harmonica." *History of Psychology* 3: 326–343.

Garland-Thomson, Rosemarie. 1997. *Extraordinary Bodies: Figuring Physical Disability in American Culture and Literature*. New York: Columbia University Press.

Garland-Thomson, Rosemarie. 2004. "The Cultural Logic of Euthanasia: 'Sad Fancyings' in Herman Melville's *Bartleby*." *American Literature* 76: 777–806.

Goldschmidt, Harry. 1983. "Schubert und kein Ende." *Beiträge zur Musikwissenschaft* 25: 288–292.

Gouk, Penelope, ed. 2000. *Musical Healing in Cultural Contexts*. Aldershot: Ashgate.

Grave, Floyd. 2008. "Recuperation, Transformation, and the Transcendence of Major over Minor in the Finale of Haydn's String Quartet, Op. 76, No. 1." *Eighteenth-Century Music* 5 (1): 27–50.

Gur, Golan. 2008. "Body, Forces, and Paths: Metaphor and Embodiment in Jean-Philippe Rameau's Conceptualization of Tonal Space." *Music Theory Online* 14 (1).

Hawkins, Anne Hunsaker. 1999. *Reconstructing Illness: Studies in Pathography*. 2nd ed. West Lafayette, IN: Purdue University Press.

Hogan, Brian. 2016. "'They Say We Exchanged Our Eyes for the Xylophone': Resisting Tropes of Disability as Spiritual Deviance in Birifor Music." In *The Oxford Handbook of Music and Disability Studies*, edited by Blake Howe, Stephanie Jensen-Moulton, Neil Lerner, and Joseph N. Straus, 112–130. New York: Oxford University Press.

Horden, Peregrine, ed. 2000. *Music as Medicine: The History of Music Therapy since Antiquity*. Aldershot: Ashgate.

Howe, Blake. 2009. "The Allure of Dissolution: Bodies, Forces, and Cyclicity in Schubert's Final Mayrhofer Settings." *Journal of the American Musicological Society* 62: 271–322.

Howe, Blake. 2016. "Saul, David, and Music's Ideal Body." In *The Oxford Handbook of Music and Disability Studies*, edited by Blake Howe, Stephanie Jensen-Moulton, Neil Lerner, and Joseph N. Straus, 539–559. New York: Oxford University Press.

Katz, Ruth, ed. 1987. *Substance. Vol. 1 of Contemplating Music: Source Readings in the Aesthetics of Music*. New York: Pendragon.

Keith, Lois. 2001. *Take Up Thy Bed and Walk: Death, Disability and Cure in Classic Fiction for Girls*. New York: Routledge.

Keller, Peter E., and Martina Rieger, eds. 2009. "Musical Movement and Synchronization." Special issue, *Music Perception* 26 (5).

Kennaway, James. 2012. "Musical Hypnosis: Sound and Selfhood from Mesmerism to Brainwashing." *Social History of Medicine* 25: 271–289.

Kerner, Justinus. 1845. *The Seeress of Prevorst, Being Revelations Concerning the Inner-Life of Man and the Inter-Diffusion of a World of Spirits in the One We Inhabit*. Translated by Mrs. Crowe. London: J. C. Moore.

Kramer, Cheryce. 2000. "Music as Cause and Cure of Illness." In *Music as Medicine: The History of Music Therapy since Antiquity*, edited by Peregrine Horden, 338–352. Aldershot: Ashgate.

Krell, David Farrell. 1998. *Contagion: Sexuality, Disease, and Death in German Idealism and Romanticism*. Bloomington: Indiana University Press.

Kümmel, Werner Friedrich. 1977. *Musik und Medizin: Ihre Wechselbeziehungen in Theorie und Praxis von 800 bis 1800*. Freiburg: Alber.

Larson, Steve. 2012. *Musical Forces: Motion, Metaphor, and Meaning in Music*. Bloomington: Indiana University Press.

Linton, Simi. 1998. *Claiming Disability: Knowledge and Identity*. New York: New York University Press.

Lüdtke, Karen. 2009. *Dances with Spiders: Crisis, Celebrity, and Celebration in Southern Italy*. New York: Berghahn Books.

Makari, George. 2015. *Soul Machine: The Invention of the Modern Mind*. New York: W.W. Norton.

Mead, Richard. 1762. *The Medical Works of Richard Mead*. London.

Mitchell, David T., and Sharon L. Snyder. 2000. *Narrative Prosthesis: Disability and the Dependencies of Discourse*. Ann Arbor: University of Michigan Press.

Monroe, John Warne. 2008. *Laboratories of Faith: Mesmerism, Spiritism, and Occultism in Modern France*. Ithaca, NY: Cornell University Press.

Owen, Charles. 1742. *An Essay towards a Natural History of Serpents*. London.

Quayson, Ato. 2007. *Aesthetic Nervousness: Disability and the Crisis of Representation*. New York: Columbia University Press.

Rankin, Susan. 2005. "*Naturalis concordia vocum cum planetis*: Conceptualizing the Harmony of the Spheres in the Early Middle Ages." In *Citation and Authority in Medieval and Renaissance Musical Culture: Learning from the Learned*, edited by Suzannah Clark and Elizabeth Eva Leach, 3–19. Woodbridge: Boydell.

Regnault, Père (Noël). 1731. *Philosophical Conversations; or, A New System of Physics*. Translated by Thomas Dale. London.

Richards, Annette. 2001. *The Free Fantasia and the Musical Picturesque*. Cambridge: Cambridge University Press.

Schneider, Peter Joseph. 1835. *System einer medizinischen Musik*. Bonn: Carl Georgi.

Schubart, Christian Friedrich Daniel. 1806. *Ideen zu einer Ästhetik der Tonkunst*. Edited by Ludwig Schubart. Vienna: J. V. Degen.

Siraisi, Nancy G. 1975. "The Music of Pulse in the Writings of Italian Academic Physicians (Fourteenth and Fifteenth Centuries)." *Speculum* 50: 689–710.

Solomon, Maynard. 2003. *Late Beethoven: Music, Thought, Imagination*. Berkeley: University of California Press.

Steblin, Rita. 2002. *A History of Key Characteristics in the Eighteenth and Early Nineteenth Centuries*. 2nd ed. Rochester: University of Rochester Press.

Straus, Joseph N. 2011. *Extraordinary Measures: Disability in Music*. New York: Oxford University Press.

Strauss, Richard. 1979. *Tone Poems: Series 1*. New York: Dover.

Tomlinson, Gary. 1993. *Music in Renaissance Magic: Toward a Historiography of Others*. Chicago: University of Chicago Press.

Tomlinson, Gary, ed. 1998. "The Renaissance." In *Strunk's Source Readings in Music History*, revised ed., edited by Leo Treitler, 281–508. New York: W.W. Norton.

Tyson, John R. 2006. "The Methodist National Anthem: 'O for a Thousand Tongues to Sing' and the Development of American Methodism." In *Sing Them Over Again to Me: Hymns and Hymnbooks in America*, edited by Mark A. Noll and Edith L. Blumhofer, 20–42. Tuscaloosa: University of Alabama Press.

Walker, Alan. 1989. *The Virtuoso Years, 1811–1847. Vol. 1 of Franz Liszt*. Revised ed. Ithaca, NY: Cornell University Press.

Watson, J. R. 2010. "Charles Wesley and the Music of Poetry." In *Music and the Wesleys*, edited by Nicholas Temperley and Stephen Banfield, 26–40. Urbana: University of Illinois Press.

Wesley, Charles. 1849. *The Journal of the Rev. Charles Wesley*. Edited by Thomas Jackson. London: John Mason.

Wesley, John, and Charles Wesley. 1740. *Hymns and Sacred Poems*. London: Strahan.

Wheatley, Edward. 2010. "Medieval Constructions of Blindness in France and England." In *The Disability Studies Reader*, 3rd ed., edited by Lennard J. Davis, 63–76. New York: Routledge.

CHAPTER 15

VIRTUOSITIES OF DEAFNESS AND BLINDNESS

Musical Performance and the Prized Body

STEFAN SUNANDAN HONISCH

THE terrain of "music and the body" has become more densely populated with the recent emergence of music and disability studies (Lubet 2010, 2015; Straus 2006, 2008, 2011; Lerner and Straus 2006). To be sure, race, gender, and sexuality—with their complex histories, politics, and aesthetic practices—had already entered into the collective consciousness of music studies, bearing out Richard Leppert's assertion that "music's aural and visual presence constitutes both a relation to and a representation of the body" (1993, xx). In this sense, disability studies is not so much an interloper as a latecomer to critical work that explores the relationship between music, embodiment, and, relatedly, questions of power, privilege, oppression, and resistance (McClary 2002; Brett, Wood, and Thomas 1994; Kerman 1985; Southall, 1999). Uniting their diverse critical projects is a shared concern with how music embodies gender, race, sexuality, and disability through composition, performance, and reception. Their collective task lies in demonstrating how musical representations of bodily difference constituted within each of these spheres of musical activity assume the existence of a normal body against which bodies that depart from purportedly normal characteristics become the objects of normalizing interventions, both metaphorical and actual.

Mining the critical veins of music and the body with the tools of disability studies, this chapter explores how the convergence of disability and virtuosity in competitive music performance displaces the normal body from its privileged status. What Adam Overton (2006) calls "the virtuosic body" prized in competition is, like the disabled body, abnormal, sometimes extraordinarily so.[1] The difference is that in competitive musical performance, the virtuosic body takes pride of place as the body to which both normal and disabled bodies must aspire. When the disabled body manifests symptoms of virtuosity in a competitive environment, however, musicians, critics, and the public carry out a variety of diagnostic procedures in order to make sense of this seeming paradox.

Two illustrative case studies show how conceptual distinctions between able and disabled bodies fall apart when the works of a deaf composer are performed by a blind pianist in the competitive arena. More so than in an ordinary performance setting, competitive music performance forefronts virtuosity to the extent that anything less, including mere ability, is taken as a sign of disability. The music discussed in the two case studies is the *Appassionata* and *Hammerklavier* sonatas of Ludwig van Beethoven, the pianist is Nobuyuki Tsujii, and the competitive arena is that of the Cliburn competition, in which Tsujii shared the top prize in 2009 (Honisch 2015, 2016; Johnston 2015; McCormick 2015; Yoshihara 2009). The encounter between a "deaf composer" and a "blind pianist" imagined in Tsujii's critical and popular reception is of interest from a disability studies perspective not because of the simultaneous presence of two musicians with disabilities within a single performance context but because of how competitive performance, in particular, brings into play representations of virtuosity that exist in unfixed relationships to both the normal and disabled body.

The visual and sonic "encounter" between Beethoven's deafness and Tsujii's blindness can be understood in the spectacular arena of the Cliburn competition as a convergence of music and the body in which the disparities between the virtuoso's prized body[2] and the devalued sensory experiences of deafness and blindness are stark. Normatively defined as barriers to virtuosity, sensory disabilities are often reduced to the role of antagonists within musical narratives of the body, a framing that disability studies approaches counteract (Deaville 2013; Straus 2011). The staging of the virtuosic body takes place in an arena in which the normative body—that is, the body against which all others are judged—is one defined not by its "normalcy," that is, by its similarity to other bodies capable of seeing, hearing, and moving, but rather by its extraordinary capacity to command itself, the musical instrument, and the attention of the audience. At the same time, however, the communicative medium of music performance, competitive or not, requires the bodies of disabled musicians simultaneously to transcend and to perform disability (Straus 2011, 126). While the able-bodied performer likewise has to create a dual performance of music and ability, the taken-for-granted status of the normal, able body obscures the degree to which corporeal normalcy—like disability—must be overcome in order to attain mastery over the instrument (Straus 2011; Howe 2010; Lerner 2006). Complicating matters further within the competitive sphere is that able-bodied and disabled performers must also perform virtuosity. Tsujii's performances of the *Appassionata* and *Hammerklavier* sonatas during the competition raise questions about the simultaneous presence and absence of disabled virtuosities in a context that prizes bodies capable of transcending the limits of normal ability. Disabled virtuosities are defined here as relational, signifying the interaction and interdependence of bodies with each other, rather than mastery over an individual body. Conceived in this way, embodied musical practices that unfold through encounters between historical accounts of Beethoven as a deaf pianist and composer, and present-day responses to Tsujii as a blind pianist can be interpreted using the critical apparatus elaborated by Tobin Sieberss (2010) as *disability aesthetics*, in which the focus of analysis is on how "some bodies make other bodies feel" (20).

The aesthetic significance of a blind pianist performing the music of a deaf composer can also be situated in relation to both the reception of virtuoso performers and performers with disabilities, digging below the surface engagement in critical and popular responses to Tsujii's performances of Beethoven, which often tread familiar ground. While connections between "the deaf composer" and "the blind pianist" were a distinct feature of Tsujii's reception during the 2009 Cliburn competition, they obscure disability's semiotic contradictions within the popular imagination as a sign of loss and of contact with the extraordinary. It is important to emphasize that the purpose of this chapter is not simply to argue that deafness and blindness have been framed negatively, and that what is needed is to celebrate deafness and blindness as wellsprings of exceptional musical insight and technical brilliance. Such an approach would romanticize disability, and disregard the practices of exclusion that face disabled bodies at every turn, on and off stage. Instead by approaching the issue through engagement with the world of competitions, this chapter moves away from pat solutions, directing analytical attention to how both able and disabled bodies can be displaced by the virtuosic body. This displacement challenges—even as it asserts—the rules and conventions associated with music performance and reception.

The landscape of "music and the body" contains within it, then, a region defined by the convergence of disability and virtuosity, an analytical space narrowed in this chapter to "musical performance and the prized body." By arguing that the normal, able body is effectively disabled by the normalization of virtuosity in competition, I do not mean that "everyone is disabled," the well-intentioned, but ultimately futile platitude castigated within disability studies for its failure to consider the real structural and attitudinal inequalities faced by disabled people (Titchkosky 2011, 56). Instead, virtuosity, ability, and disability are interlocking corporeal performances. Because ability and disability each constitute different articulations of music to the virtuosic body in a competitive environment, a straightforward opposition between normal ability as that which serves as the necessary basis for developing virtuosity, and disability as that which *prevents* virtuosity, fails to withstand empirical realities of reception practices that are ambiguous at best.

What musicians and audiences individually see and hear, and what they do not see and hear are inseparable parts of a sensory and corporeal fabric that makes disability simultaneously absent and present as virtuosic. For this reason, Leppert's (1993) assertion of the inescapable presence of the performative body in music (xx) offers a useful point of entry into this chapter because of the immediate questions it raises about deafness and blindness as forms of virtuosity: how does music's visual and sonic presence and absence make disabled, normal, and virtuosic bodies visible and audible? Furthermore, whereas the traditional binary separation of ability and disability assumes a hierarchy, what are we to make of music's sensory absences in constituting different types of bodies? If the normal body is no longer a stable reference point between disability and virtuosity, what are the implications for understanding music and the body?

Constructions of the virtuoso pianist's "prized body" through competitive music performance mirror the construction of what Paul Longmore has memorably termed "the severely able-bodied" (as quoted in Garland-Thomson 2005, 33). This is because the competitive arena is not merely an institution that disables and excludes, as recent scholarship reminds us (McCormick 2009, 2015). Indeed, at the entry level, competitions are "inclusive in that musicians of every nationality, ethnic origin, religion and gender can compete; competition organizations also seek to expand the audience and the available forms of participation through ancillary programming, outreach initiatives and communications technologies" (McCormick 2014, 2264). On one level, this observation is unproblematic, although it is worth noting, at this initial level, that disability is missing from the multiple, intersecting identity categories that McCormick enumerates. On a deeper level, as McCormick herself has argued, institutional postures of inclusiveness are encumbered by the weight of the normal body, whose invisible and silent privilege circumscribes how music relates to the body in performance and reception (2015). To illustrate this paradox between inclusive and exclusionary practices, it is helpful to understand the ways in which the relationship between music and the body in competition is laden not merely with aesthetic judgment but also with ethical and moral judgments.

Competitive music performance articulates music to the prized body, granting virtuosity significance beyond mere ability, in effect, making a virtue out of virtuosity (Deaville 2013; O'Dea 2000). Analogies between musical and athletic performance are perhaps overly familiar, as Naomi Cumming observes in her book *The Sonic Self: Musical Signs and Subjectivity* (2000, 20). Yet such analogies can also reveal much about sociocultural practices that mark particular kinds of bodies as deserving of recognition, linking ability to character: "mundane sports metaphors . . . invite gendered interpretations or heroic tropes that emphasize traits such as age or disability to amplify the significance of achievements and underscore moral worth" (McCormick 2015, 84).

Responses to Tsujii's performances of the *Appassionata* and *Hammerklavier* sonatas discussed in the following sections of this chapter express a fascination with the imagined encounter between "the deaf composer" and "the blind pianist" that runs swift and deep, yet the movements of our imagination in the present cannot be understood independently of the slower pulse of history. Virtuosity and disability have often been framed as incongruous features, the quizzical reception of disabled virtuosity intermingled with awe and disapproval at mere technical display. Uncovering the reasons behind the puzzled attitudes that implicitly ask how a body can be simultaneously disabled and virtuosic brings into play these multiple ways of responding to artistic depth, technical skill, as well as social and cultural attitudes toward the body that shape the dynamics of performance and reception (Lubet 2015).

As a forum for "exemplary able-bodiedness," musical performance blurs the edges between ability and disability. Taken together, the performative modes of ability and disability contrast "what a performer can do (with one part of the body) with what a performer cannot do (with another part of the body)" (Howe 2015, 191). In the next section of this chapter, I take up Howe's notion of "disabled performance," pursuing the contradictory figure of the disabled virtuoso, a figure that recent scholarship on disability and musical performance has looked at, and listened to in new ways (Howe 2015; Deaville 2013; Burstein 2006).

Virtuosity and the Body: A Disability Studies Approach

Musical performance can be understood in general terms as an encounter between senses, minds, and bodies, in which "the visibility and audibility of a disability seldom align, producing revealing and interesting incongruities" (Howe 2015, 192). Beethoven's deafness and Tsujii's blindness are simultaneously visible and invisible, musically audible and inaudible, depending on the performance context and on the sensibilities of a given audience. In the mediated formats of sound recordings, for example, Tsujii's performances of the *Appassionata* and *Hammerklavier* sonatas during the Cliburn competition render his blindness relatively invisible. Outside of the competition arena and concert circuit, Tsujii's blindness is kept visible through its circulation in media images of the pianist and in critical and popular commentary on his Cliburn competition performances and in his postcompetition reception (Honisch 2016).

From a disability studies perspective, musical performances by disabled bodies jar the senses and unravel the fabric of musical space and time. At the same time, however, performances by disabled bodies can reshape the way audiences experience a familiar repertoire, not because of their interpretive, technical, or affective singularity, but because their inevitable transgression of expectations placed on the virtuosic body. The physical separation of the performer and audience is metaphorically dissolved in the concert hall, and even in the technological mediation of sound and video recordings "the virtuoso cannot keep his body to himself" (Kramer 2002, 92). Media representations of the figure of the "disabled virtuoso" transmit, with varying degrees of strength, the assumption that disability and virtuosity are paradoxical, or at the very least, that the convergence of the two is worth noting (Marsallis, 2014; Onions 2015).

The precariousness of being simultaneously disabled and virtuosic is vividly captured in Howe's (2010) study of Paul Wittgenstein and his performances of one-handedness. As Howe explains, "Wittgenstein participated in this quest for transcendental virtuosity while his disability was simultaneously marked by the ideology of ability that fueled it" (144). Pianists who perform disabled virtuosities in the competitive arena must therefore deal with, if not resolve, a dissonance between transcendent musical performance and an ideology of corporeal mastery, which all too often assumes that disabled virtuosity does not belong. At the same time, however, while institutional practices stipulate a tightly choreographed set of responses, actual critical and popular reception often departs from this script. To the extent that disabled bodies are seen, heard, and felt to do things that normal, able bodies cannot do in competition, the disabled body, too, becomes virtuosic at the level of reception.

The intertwining of Beethoven's deaf virtuosity and Tsujii's blind virtuosity in the pianist's competition performances of the *Hammerklavier* and *Appassionata* sonatas complicate efforts to separate virtuosity's public exterior and the virtuoso's embodied, subjective interior. Like disability, virtuosity is both stigmatized and valorized (Hennion 2012). Disabled virtuosities intensify this cultural ambivalence toward performances of virtuosity understood as inherent in the performer's command of the instrument rather than in particular traits of the virtuoso's musical sensibilities (Cumming 2000; Bernstein 1998). Exceeding the sum of its stigmatized and valorized forms, virtuosity connotes "performance aimed at a public that outwardly demonstrates exceptional agility . . . especially as [it] relates to speed of execution, both admired as demonstrations of a divine gift, and condemned as a diabolical perversion of musical performance" (Hennion 2012, 126). Virtuosity in this way comes to represent a "corporeal and risky litmus test of the mysterious power of music and musician's talents . . . primary emotion, collective intoxication, and moments of trance" (Hennion 2012, 126).

It is important not to overstate the purported similarities between virtuosity and disability. Unlike virtuosity, disability does not require a circumscribed aesthetic context in order to register as extraordinary, a reality experienced by people with disabilities every day: "As soon as the disabled body enters the public sphere—on the street, in social institutions, in medical institutions or in popular theatre, television, film or literature—it becomes a spectacle" (Hadley 2014, 2; see also Sandahl and Auslander 2005). As a consequence, the element of spectacle that attends the disabled body in musical performance weakens the autonomy of musical performance from the exigencies of daily life (Howe 2010, 2015; Straus 2011; Lubet 2010; Lerner 2006).

The first half of the twentieth century witnessed the ascent of several blind pianists to the status of international piano competition laureates (Honisch 2015). Yet the medical and sociocultural markings of blindness as the absence of sensory experience, and the diminishment of selfhood, have been sufficiently powerful to render the figure of the blind virtuoso pianist as anomalous. Consider, for example, the following excerpt from a report on music pedagogy for blind students during this period: "In spite of the fact that a high degree of virtuosity is seldom acquired by the blind pianist, the piano is nevertheless the most practical instrument for those without sight" (Hough 1940, 65). There is a subsequent acknowledgment that "the blind pianist has an advantage" in being able not only to play recreationally but also to earn money "by piano teaching on the side" (65). Apart from teaching, tuning pianos was also considered a viable source of employment for blind people, as indicated by the training programs in schools for the blind in Europe and North America during the nineteenth and twentieth centuries. Despite publicized instances of blind pianists proving their mettle in competition against sighted pianists, however, blind pianism was reduced largely to the domains of necessity and recreation. Blindness has therefore often been rendered invisible and inaudible within the high-stakes world of publicly consumed virtuosity. Before Imre Ungár's second prize at the 1932 Chopin competition (Honisch 2015), Alfred Hollins had attained recognition as a pianist of stature, although the latter himself suggests in his 1936 memoirs that the advancement in twentieth-century compositional styles made virtuosity inaccessible to blind pianists (Hollins [1936] 1998; see also Honisch 2015).

Beethoven and the Performance of Virtuosity: The Skills of Deaf Hearing

Historically, the term "virtuoso" has been associated more frequently with composers than with performers. Beethoven's reception, including the reception of his late-period music relevant to the concerns of the present chapter, might therefore be situated more promisingly in relation to his contemporaneous reception as a virtuoso composer rather than as a virtuoso pianist (Jander 2001). However, while Beethoven's pianistic activities faded away, there are nonetheless contemporaneous accounts that enrich our understanding of how deafness shaped his compositional and pianistic virtuosities (Skowroneck 2010).

The analysis of deafness as disabled virtuosity in Beethoven's music begin by, but does not end with, understanding Beethoven's musical experiences as a composer and pianist who did not hear physically embodied music but who was able mentally to embody music to a degree that may be considered virtuosic (Straus 2011, 170). This kind of interior musical activity is fundamental to the musician's toolkit; what necessarily differs is the degree of its cultivation. Its very necessity carries important implications for connecting deafness to virtuosity: "Inner hearing—the ability to conceptualize music in its full particularity in the absence of audible sound is a distinguishing ability of trained musicians.... It is obviously a crucial ability for performers... and composers. It is also characteristic of a certain mode of contemplative hearing in which music, in either the presence or absence of musical notation, is perceived and understood in the musical mind" (Straus 2011, 170). During the late period of his life, Beethoven's absorption in an inner world of musical experience, destined to remain inaudible without notation, reaches a level of virtuosity that displaces, or perhaps silences the inner hearing identified by Straus as essential to the musician's art.

As musical expressions of deaf virtuosity, the *Appassionata* and *Hammerklavier* sonatas as performed by Tsujii afford audiences the opportunity to transcend the humdrum activities of simply marveling that a deaf composer can write masterpieces, and of simply puzzling over how a blind pianist can learn such difficult music. The possibilities for audiences to transcend their own understanding of deafness and blindness lies in imagining inaudible and invisible musical experience through virtuosic acts of identificatory projection. Through their own cultivation of virtuosic reception practices, audiences can imagine deaf and blind virtuosities not as the loss of sound and sight, but rather as the inner presence of music, a communicative experience that far exceeds the normal visual, aural, and even tactile abilities that ordinarily limit our musical encounters.

The generative possibilities opened up by this kind of virtuosic reception can be understood by reflecting on the ongoing struggles of the percussionist Evelyn Glennie, who resists the tendency of critics, journalists, and audiences to treat deafness as paradoxical and defining features of her performances: "A common and ill-informed question from interviewers is 'how can you be a musician when you can't hear what you are doing?' The answer is of course that I couldn't be a musician if I were not able to hear. Another often asked question is 'how do you hear what you are playing?' The logical answer to this is 'how does anyone hear?' " (2015). In charting a move from her personal experiences to a philosophical questioning of what it means to "hear," Glennie stages an intervention that simultaneously normalizes deafness in music and estranges normal hearing from its familiar surroundings as merely what everyone does. Hearing can no longer be readily accepted as a common sense, with a common-sense meaning (Neill and Stein 1996; see also the chapter on Glennie).

Two contemporaneous accounts of Beethoven as both pianist and composer exemplify how Beethoven embodied the kind of inner hearing delineated by Straus (2011), creating a deaf virtuosity. These accounts, from different stages of Beethoven's career, illustrate Beethoven's virtuosic performance of inner hearing as both composer and pianist. First, here is how Ferdinand Ries describes Beethoven's initial conception of the last movement of the *Appassionata* during a walk the two men took in Doebling during the summer of 1804: "He had been all the time humming and sometimes howling, always up and down, without singing any definite notes. In answer to my question what it was, he said: 'A theme for the last movement of the [Appassionata] sonata' has occurred to me.' " Upon their return to Beethoven's rooms, "he stormed for at least an hour with the beautiful finale of the sonata" (Thayer [1967] 1991, 356; see also Kinderman 1995, 101).

The second vignette comes from the fifth chapter of Sir John Russell's *A Tour in Germany and Some of the Southern Provinces of the Austrian Empire in 1820, 1821, and 1822* (1825). Russell's description parallels Straus's definition of inner hearing. Russell begins by framing Beethoven, "the deaf pianist," according to a trope that equates disability with social isolation: "Though not an old man he is lost to society in consequence of his extreme deafness" (392). However, the subsequent description of Beethoven's deaf pianism is sensitively tuned to the possibilities of music experienced within an absorbing inner world:

> The moment he is seated at the piano, he is evidently unconscious that there is anything in existence but himself and his instrument; and, considering how very deaf he is, it seems impossible that he should hear all he plays. Accordingly, when playing very *piano*, he often does not bring out a single note. He hears it himself in the "mind's ear." While his eye, and the almost imperceptible motion of his fingers, show that he is following out the strain through all its dying gradations, the instrument is actually as dumb as the musician is deaf. (393; emphasis in original)

Rather than Beethoven's deafness providing an antidote to meretricious virtuosity, it adds experiential heft to virtuosity by constituting a form of virtuosity in itself.

Nobuyuki Tsujii and the Skills of Blindness: Virtuosity as Not Seeing

In order not to exaggerate the singularity of Tsujii's encounter with Beethoven (as has sometimes happened in the critical and popular response to the 2009 Cliburn competition), it is helpful to consider how performances of Beethoven's music by other blind pianists have been received. A notable example in this regard is Bernard d'Ascoli, who took the third prize at the 1981 Leeds International Piano Competition: "Millions of viewers watched his performance, and his youth and his blindness touched a special chord just as much as the quality of his playing of the Beethoven Fourth" (Jones 1983, 8). To Edward Greenfield is attributed the following observation: "while it was inevitable that d'Ascoli should be known as 'the blind pianist *whose virtuosity is hardly credible*,' his blindness would come in time to seem an irrelevance to his musicianship" (Jones 1983, 8; emphasis added).

D'Ascoli has acknowledged that audiences find his blind pianism remarkable. A 1983 article in the *Glasgow Herald* titled "In Sight of Greatness" quotes D'Ascoli as follows: "Yes, of course, I wish people would ignore it and concentrate on my music—but I know I can't censor newspapers, or stop people being interested. To play the piano well without being able to see the keys must seem like a kind of virtuosity in itself . . . but as time goes on, people will pay less and less attention to it" (D'Ascoli, as quoted in McMillan 1983, 4). To accept D'Ascoli's framing of blindness as "a kind of virtuosity in itself," rather than as a barrier, is not uncritically to accept D'Ascoli's own dismissal of the popular attitude that there is, indeed, a kind of virtuosity in blind pianism, nor to endorse, without question, his suggestion that blind pianism gradually loses its novelty through repeated encounters. To imagine deafness and blindness as forms of virtuosity is to undertake an exploration of the edges and gaps, the assumptions behind the discourses that frame sensory disability as the absence of musical experiences deemed normal. In this regard, theoretical debates about disability do not always intersect with the real-world concerns of musicians with disabilities navigating the competitive arena and the actual reception practices of critics and audiences.

D'Ascoli's statement rests on the assumption that, given time, the relevance of, and interest in the pianist's blindness will disappear, an assumption that has been voiced most recently as a wish by Nobuyuki Tsujii (Oda 2009). Efforts to make blindness seem unremarkable idealize reception, and, indeed, performance, as autonomous from the body. If, instead, both musicians and audiences were to be called on to do something other than forget a performer's blindness, namely, to engage in a kind of experiential learning of blindness, a pedagogical strategy not readily accepted within disability studies, then the privilege accorded to sight, which both D'Ascoli and Tsujii cannot dislodge through their public statements, might begin to fall away, to be replaced by a more inclusive understanding of virtuosity.

The Skills of Blindness and Deafness: Performing the Invisible and Inaudible

Let us now consider the multiple responses that Tsujii's performances of the *Hammerklavier* and *Appassionata* sonatas at the 2009 Cliburn competition drew from musicians, critics, and the larger public for the competition. As the modes of reception discussed here show, relationships between virtuosity and disability are often conceived in ways that accentuate the fundamentally abnormal status of both forms of embodiment. In the following excerpt, technical excellence of the rigorous standard expected of competitors is accorded to Tsujii, while artistic depth and interpretive sensibility are identified as absent from the pianist's highly qualified, that is "blind," virtuosity: "It's amazing that a 20-year-old blind from birth has acquired so sure-fire a technique and so basically secure a musicality. But finer nuances of tone and shape sometimes elude him, and sometimes at crucial points. His Beethoven Appassionata began with real authority, but the second-movement variations ambled along on auto-pilot, and there was no sense of cumulative energy in the finale" (Cantrell 2009).

Wholly different are responses to the pianist's *Appassionata* in which both technical brilliance and musical sophistication are found to be present: "Tsujii proved once again to be a natural musical dramatist with an exciting performance of Beethoven's *Appassionata* Sonata. The finale was breathtaking" (Chism 2009).

Perhaps the most scathing response to Tsujii came from the *Wall Street Journal*'s Benjamin Ivry, who not only questioned Tsujii's pianistic abilities but also faulted the jury for prizing a disabled body that, in his view, did not attain the level of virtuosity expected of a competition laureate: "Mr. Tsujii was highly uneven even in solo music, such as a jejune version of Beethoven's 'Appassionata' Sonata on June 7, yet the jury . . . awarded him first place" (2009).

At times, the reaction to Tsujii's Cliburn performances suggest that even though his command of the instrument does not qualify as that of a virtuoso, his musicality nevertheless legitimizes his status as an emerging pianist on the international concert circuit: "It will surely be worthwhile to observe Tsujii further, *even if he did not succeed so spectacularly in other, more muscular works such as Beethoven's Appassionata*, or the Rachmaninoff concerto" (Arndt 2009; emphasis added).

Audiences aware of both Tsujii's blindness and Beethoven's deafness might respond to Tsujii's rendering of the *Appassionata* and *Hammerklavier* sonatas in ways that reflect greater and lesser degrees of awareness of the changing meaning of disability over time. Virtuosic reception, in this regard, may consist in synthesizing these various dimensions, going beyond inert references to "the deaf composer" and "the blind pianist," and situating analyses of the compositional processes at work in both sonatas in relation to the interpretive and technical features of Tsujii's performances. Going further, virtuosic reception so conceived would entail making larger connections between the performance of disability, the composition of disability, and the changing social constructions of the disabled body over time. Listening to music involves a performative dimension suggesting the possibility of adding a third strand to conversations about the performance of music and the performance of disability; namely, the performance of hearing (Straus 2011).

Tsujii's performance of the *Hammerklavier* sonata at the 2009 Cliburn competition was held up by several observers as an example of prodigious memory and sheer force of will. Discourses of virtuosity are implicit in this way of responding to Tsujii's performance of the *Hammerklavier*. "It's hard to watch Nobuyuki Tsujii play without considering the miraculous gift he has, being blind . . .", wrote Gregory Allen,"and yes, I mean that the way it reads: *his handicap is the springboard of his creativity, in much the same way that a deaf composer managed to produce the Hammerklavier Sonata*. Whether or not you respond positively to his artistic intention (or Beethoven's, for that matter), the fact remains that this young man has phenomenal abilities that the rest of us can only dimly comprehend" (Allen 2009, para. 1; emphasis added). At this point, the commentator, himself a pianist and teacher, modulates into a register that focuses more clearly on issues of musical interpretation:

> The first two movements of the Beethoven [*Hammerklavier*] had finely controlled rhythmic energy and clean, unadorned sonorities. In the magnificent slow movement, he didn't quite blow me away; other pianists have plumbed its depths more meaningfully, though probably few of them are only 20. During the Fugue I confess I had in my head the image of the proverbial roomful of monkeys tapping out the complete works of Shakespeare—but come on, doesn't it always sound rather like that, even in the best performance? (Allen 2009, para. 1)

For this commentator, then, Tsujii's virtuosic memorization and technical command of the sonata was not sufficiently accompanied by a truly virtuosic grasp of its structural design and expressive world. The popular image of an heroic pianist grappling with this lengthy and taxing music overtakes the equally familiar image of a blind musician heroically struggling to overcome disability.

In D'Ascoli's statement discussed earlier, perceptions of blindness as virtuosity by sighted audiences do not easily square with the realization that a blind virtuoso musician actually considers such virtuosity as simply normal, a mode of self-representation that turns normalcy itself into contested terrain. Consider, for example, how Nobuyuki Tsujii responds to a question about navigating the opening left-hand leap in Beethoven's *Hammerklavier* sonata: "People always ask me about leaps, but they are not difficult. For me the piano is an extension of my own body, so I know exactly where everything is'" (Hewett 2013; see also Honisch 2016). In contrast to Beethoven's deafness, Tsujii's blindness is hyper visible, even if an ideal mode of reception is often invoked by fellow musicians, critics, and indeed by Tsujii himself, to resist this visibility. According to this idea, his blindness is, or should be, inaudible in his performances. By way of comparison, Evelyn Glennie suggests that if an audience comes away from one of her performances with the impression that it is miraculous for a deaf musician to be able to make music, then she considers herself to have "failed as a musician" (2015).

The critical and popular responses to Tsujii's performances of the *Appassionata* and *Hammerklavier* sonatas emphasized the encounter between a deaf composer and a blind pianist. Tsujii himself has spoken of an affinity that he perceives between himself and Beethoven, through an imagined sharing of disability experience. Tsujii himself pulls away from the label "the blind pianist," and elsewhere has expressed frustration that just as a composer like Beethoven is not perennially marked by his disability, so, too, should he (Tsujii) be accorded the status of "pianist," without the accompanying descriptor "blind." At the same time, this framing of Tsujii's performances of Beethoven's music conceals more than it reveals about the relationship between musical performance and the prized body. The generalities that follow from blithe references to "the deaf composer" and "the blind pianist," ignore the physical embodiments of Beethoven and Tsujii within different historical, social, cultural, and geographic contexts. In reimagining the aesthetic significance of the encounter between the deaf virtuosity of Beethoven and the blind virtuosity of Tsujii, it is therefore vital to remember that musical performance is both the intertwining of composition, performance, and reception, as well as the meeting of different bodies within, but not necessarily across, difference: "the irreducible specificity of these bodies becomes clear when technique and the particularities of performance overshadow the musical message" (Bernstein 1998, 59).

Beethoven's status as the archetypal musician with a disability confers on his compositional narratives of disability a paradoxical transcendence. On the one hand, popular discourses ascribe to his music the capacity for leaving historical context behind, a vaunted universality that may be understood as dissonant relative to the spatial and temporal rootedness of individual performances by present-day disabled performers. On the other hand, however, as scholars of music and disability have shown, shifts in historical understandings of disability, in general, and deafness, in particular, cannot be simply ignored in our experiences of Beethoven's music. By grounding the analysis of virtuosity or, perhaps better, virtuosities in the embodiment of disability, an analytical gesture fundamental to the production and reception of texts, the body is accorded "a place equal rather than in subordination to the text" (Fischer-Lichte 2008, 89)—in this case the musical text. Lines dividing the musical composition of disability as text from the musical performance of disability as embodiment evaporate. Reframing this experience thus brings us into contact with the "constitutional ambivalence of virtuosity" (Hennion 2012, 125), and what demands our engagement is not "the specific property some music or artists possess" but rather something more nebulous, even dangerous: "a distanced attraction—a dangerous temptation, an impossible junction of which [we] need to be wary" (125).

The virtuosities demanded of the pianist in the *Hammerklavier* and *Appassionata* sonatas are not of the kind readily associated with competitive musical performance. Tsujii's inclusion of the *Hammerklavier* sonata in his recital repertoire for the 2009

Cliburn competition, in particular, was nonetheless seen and heard as an act of virtually superhuman effort, a transcendence not only of the body but also of the mind, given the intense strain of performing this work in public, and the navigational perils entailed in the performance of this sonata. That Tsujii committed such a large work to memory was perceived as a kind of virtuosity in itself, and here I invoke D'Ascoli's formulation cited earlier, even if reviewers did not precisely describe it as such. This mode of reception shifts focus away from his actual performance of the sonata, directing attention, instead, to Tsujii's personal attributes. In this regard, the reception of Tsujii's blind virtuosity in the *Hammerklavier* sonata embodies a history of representation that confers on blind people virtuosity not in spite of but rather precisely because of their inability to see. As individuals deemed extraordinary within a normal, that is sighted society, those without sight have been considered to be geniuses, a representation of blindness that reaches as far back as antiquity (Straus 2011). We might observe here the faint imprint of this trope of blindness as genius in the present-day reception of Tsujii's *Hammerklavier*. Perhaps the trope's presence is unintentional, perhaps it is dimly felt, long since buried underneath the fictive unity brought about by common sense. Historically, the "inner vision" to which a blind virtuoso purportedly has access is made possible by a "highly developed memory and a highly structured system of technical knowledge" (Metzner 1998, 260).

Conclusion: Virtuosities "Sometimes Fallible, Sometimes Strong, Sometimes Inspired, Sometimes Distorted, Sometimes Mundane, Sometimes Extraordinary"

Virtuosity, writes Jim Samson, "brings into sharp focus the relationship between music's object-status and its event-status. It marks out a relational field in which text, instrument, performer and audience are all indispensable to defining significance" (2003, 2). The critical task of rethinking deafness and blindness as virtuosities expressive of the richness of aural, tactile, and kinesthetic musical experiences can be furthered by considering scholarship outside of musical performance. As a relational embodied encounter, competitive music performance featuring disabled bodies opens up the possibility for further undisciplining the body (Sandahl and Auslander 2005). Constructions of the virtuosic body can be brought into productive dialogue with representations of disability, to make sense of competitive performance, in all of its psychological, emotional, and physical complexity, demanding an analytical balance which can, itself, be considered as a virtuosity. Bodily abilities, such as hearing and seeing, to take the relevant examples, serve not as taken-for-granted frames of reference, a kind of somatic lingua franca, for audience reception, but rather as contingent interpretive frames for rethinking virtuosic practices.

If musical performance becomes tangible through the expressive play of bodily movement, then widening the range of movements allowed in cultivating the virtuosic body need not perpetuate unimaginative representations of disability as antithetical to virtuosity. Going further, reception practices can offer an aesthetically coherent foundation for including a repertoire of movements too easily dismissed as the symptoms of abnormal musical experience. A host of pedagogical and cultural systems have restricted the parameters of the virtuosic body to what the normal body can do, positioning normal bodily ability as the essential corporeal state out of which virtuosities are fashioned (Sandahl and Auslander 2005). Prioritizing virtuosity as a textual, performative, and practice-oriented discourse, in contrast, emphasizes what senses, bodies, and minds can do outside of the boundaries defined by the normal. At the same time, this enlarged scope cannot mark the beginning of the end of critical judgment, especially when the careers of young musicians are at stake, as they are in international competitions. For this reason, it may be argued that international competitions would do well to emulate the inclusivity to which disability studies in music gestures: "Certainly, we need standards in order to judge and appreciate accomplishments, musical or otherwise. What is needed in music is a more expansive range of acceptable practices [and] various criteria for judging musical abilities—and the tolerance to accept and appreciate their differences" (Howe n.d., 3).

Through the convergence of disability and virtuosity in competitive music performance, sound barriers between a fixed, notated musical text and an ephemeral music performance fade into silence. Analyzing virtuosity in musical performance using a disability studies framework holds the promise of breaking free of restrictive dichotomies not only between ability and disability but also between technique and expressivity, form and content, creation and interpretation. As Rosemarie Garland-Thomson reminds us, "performance lends itself especially well to the project of renarrating disability because the body is the artistic medium of performance" (2005, 33; see also Honisch 2009).

As in dance, so too in musical performance, the disabled body can be virtuosic, and sometimes unexpectedly so (Polson 2013). A disabled body in virtuosic musical performance, like its dancing counterpart, has the potential to cause audiences to reconsider beauty in relation to "our internal and external constructs of time and space" and "to question the origins and effects of our own perspectives and prejudices" (Benjamin 2002, 40). These speculations are all voiced in the language of possibility. However, empirical research involving ethnographic methods with present-day audiences, or historical approaches that bring out past reception practices may very well present a range of obstacles to the smooth translation of possibility into actuality. As we tug at the edges of common sense, the fabric of our ideas about what it means to be a virtuoso composer or performer starts to come undone. As demonstrated in the case studies of how Tsujii's performances of the music of Beethoven were received during the 2009 Cliburn competition, musical encounters between blindness and deafness can be reimagined as an encounter between different manifestations of sensory virtuosity. It is in this work of unraveling that disability studies can provide an organizing theoretical framework for revealing the fabrication of a normative body against which disability and virtuosity become "corporeal others," even while becoming other to each other. Binary conceptions of corporeality are inadequate for understanding how these processes occur at the nexus of performance and reception, since the bodies that virtuosity prizes are bodies constantly on the move, neither fully normal nor wholly abnormal, bodies constantly straining the limits of what virtuosity has meant, continues to mean, and may mean in futures yet unseen and unheard.

Notes

1. I thank Reetika Khanna for varied and fascinating conversations about the connections between lived experience and responses to music. I have likewise benefitted from discussing arguments presented in this chapter with Alex Fisher, Erika Supria Honisch, Mary Ingraham, Laura Kinderman, Hedy Law, David Metzer, and August Sheehy. My thanks, also, to colleagues in the Engaging Vulnerability program, Department of Cultural Anthropology and Ethnology, Uppsala University, I wish to acknowledge colleagues in Uppsala University's Department of Musicology, with whom I shared an earlier draft of this chapter during a research seminar.
2. The formulation "prized body" has been used in several areas of scholarly inquiry. My own usage in this chapter encompasses the figurative and literal: the prized body that concerns me here is the virtuosic body that is both figuratively prized (highly valued) and often literally prized as well (in the sense of being awarded prizes). Disability upsets this hierarchy because of the ways in which the sensory, bodily, and cognitive differences that constitute disability are devalued.

References

Allen, Greg. 2009. "Last of the Semis." *American Public Media*. http://www.publicradio.org.

Arndt, Stefan. 2009. "Der blinde Pianist Nobuyuki Tsujii gewinnt Cliburn Wettbewerb." *Hannoverische Allgemeine*, May 2. Accessed May 10, 2016. http://www.haz.de/Nachrichten/Kultur/Uebersicht/Der-blinde-Pianist-Nobuyuki-Tsujii-gewinnt-Cliburn-Wettbewerb.

Benjamin, Adam. 2002. *Making an Entrance: Theory and Practice for Non-Disabled Dancers*. London: Routledge.

Bernstein, Susan. 1998. *Virtuosity of the Nineteenth Century: Performing Music and Language in Heine, Liszt, and Baudelaire*. Stanford: Stanford University Press.

Brett, Philip, Elizabeth Wood, and Gary C. Thomas. 1994. *Queering the Pitch: The New Gay and Lesbian Musicology*. New York: Routledge.

Burstein, L. Poundie. 2006. "Les chansons de fous: On the Edge of Madness with Alkan." In *Sounding Off: Theorizing Disability in Music*, edited by Neil Lerner and Joseph N. Straus, 278–287. New York: Routledge.

Cantrell, Scott. 2009. "Cliburn 2009: The Last Performances." *Dallas Morning News*, June 7. Accessed May 10, 2016. http://artsblog.dallasnews.com/2009/06/cliburn-2009-the-last-performa.html/.

Chism, Olin. 2009. "Three Fine Performances Conclude Cliburn." *Art Seek*, June 7. Accessed May 10, 2016. http://artandseek.org/2009/06/07/three-fine-performances-conclude-cliburn/.

Cumming, Naomi. 2000. *The Sonic Self: Musical Subjectivity and Signification*. Bloomington: Indiana University Press.

Deaville, James. 2013. "Virtuosity and the Virtuoso." In *Aesthetics of Music: Musicological Perspectives*, edited by Stephen Downes, 276–296. New York: Routledge.

Fischer-Lichte, Erika. 2008. *The Transformative Power of Performance: A New Aesthetics*. New York: Routledge.

Garland-Thomson, Rosemarie. 2005. "Dares to Stares: Disabled Women Performance Artists and the Dynamics of Staring." In *Bodies in Commotion: Disability and Performance*, edited by Carrie Sandahl and Philip Auslander, 30–42. Ann Arbor: University of Michigan Press.

Glennie, Evelyn. 2015. *Hearing Essay*. Accessed June 5, 2016. https://www.evelyn.co.uk/hearing-essay/.

Hadley, Bree. 2014. *Disability, Public Space Performance and Spectatorship: Unconscious Performers*. New York: Palgrave Macmillan.

Hennion, Antoine. 2012. "'As Fast as One Possibly Can . . . ': Virtuosity, A Truth of Musical Performance?" In *Critical Musicological Reflections: Essays in Honour of Derek B. Scott*, edited by Stan Hawkins and Derek B. Scott, 125–138. Burlington, VT: Ashgate.

Hewett, Ivan. 2013. "Nobuyuki Tsujii: 'The piano is an extension of my body.'" *The Telegraph*. July 16. Accessed May 2, 2016. http://www.telegraph.co.uk/culture/music/proms/10182751/Nobuyuki-Tsujii-The-piano-is-an-extension-of-my-own-body.html.

Hollins, Alfred. [1936] 1998. *A Blind Musician Looks Back: An Autobiography*. Portsmouth: Bardon.

Honisch, Stefan Sunandan. 2009. "Renarrating Disability through Musical Performance." *Music Theory Online* 15.

Honisch, Stefan Sunandan. 2015. "Blindness and the Performing Self in Imré Ungár's Chopin." In *The Oxford Handbook of Music and Disability Studies*, edited by Blake Howe, Stephanie J. Moulton, Neil Lerner, and Joseph N. Straus, 246–271. New York: Oxford University Press.

Honisch, Stefan Sunandan. 2016. "Different Eyes, Ears, and Bodies: Pianist Nobuyuki Tsujii and the Education of the Sensorium through Musical Performance." PhD diss., University of British Columbia.

Hough, Basset W. 1940. "Musical Education of the Blind." *Thirty-Fifth Biennial Convention: Proceedings of the American Association of Instructors of the Blind June 24th to 28th 1940, Western Pennsylvania School for the Blind*. Accessed May 2, 2016. https://archive.org/details/proceedingsofameoounse.

Howe, Blake. n.d. "Enforcing and Recasting Disability through Music: Two 'Morals in a Ruin.'" Unpublished paper. Accessed May 2, 2016. https://musicdisabilitystudies.files.wordpress.com.

Howe, Blake. 2010. "Paul Wittgenstein and the Performance of Disability." *Journal of Musicology* 27 (2): 135–180.

Howe, Blake. 2015. "Disabling Music Performance." In *The Oxford Handbook of Music and Disability Studies*, edited by Blake Howe, Stephanie J. Moulton, Neil Lerner, and Joseph N. Straus, 191–209. New York: Oxford University Press.

Ivry, Benjamin. 2009. "What Was the Jury Thinking?" *Wall Street Journal*. Accessed June 6, 2016. http://www.wsj.com/articles/SB124458728669699751.

Jander, Owen. 2001, January 20. "Virtuoso." *Grove Music Online*. Oxford University Press. Accessed May 2, 2016. http://www.oxfordmusiconline.com.

Johnston, Kirsty. 2015. *Stage Turns: Canadian Disability Theater*. Toronto: University of Toronto Press.

Jones, M. 1983. "Blindness a Spur to Visiting Young Virtuoso." *Sydney Morning Herald*. February 11.

Kerman, Joseph. 1985. *Contemplating Music: Challenges to Musicology*. Cambridge, MA: Harvard University Press.

Kinderman, William. 1995. *Beethoven*. Berkeley: University of California Press.

Kramer, Lawrence. 2002. *Musical Meaning: Toward a Critical History*. Berkeley: University of California Press.

Leppert, Richard. 1993. *The Sight of Sound: Music, Representation, and the History of the Body*. Berkeley: University of California Press.

Lerner, Neil. 2006. "The Horrors of One-Handed Pianism." In *Sounding Off: Theorizing Disability in Music*, edited by Joseph N. Straus and Neil Lerner. New York: Routledge.

Lerner, Neil, and Joseph N. Straus, eds. 2006. *Sounding Off: Theorizing Disability in Music*. New York: Routledge.

Lubet, Alex. 2010. *Music, Disability, and Society*. Philadelphia, PA: Temple University Press.

Lubet, Alex. 2014. "Tunes of Impairment: An Ethnomusicology of Disability." *Review of Disability Studies* 1 (1): 133–156.http://www.rds.hawaii.edu/ojs/index.php/journal/article/view/20.

Lubet, Alex. 2015. "Social Confluence and Citizenship: A View from the Intersection of Music and Disability." In *Civil Disabilities: Citizenship, Membership, and Belonging*, edited by Nancy J. Hirschmann and Beth Linker, 123–142. Philadelphia: University of Pennsylvania Press.

Marsalis, Wynton. 2014. "The Virtuoso: Marcus Roberts." *CBS News 60 Minutes*. March 30. Accessed May 10, 2016. http://www.cbsnews.com/news/the-virtuoso-marcus-roberts-2/.

McClary, Susan. 2002. *Feminine Endings: Music, Gender, and Sexuality*. Minneapolis: University of Minnesota Press.

McCormick, Lisa. 2009. "Higher, Faster, Louder: Representations of the International Music Competition." *Cultural Sociology* 3: 5–30.

McCormick, Lisa. 2014. "Tuning in or Turning Off: Performing Emotion and Building Cosmopolitan Solidarity in International Music Competitions." *Ethnic and Racial Studies* 37: 2261–2280.

McCormick, Lisa. 2015. *Performing Civility: International Competitions in Classical Music*. Cambridge: Cambridge University Press.

McMillan, J. 1983. "In Sight of Greatness." *Glasgow Herald*. November 1, 1983.

Metzner, Paul. 1998. *Crescendo of the Virtuoso: Spectacle, Skill and Self-Promotion in Paris during the Age of Revolution*. Berkeley: University of California Press.

Neill, Michael, and E. Stein. 1996. "Sounds of Silence: Deaf Virtuoso Evelyn Glennie Feels the Beat of a Different Drummer—Herself." *People* 45 (12): 109.

Oda, Yuki. 2009. "'Nobu' Fever: Japan Falls for a Blind Piano Prodigy." *TIME*. Accessed June 5, 2016. http://content.time.com/time/world/article/0,8599,1940215,00.html.

O'Dea, Jane. 2000. *Virtue or Virtuosity? Explorations in the Ethics of Musical Performance*. Westport, CT: Greenwood.

Onions, I. 2015. "Disabled Virtuoso Musicians Will Be Giving a Concert at Bristol's Colston Hall." *Bristol Post*, June 25, 2015. Accessed May 10, 2016. http://www.bristolpost.co.uk/Disabled-virtuoso-musicians-giving-concert/story-26760021-detail/story.html.
Overton, Adam. 2006. "Invisible Performance and the Virtuosic Body." *Contemporary Music Review* 25 (1–2): 173–182.
Russell, John. 1825. *A Tour in Germany, and Some of the Southern Provinces of the Austrian Empire, in the Years 1820, 1821, 1822.* Edinburgh: Archibald Constable and Co.
Samson, Jim. 2003. *Virtuosity and the Musical Work: The Transcendental Studies of Liszt.* Cambridge: Cambridge University Press.
Sandahl, Carrie, and Philip Auslander, eds. 2005. *Bodies in Commotion: Disability and Performance.* Ann Arbor: University of Michigan Press.
Siebers, Tobin. 2010. *Disability Aesthetics.* Ann Arbor: University of Michigan Press.
Skowroneck, T. (2010). *Beethoven the Pianist.* New York: Cambridge University Press.
Southall, Geneva H. 1999. *Blind Tom, the Black Pianist-Composer (1849–1908): Continually Enslaved.* Lanham, MD: Scarecrow.
Straus, Joseph N. 2006. "Normalizing the Abnormal: Disability in Music and Music Theory." *Journal of the American Musicological Society* 59: 113–184.
Straus, Joseph N. 2008. "Disability and 'Late Style' in Music." *Journal of Musicology* 25: 3–45.
Straus, Joseph N. 2011. *Extraordinary Measures: Disability in Music.* New York: Oxford University Press.
Thayer, Alexander Wheelock. [1967] 1991. *Thayer's Life of Beethoven.* Vol. 1. Revised and edited by Elliott Forbes. Princeton, NJ: Princeton University Press.
Titchkosky, Tanya. 2011. *The Question of Access: Disability, Space, Meaning.* Toronto: University of Toronto Press.
Whatley, Sarah, Charlotte Waelde, Abbe Brown, and Shawn Harmon. 2015. "Validation and Virtuosity: Perspectives on Difference and Authorship/Control in Dance." *Choreographic Practices* 6 (1): 59–83.
Yoshihara, Mari. 2009. "Pianist Tsujii Takes the Global Stage." *Japan Echo: An Interactive Journal of Informed Opinion* 36 (5): 46–49.

Further Readings

Cook, Nicholas. 2013. *Beyond the Score: Music as Performance.* New York: Oxford University Press.
Cook, Nicholas, and Richard Pettengill, eds. 2013. *Taking It to the Bridge: Music as Performance.* Ann Arbor: University of Michigan Press.
Henderson, Bruce, and Noam Ostrander, eds. 2010. *Understanding Disability Studies and Performance Studies.* New York: Routledge.
Montgomery, M. R. 1992. "The Insight and Passion of Bernard D'Ascoli." *Boston Globe*, April 2.
Pincherle, Marc. 1949. "Virtuosity." *Musical Quarterly* 35 (2): 226–243.
Raykoff, Ivan. 2014. *Dreams of Love: Playing the Romantic Pianist.* New York: Oxford University Press.
Sandahl, Carrie. 2001. "Performing Metaphors: AIDS, Disability, and Technology." *Contemporary Theatre Review* 11 (3–4): 49–60.

Tommasini, Anthony. 2011. "Virtuosos Becoming a Dime a Dozen." *New York Times*, August 12, 2011. Accessed April 28 2018. http://www.nytimes.com/2011/08/14/arts/music/yuja-wang-and-kirill-gerstein-lead-a-new-piano-generation.html?_r=0.

Treffert, D. A. 1989. "An Unlikely Virtuoso." *The Sciences* 29 (1): 28–35.

Virtuoso Voices. 2006. "Evelyn Glennie (Playing Barefoot)." *Public Radio Exchange*, August 1 2006. Accessed May 2, 2016. https://beta.prx.org/stories/13401.

Wasserman, David. 1998. "Distributive Justice." In *Disability, Difference, Discrimination: Perspectives on Justice in Bioethics and Public Policy*, edited by Anita Silvers, David Wasserman, and Mary B. Mahowald, 189–207. Lanham, MD: Rowman & Littlefield.

CHAPTER 16

EMBODIED REPRESENTATION IN STAGED OPERA

LINDA HUTCHEON AND MICHAEL HUTCHEON

THE exuberantly dramatic musical art form called opera offers what Peter Brooks has described as "stories on the body, and the body in story" (1993, xi). This is in part because it came into existence at the start of the seventeenth century, at the very moment in history that Michel Foucault pinpointed as marking the rise of a new concept of the human body in Western culture (1980, 17). The body, for Foucault, is not a simple and unproblematic entity, not a given or a universal: the body is given meaning by culture. That cultural bestowing of meaning on the singing, acting body is tellingly manifest in staged live opera, though it has not often been studied from this perspective.

It was feminist musicology that brought the body into opera studies in the late 1980s and early 1990s through its focus on the (once ignored) gendered bodies of performers, creators, and characters of opera (see Blank's chapter for more on this subject). Studies of masculinity in opera followed (see Purvis 2013). Susan McClary noted that while studies of classical music in the West had tried to transcend the body to concern themselves with "nobler domains of imagination and even metaphysics," the fact was that opera—as music set to a text, usually one about love and death—made transcendence considerably more difficult (1991, 57). But there is another reason for that difficulty: opera is not only texted; it is staged, and therefore the body is not only verbally represented, it is literally and physically embodied. Opera is an art form to watch as well as listen to.

In studies of singers' bodies, corporality has indeed been acknowledged. Susan J. Leonardi and Rebecca A. Pope introduce *The Diva's Mouth: Body, Voice, Prima Donna Politics*, with these words: "this is very much a book about bodies. A book about body parts. Fragile throats, open mouths, powerful lungs, swollen epiglottises, larynxes on the edge, pierced ears, notes from the chest, notes from the head, heaving breasts, bared breasts, bound breasts, snipped testicles, clipped frenums. It's a book about muscles, hair, fat, blood, sweat, and secretions" (1996, 8). But even here, it is the *singing* body, not the singing and *acting* body that is given center stage. It is the voice that takes precedence, as in this typical statement: "like the body from which it emanates, the female voice is construed as both a signifier of sexual otherness and a source of sexual power, an object at once of desire and fear" (Dunn and Jones 1994, 3). Queer theory has further embedded the voice—as sung and as listened to—as the focus of attention in opera studies, from Elizabeth Wood's "Sapphonics" (1994, 27) to the work of Wayne Koestenbaum (1993) and Sam Abel (1996).

In fact, much of the "body" focus in opera studies has been on precisely these two things: the "voice in performance" (Risi 2011, 49) and the responses of listeners. While it is obvious that the voice is housed in a body and comes out of a body, somehow it is deemed separate from it: we are told that beauty "and flexibility of voice in opera are . . . roughly equivalent to physical beauty in literature and art" (Locke 1995, 65). The voice somehow transcends the body of its producer but makes a direct link to the listener: "The singer destroys the division between her body and our own, for her sound enters our system" (Koestenbaum 1993, 43). Michelle Duncan notes, "As for the body of the singer, opera studies has tended to ignore it altogether unless it possesses currency as the object of desire or of a fetish. And when this happens, both the body and voice of the singer become secondary to the affect or erotic desire of the spectator" (2004, 285). Somehow, the corporeal body on stage is ignored, as if, in Mary Ann Smart's ironic words, it "could overwhelm analysis with an ungraspable excess of materiality" (2000, 9).

The fact is, however, that the represented body on the opera stage is more than a voice and thus does more than sing musical notes: it also moves, it mimes, it dances, it gestures, it laughs, it cries. It is dressed in a certain way. It enacts a drama along with other bodies. It has physical characteristics: its age, size, shape, physiognomy, race, and sex are all signifiers. In short, the staged body is a sign: in any art form, the body "cannot but represent both itself and a range of metaphorical meanings, which the artist cannot fully control, but only seeks to limit by the use of context, framing and style" (Mirzoeff 1995, 3). In the collaborative art of live opera production, it is the task not only of the artist/performer but of the director or producer at once to limit that range of meanings and to stage the represented body within the dramatic action that it is embodying live on stage: "The process [of creating a role and drama] is also thoroughly embodied, as producers and performers attempt to perform physically and make visible the motives and intentions of the opera's characters. Plausible drama depends on the legibility of performers and their embodied presence on-stage" (Atkinson 2006, 99). This is a conscious construction or literalization of what sociologists call the "dramaturgical body": people do not merely *have* a body, they "actively *do* a body" (Waskul and Vannini 2006, 6).

Like all theater, opera has to deal with the fact that bodies on stage are never "natural" but are always historical and cultural: "No body ever simply appears on stage. Bodies are, rather, made to appear in performance, rendered visible as the encoded tissues interwoven by systems of ideological representation that mediate the anxieties and interests at play in specific historical moments. On stage, even the naked body represents not natural material, but the matter of history" (Rouse 1997, 394). This was most certainly made obvious by the audience's shocked response to the director Peter Hall's 1986 production of Richard Strauss's *Salome* (1905), in which his wife, the soprano

Maria Ewing, revealingly removed every one of her seven veils during her infamous dance (Fuller 2009, 254).

As Simon Shepherd argues in *Theatre, Body, Pleasure* (2006), theater—and opera is a form of theater as well as music—is a medium in which bodies watch bodies. It is "a place which exhibits what a human body is, what it does, what it is capable of" (1). Opera is indeed a medium, but so is the body as sign on stage, as Fuller has explored: "from the vogue for the castrato in heroic roles in eighteenth-century opera, to the so-called 'trousers' roles . . . the underlying issue common to all these forms of the body as a medium concerns the nature and importance in opera of dramatic illusion" (2009, 263–265). For directors, then, the issue is one of matching the physical demands of the music and the implications of the drama: this is where a body's age, size, physique, and even race can enter the picture.

When the director of Covent Garden's 2004 production of Richard Strauss's *Ariadne auf Naxos* (1916) infamously decided that the hefty soprano Deborah Voigt would be replaced—for reasons of appearance—by the slimmer Anne Schwanewilms, the opera community was up in arms. Two years later Voigt was rehired after losing 135 pounds. The aesthetics of film/television realism clearly won out over opera's tradition of the voice reigning supreme. This was but the most public of such conflictual moments around realism or body-matching-role: comments have certainly been heard circulating among audience members about sixty-year-old tenors of a certain size playing the young, starving student Rodolfo in Puccini's *La Bohème*. And Yevgeny Nikitin was not allowed to sing Richard Wagner's Dutchman in *Der fliegende Holländer* at Bayreuth in 2012 because of his "ill-advised" swastika tattoo. This is the operatic body as medium.

There are, however, times when the opera's subject directly involves the body, as the example of Salome suggests. The diseased body, the disabled body, the wounded body, the beautiful body: all figure in operatic plots. Because of the plot, Giuseppe Verdi's Falstaff must be obese (see Gilman 2005), even if Ariadne at the Royal Opera was not allowed to be so. Verdi's Otello should be dark-skinned, though he often is not. Like size, race marks the staged body, either as medium or subject (see André, Bryan, and Saylor 2012). So too does disability and disease. Opera was born in the Neo-Platonic Renaissance with its idealization of the beautiful body as the externalization of the good soul. Not surprisingly, this philosophical truism was translated into operatic terms, but it is most often the marked body of the staged male that signifies "deformation"—that is a lack of symmetry and order, both aesthetic and moral (see Hutcheon and Hutcheon 2000, 53). However, the narratives of these operas—from *Pagliacci* to *Rigoletto*—make clear that, as Erving Goffman (1963) and others have argued, it is the stigma of social rejection that induces hatred and anger and causes these characters to act as they do.

Sometimes, all of these different staged representations of the body—the embodied "voice in performance," and the actual body as both medium and subject—come together within one staging, offering a rich case study through which to investigate the aesthetics and ethics of the staged operatic body. One such production is that of Gaetano Donizetti's *Lucia di Lammermoor* (1835), directed by David Alden in London (English National Opera) in 2008 and 2010, Washington (Washington National Opera) and Goteborg (Goteborg Opera) in 2011, and Toronto (Canadian Opera Company) in 2013. A bel canto opera now part of the standard repertoire, *Lucia* adapts Walter Scott's 1819 novel, *The Bride of Lammermoor*, based on a true tale of star-crossed lovers set in what was seen as the romantic Scottish Lowlands of the seventeenth century. The librettist, Salvadore Cammarano, condensed the novel's plot and altered the emphasis, but retained the general intrigue. Lucia Ashton has been promised in marriage to the wealthy and well-connected Arturo Bucklaw in a bid by her brother, Enrico, to save the family estate in both financial and political terms. She, however, has already secretly pledged her troth to Edgardo, who unfortunately is a Ravenswood—that is, of a family with whom the Ashtons have long feuded. Through fraud (a forged letter) and appealing to her sense of family honor, Enrico persuades Lucia to marry Arturo, but at the wedding Edgardo appears—enraged that Lucia has betrayed him. A shattered Lucia retires to the bridal chamber with her new husband, where she kills him with his own sword. In what has been called opera's "quintessential Mad Scene" (McClary 1991, 92), she appears before the wedding guests, hallucinating her fantasy marriage to Edgardo before dying; he, when he is told Lucia has died, commits suicide.

It is Lucia's madness—in its vocal and corporeal manifestations—that has captured the attention of critics and audiences. Indeed, in its early performances, the opera was often shortened in order to end with the "Mad Scene." Star soprano Nellie Melba also performed the scene to great acclaim in concert (Ashbrook 1990). The embodied "voice in performance" aspect of this opera, however, has never been separable from the larger historical issue linking women with madness. Tellingly, David Alden's production deliberately set the staging in the nineteenth century in order to capture the gloom and claustrophobia of Victorian Britain and post-Napoleonic Europe—in other words, right after the time when Scott and Donizetti/Cammarano were writing. In these same years, as Elaine Showalter (1985) has explored in detail, madness came to be seen as "the female malady." And Scott's Lucy Ashton has been understood as the most dangerous manifestation of "female sexuality as insane violence against men" (14). Madness was not only feminized at this time, however, but also medicalized, thereby linking insanity to the female biological life cycle, especially in the adolescent years. This was a society that saw all women as "childlike, irrational, and sexually unstable," and so rendered them "legally powerless and economically marginal" (73)—as was true of both Lucy and Lucia's family situation. And it was *young* women who were deemed the most vulnerable.

Scott's Lucy is practically mute in her docility and passivity, but she finally does speak in her moment of "demoniac" insanity. With "a sort of grinning exultation" as they take away the body of her wounded husband, she utters the ironic words: "So, you have ta'en up your bonny bridegroom?" (1850, 455). The operatic Lucia's voice cannot be so easily silenced, of course. Catherine Clément (1989), Susan McClary (1991), and Naomi André (2006) see Lucia's mad scene as proof of her liberatory transgression of patriarchal social limits through its "awesome and frightening" exuberance (McClary 1991, 96). But Mary Ann Smart has argued, instead, that while Lucia certainly "stretches the boundaries of her musical world—by flights of coloratura, distortion of conventional forms, and quotations of melodies used earlier in the opera" (1992, 122), she is in the end "defeated both by the forces of musical language and by the oppressive power of the plot" (141).

Whichever way we interpret the scene, what is significant for our argument here is what Alden's particular production does with these excessive and embodied "voice in performance" aspects of the opera. In the original English National Opera version, the tessitura of Lucia's vocal role was raised from its usual lowering back to the original higher tessitura conventionally associated with female stage madness (Willier 1992, 145). What was also restored was the use of the glass harmonica (in a passage later revised by the composer for obbligato flute) in the mad scene. Its eerie, otherworldly sound not only recalls its reputation for causing insanity in those who play it, but further suggests Lucia's less than firm grasp on reality. As Alden asserted in the 2013 program of the Canadian Opera Company production, it "creates an amazing soundscape in the mad scene which is very unusual and deeply disturbing." A "poetic sign of liminality" (Hadlock 2000, 509), the glass harmonica—with its range, timbre, and tone quality—came to be associated with the vulnerable feminine (538). This musical imagining of female insanity could work on a contemporary twenty-first-century audience as readily as it could on its original spectators, but what could *not* work is that (period-specific) nineteenth-century pathologization of the sexualized adolescent female.

How then might Alden psychologically motivate Lucia's murderous madness to make it emotionally believable for a modern audience? A common narrative and dramatic device that provides precisely such motivation today is that of sexual abuse, as strikingly evoked in Atom Egoyan's 1996 production of *Salome*. Alden ups the ante in a variety of ways by taking this plot device to extremes. First, it is Enrico, Lucia's brother, who takes violent and incestuous advantage of his sister—and does so on stage. The musical link to her later madness is made through her music's use of melismas and her coloratura singing that we hear as her brother abuses her. But this is not just incest; it is the pedophilic abuse of a much younger child. Lucia in this production is infantilized: she is dressed in young girl's clothing, is surrounded by dolls, and sleeps in a kind of crib (to which her brother, at one point, ties her). In responding to the production, critics everywhere reacted strongly to this addition to the narrative, repeatedly noting that the singer for whom Alden always claimed he had specifically designed the production, Anna Christy, looked vulnerable, young, "doll-like." She is a "child-woman," as the Director's Note in the North American productions put it.

Anna Christy, a talented singer more than capable of the vocal demands of this role, did indeed look young and thus made Lucia's physical vulnerability all the more striking. The body as medium here reinforces the interpretation of the body as "voice in performance." Lucia's young body is very much also the subject of the opera, of course, and the stage directions of the librettist make clear that her staged visual presence is important. When she appears to the wedding guests at the start of her mad scene, we are told that she is dressed in white, with disheveled hair. Her pale face ("coperto da uno squallore di morte") makes her appear more a ghost than a living person: "Her stony gaze, her convulsive movements and her malevolent smile disclose not only a frightening insanity, but the signs that her life is approaching its end" (quoted in Smart 1992, 125). What Lucia looks like on stage, in short, is important.

We emphasize this because the Washington and Toronto Director's Note (written by the assistant director, Ian Rutherford) goes on to make a comment that seems to come out of nowhere: "In the same Victorian era, mental health pioneer John Langdon Down was understanding and naming Down's syndrome in his sanatoriums." (This information did not appear in the English National Opera programs—and may help explain why reviewers were so puzzled by the production.) What happens when audience members do indeed read this information before the curtain rises, look at Anna Christy on stage (Figure 16.1), and see a Eurasian singer in the role of Lucia (and in Toronto, the cover,

FIGURE 16.1 Anna Christy as Lucia di Lammermoor, English National Opera 2010.
Photo by Robert Workman.

Keiya Kasahara, was also Eurasian)? Has Enrico's abuse of his young sister just been raised up one more notch in the scale of evil: is he abusing a Down syndrome child?

In case said audience members think taking this step was going too far in interpretation, the Director's Note continues to point them further in this direction: "As part of his treatment regimen, [Dr. Down] incorporated theatres into the sanatoriums, and the public was invited to see performances by the patients. The belief was that it helped the patients, but audiences were fascinated by hysteria and madness." And indeed, the stage and theater at Down's Normansfield institution are still intact today, and we are told in the program that they provided the visual model for the stage-within-the-stage set on which Alden allows his Lucia to descend into madness. This embedded stage faces a silent chorus, portraying a clearly engaged audience—thereby implicating those in the theater watching, of course.

Some reviewers picked up on this information, but only insofar as to link it to the well-known fact of asylum visitation by the curious in the nineteenth century and the period's much documented fascination with female hysteria. But the fact is that Dr. Down did *not* run an asylum for hysterical women. This well-known British physician (1828–1896) was interested in helping learning-disabled children, working first at the Royal Earlswood Asylum for Idiots (as it was known at the time) before setting up his own institution for the "feeble-minded," as they were called, at Normansfield in 1868. (This house is currently the headquarters of the Down's Syndrome Association.) A decade later, he opened that theater in the building to provide entertainment for residents and staff. In Down's words: "music and dancing may for all alternate with dramatic entertainments, which are most useful in appealing to both the eye and the ear" (Down [1887] 1990, 83–84). Working with learning-disabled children, he believed he was offering not the withdrawal of an asylum, but training within a social environment aimed at making "life joyous" (78).

After exposure to the ideas of the German anthropologist, naturalist, and physician Johann Friedrich Blumenbach (1752–1840), Dr. Down had made an observation that would eventually lead to the naming of Down syndrome. Blumenbach's 1795 *Anthropological Treatises* (translated into English in 1865) argued that on the evidence of craniometry and multiple other physical characteristics (drawn from his collections of skulls, hair, fetuses, anatomical preparations, and pictures), the human species could be divided into five races: Aztecs (Americans), Caucasians, Malayans, Ethiopians, and Mongolians. Down undertook to classify all the residents at Earlswood by "race" using measured diameters of their heads and specific facial features (taken from photographs of his patients). In an 1866 article published in *London Hospital Reports* (259–262) called "Observations on an Ethnic Classification of Idiots," Down drew attention to the fact that "a very large number of congenital idiots are typical Mongols" and thus look more like each other than like members of their families. He described their physical features, their capacity for humor and imitation, and their reduced life span. He further noted their capability of learning with training and went on to offer this ethnic classification of "Mongolian" as proof of, not racial difference, for these Mongolian children were Caucasian, but rather of "arguments in favor of the unity of the human species" (1866, 262). This clearly challenged an important racial theory of the day held by proponents of slavery who asserted both racial difference and hierarchy.

It was Dr. Down who was responsible for the term "Mongoloid idiot" and for the use of the terms "Mongoloid" and "Mongolism." In 1961, a group of eminent geneticists wrote to the editor of the *Lancet* offering four options for renaming this syndrome: Langdon Down anomaly, Down's syndrome, Congenital acromicria, and Trisomy 21. The editor chose the second, but today, "Down syndrome" is the preferred term. The new name was confirmed by the World Health Organization in 1965 after the director general had been approached by representatives of the People's Republic of Mongolia who objected to the earlier term as derogatory. As Ann Gath points out in the foreword to a recent publication of Down's Lettsomian lectures: "Many labels have been applied to those regarded as mentally retarded. At first believed to be free from stigma, each label becomes tarred with prejudice and is soon superseded by another then another in rapid succession, each only acceptable for a brief period" (Down [1887] 1990, iv).

Could it have been this difficulty with the terminology or perhaps the "political correctness" fears of talking about this kind of "developmentally disabled" (USA) or "learning disabled" (UK) child that stopped reviewers from even mentioning the Eurasian casting and the possibility of this Lucia having Down syndrome? They had no difficulty noting and often protesting the sexual abuse by Enrico of his young sister, but had Alden gone too far in making that child an even more vulnerable victim? Or did people not "get" the reference, despite the repeated pointing to Dr. Down in the Director's Note? The dramaturgical effect for those in the audience who did "get" it, was to increase, to an almost painful degree, the trauma of the abuse and the helplessness of the victim. In subsequent discussions of this production, the director has never openly addressed this point, but the set designer, Charles Edwards, has clearly asserted, "in using Normansfield as a touchstone we were able to give Lucia a slightly more disturbing psychological angle to match what lurks in the music" (quoted in Vanstone 2013).

Affectively, therefore, the casting was immensely powerful because of this disturbing addition. But, in ethical or even political terms, some audience members might well find themselves asking whether the dramatic effect justifies the taboo-breaking means, for it must be said that this production does go against new social conventions. The current view of Down syndrome children includes their learning disability, of course, and thus Alden's particular staging consolidates the "sign" of the vulnerable child. But, in stark contrast to his representation, the public presentation of the various national Down syndrome associations is consistently of people with the syndrome as capable, functioning members of society. As Herbert Blau argued, "Like all symbolic activity, the theater event both excludes and disguises whatever exists outside the margins of the acceptable. What is left out is substituted for by the appearance, which seems more acceptable. The taboo gets into performance . . . in a kind of incognito, which slips past the censor and through the work of representation" (1990, 196–197). This is what likely happened with casting and thus the inferred Down syndrome representation on stage.

Just as the opera's use of coloratura is a marked musical gesture of madness in dramaturgical terms for the embodied "voice in performance" (see Parr 2009), so the specifically marked corporeal body of Lucia here functioned powerfully as both medium and subject: in the context of this particular stage production, Anna Christy literally embodied the excruciating vulnerability of Lucia as subject through the medium of her youthfulness and her mixed race. While Richard Leppert has asserted that all music, "despite its phenomenological sonoric ethereality, is an embodied practice, like dance and the theatre" (1993, xxi), opera is much more obviously precisely such an embodied practice. Thanks to the pioneering theoretical work of scholars working in fields such as feminist and disability studies, the visual as well as aural dimensions of opera's staged representations of the body have come into critical focus, and with that new attention has come an increased concern about the politics and ethics of such representations. As Leppert points out, the connection between music and the body throughout Western history may well be "highly problematic and contradictory, the product of deep socio-cultural anxieties and antagonisms" (xx). But in opera, those very anxieties and antagonisms are what are often literally played out on the stage by bodies representing bodies to be watched (as well as listened to) by still other bodies.

The living body of the singing actor on the operatic stage both *has* and *performs* a body, and that body is more than a biological entity. It is also an ideological construct, a product of its particular historical and cultural time and place. The performer's body is assigned the task by the director of presenting a role on stage in such a way as to "speak" to the particular audience listening and watching. Thus the performance is going to be presented and interpreted within a particular historical and social context. As Shepherd explains, "the preparation of the performing body is undertaken within a context of assumptions about 'body' in society and thus a relationship—conscious or not, critical or not—with what are perceived to be dominant norms" (2006, 11). Opera, like all theater, is always going to be "a practice in which societies negotiate around what the body is and means" (1).

Note

Our thanks to the Canadian Opera Company, especially Claudine Domingue, for assistance in accessing materials, and to Kathy Chung for her keen research skills. Our gratitude to Emory University's Disability Studies group, especially Sander L. Gilman and Rosemary Garland-Thomson, for giving us the initial opportunity to examine this subject.

References

Abel, Sam. 1996. *Opera in the Flesh: Sexuality in Operatic Performance*. Boulder, CO: Westview Press.

André, Naomi. 2006. *Voicing Gender: Castrati, Travesti, and the Second Woman in Early-Nineteenth-Century Italian Opera*. Bloomington: Indiana University Press.

André, Naomi, Karen M. Bryan, and Eric Saylor, eds. 2012. *Blackness in Opera*. Champaign-Urbana: University of Illinois Press.

Ashbrook, William. 1990. "Popular Success, the Critics and Fame: The Early Careers of *Lucia di Lammermoor and Belisario*." *Cambridge Opera Journal* 2: 65–81.

Atkinson, Paul. 2006. "Opera and the Embodiment of Performance." In *Body/Embodiment: Symbolic Interaction and the Sociology of the Body*, edited by Dennis D. Waskul and Phillip Vannini, 95–107. Aldershot: Ashgate.

Blau, Herbert, 1990. *The Audience*. Baltimore, MD: John Hopkins University Press.

Brooks, Peter. 1993. *Body Work: Objects of Desire in Modern Narrative*. Cambridge, MA: Harvard University Press.

Clément, Catherine, 1989. *Opera, or the Undoing of Women*. Translated by Betsy Wing. London: Virago.

Donizetti, Gaetano, and Salvatore Cammarano. [1835]. 1941. *Lucia di Lammermoor*. Milan: E. Bestetti.

Down, John Langdon. 1866. "Observations on an Ethnic Classification of Idiots." *London Hospital Reports* 3: 259–262.

Down, John Langdon. [1877] 1990. *On Some of the Mental Affections of Childhood and Youth*. Philadelphia, PA: J.B. Lippincott.

Duncan, Michelle. 2004. "The Operatic Scandal of the Singing Body: Voice, Presence, Performativity." *Cambridge Opera Journal* 16 (3): 283–306.

Dunn, Leslie C., and Nancy A. Jones. 1994. *Introduction to Embodied Voices: Representing Female Vocality in Western Culture*, edited by Leslie C. Dunn and Nancy A. Jones, 1–13. Cambridge: Cambridge University Press.

Egoyan, Atom, dir. 1996. *Salome* by Richard Strauss. Canadian Opera Company.

Foucault, Michel. 1980. *The History of Sexuality*. Vol 1. Translated by Robert Hurley. New York: Vintage.

Fuller, David. 2009. "The Erotic and the Sacred Body in Opera: The Venusberg to Monsalvat—and Beyond." In *The Body and the Arts*, edited by Corinne Saunders, Ulrika Maude, and Jane Macnaughton, 254–269. Basingstoke: Palgrave Macmillan.

Gilman, Sander L. 2005. "Obesity and Diet in the Nineteenth Century: Framing Verdi and Boito's Healthy 'Falstaff.'" *University of Toronto Quarterly* 74 (2): 759–775.

Goffman, Erving. 1963. *Stigma: Notes on the Management of Spoiled Identity*. Englewood Cliffs, NJ: Prentice-Hall.

Hadlock, Heather. 2000. "Sonorous Bodies: Women and the Glass Harmonica." *Journal of the American Musicological Society* 53 (3): 507–538.

Hall, Peter, dir. 1986. *Salome* by Richard Strauss. Los Angeles Opera.

Hutcheon, Linda, and Michael Hutcheon. 2000. *Bodily Charm: Living Opera*. Lincoln: University of Nebraska Press.

Koestenbaum, Wayne. 1993. *The Queen's Throat: Opera, Homosexuality, and the Mystery of Desire*. New York: Poseidon Press.

Leonardi, Susan J., and Rebecca A. Pope. 1996. *The Diva's Mouth: Body, Voice, Prima Donna Politics*. New Brunswick, NJ: Rutgers University Press.

Leppert, Richard. 1993. *The Sight of Sound: Music, Representation, and the History of the Body*. Berkeley: University of California Press.

Locke, Ralph P. 1995. "What Are These Women Doing in Opera?" In *En Travesti: Women, Gender Subversion, Opera*, edited by Corinne E. Blackmur and Patricia Juliana Smith, 59–98. New York: Columbia University Press.

Loy, Christof, dir. 2004. *Ariadne auf Naxos* by Richard Strauss. Royal Opera House, Covent Garden.

McClary, Susan. 1991. *Feminine Endings: Music, Gender, and Sexuality*. Minneapolis: University of Minnesota Press.

Mirzoeff, Nicholas. 1995. *Bodyscape: Art, Modernity and the Ideal Figure*. New York: Routledge.

Parr, Sean M. 2009. "Melismatic Madness: Coloratura and Female Vocality in Mid-Nineteenth-Century French and Italian Opera." PhD diss., Columbia University.

Purvis, Philip, ed. 2013. *Masculinity in Opera*. New York: Routledge.

Risi, Clemens. 2011. "The Diva's Fans: Opera and Bodily Participation." *Performance Research* 16 (3): 49–54.

Rouse, John. 1997. "Comment." *Theatre Journal* 49: 394–395.

Scott, Walter. 1850. *The Novels of Walter Scott*. Vol. 2. Edinburgh: R. Cadell.

Shepherd, Simon. 2006. *Theatre, Body, Pleasure*. London: Routledge.

Showalter, Elaine. 1985. *The Female Malady: Women, Madness, and English Culture, 1830–1980*. New York: Penguin.

Smart, Mary Ann. 1992. "The Silencing of Lucia." *Cambridge Opera Journal* 4 (2): 119–141.

Smart, Mary Ann. 2000. *Introduction to Siren Songs: Representations of Gender and Sexuality in Opera*, edited by Mary Ann Smart, 3–16. Princeton, NJ: Princeton University Press.

Strauss, Richard. [1905]. 1943. *Salome: Musikdrama in einem Aufzuge nach Oscar Wilde's gleichnamiger Dichtung in deutscher Übersetzung von Hedwig Lachman*. Op. 54. London: Boosey & Hawkes.

Strauss, Richard. [1916]. 1943. *Ariadne auf Naxos: Oper in einem Aufzuge nebst einem Vorspiel von Hugo von Hofmannsthal*. Op. 60. London: Boosey & Hawkes.

Vanstone, Susan. 2013. "Normansfield Inspires *Lucia*." *Parlando: The COC Blog*. http.//www.coc.ca/ExploreandLearn/NewToOpera/OnlineLearningCentre/ParlandoTheCOCBlog.aspx?EntryID=24407.

Waskul, Dennis, and Phillip Vannini. 2006. "Introduction: The Body in Symbolic Interaction." In *Body/Embodiment: Symbolic Interaction and the Sociology of the Body*. Edited by Dennis D. Waskul and Phillip Vannini, 1–18. Aldershot: Ashgate.

Willier, Stephen A. 1992. "Mad Scene." In *The New Grove Dictionary of Opera*, edited by Stanley Sadie, 3:145–146. London: Macmillan.

Wood, Elizabeth. 1994. "Saphonic." In *Queering the Pitch: The New Gay and Lesbian Musicology*, edited by Philip Brett, Elizabeth Wood, and Gary C. Thomas, 27–66. New York: Routledge.

CHAPTER 17

SEXUALITY, DIS/ABILITY, AND SUBLIMITY IN GRAND OPERA

HANNE BLANK

Devotees of opera often swoon that, at its best, opera is sublime. They generally seem to mean it figuratively, "sublime" as a bit of colloquial hyperbole meaning "extremely pleasing," or "transporting," or "of exceptional quality," not necessarily something that strikes deep awe into one's heart and psyche. But they could just as well mean real sublimity. Grand opera—which for the purposes of this chapter is defined as the post–*opera seria* operatic literature that begins roughly with Mozart's *Don Giovanni* and extends into the twentieth century—is devoted to and dependent on the body (Kramer 2004, 7). It is also, and not coincidentally, devoted to a particular type of aesthetic experience, one of ravishment, of enthrallment, a state of *overwhelmedness* in which one experiences intense yet conflicted joy. This is an experience that can, and sometimes is, driven by a very genuine engagement with the sublime, an argument that has been made in music criticism since E. T. A. Hoffman's early nineteenth-century defenses of Beethoven.[1] In the case of opera, experiencing such sublimity hinges on the operatic presence of the body in extremis, and specifically in connection with contexts and narratives of sexuality and disability.

Bodily extremes, particularly those whose representation is inimitably aided by music's ineffable capacity to invoke bodily sympathy in the audience, invite an operatic experience that resonates with the conception of the sublime developed by Arthur Schopenhauer in his popularly influential magnum opus, *Die Welt als Wille und Vorstellung* (1818/1844, published in English as *The World as Will and Representation* in 1958). Neither Schopenhauer nor his philosophical ancestors Immanuel Kant and Edmund Burke specifically articulated the sublime in connection with encountering human embodiment, preferring to conceive of it as the result of confrontations between the limited, embodied human being and the vast, impersonal presences of the natural world. Yet to these writers the sublime existed, as Schopenhauer explained, due to one's awareness of the human body and self as a constantly vulnerable "frail phenomenon." The strange and only partly intellectual thrill of the sublime lay at the intersection of witnessing forces capable of damaging or obliterating that frail phenomenon, the knowledge of what it would mean to be thus damaged or obliterated, and the physical and mental frissons of pleasure associated with contemplating all this from a position of relative safety. Sublimity was a way to characterize the experience of interpolating one's thinking, feeling, experiencing body into a natural world that exceeded the body's capacities. During the late eighteenth and nineteenth centuries, sociocultural, intellectual, technological, and even bureaucratic developments created a climate in which it became possible to experience human embodiment—at least in some forms and cases—as likewise being a manifestation of a powerful, incommensurable Nature. At the core of these experiences of embodiment we find the same bodily experience: that of extremity and the exceeding of human capacity and will (Schopenhauer 1958, 201–207).

In opera's polysemic temple of exquisite disasters, then, it may be more than superficially relevant, more than historically and symbolically relevant, that Lucia goes mad, that Rigoletto is deformed, that Amfortas's leg refuses to heal, that Violetta's consumptive lungs gasp out their death-rattle. The aesthetic experiences opera can offer its audiences hinge on the ways in which women's bodies struggle and suffer when they play out their roles as sacrificial lambs, vicious manipulators, or relentless, deliberate *femmes fatales*, and in the ways that men yearn, or rape, or stand, as Rodolfo does at the end of *La Boheme*, gutted and dumbstruck at their utter inability to protect women from the intransigencies of their own bodies. It matters as well that all this happens in extraordinary fashion, larger than life, opulently costumed, staged, and lit especially for the eyes of the audience. Most importantly and most potently, it matters that all of this is carried in and through music that is played, acted, and, crucially, sung, by singers on whose literal backs, breasts, and throats the enactment of opera's embodied disasters rests (see Hutcheon and Hutcheon's chapter for more on this subject).

Scholars have become accustomed, over the past generation or so, to considering sexuality in opera as a characteristic feature. Opera is, infamously as well as famously, the genre of love duets and love-deaths, women driven mad on their wedding night, castrated men embodying and giving voice to goddesses and queens, and low-voiced women embodying and singing serenades in the guise of romantic young men. Within opera, a panoply of sex and gender expectations are demonstrated via both example and transgression, communicated through narrative, costume, gesture, text, and music itself. Feminist musicology of opera, decisively launched by Catherine Clément and Susan McClary with their deservedly classic *Opera, or the Undoing of Women* (1988) and *Feminine Endings: Music, Gender, and Sexuality* (1991), was rapidly taken up by other scholars including Carolyn Abbate (1991), Mary Ann Smart (2000), and Wendy Heller (2004), all of whom have used feminist analysis to look at the ways in which ideas and expectations of gender and sexuality have been communicated through opera, musically and otherwise. Particularly relevant was McClary's essay "Excess and Frame" (1991, 80–111), which applied feminist historiographical and musicological analysis to perhaps the quintessential mad scene of the grand opera repertoire, "Il dolce suono" from

Gaetano Donizetti's 1835 *Lucia di Lammermoor*. In it, McClary showed how musical structure, ornamentation/fioritura, and tessitura helped to inscribe and indeed simulate central ideas about gender, sexuality, sanity, femaleness, and the nature of delirium as a form of embodied extremis.

Around the same time as the emergence of these feminist scholars and commentators a related array of writing about the sexuality of opera performance and reception also appeared, this one heavily influenced by gay male opera queen culture and the nascent discipline of queer studies, as well as more traditional reception and performance studies. These have been framed, and even titled, in connection with embodiment and extremis. Ethan Mordden's *Demented: The World of the Opera Diva* (1984) takes its title from the one-time use of the word "demented" to mean a superlative performance by an opera diva, a usage with clear connections to the trope of the sexually hyperdetermined madwoman. Direct references to bodies and body parts abound, for instance Wayne Koestenbaum's *The Queen's Throat: Opera, Homosexuality, and the Mystery of Desire* (1993); Samuel Abel's *Opera in the Flesh: Sexuality in Operatic Performance* (1996); and Rebecca Pope and Susan Leonardi's *The Diva's Mouth: Body, Voice, Prima Donna Politics* (1996). Delving yet more closely into the connections between embodiment and opera, and likewise beginning in the same period, is the small but important literature on opera and the medicalized body, of which Linda Hutcheon and Michael Hutcheon's *Opera: Disease, Desire, Death* (1996) is emblematic.

In these works, as in the feminist scholarship on sexuality in opera, the means by which opera produces physical and emotional cathexis in its audiences is a key concern. At one point, musicologically and historiographically speaking, this question of affective means would have been considered exclusively from the standpoint of theoretical musicology's analyses of structure and harmony, leaning hard on a notion of "pure music" having transcendent, verbally incommensurable meanings and effects that transcend and exceed the capacities of mundanities such as words, plots, and costumes.[2] However, the cultural turn in historical musicology since the 1980s—of which feminist and sex/gender analyses are emblematic—has broadened the scope of this discussion to include considerations of the affective and aesthetic roles of narrative tropes, physical acting, and costuming as well as materialist, biomedical analyses of the observable effects of factors like musical rhythm and frequency on the body (Hutcheon and Hutcheon 2000, 153–181).

But opera is a combinatory, synthetic art form, at minimum consisting of both instrumental and vocal music plus text. It is, whether we like it or not, difficult to speak of opera without engaging the Wagnerian concept of the *Gesamtkunstwerk* or all-embracing artwork. There is no, and perhaps can be no, succinct, clear, direct schedule of one-to-one correspondences that explain the affective power of such an integrally multilayered, polysemic genre. In the face of the stubborn evasiveness of opera's mechanisms of affect, however, scholars end up writing not about what produces the audience's responses to opera but about the responses themselves. They write of overwhelming bodily sensation, of tears and gasps, intoxication, and even sexual consummation. Tasked with explaining their own complex bodily/cognitive/emotional responses to opera, scholars write, as Judith Peraino does in her essay "I Am an Opera: Identifying with Henry Purcell's *Dido and Aeneas*," of experiencing

> literally sympathetic vibrations resulting from the music and the vision of the singers. This catharsis combines with the engagement of the emotions and the aesthetic experience of identification—that is, a vicarious/empathetic experience of the sentiment displayed by one or more characters. Opera is a unique art form that provides for the intertwining of emotional/physical rapture—a complete catharsis that unites the mind and the body. (1995, 127–128)

What Peraino and other writers thus describe are moments where the mind is not in control of how one reacts, intervals where the ego or the will submits to the encompassing onslaught of opera's sensory and narrative elements and involuntary physiological and emotional responses become at least as powerful as the cognitive, if not more so (Hutcheon and Hutcheon 2000, 28).

One of the reasons this happens, and the crucial component in the operatic synthesis, is that it is an integrally embodied performance genre. Without voices and the bodies that produce them, without the breath and the muscle to support it, opera loses its most distinctive characteristic: song. The centrality of song to opera, and our tendency as human beings to react sympathetically to human breathing and vocalization, renders opera a member of the class of performance genres that the film scholar Linda Williams calls "body genres" (1991, 2–13). Like horror, pornography, and—not coincidentally—melodrama, opera's effectiveness is reliant on the bodily sympathy of its audiences to the bodily activity of its performers.[3] This bodily sympathy contributes to the effectiveness of embodied states as suggested or symbolized in song, as Susan McClary shows in her analysis of the wild flights of Lucia's delirium.

We see analogous (and sometimes similar) composerly attention to the scoring of embodied states in a range of operas. In Verdi's *La Boheme*, for example, careful notations of Violetta's coughs and shortness of breath begin to appear in the score as early as act 1, scene 2, in the run-up to Violetta and Alfredo's first duet. Similar attention to embodied detail runs throughout *La Boheme*, especially in Puccini's painterly rendering of Mimi's last breaths in notated gasping, slowing phrases against a lean accompaniment of strings, flute, clarinet, and harp prior to the fateful *lunga pausa* that announces her death. The tendency for instrumental accompaniment to drop out beneath Peter Grimes's vocal line is Britten's way of expressing not only Grimes's social alienation but also his cognitive and emotional distance from reality, and in the penultimate scene, the evocative alienated vocalises of his name that Britten gives Grimes (first over a choral pedal tone, then over silence) not only echo the same sort of chromatic melismatic excess to which McClary notes Lucia is prone but also render Grimes's psychic and cognitive inability to recognize himself fully audible.

These are bodies whose existence, functioning, and malfunctioning are being expressed through vocalization, through the very noise of breath moving in and through the body. The operatic audience is aware of embodied presence, indeed cannot help but be aware of it, in a way that is simply not true for audiences of instrumental music. Just as stylized fear reactions and arousal seen on movie screens provoke bodily reactions in horror and porn movie viewers, stylized vocal exertion and expression are part of what provoke opera audiences' curiously and wondrously intense somatic reactions.

Such involuntary, even unavoidable bodily response is part of what produces in the operagoer the preconditionary sensibility for the experience of the sublime: the sense of being confronted with a presence and a force larger than the self and beyond the influence of one's will. Voice, first among other musical and dramaturgical components, is the medium that translates somatic presence onstage into somatic reaction in the audience. And it is not just any voice, but athletic, cultivated, prolonged, and physically extreme vocalism that achieves this. When the audience member receives an operatic performance, their body becomes not secondary to cognition in response to the performance, but more of an equal responder by virtue of empathetic and sympathetic response to vocalism and breath as well as music, narrative, and stage action. This somatic susceptibility helps make some sense of why opera is the genre of consumptives who nevertheless die at the tops of their lungs and where the staggering and delusional are somehow capable of long spans of physically punishing, technically complex singing, why dwarves and hunchbacks and erotomaniacs and syphilitics thrive on the operatic stage.

Indeed, it makes it something of a wonder there are not more of them, given the nature of the shifts that took place with regard to expectations and meanings of embodiment during the long nineteenth century. Across the century, Enlightenment leanings toward objectification and quantification manifested in intense concern with the rationalization of human life through new concepts like "average" and "normal." The Belgian statistician Adolphe Quetelet's 1845 *Sur l'homme et le developpement de ses facultés, essai d'une physique sociale* (A Treatise on Man and the Development of His Faculties in its English translations) gave us the concept of the "average man" as common denominator. This idea, filtered through nationalist and medical imaginaries that held up the average as the desirable, properly constituted body of the healthy person and good citizen, was further entrenched by workplace and commodity standardizations imposed by industrialization's regularized built environments and economies of scale; Lawrence Kramer calls this "the rise of the regime of the norm as the paramount regulatory mechanism of middle-class modernity" (2004, 7).

Thus, as the century wore on, bodies that in some way or another were not average came to represent a newly significant, intensified kind of difference. Because the average or "normative" was acquiring a new and unprecedentedly thoroughgoing value in the political and social economy, nonaverage, nonnormative bodies were less assimilable, harder to make exceptions for.[4] Adding to and deepening this new cultural rhetoric of the average and normative body was the increasing acceptance of the notion that bodies might and indeed should be susceptible to human control. Paradigms of medical and bioscientific understanding were changing. Humors and miasmas were giving way to chemical analysis and germ theory, Lamarck yielding to Darwin and Mendel. Increasingly, physicians were able to offer the miracle of painless surgery, and some of their patients even had the benefit of new aseptic techniques, making invasive therapeutic interventions not only possible but also successful in ways they had never been (Hutcheon and Hutcheon 2000, 25). Eugenics, that seductive and pernicious idea that human perfection might be bred for, was formally articulated in the early 1880s. It was an age of wonders in which it seemed as if man (this noun is used advisedly) might finally really become his own bodily master.

Against this backdrop, bodies that did not conform to newly important norms of formation and function acquired new resonance and meaning. The windows of acceptable human variation narrowed, and the connotations of the bodies that would not fit through them came to include not only older customary notions of misfortune and sin but also a new, complex sensibility of imperviousness to reason, science, and similar modern priorities. Such bodies flagrantly refused control and were unmoved by economic and social incentive. The biopower of machines, laws, and statistics could not un-hunch a hunchback, nor restore sanity to the insane. Neither could the ingenuity of medical men and scientists.[5] These nonconformist, excessive bodies were intransigent, and at the core of their intransigence was nothing less than nature itself.

Insofar as they appeared spontaneously physical and physiological deformities, asymmetries, absences, and superfluities were clearly natural, and insofar as they were natural, they were a reminder that nature tends to surpass our human capacities. Nature, not man, was the ultimate master of man's embodiment. A disabled, sick, or dysfunctional body that could not be brought into line represented not just an individual's inability to discipline their own body into conformity but also science and medicine's failure, in spite of dramatic and suggestive new advances and ideas, to be able to bring it to heel. Nonnormative embodiments presented uncomfortable reminders to the normative and the average of the arbitrariness and unenforceability of the expectations whose fulfillment gave them a certain degree of status. (The word "normative," meaning a thing that serves as a prescriptive norm or standard, dates from 1852. The timing is suggestive.) They additionally proclaimed, by their very lack of normativity, the limits of human mastery over embodiment. Nonnormative embodiments signaled the continued presence of nature in the human animal, but not Nature in its guise of benevolent provider or spectacular scenery. Reflected in these troublemaking bodies was the same dynamic, forceful, threatening Nature the philosophers invoked in their classic examples of the sublime: the jagged cold rock faces of mountains, the lifeless wastes of deserts, the crashing waves and swallowing depths of the sea, the motiveless raging of storms, all of them vast, inhuman, and uncontrollable.

It was not only the physical body that might provide glimpses of a feral nature unamenable to human control but also the behavioral and emotional body. Such instances, of course, are more complex: it frequently cannot be known whether emotional or behavioral acting-out is a matter of malfunction, misbehavior, or some mixture of the two. Yet particularly in the realm of romantic and erotic extremity and excess, there was historical precedent for interpreting emotional and behavioral excessiveness as symptomatic of dysfunction, and not necessarily under the individual's control. Lovesickness, particularly, had a centuries-long history of being more than a metaphor. Until the time the materialist and mechanistic turn of post-Enlightenment diagnostics complicated the nosological picture, humoral medicine made "amorous melancholy" and "erotomania" fairly straightforward as disorders of the body (Small 1996, 14–18; see Kennaway's chapter for more on this subject). Like fevers, they involved an excess of humoral heat. For some writers, such as the German Jerome David Gaub, who taught and wrote in Leiden in the second half of the eighteenth century, such fevers could provoke delirium and bodily disease; a stream of medical thought represented in the sixteenth century *Theatrum Mundi* among other sources held that the heat of excessive love and desire actually burned and desiccated internal organs such as the liver and intestines (Dawson 2008, 14–17; Duffin 2005, 60–65). The absence of love and sexual consummation could also create illness. In virginal women, greensickness or chlorosis was a condition in which the absence of carnal love could lead to fainting, feebleness, madness, and death. Such notions of love and desire causing bodily dysfunction were on their way out in the nineteenth century; by century's end they would be rethought and repackaged as problems of purely psychological origin. But this was a slow and late transition. During grand opera's heyday the older ideas coexisted with the developing new paradigms, and thus we must allow for the possibility, in nineteenth-century opera particularly, that lovesickness and sexual desire are part of the array of literally embodied disorder and extremis that gets represented on opera stages. Thus they are part of that which represents the natural body in all its uncontrollability—in addition to possibly representing emotional derangement or social misbehavior—on the operatic stage.

Lovesickness and erotomania are no longer formally diagnosed, neither exists in the current editions of the *Diagnostic and Statistical Manual* (DSM) or the *International Statistical Classification of Diseases and Related Health Problems* (ICD). Cases resembling what used to be known by those names are now more likely to be couched in terms of delusional, obsessive-compulsive, or histrionic disorders. Whatever names they may go by, the dual nature of these conditions, at once psychological and bodily, sustains a connection to the aesthetic experience of sublimity their representation can help to invoke. As James Luchte notes in his analysis of Schopenhauer's aesthetic phenomenology, the sublime is "a state of exaltation that is almost a dare—in remembrance and dreadful recognition of the threat to oneself" in which "the terrible becomes susceptible to contemplation" (2009, 235). This experience, reacting to (dreadful recognition) but also contemplating a threat to the self, is the heart of the experience of the sublime.

Such deep cognitive and experiential tension finds echoes throughout the operatic experience. We find it in the distinctions and unities between what Schopenhauer called the "sacred, mysterious, intimate language of music" and the visual and narrative aspects of opera—especially the "insipid love story" elements—he found so distracting (Schopenhauer 1970, 163). It exists in the discrepancies between the eminently abled and highly trained bodies and minds of singers, without which the opera could literally not be performed, the disabled or diseased or dysfunctional embodiments they occupy on stage, and the audience's simultaneous awareness of both statuses. There are multiple simultaneous levels at which embodiment and meaning (sociocultural, narrative, representational, affective) are held in productive, even ravishing, tension.

When we look to the composers and librettists, we find that this is no accident. Opera's creators were fully aware of these tensions, and of the affective and aesthetic power in juxtaposing the vulnerability created by opera's relentless embodiment with the vulnerability created by disease, dysfunction, and disability, using each to heighten the other. Indeed, they exploited the paradoxes purposefully. Giuseppe Verdi, creator of Rigoletto, opera's most famous hunchback, recognized the aesthetic value of the tension between his character's embodiment and his emotionality in a letter to the director of Teatro La Fenice in 1850: "I thought it would be beautiful to portray this extremely deformed and ridiculous character who is inwardly passionate and full of love. I chose the subject precisely because of these qualities and these original traits" (Verdi 1971, 76). Wagner opens *Das Rheingold* with a protracted dissertation—carried out by the dwarf Alberich in conversation with the three beautiful Rhinemaidens—on the same apparent paradox, the passions and desires that dwell within the physically deformed (and thus erotically inconsiderable, at least to the mocking Rhinemaidens) body. In yet another similar case, Leoncavallo's hunchbacked "half-wit" clown Tonio (*Pagliacci*) begins the opera with an introductory prelude, explaining that the artist is a man and as such, his mission is to depict something human for his fellow men—"Egli ha per massima sol che l'artista / È un uom, e che per gli uomini scrivere ei deve." It is his plea, from a player to his audience, that they see him as a valid and fully human person, the very thing his love interest, Nedda, will violently refuse him when she lashes out at his advances with "Hai l'animo siccome il corpo tuo difforme, lurido!" (Your soul is deformed and filthy, just like your body!)

Operatic composers and librettists thus pit humanity (represented by love and relationship) against nonhuman nature (represented by disability and deformity) in open contest. They do so not only in dialogic form—Nedda and Tonio, Alberich and the Rhinemaidens, Rigoletto and Gilda—but also in monologue and soliloquy as well. Alberich renounces love and curses the gold, Tonio asks to be thought of as a man as well as the caricature he knows he inevitably represents. In his act 1 soliloquy, Rigoletto puts it quite baldly: "O uomini! O natura! Vil scellerato mi faceste voi!" (Oh mankind, oh nature, it is you that made me vile and corrupt!). Librettists and composers produce and foreground a fundamental and insurmountable conflict having to do with humanity and nature, conformity and deformity, not only as a dialogic reflection of naive physical disgust or an increasing cultural impermissibility of allowing full humanity to the disabled, although these are certainly in evidence, but also as self-aware internal schism.

In these moments there is a sensibility of actual incommensurability between bodies that are tangibly marked by inhuman nature and bodies whose conformity or conformability to norms marks them as belonging to civilization. This incommensurability is marked by inclusion or exclusion in human emotional experience that is linked to sexuality either directly (Lucia, Violetta, Tonio, Mimi, Alberich, Aschenbach, Lulu, Salome) or as in Rigoletto's case, or Peter Grimes's, at some remove (Brett 2006, 11–53). Raised up by idealized expressions of desire and love, whether passionate or paternal, and simultaneously cast down or out by embodiment that does not conform (whether in shape or function), the character onstage is both exalted and humbled by the conflict, as is the operagoer who sympathetically responds with cognitive, emotional, and bodily reactions.

This ambivalence, the philosopher Bart Vandenabeele argues, is at the core of the Schopenhauerian sublime, which he writes is "not an experience which is first painful or distressing and subsequently uplifting, but which is simultaneously painful and pleasurable or characteristically oscillating between feelings of tranquility and turbulence, elation and humiliation, pleasure and pain, and so on. The sublime is typically an unstable emotion in which the mind seems to go back and forth from negative to positive feelings and back again" (2015, 115). This sublimity, more than just an awareness of forces and factors larger than the self, and more than just the reception of a presence that beggars the force of one's will, is also something that invites the simultaneous efforts of cognition and sensation to engage fully with the environment at hand. In nineteenth- and early twentieth-century grand opera, disability writ large—including deformity, disease, and dysfunction—and sexuality combine with opera's musical, literary, and dramatic elements to engender a profoundly humanistic experience of the aesthetic sublime. The repeated deliberate cultivation of this particular mode of sublime experience speaks volumes about the cultural priorities of a nineteenth century increasingly concerned with internal tensions between passion and control, individualism and conformity, the pathological and the normal.

These moments of a Schopenhauerian humanistic sublime are not, by far, the only ways in which opera is deepened and enriched by the presence of both sexuality and disability. As the writings of scholars like Sander L. Gilman (1988), Lawrence Kramer (2004), and Linda Hutcheon and Michael Hutcheon (1996, 2000) have notably shown, sexuality and embodiment carry multiple meanings and exist in the frames of multiple rhetorics on the opera stage. Considering sublimity as an extraordinarily powerful aesthetic experience affords us a way to understand the depth and immediacy with which the images and ideals opera communicates affect the operagoer. Certainly sublimity can be a useful, perhaps indexical, consideration in the ongoing effort to tease out the methods by which opera generates the spiritual, emotional, and physical excitement and exaltation that it affords those who pay attention to it. It bears noting that this use of disability—as Joseph Straus and others note is historically true of most such representations—creates this aesthetic experience at the expense of disability and the disabled, and is thus in itself aligned with a broadly normative aesthetics and erotics of human embodiment (Straus 2011, 46–48). This is hardly surprising in an art form that canonically serves an elite audience, one whose power and status derive in part from particularly stringently conformist embodiments in terms of beauty, health, strength, gender, and other attributes. But perhaps audiences who particularly prize conformity and control may be—and may historically have been—all the more susceptible to the shock and threat represented by the onstage consumptive, the dwarf, the hunchback, and hence all the more likely to find themselves caught up in the sublime operatic moments when love and sexuality collide and entangle with the ineluctable, irreducible presence of bodies whose varied extremes refute the norms.

Notes

1. Two recent works offer insightful overviews of the trajectory of "sublime" in our understanding of music. For a brief synopsis, see Downes (2014). For a monograph, Wurth (2009).
2. Indeed, this intense prejudice in favor of wordless, primarily instrumental "pure music" is also derived at least in part from Schopenhauer, whose championing of music as the supreme art rested on the expressive power of untexted music. Richard Wagner, Gustav Mahler, and Thomas Mann were among the popularizers of this ultimately widely popular and influential idea, as Lydia Goehr discusses in "Schopenhauer and the Musicians" (1996, 200–228).
3. I am not the first or only writer to add opera to the roster of body genres or, for that matter, to note the indexical relation between body genres and the effect of the presence of dis/ability in performance (see, e.g., Snyder and Mitchell 2006, 159–162; Hutcheon and Hutcheon 2000).
4. There are many relevant discussions of this trajectory in, among others, the very useful themed issue of *Psychology and Sexuality* 1, no. 3 (September 2010), Garland-Thomson (2009, 30–35), Snyder and Mitchell (2006, 3–7), and Davis (1995, 129–131).
5. Medicine and science could not unsex a woman nor make brown skin white, for that matter, and it is no coincidence that femaleness and nonwhiteness were also considered, at least by some, to be physical inferiorities or disabilities as well by comparison to whiteness and maleness.

References

Abbate, Carolyn. 1991. *Unsung Voices: Opera and Musical Narrative in the Nineteenth Century*. Princeton, NJ: Princeton University Press.

Abel, Samuel. 1996. *Opera in the Flesh: Sexuality in Operatic Performance*. Boulder, CO: Westview Press.

Brett, Philip. *Music and Sexuality in Britten: Selected Essays*. Edited by George E. Haggerty. Berkeley: University of California Press, 2006.

Clément, Catherine. 1988. *Opera; or, The Undoing of Women*. Minneapolis: University of Minnesota Press.

Davis, Lennard J. 1995. *Enforcing Normalcy: Disability, Deafness, and the Body*. New York: Verso.

Dawson, Lesel. 2008. *Lovesickness and Gender in Early Modern English Literature*. Oxford: Oxford University Press.

Downes, Stephen. 2014. "Beautiful and Sublime." In *Aesthetics of Music: Musicological Perspectives*, edited by Stephen Downes, 84–110. New York: Routledge.

Duffin, Jacalyn. 2005. *Lovers and Livers: Disease Concepts in History*. Toronto: University of Toronto Press.

Garland-Thomson, Rosemarie. 2009. *Staring: How We Look*. Oxford: New York: Oxford University Press.

Gilman, Sander L. 1988. *Disease and Representation: Images of Illness from Madness to AIDS*. Ithaca, NY: Cornell University Press.

Goehr, Lydia. 1996. "Schopenhauer and the Musicians: An Inquiry into the Sounds of Silence and the Limits of Philosophizing about Music." In *Schopenhauer, Philosophy, and the Arts*, edited by Dale Jacquette, 200–228. Cambridge: Cambridge University Press.

Heller, Wendy. 2004. *Emblems of Eloquence: Opera and Women's Voices in Seventeenth-Century Venice*. Berkeley: University of California Press.

Hutcheon, Linda, and Michael Hutcheon. 1996. *Opera: Desire, Disease, Death*. Lincoln: University of Nebraska Press.

Hutcheon, Linda, and Michael Hutcheon. 2000. *Bodily Charm: Living Opera*. Lincoln: University of Nebraska Press, 2000.

Koestenbaum, Wayne. 1993. *The Queen's Throat: Opera, Homosexuality, and the Mystery of Desire*. New York: Poseidon Press.

Kramer, Lawrence. 2004. *Opera and Modern Culture: Wagner and Strauss*. Berkeley: University of California Press, 2004.

Lerner, Neil William, and Joseph Nathan Straus. 2006. *Sounding Off: Theorizing Disability in Music*. New York: Routledge.

Luchte, James. 2009. "The Body of Sublime Knowledge: The Aesthetic Phenomenology of Arthur Schopenhauer." *Heythrop Journal* 50: 228–242.

McClary, Susan. 1991. *Feminine Endings: Music, Gender, and Sexuality*. Minneapolis: University of Minnesota Press.

Mordden, Ethan. 1984. *Demented: The World of the Opera Diva*. New York: F. Watts.

Peraino, Judith. 1995. "I Am an Opera: Identifying with Henry Purcell's *Dido and Aeneas*." In *En Travesti: Women, Gender Subversion, Opera*, edited by Corinne E. Blackmer and Patricia Juliana Smith, 99–131. New York: Columbia University Press.

Pope, Rebecca, and Susan Leonardi. 1996. *The Diva's Mouth: Body, Voice, Prima Donna Politics*. New Brunswick, NJ: Rutgers University Press.

Schopenhauer, Arthur. 1958. *The World as Will and Representation*. Translated by E. F. J. Payne. Indian Hills, CO: The Falcon's Wing.

Schopenhauer, Arthur. 1970. *Schopenhauer: Essays and Aphorisms*. Translated by R. G. Mollingdale. Harmondsworth: Penguin.

Small, Helen. 1996. *Love's Madness: Medicine, the Novel, and Female Insanity, 1800–1865*. New York: Oxford University Press.

Smart, Mary Ann. 2000. *Siren Songs: Representations of Gender and Sexuality in Opera*. Princeton, NJ: Princeton University Press.

Snyder, Sharon L., and David T. Mitchell. 2006. *Cultural Locations of Disability*. Chicago: University of Chicago Press.

Straus, Joseph N. 2011. *Extraordinary Measures: Disability in Music*. New York: Oxford University Press.

Vandenabeele, Bart. 2015. *The Sublime in Schopenhauer's Philosophy*. New York: Palgrave Macmillan.

Verdi, Giuseppe. 1971. *Letters of Giuseppe Verdi*. Translated and edited by Charles Osborne. London: Victor Gollancz.

Williams, Linda. 1991. "Film Bodies: Gender, Genre, and Excess." *Film Quarterly* 44: 2–13.

Wurth, Kiene Brillenburg. 2009. *Musically Sublime: Indeterminacy, Infinity, Irresolvability*. New York: Fordham University Press.

Further Readings

Baum, Günther, and Dieter Birnbacher. 2005. *Schopenhauer und die Künste*. Göttingen: Wallstein.

Blackmer, Corinne E., and Patricia Juliana Smith. 1995. *En Travesti: Women, Gender Subversion, Opera*. New York: Columbia University Press.

Brett, Philip, Elizabeth Wood, and Gary C. Thomas. 1994. *Queering the Pitch: The New Gay and Lesbian Musicology*. New York: Routledge.

Garland-Thomson, Rosemarie. 1997. *Extraordinary Bodies: Figuring Physical Disability in American Culture and Literature*. New York: Columbia University Press.

Henson, Karen. 2015. *Opera Acts: Singers and Performance in the Late Nineteenth Century*. New York: Cambridge University Press.

Jacquette, Dale. 1996. *Schopenhauer, Philosophy, and the Arts*. New York: Cambridge University Press.

Rutherford, Susan. 2013. *Verdi, Opera, Women*. New York: Cambridge University Press.

CHAPTER 18

IS THERE DISABLED MUSIC?

Music and the Body from Dame Evelyn Glennie's Perspective

EVELYN GLENNIE, SANDER L. GILMAN, AND YOUN KIM

OFTEN the discussion of disability and music rests on either the question of the representation of disability in music (in opera for example) or the question of the role of the disabled as performer or listener/viewer. The former is a means of tracking meanings associated with historical categories of disability in music; the latter in the twenty-first century rests on a rethinking of what disability means within both performance and its reception. That this too has a historical dimension is without a doubt true. But the question of the intersection of representation, performance, and creativity is one that is central to any reading of the disabled (non-normate) body in the public sphere. This was in complex ways precluded in the recent history of disability.

In a radical attempt to create an advocacy movement for the disabled, the World Health Organization, in its 1980 *International Classification of Impairments, Disabilities, and Handicaps*, made a seemingly clear distinction between impairment, disability, and handicap. Impairment is an abnormality of structure or function at the organic level, while disability is the functional consequence of such impairment. A handicap is the social consequence of impairment and its resultant disability. Thus, cognitive or hearing impairments may lead to communication problems (disability), which in turn result in isolation or dependency (handicap). Such a functional approach (and this approach was long the norm in American common and legal usage) seemed to be beyond any ideological bias.

The most recent rethinking of disability redefines disability on a scale of "human variability" that postulates the difficulties of the disabled as the result of the inflexibility of social institutions rather than resulting from impairment. Yet even with the passing of the Americans with Disabilities Act of 1990 and the United Nations Convention of

Rights of People with Disabilities of 2009, the World Health Organization's 1980 definition of disability as "impairment, restriction, dysfunctionality, abnormality" has remained very much in place in the public sphere (Garland-Thomson 2012, 341). Music and disability studies in the academy has begun to radically rethink this. One of the creators of the field, Rosemarie Garland-Thomson, understands "disability as a way of being in an environment" produced by the "discrepancy between body and world, between that which is expected and that which is" (342). Not only is disability the disjuncture between the world and the individual, but as all who age become disabled, "disability is thus inherent in our being: What we call disability is perhaps the essential characteristic of being human" (342). The "ableist" view defines the disabled as their antithesis; the "normate," to use Garland-Thomson's helpful term, comes to define the healthy and the beautiful.[1]

Thus the debate entered into by Erving Goffman in the 1960s about stigmatization is joined in another way. What Goffman had seen as the internalization of a "spoiled" identity as defined by society's drive for "normalcy" on the part of the disabled comes to be understood not as a "special kind of relationship between attribute [condition] and stereotype" but as an antithesis between the experienced reality of each (1968, 4). The argument now is that these very categories perpetuate the social isolation and stigma of those differently able, but that these also have an autonomous manner of dealing with the external world through their own disability perspective.

Yet stigma, as Goffman imagines it, is merely the projection of the feared loss of the self on to others in the world. Thus disability is a social category of stigmatization. But it is always defined by the outsider, the marginal, the proscribed: these are defined through this filter as "less than" the normal. For postmodern theory, Davis argued:

> the body is seen as a site of *jouissance*, a native ground of pleasure, the scene of an excess that defies reason, that takes dominant culture and its rigid, powerladen vision of the body to task. . . . The nightmare of that body is the one that is deformed, maimed, mutilated, broken, diseased. . . . Rather than face this ragged image, the critic turns to the fluids of sexuality, the gloss of lubrication, the glossary of the body as text, the heteroglossia of the intertext, the glossolalia of the schizophrenic. But almost never the body of the differently abled. (1995, 5)

At the core of Davis's vision of the disabled are the classical figures of the deaf, the blind, and, as Michel Foucault suggested, building on half a century of cultural fascination with difference, the "mad" (Gutting 1989, 82; see also Gilman 1985). It is the abject body not the superior body. For the bell curve of stigma seems to focus primarily on the "lower end." What about the higher end of the bell curve? What do you do with the highly polished professional performer who provides great pleasure but is disabled? What do we do with the disabled body that gives pleasure without the visible clues enumerated above to their disability? What do we do with the disabled performer whose disability is not visible, such as the deaf percussionist? What do we do when that person is an exemplary performer: neither marginalized nor stigmatized? That is, what do we do when the desire to level the difference between the normate and the disabled is not possible because there really is a radical difference between the "normal performer" and the disabled performer, but it is a category of superiority not inferiority?

For the definition of the disabled performer one needs to rethink the question of stigma as well as that of universals. For the claim of at least professional musicians is that they are by definition not of the common garden variety, they are different from everyone else. The acclaimed performer is defined as inherently different from, perhaps as greater than, but certainly different from all other human beings: the distinction between the professional and the amateur, unlike in sports, is not based on whether one is paid for one's expertise, but whether one is considered to have greater proficiency than the norm. What happens when the image of the superior performer is merged with that of the disabled performer? Is there an increased "Romance of Disability," the notion well embraced by popular culture (including music) that the disabled performer has "overcome" a disability to become proficient as a musician (Morris 1991, 100)? Often called the myth of the "supercrip" in performance, it hints not at excellence in performance but rather the overcoming of disability. According to Jenny Morris, a supercrip is a person with a disability who lives out the popular representation of disability as adversity to be overcome (1991, 100). The classical categories that defined disability, the blind and the deaf, have represented disability within the greater society: from the teacher of the blind, Louis Braille, to the icon of disability in midcentury America, the blind and deaf Helen Keller, these disabled icons were the public image of disability (Mirzoeff 1995). It is these figures that inhabit the myth of self-overcoming. Certainly the greatest nineteenth-century novel on the topic, Victor Hugo's 1831 *The Hunchback of Notre Dame*, is the story of a figure with deformed posture working against type. It is the tale of Quasimodo, the deaf and half-blind bell ringer of the Cathedral, whose love for the gypsy dancer Esmeralda transfigures him. This image continues on the stage, in film, in ballet, and indeed even in video games to shape our image of the ability to transcend the limitations of posture. Such an image of posture as moral force continues through Verdi's 1851 *Rigoletto*, based on Hugo's *Le roi s'amuse* (1832), through Dalton Trumbo's antiwar film *Johnny Got His Gun* (1971), based on his 1938 novel. Yet historically these are immediately visible disabilities, unlike deafness, which is by its very nature invisible. This myth of self-overcoming is inherently the argument dismissed by disability studies that wish to level the playing field and see all human beings as able to function in the world, if the question of access is answered. Thus the disabled performer is essentially better than the disabled in this narrative.

There is a difference in the reception of disability within music as the traditional "visible" disabilities, with their modern origin in the concerns of the Enlightenment for the education and integration of the disabled into the economic system, give way to markers of "access," of "integration," of "flexibility" (Hamraie 2013). The three-dimensionality of mapped space, what Isabel Dyck calls "conceptualizing the environment," comes to be "crucial to the politics of disability research in delineating issues of access, a crucial dimension of a *socio-spatial model* of disability" (2010, 254; emphasis in original). Here disability dance becomes a model for a changed understanding of how disability studies come to define such new models of access also as an aesthetic question. As we shall see, "space" too figures heavily in other manners of performance, both three-dimensional space and acoustic space. Is the disabled musician, whether instrumentalist or vocalist, to be included in this new manner of seeing disability as a reflection of a universal embeddedness in the phenomenological world?

The question of placing oneself in the position of the disabled performer is encapsulated in the feminist disability concept of "sitpoint theory," which stressed the postural impact on "knowing" from the position of sitting in a wheelchair, as opposed to standing, as productive of a different epistemology. Indeed, it was the late Emory University theologian Nancy Mairs, who had multiple sclerosis and used a wheelchair, who evolved the notion that human beings came to be defined by their literal view of the world as they acquired upright posture. She noted that human grandiosity has its roots in the ability of the child to move from crawling to walking, "and at this momentous point in the practicing stage, upright locomotion permits such vista, such possibilities, such triumphs, that a child can grow drunk on omnipotence and grandeur" (1997, 68). This is the claim of the "normal" seeing the world from a new perspective, that of upright posture. The normate was defined by the difference between those who were able to stand and for whom "the very thought of ever being like me so horrifies them that they can't permit themselves to put themselves on my wheels even for an instant" (32). Thus the question that we can address to the disabled performer is twofold: what is it like to be disabled and the best in one's field?

The editors of *The Oxford Handbook on Music and the Body* addressed a number of these queries to the percussionist Dame Evelyn Glennie. As perhaps the most recognized deaf professional musician, her work stands at the center of many of these questions. Born in 1965 in Scotland, she became profoundly deaf at the age of 12, having begun to lose her hearing at the age of 8. A trained classical musician (both at the Ellon Academy and the Royal Academy of Music), she is equally at home in the Scots indigenous musical tradition of her father, an accordionist in the country dance band. Thus she has played with leading classical orchestras worldwide but has also collaborated with diverse artists such as Björk and Mark Knopfler. Awarded an Officer of the Most Excellent Order of the British Empire (OBE) in 1993 she became a Dame Commander of the Most Excellent Order of the British Empire (DBE) in 2007. She plays barefooted and has noted that she has taught herself to "distinguish the rough pitch of notes by associating where on my body I felt the sound with the sense of perfect pitch I had before losing my hearing. The low sounds I feel mainly in my legs and feet and high sounds might be particular places on my face, neck and chest" (Glennie 1993). Her substantive "Hearing Essay" outlines her philosophy of "hearing":

> my hearing is something that bothers other people far more than it bothers me. There are a couple of inconveniences but in general it doesn't affect my life much. For

> me, my deafness is no more important than the fact I am female with brown eyes. Sure, I sometimes have to find solutions to problems related to my hearing and music but so do all musicians. Most of us know very little about hearing, even though we do it all the time. Likewise, I don't know very much about deafness, what's more I'm not particularly interested. I remember one occasion when uncharacteristically I became upset with a reporter for constantly asking questions only about my deafness. I said: "If you want to know about deafness, you should interview an audiologist. My speciality is music." (1993)

The editors sent a number of follow-up questions to Dame Glennie in the fall of 2015.

Interview

Editors: How would you define "hearing"? I would like to ask you to explicate what you meant by "the body as a resonating chamber."

Evelyn Glennie

Hearing is the involvement of all our senses to some degree but three in particular—sight, touch, and hearing. The eye sees something move, it then registers the relationship of that movement to the body which I believe gives us a feeling of vibration through parts of the body and sometimes the ear itself, that is, the air movement over the skin. I believe there is an assumption in the brain, possibly created from memory, defining what kind of feeling or sound is occurring. For example, if I see the branches of a tree move I assume the branches, leaves, birds, wind, and the immediate environment are creating sound regardless. My senses are triggered by what my eyes see. The ripples from my visual experience is what my imagination is using as a basis for assumption.

The body can act as a resonating chamber because vibrations are felt throughout its parts. In my case I have learned to recognize and associate particular vibrations and connect them to the visual image I see. For example, when I strike an instrument the sound vibrates through parts of my body, which I have repeatedly experienced. I feel some vibration from higher pitches in the upper part of my body such as my cheekbones. I also identify the lower pitches when I feel vibration in my abdomen. Through recognition of these sensations I have trained myself to think of my body as a huge ear. As a musician I believe too much prolonged sound can numb the listening experience, which can result in the whole body being exposed to a cacophony of sound. The concentration required to digest resonance is much more of a focused activity—remember, resonance extends sounds.

Editors: Do you think people can learn your way of engaging the "body as a huge ear" in listening? If it can be taught, do you have any suggestions?

Glennie:

We can learn to distribute sound by observing our environment and through listening and connecting sounds with sensations we experience in the body. We can use opportunities around us to experience resonance felt in the body in many ways. For example, as a child I was taught to press my hand against the side of a drum or a classroom wall when a drum was struck. You can learn to locate where sound is coming from by closing your eyes and listening for the direction sound is coming from. As you gradually build up these experiences you can refine them to a point when your brain recognizes the sensation because of the feeling in the hand, face, or other parts of the body.

To begin getting used to this practice one can literally declutter and pay attention. Having the notion of giving your environment (home or workplace) a sound spring clean will help. Get rid of any sound that is not required or perhaps by tempering the amount of sound penetrating through your environment is a good start. Sound is like food, we are often very careful about what we eat and the quantity and quality of the food itself. The same philosophy can be applied to sound. By giving ourselves time to digest one sound at a time we are allowing our bodies to process the sounds. Listening, after all, is about focus, concentration, and patience.

Editors: Do you recognize a category of "disabled music"? If it exists would this refer only to the performer(s) or can there be a "disabled" listening practice?

Glennie:

I do not believe there is any such thing as "disabled music" or indeed disabled listeners. I do believe we are all capable of experiencing sound and vibration, therefore music.

Listening is a living experience, not necessarily done the same way every time. Of course we can be lazy listeners, inattentive listeners, short-spanned listeners, pompous listeners, arrogant listeners, and so on.

Editors: Can you see your musical performances as supporting or negating the idea of a "deaf culture"?

Glennie:

My audiences are varied and not exclusive to people with hearing or without. My performances are there for all to relate to in their own way according to their own physical, mental, and emotional journey. I don't categorize the music I play—it is simply a painted aural sensation that can be experienced or felt as a physical medium or simply heard through the ears should the listener choose to. Each person, deaf or hearing with all its variations, has the capability to experience the performance in more diverse ways than they think should they allow their minds to open up to those possibilities. The main point is how an individual categorizes themselves as this will have an impact on how they think they can relate to sound.

Editors: You are known both for virtuosic solo performance and for collaborations and ensemble performances. Can you tell us about how you engage your body differently in these two cases?

Glennie:

When I perform as a soloist my body becomes the tool to create the sound. I think of a much bigger sound experience and relate to the ambient environment around me. The sound in my mind is not yet possible to project from the tools I have but I still strive for that. For example, the marimba in my head does not yet exist in reality but I perform as though it does. Therefore, everything is more exaggerated when playing alone and I can take my time to digest the whole process of creating the sound. This process alters the dynamic between myself and the way I embrace my audiences.

When playing as part of an orchestra or with other musicians there has to be more give and take between the players, which also means more compromises. There needs to be a greater sense of balance and respect for each other's aural space. As a result, all the senses are heightened and engaged. There is much more of a feeling of compartmentalizing, and the players are bringing their own performance styles together to combine as one. The listening preparation before coming together is more academic; each musician needs to listen out for cues from the other performers. For example, by looking at a full score I would be analyzing what the trumpets are playing and interpreting how they might play each segment. This type of preparation is also something a conductor would be studying. Your listening palette is open to sounds even before striking your own instruments. The expectation is slightly different—it's a bit like driving in a busy city where you know the rules, you know where you want to go, but you also know there is traffic coming from all directions, so your awareness is much more heightened.

Editors: In your "Hearing Essay," you wrote about the performer's role as follows:

Glennie:

"Music represents life. A particular piece of music may describe a real, fictional or abstract scene from almost any area of human experience or imagination. It is the musician's job to paint a picture which communicates to the audience the scene the composer is trying to describe. I hope that the audience will be stimulated by what I have to say (through the language of music) and will therefore leave the concert hall feeling entertained."

Editors: Can you tell us more about the role of a performer in musical communication? Is there such a thing as a problem of communication or interpretation? How do you understand the way in which the performer's body is involved in this communication process?

Glennie:

A performer's job is to explore sound and to process their sound creations into an emotional journey for the audience, who then digest the performer's journey according to their own state of mind at the time. Ultimately, a performer knows there are infinite ways of interpreting a piece of music, yet at the time of the performance he/she has to decide on one way. When making a recording this is known as a "frozen moment in time." During a live performance the performer decides which pathways he/she might choose to express the interpretation. This is usually in relation to their body movement. Questions they would ask themselves might be whether or not to close their eyes; consider if there's excessive movement; should they create vocal noises whilst playing; do they look at the audience; is the body showing signs of not giving any more than producing the sound, and so on? It's really then up to the audience to decide how they are moved by the performance. Some will like the excessive movement, and some will not. Some audience members will be hooked on tradition, whilst others are not concerned. I believe these factors all have a bearing on how we listen. There is no right or wrong way, it's personal to each performer and every member of the audience. We are all part of the performance because our dynamic affects the actions of the other. The presence of an audience ultimately changes the performer's interpretation of a piece and vice versa.

There are many factors to consider, spanning from the type of venue to the size of the audience, and they come into play when putting up a performance. Playing in a wet acoustic, like a church, allows the body to wait and be patient with sound. In reverberate acoustic environments the performer needs to recognize and be aware of the resonance. Therefore, subtlety is everything. Time takes on a different dimension as one plays with resonance, which is much more easily felt through the body than in an outdoor venue.

Editors: You have performed in various venues, not limited to the traditional concert halls but also in everyday places. How does the musical space in which you perform affect the use of your body in listening and performing?

Glennie:

The different performance spaces and environments do affect the way I use my body to listen and perform. An outdoor venue injects less subtlety in my playing as all dynamics rise several notches to cope with the environment. Therefore, the physical projection becomes more profound. Everything is larger and broader with less emphasis on what you're feeling through the body but more emphasis on what is being seen. For example, I could not have played in the London 2012 Opening Ceremony as though I was striking baroque timpani; I had to use the physical exaggeration to reach to the many millions of people around the world. I wanted them to have the illusion of *feeling* the sound through what they were seeing. That type of exaggeration would be totally inappropriate if executed in a church or cathedral because here the environment acoustic allows the performer and audience to be more connected to the physical subtlety of the sound.

Editors: Hearing and seeing are intertwined closely in today's listening mode. Do you think seeing the performer's body will facilitate communication or not?

Glennie:

I always enjoy a performance more when I see the performer(s). There is an added wonderment when seeing someone at work creating a sound meal just as it's an experience seeing a great chef at work in a restaurant. The physical body movement and facial expression can all be heightened from the performer in the presence of an audience, unlike a recording, which is usually tweaked from a number of "takes." A recording is about the quality of sound projection, which can affect an interpretation, whereas a live performance can be more about interpretation than sound projection.

Editors: Do online experiences enhance or diminish the communication of the performer with the audience?

Glennie:

This is hard for me to say. I feel the relationship to technology today is different with younger generations. Technology seems to have become an extension to their limbs and they may have a different viewpoint to mine. I was not brought up listening to music other than from music created using the raw hand, and my experiences are more of an organic nature. When I listen to sounds produced in the moment my listening skills differ from those I experience when I am listening to sounds created through a live performance. I know the physical experience I get from listening to digital media is diminished. I believe this is due to the reduction of tactile communication when the sound is coming from a radio or computer. It becomes much more to do with the ear digesting the sound rather than the body, which for me is not so satisfying. I believe digital listening experiences are more to do with being satisfied quickly on a "hearing" level as opposed to an actual "listening" experience.

Editors: I am particularly interested in what you said about silence in music. Can you tell us about how you use your body to perform the silence in music?

Glennie:

Silence in music is open to interpretation similar to the physical sounds. Silence in music has the potential to be the most profound sound of all that binds us together. I believe we listen to silence differently—the attention is more personal and yet binding, whereby both audience and performer change physically and mentally. Silence is the sound of the environment we are in, the canvas we are painting other sounds on. Silence engulfs the whole being. The body can present any stance whilst respecting silence as a sound, and the audience will accept the chosen stance.

Dame Evelyn Glennie's responses, rather than being confined to the topic of music and disability, more comprehensively illuminate various aspects of music and the body from a practitioner's perspective. She speaks, for example, about how she uses her body as an instrument not only to produce sounds but also to communicate effectively with her audience and co-performers. In speaking of the different preparations needed for solo versus ensemble performance, she addresses the relationship between music analysis and performance and seems to validate what Rothstein (2005) calls "playing with the work" as opposed to "playing the work" (218). Her statement also corroborates the significance of the ambient space for musical performance in terms of both acoustics and ecology. The venue of musicking involves the sociological dynamics between the performer(s) and the audience, the cultural significance of media, and the architectural structure and physical condition. As argued in the ecological psychology pioneered by Gibson (1979), our perception, cognition, and actions are embedded and situated in the context, and the act of musicking (whether it be composition, performance, or reception) arises from the interaction between the body (or bodies) and the world.

Our bodies are all different, and we hear music differently with our different bodies. Glennie's listening attests to yet a different, sensitive, and focused way of hearing music.

Her entire body serves as an instrument for listening, which involves the act of seeing, hearing, and feeling—a truly multimodal experience.

Pointing out the problematic nature of the terms "disability" and "disabled," Straus (2011) proposes "disablist hearing." In contrast with "prodigious hearing" and "normal hearing," each of which is promoted in the fields of music theory and music cognition/education, respectively, "disablist hearing" refers to "the ways that people whose bodily, psychological, or cognitive abilities are different from the prevailing norm might make sense of music" (150). Autistic hearing, for example, emphasizes the parts over the whole and creates associative networks rather than hierarchies. Such a mode of hearing is enabled when one listens to atonal music, which, unlike tonal music, disables normal listeners and hierarchical listening. Glennie's listening is an example of "deaf hearing": it emphasizes the tactile dimension of music listening, but it can also involve seeing and kinesthetics. Deaf hearing is also closely tied to the idea of "inner hearing": we hear music in the mind's ear in the absence of the acoustic stimuli (see Kraemer et al. 2005). Via this connection to musical imagery, the idea of deaf hearing not only illustrates diverse possible modes of hearing but also carries serious implications for music education. After all, inner hearing is a vital ability for all composers and performers.

Dame Evelyn Glennie's sense of the silence of music is analogous to Friedrich Nietzsche's observation that one is aware of one's own body only when it is ill or perhaps here when it is discordant with the world in which it must function. He writes in *Human, All Too Human* about the "usefulness of sickliness":

> He who is often sick does not only have a much greater enjoyment of health on account of the frequency with which he gets well: he also has a greatly enhanced sense of what is healthy and what is sick in works and actions, his own and those of others: so that it is precisely the sickliest writers, for example—and almost all the great writers are, unfortunately, among them—who usually evidence in their writings a much steadier and more certain tone of health, because they understand the philosophy of physical health and recovery better and are better acquainted with its teachers—morning, sunshine, forests and springs—than the physically robust. (1996, 293)

We live in a world of silence broken, as Evelyn Glennie noted, by moments of sound. This is the world that we all can experience but few ever become aware of. As Lennard Davis noted, "the fact is that no one is or can be normal, as no one is or can be equal. Everyone has to work hard to make it seem they conform, so the person with disabilities is singled out as a dramatic case of not belonging. This identification makes it easier for the rest to think they fit the paradigm" (2002, 117). Shaping our acoustical world to be heard from the perspective of "sitpoint theory" seems to be possible. We can imagine a world of silence ruptured by sound. Yet as listeners we are still left with the impossibility of seeing superior performance from our position in the audience. If we can dismiss the supercrip "Romance of Disability," as Evelyn Glennie does, there remains a mystery of excellence, the infinitely more complex one of the creativity of the performing artist. The question that remains is not whether we admire the disabled performer for their aesthetic ability but whether their extraordinary ability is indeed a product of their difference. And that question is unresolvable.

Note

1. Garland-Thomson (1996, 8) coined the term "normate" to indicate the nonstigmatized figure body—a body that is "normal" in society in that it is the unspoken standard against which society measures difference. The "normate" is a dynamic concept and changes over time.

References

Davis, Lennard J. 1995. *Enforcing Normalcy: Disability, Deafness, and the Body*. New York: Verso.
Davis, Lennard J. 2002. *Bending Over Backwards: Disability, Dismodernism and Other Difficult Positions*. New York: New York University Press.
Dyck, Isabel. 2010. "Geographies of Disability: Reflecting on New Body Knowledges." In *Towards Enabling Geographies: "Disabled" Bodies and Minds in Society and Space*, edited by Edward Hall, Vera Chouinard, and Robert Wilton, 253–264. Burlington: Ashgate.
Garland-Thomson, Rosemarie. 1996. *Extraordinary Bodies: Figuring Physical Disability in American Culture and Literature*. New York: Columbia University Press.
Garland-Thomson, Rosemarie. 2012. "The Case for Conserving Disability." *Journal of Bioethical Inquiry* 9 (3): 339–355.
Gibson, James J. 1979. *The Ecological Approach to Visual Perception*. Boston: Houghton Mifflin.
Gilman, Sander L. 1985. "The Mad Man as Artist: Medicine, History and Degenerate Art." *Journal of Contemporary History* 20 (4): 575–594.
Glennie, Evelyn. 1993. "Hearing Essay." https://www.evelyn.co.uk/hearing-essay/.
Goffman, Erving. 1968. *Stigma: Notes on the Management of Spoiled Identity*. Harmondsworth: Penguin.
Gutting, Gary. 1989. *Michel Foucault's Archaeology of Scientific Reason*. New York: Cambridge University Press.
Hamraie, Aimi. 2013. "Designing Collective Access: A Feminist Disability Theory of Universal Design." *Disability Studies Quarterly* 33 (4). http://dsq-sds.org/article/view/3871/3471.
Kraemer, David J. M., C. Neil Macrae, Adam E. Green, and William M. Kelley. 2005. "Musical Imagery: Sound of Silence Activates Auditory Cortex." *Nature* 434 (7030): 158.
Mairs, Nancy. 1997. *Waist-High in the World: A Life among the Nondisabled*. Boston: Beacon.
Mirzoeff, Nicholas. 1995. *Silent Poetry*. Princeton, NJ: Princeton University Press.
Morris, Jenny. 1991. *Pride against Prejudice: Transforming Attitudes to Disability*. Philadelphia, PA: New Society.
Nietzsche, Friedrich. 1996. *Nietzsche: Human, All Too Human*. Translated by R. J. Hollingdale. New York: Cambridge University Press.

Rothstein, William. 2005. "Analysis and the Act of Performance." In *The Practice of Performance: Studies in Musical Interpretation*, edited by John Rink, 217–240. New York: Cambridge University Press.

Straus, Joseph N. 2011. *Extraordinary Measures: Disability in Music*. New York: Oxford University Press.

Further Readings

Glennie, Evelyn. 1990. *Good Vibrations: My Autobiography*. London: Hutchinson.

Glennie, Evelyn. 1994. "Evelyn Glennie: Teach the World to Listen." Accessed August 24, 2016. https://www.evelyn.co.uk/biography/.

Holland, James. "Glennie, Evelyn." *Grove Music Online*. Oxford University Press. Accessed August 3, 2015. http://www.oxfordmusiconline.com/subscriber/article/grove/music/44870.

Riedelsheimer, Thomas. 2006. *Touch the Sound—A Sound Journey with Evelyn Glennie*. New Video Group.

PART V

MUSIC AS MEDICINE

CHAPTER 19

MUSIC AND THE BODY IN THE HISTORY OF MEDICINE

JAMES KENNAWAY

ATTEMPTS to understand the relationship between music and the body often draw on medicine in ways that go much deeper than the sometimes-marginalized phenomenon of music therapy. Today's boom in interest in music and the brain is only the latest example of medical conceptions of how listening works at the level of the body. Especially since the Enlightenment, anatomy, physiology, and neurology have been used to try to explain the aesthetic, emotional, moral, therapeutic, and even pathological power of music (Kennaway 2012a, 2010; Gouk and Sykes 2011; Gouk 2004). Approaching the topic from a history of medicine perspective can reveal some of the fundamental assumptions about how listening functions. Along with this acknowledgment that music's sensual aspect might be explained in medical terms, there has also been a profound recurring moral hostility to the association of that physicality with lasciviousness. At the same time, there has also often been a powerful countercurrent that has sought to deny that medicine has anything to tell us, rejecting any significant link between music and the body, and asserting music's abstract, cosmic, and quasi-disembodied character. Yet even for those who deny or denounce the role of the physical and sensual in music, the body and its concomitant medical framework often seem inescapable, cropping up in unexpected ways.

This chapter sets out the ways that changing medical ideas of listening and of the embodied self have influenced theories of music, not only in medical circles but also in aesthetics, criticism, and culture more broadly. After briefly setting out the role of medicine and the body in the cosmological speculation and humoral conceptions of the body from antiquity until the seventeenth century, it examines the radical shift to understanding music in terms of the stimulation of the nerves. By locating music's effects in the nervous system, many Enlightenment physicians and music theorists created an essentially materialist conception of music. The chapter then looks at the way the nineteenth-century idealist aesthetic of music based on an idea of the transcendental listening subject and a positivist science skeptical of music's physical powers led to a dramatic decline in medical interest in music therapy and in the influence of medical conceptions of the body in music. Nevertheless, even here, medical views of music continued to be highly influential in a number of contexts, such as psychiatry, gynecology, and in debates on "mad genius" and race.

Starting in the 1890s, thinking on music began to reflect Sigmund Freud's emphasis on the individual symbolic and psychological meanings of music, apparently leaving the body, already marginalized in aesthetics and somatic medicine, in some ways increasingly irrelevant in psychiatry, too. Likewise, mainstream medical critiques of music's perils for the body declined in the first half of the twentieth century, even as the idea was being instrumentalized for political reasons in Nazi Germany and the Soviet Union. A serious revival of interest in music therapy began in the 1950s, as the field was gradually professionalized, but the real shift in the relationship between music and the body has occurred in the "music and the brain" boom since the 1990s. As this chapter shows, recent decades have seen a dramatic return of medical conceptions of the body in neuroscience, music therapy, and aesthetics. Social change and scientific and technological developments have undermined the Freudianism and the idealist musical aesthetics of the past and taken us to a position where medicine and the body again provide vital paradigms for understanding music.

The Body and Cosmic Harmony

The link between music and the body in medical terms goes back a long way, with an intermittent tradition of using music in treating disease since antiquity and presumably back to prehistory (West 2000). As far as the West is concerned, thinking on the subject can be traced back to Orphic and Pythagorean traditions focused on bringing the soul into harmony in terms of numbers. However, a view of music based on more concrete medical thinking on the body is also apparent. Even Plato, whose Pythagorean cosmological thinking and ethical and political conceptions of music have proved influential ever since, also acknowledged the role of body in listening—describing how sound was transmitted to the soul via the ears, blood, and liver (Pelosi 2010; Barker 2000). And the later Greek author Aristides Quintilianus, whose compendium of classical arguments on music contains much on the role of the abstract number, also linked the role of ratio in music to medical explanations of disease (Barker 1989, 2:506).

The early modern era saw a remarkable flowering of speculative semimedical writings on music and the body. In particular, the Florentine philosopher Marsilio Ficino (who trained as a physician) set out a neo-Platonic conception of music therapy related to natural magic and fitting the body into the cosmic order (Voss 2000). Music's apparent power to help with the regulation of the "passions of the mind" meant that much of the discussion of the subject related to "diseases of the soul" such as melancholy or lovesickness. But despite that, and the magical and cosmological elements in these theories of music's therapeutic potential, the body is very much present, both as the means by which sound reaches the soul and as part of the microcosm related to the cosmic macrocosm. For Ficino, for example, the spirit affected by music, as "defined by the physicians," is "a certain vapour of the blood, pure, subtle, hot and lucid. And, formed from the subtler blood by the heat of the heart, it flies to the brain, and there the soul assiduously employs it for the exercise of both the exterior and interior senses" (Gouk 2004, 102). Likewise, the English physician, music theorist, and esoteric philosopher Robert Fludd may have written at length on the celestial aspects of music, but he was also clear that "the soul is linked to the body of microcosm" (Figure 19.1). This connection between music and the "humors" of the body meant that music could lead "to both vivification and invigoration" and thereby remove headaches, illnesses, and evil humors (see Hauge 2011, 43–45).

However, the so-called Scientific Revolution of the seventeenth century recast the whole question of music and the body and its medical basis in radical new ways. In what was arguably the most significant shift in thinking about music in Western history, there was a gradual rejection of cosmic harmony and a move toward materialist explanations based on the physics of acoustics and, crucially, on new medical understandings of hearing and of the listening self in terms of the nervous system (Cohen 1984). The "Untuning of the Sky" had enormous consequences for thinking on music, as it began to be treated in the context of a Newtonian mechanical conception of nature, suitable for experimentation and quantification in terms of questions such as the role of irrational numbers in musical temperament, recasting music as essentially a form of stimulation of the body rather than a numerological phenomenon of the soul that could have physical consequences.

However, while music had been in some ways at the forefront of the Scientific Revolution, it is striking that it took some time for medical theories of music to reflect this shift. As late as the early eighteenth century, thinking on the medicinal power of music still tended to focus on its supposed ability to bring patients into line with cosmic order. For instance, the Jena lawyer and composer Friedrich Erhardt Niedt's *Veritophili* (1717) talked about the uses of music in therapy for physical illnesses such as gout in terms of bringing the *Gemüth* (mind or spirit) into harmony (35–40). It was only from the mid-eighteenth century that a fundamentally materialist argument for music therapy based on the nervous system was put forward. Looking back, Peter Joseph Schneider's *System einer medizinischen Musik* (System for a Medical Music) distinguished between the (old-fashioned) followers of the German physician Georg Stahl, who, he wrote, "believed that music gives an arithmetical example, that the soul counts the individual vibrations and delights when it perceives a correct relationship between them," and those who followed the Scottish physician George Cheyne and others of the period who ascribed music's power to the impact of music on the "nervous fibres" and its sympathetic resonance throughout the nervous system (1835, 1:63).

The role of the nerves went way beyond the ear to include a whole model of the listening self. As historians such as George Rousseau and Anne Vila have shown, in the wake of the work of René Descartes and Thomas Willis, the nervous system became the center of new philosophical conceptions of selfhood and of a whole culture and aesthetic of sensibility based on feeling in the sense of both sensation and emotion (Rousseau

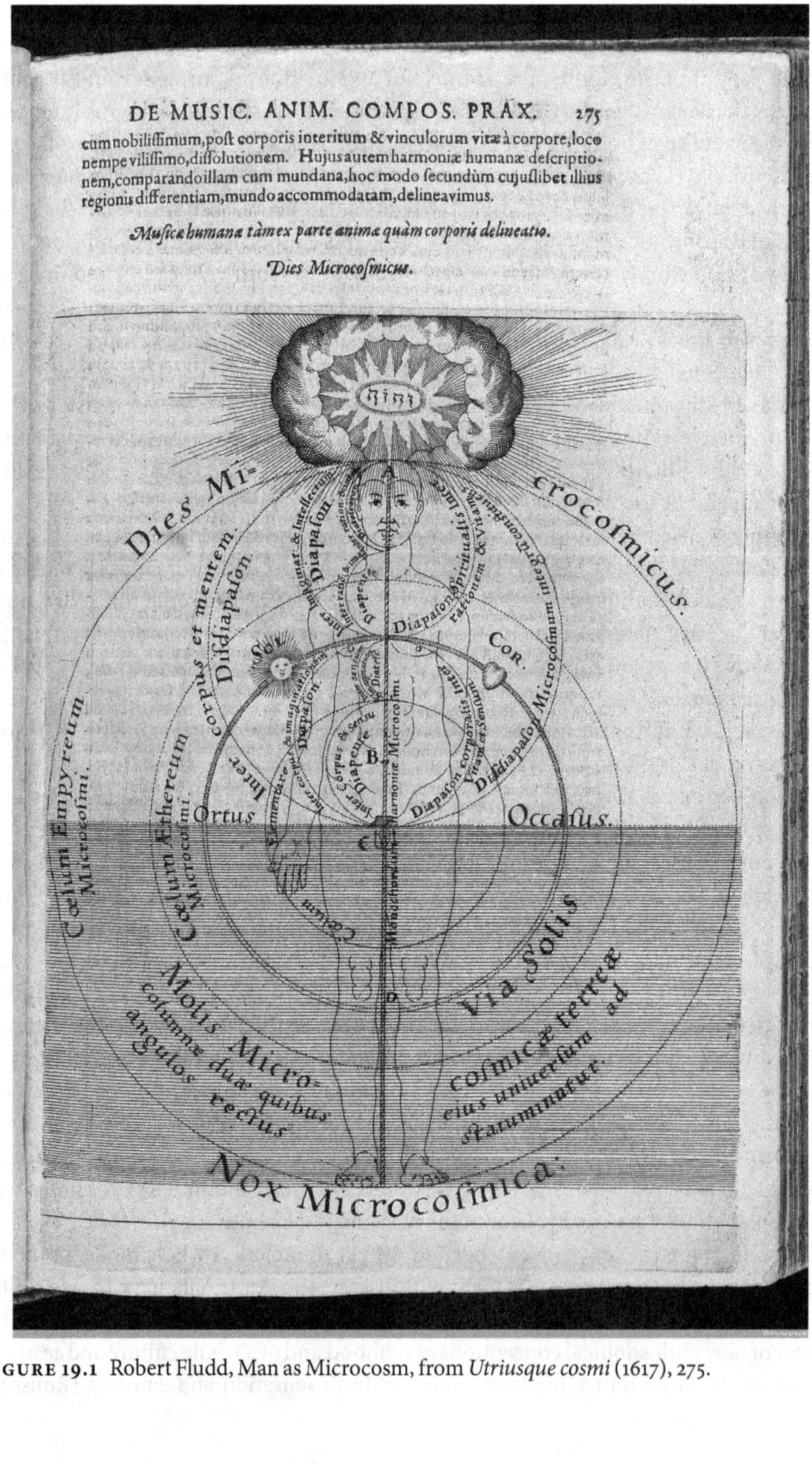
DE MUSIC. ANIM. COMPOS. PRAX. 275

cum nobilissimum, post corporis interitum & vinculorum vitæ à corpore, loco nempe vilissimo, dissolutionem. Hujus autem harmoniæ humanæ descriptionem, comparando illam cum mundana, hoc modo secundùm cujuslibet illius regionis differentiam, mundo accommodatam, delineavimus.

Musicæ humanæ tàm ex parte animæ quàm corporis delineatio.

Dies Microcosmicus.

FIGURE 19.1 Robert Fludd, Man as Microcosm, from *Utriusque cosmi* (1617), 275.

2004; Vila 1998). From the 1720s the nerves began to be used as the fundamental basis for explanations of music's effects, as an intermediate substance between body and soul, for a great number of books on the topic, such as Richard Browne's *Medicina musica* (1729), Richard Brocklesby's *Reflections on Antient and Modern Music* (1749), Louis Odier's *Epistola physiologica inauguralis de elementariis musicae sensationibus* (Inaugural Physiological Letters on the Elements of Musical Sensations) (1770), David Campbell's *Disquisitio inauguralis de musices effectu* (Inaugural Inquiry on the Effects of Music) (1777), or Jean-Louis Lasserre's *Tentamen medicum de musica therapeutice considerata* (Treatise on the Therapeutic Use of Music) (1779), to name just a few of many examples of a genre that persisted strongly into the 1830s.

There was, however, little consensus about the nature and function of the nerves that were being stimulated by music, with theories of animal spirits, nervous fluid, vibrating, elastic, and electric nerves all widely represented. Others compared the nerves to the resonating strings of a musical instrument, sometimes drawing on the idea of nervous sympathy promoted by the Montpellier anatomist Raymond Vieussens's work on the seventh and eighth pairs of nerves (Vieussens [1716] 1998; Hiller 1781). For a generation from the 1790s, the theories of the Scottish physician John Brown (known as Brunonianism) provided the language for a wave of books on music therapy and the medical dangers of music (Brown 1814). The strength of Brunonian ideas in Germany was reflected in books on music therapy such as Wenceslaus Schmolznop's *Dissertatio inauguralis medica de Musae Polyhymniae in sanos et aegrostantes effectu* (Inaugural Medical Dissertation on the Effect of the Muse Polyhymnia on the Healthy and the Sick), which set out principles on which a physician could prescribe doses of music depending if the case demanded an addition or reduction of stimulation (1836, 18–21).

Alongside the debate on the therapeutic value of stimulating the nerves with music, there was also a lot of discussion of the dangers of excessive music for health (Kennaway 2012a). Indeed, much of the time, the discourse of pathological music was considerably more substantial than that on music therapy. It is striking that it developed rather later than the therapeutic angle on music and the nerves, only really becoming a systematic attack on the perils of music from the 1790s, often in an explicitly Brunonian context. For instance, Carl Schöne's 1806 *Versuch eines systematischen Entwurfs der gesammten Medicin* (Attempt at a Systematic Outline of the Whole of Medicine) warned of the potential danger of getting a dose of music in terms of the Brunonian concept of asthenia (155). Robinot de la Lande's 1835 *Dissertation sur les effets avantageux de la musique dans les maladies* (Dissertation on the Advantageous Effects of Music in Disease) compared music to "high doses of narcotics" and used the Brunonian phrase "barometer of sensibility" in his discussion of the dangers of nervous women suffering spasms after listening to the glass harmonica (19–20) (see Hutcheon and Hutcheon's chapter for more on this subject). Literature soon took up these ideas. For instance, E. T. A. Hoffmann's 1818 "Rat Krespel" ("Councillor Krespel") relates the story of Antonia, whose father is told that because she may have "exerted herself too much in singing," music would probably kill her within six months (Hoffmann 2004, 179–180).

The medical approach to explaining the effect of music offered both great advances and real challenges to Enlightenment musical aesthetics. Already in the early eighteenth century the music theory and aesthetics in Mattheson and Rameau clearly took for granted an essentially materialist conception of music and showed little interest in cosmic harmony. The mid-eighteenth-century aesthetic of sensibility (or *Empfindung*) that incorporated a cult of feeling with a sense of gentility and order often drew on medical views of the nerves. In this context, the final quarter of the eighteenth century saw the emergence of an aesthetic of music directly based on medical views of the nerves, for instance in the work of Johann Georg Sulzer (1777) and Nikolaus Forkel (1788). Another proponent of this approach, the Berlin philosopher Johann Jakob Engel, in his book on "musical painting," explicitly based his rejection of the mimesis aesthetic in music on the nerves. "All passionate ideas of the soul," he wrote, "are inseparably connected to corresponding movements in the nervous system, and are supported and strengthened by the perception of these movements," locating musical emotions firmly in the body (1802, 313).

Music and the Body in the Age of Idealist Aesthetics and Positivist Science

The view of music based on the idea of nervous stimulation reached its apogee in the years around 1800, but was undermined by developments in aesthetics and science in the nineteenth century. The number of books on music and medicine, most of which depicted music as a kind of nervous stimulant, reached an apogee in the Brunonian context of the 1830s, only to fall back dramatically in the 1840s (Friedreich 1830, 371–372). A principal reason for this shift was the rise of German positivist science, which led a new generation of physicians to have less patience for the lack of experimental basis and the Brunonian and Romantic *Naturphilosophie* overtones of arguments for music therapy. As the notion of music as a therapeutic nerve stimulant declined, the use of music therapy appears to have focused on mental conditions and to have expanded in the emerging context of nineteenth-century psychiatric institutions (Kramer 2000).

In aesthetics, parallel developments also seemed to relegate the body to a marginal position. An idealist aesthetics of "Absolute Music" emerged that saw music as essentially a matter for the listening subject, with the body serving merely as the means by which it got there. As Alexandra Hui has shown, nineteenth-century research in physics and psychophysics related to the body, conducted by the likes of Hermann Helmholtz, Gustav Fechner, and Ernst Mach, continued to strongly influence musical aesthetics (2013, 2014; see also Burford 2006). They were able to point to work in anatomy that suggested that earlier arguments about music working via sympathy between the nerves without reaching the mind were misleading. More importantly, they drew on Immanuel Kant's rejection of physical pleasure in aesthetics, though he himself regarded music almost entirely in bodily terms as akin to cooking or perfume ([1790] 1951). Already around 1800, Christian Friedrich Michaelis strongly rejected the notion that music could be reduced to the "mechanical shakings of the air and the nerves" (1800, 4; see also Michaelis 1799, 1806). Real listening was not a question of passive stimulation that could be explained in medical terms. Instead, it was a matter of the mind and imagination actively grasping the whole form of the piece, something that made the physical impact of music more or less irrelevant. A whole enormously influential tradition of musical aesthetics from A. B. Marx to Eduard Hanslick was to draw on this rejection of a medical approach to music (Marx 1855; Hanslick 1922).

It is striking that although the sense of music as nervous stimulation was marginalized in aesthetics and music therapy, it continued to thrive in the debate on the dangers of music. Music was incorporated into changing nineteenth-century theories of nervousness, such as George Beard's neurasthenia diagnosis, which shifted the focus to the perils of hard work and the stress of modern life. For instance, the French psychiatrist Alexandre Cullerre's *Nervosisme et nervoses* (1887) warned, "In some very impressionable people, music, instead of calming, may increase the nervous erethism, and in this case it is better to abstain." After producing "voluptuous enjoyment," it might lead to "spasmodic contractions of the muscles, trembling of all members, then a gradual numbness in the extremities, dizziness and syncope, in a word, a full hysterical crisis" (256–258). Drawing on Beard, many physicians depicted musical fatigue as a medical problem. As L. Levillian's 1901 *Hygiene des gens nerveux* put it: "Excitation of the ear . . . [is] often the starting point of a real nervous fatigue or can aggravate the neuropathic state of excitable patients" (142).

Some observers who fretted about the dangers of music drew on the work of Jean-Martin Charcot and his colleagues at the Salpêtrière hospital in Paris, which often portrayed music as a form of bodily neurological stimulation that could provoke hypnosis, hysteria, and full cataleptic fits, entirely bypassing the conscious mind. Paul Richer, for instance, in his 1881 *Etudes cliniques sur l'hystéro-épilepsie ou grande hystérie* (Clinical Studies on Hystero-Epilepsy or Grand Hysteria), wrote at length about the semiautomatic response to music and its ability to trigger hysterical fits and convulsions (599–600). Similar thinking can be seen in earlier discussions of dance, which some observers depicted as a semiautomatic response to music, created by direct sympathy between the nerves over which the mind had little control. In the 1770s, Johann Georg Sulzer argued that the direct impact of music on the nervous system makes us dance because the body's motor nerves are directly stimulated (Kausch 1782, 148). Already in the 1750s, Joseph-Louis Roger's *Traité des effets de la musique sur le corps humain* (Treatise on the Effects of Music on the Human Body) had explained the phenomenon of St. Vitus dance in terms of spasms caused by an unequal distribution of nervous fluid (1803, 214). Similarly, Justus Hecker's 1832 *Die Tanzwuth* (Dance Mania), a hugely influential study of the subject, was clear that dancing mania and tarantella were related to "nervous attack" and "hysteria," and that sexual excitation played a key role (Hecker 1888, 153, 157–158).

Medical views of music also continued to be highly influential in terms of sexuality. Music had long been depicted as a spur to potentially immoral and excessive or even pathological lust, but the opposite idea, that it might subdue unruly passions by bringing harmony to the soul, also has a long pedigree. As late as the 1830s, Peter Joseph Schneider's *System einer medizinischen Musik* suggested music as a cure for nymphomania (1835, 2:193). In the past two centuries, however, the sense that music was a stimulus rather than a restraint on the sensual body has been more noticeable. Sometimes it has been regarded in a positive and therapeutic light. For instance, Friedrich August Weber in an 1802 article and Aeneas Loschi in his 1844 *De musices in organismum humanum virtute* (The Power of Music on the Human Organism) both suggested that music could function as a treatment for impotence (Weber 1802, 594; Loschi 1844, 19). However, more often medical discussion of music and sexuality has been hostile, suggesting that it could be a cause of a symptom of lust, perversion, and disease.

While there was a debate on the supposed link between music and male homosexuality, the vulnerability of the female body and nerves to the stimulation involved in music was the primary concern (Kennaway 2011). J. J. Virey's *Hygiene philosophique; ou, De la santé* (Philosophical Hygiene; or, On Health), for example, suggested, "the impressionable nervous system of women is easily agitated by pitched sounds" (1828, 118). Likewise, Robinot de la Lande's *Dissertation sur les effets avantageux de la musique dans les maladies* (1835) suggested, "immoderate music culture is harmful to vaporous women who abandon themselves without reservation to this art." He went on to explain it could lead to "hysteria, epilepsy, hypochondria, catalepsy," and, by provoking lascivious images in the mind, music could cause satyriasis and nymphomania (26–37). The rest of the nineteenth century saw a remarkably intense debate within gynecology and psychiatry about the impact of music on female sexual health. French gynecologists such as Paul Briquet tended to argue that music could overstimulate the senses, leading to premature sexualization and menstruation. In contrast, American gynecologists such as Thomas Addis Emmett suggested excessive music-making was an intellectual strain that would delay or prevent menstruation altogether, leading to infertility (Kennaway 2011).

Medical theory also linked musicality and bodily abnormality to musical genius. Drawing on a tradition that went back to the pseudo-Aristotelian argument that famously ascribed genius to an excess of black bile, eighteenth-century observers were already asserting that talented musicians had a disposition toward nervous disease, partly because of excessive music-making but also due to their underlying nature. For instance, in 1758 Joseph-Louis Roger suggested that musicians were generally thought to be crazy because of the state of their nervous fluids (1803, 213). Robinot concurred, writing in the 1830s that "among the artists and professional musicians, one meets in general the melancholic and manic: among them also were those whose sudden movements, originality, impatience, announce the extreme susceptibility of the nervous system" (1835, 36–37). The autopsy conducted by Franz Richarz when Robert Schumann died in 1856 reflected such views, concluding that his episodes of "melancholy" were related to his genius and that his death had been due to "overexertion . . . immoderate mental, especially artistic, productivity . . . exhausted the substance of physically active central components of the nervous system"—which comes very close to saying he died of music (Ostwald 1985, 299).

A systematic medical theory of musical genius really became established with Moreau de Tours's influential 1859 *La psychologie morbide* (Morbid Psychology), which suggested that Ludwig van Beethoven was naturally bizarre, that Donizetti showed signs of syphilis, that Mozart was not normal, and that Paganini was nervous (541, 550, 555). Drawing on Moreau's work and on Benedict Morel's theory of degeneration, a new genre of artistic psychopathography emerged, setting out the maladies of genius, for instance in Theodor Puschmann's *Richard Wagner: Eine psychiatrische Studie* (Richard Wagner: A Psychiatric Study) (1872) and Paul Julius Möbius's *Über Robert Schumanns Krankheit* (On Robert Schumann's Illness) (1906). Biographical evidence was by no means the only criterion. Max Nordau's compendium of sickly culture *Entartung* (Degeneration) took Richard Wagner to task not only for his notorious private life but also for his music's perversity and morbidity (1892). It is striking that when Alexander Pilcz's 1905 book *Psychiatrie und Musik* (Psychiatry and Music) divided the Austro-German musical canon into healthy and unhealthy categories, he put Bach, Haydn, Mozart, Brahms, and (perhaps surprisingly) Schubert in the former. The latter group included Donizetti, Schumann, Hugo Wolf, Beethoven, Paisiello, Weber, Wagner, Hoffmann, Auber, Gluck, Gretry, and Wagner, who, even if they were not literally insane, Pilcz suggested, showed signs of instability (1–3; see also Cabanes 1930). A brief consideration of the lives and deaths of the men involved (Schubert dead with syphilis at 41, Mozart dead at 35) suggests that accusations of sickliness were also based on aesthetic grounds. Medical categories were increasingly being used as a stick with which to beat supposedly overripe Romantics and the avant-garde.

Nineteenth-century medicine also saw the emergence of a considerable phrenological literature on musical genius, much of which also foreshadowed today's vogue for studies of the role of brain localization in the experience of listening to music. At the beginning of the century, the German physician Franz Joseph Gall identified an "organ of tune" located in the lateral part of the forehead, having distinguished musical talent from general intelligence. He listed musicians whom he had examined personally (including Beethoven, Rossini, Catalini, and Galli) or whose busts he had seen (Mozart, Haydn, Gluck, and Rameau), all of who fitted this pattern to his satisfaction (1835, 5:63–64; see also Dutton 1892, 3) (Figure 19.2). The example of child prodigies and musically talented people with mental disabilities was likewise presented as evidence of the centrality of that part of the brain in musical creativity. The nineteenth century never produced the kind of clear-cut evidence for brain localization that Paul Broca provided for speech, but a rather speculative phrenological tradition of writing about music continued.

Using medical ideas to explain aspects of music opened the door to arguments that differences between the musics of national and ethnic groups were in a sense a matter of anatomy. Some of this was no more than a continuation of simplistic climatic theories of the kind that had been doing the rounds since Herodotus. For instance, in 1775 Charles

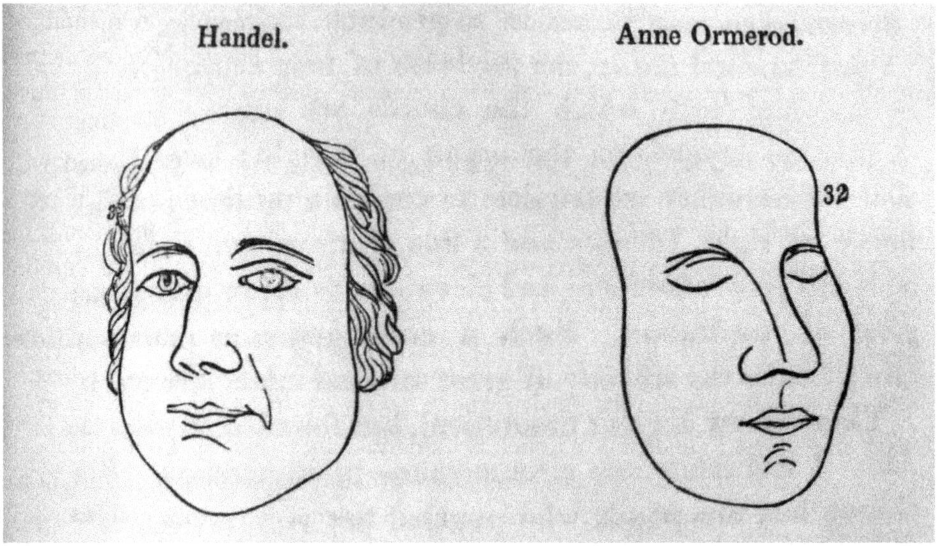

FIGURE 19.2 An image from George Combe's Lectures on Phrenology showing the face of Handel, supposedly showing a bump on his forehead indicative of musical talent.

Combe (1839, 255).

Burney suggested the supposed musicality of Mediterranean countries was due to "the auditory nerves being more irritable" there (3). However, the emergence of modern scientific racism created the circumstances for crude theories ascribing the supposed inferiority and superiority of musical cultures to the brain development of various groups. Franz Joseph Gall suggested, "negroes . . . the Spanish, the French, and the English, furnish a much smaller number of great musicians than the Italians, Bohemians, and Germans," because the latter generally had a more developed organ of tune in the forehead (1835, 5:72).

In contexts such as the antebellum United States, such medical arguments about black music took on an even more sinister tone. For instance, in 1851 the New Orleans physician Samuel A. Cartwright (notorious as the inventor of "drapetomania," a "disease" that made slaves want to escape captivity) argued in his "Diseases and Peculiarities of the Negro" that black people's "brain and nerves" were different from Europeans', citing the German anatomist Samuel Thomas von Sömmerring (Sömmerring 1785, 58). These supposed medical differences explained for Cartwright why black music was "mere sensual pleasure. . . . There is nothing in his music addressing the understanding . . . mere sounds, without sense or meaning—pleasing to the ear, without conveying a single idea to the mind; his ear is gratified by sound, as his stomach is by food" (1854, 315). The discussion of black music in a wide range of contexts from the anthropology of Africa to the debate on genres from rock 'n' roll to hip-hop has often reflected this medicalized and racialized approach (Kennaway 2012a, 117–129; see also Bastide 1972; Neher 1962; Herskovits 1943; Rodrigues 1935).

Music from the Freudian Subject to the Neuroscientific Body

Since the late nineteenth century, the relationship between music and medical conceptions of the body has gone through several dramatic changes. The focus on music's embodied character, already challenged by the dominance of idealist musical aesthetics, was further undermined by the work that Sigmund Freud and some of his contemporaries did from the 1890s, which in many ways overturned the nervous stimulation model of mental illness, looking instead to its psychological roots (see S. Gilman's chapter for more on this subject). This shift potentially relegated the bodily aspects of music to near irrelevance, and subsequent discussion of music and health tended to look at its psychological rather than physiological impact. The twentieth century saw serious steps toward institutionalizing music therapy, with the foundation of the US National Association of Music Therapy (1950) and the British Society for Music Therapy (1958), but the focus was often on well-being and moral uplift instead of the body per se.

As far as the opposite of music therapy is concerned, mainstream medicine took far less interest in the dangers of musical stimulation by the 1920s, marginalizing the body. In retrospect, it is clear that the famous Anna O. case, set out by Freud and Josef Breuer in 1895, marked a significant change. The case involved a cough triggered by hearing music, but rather than ascribing causation to overstimulated nerves, as had been normal since the eighteenth century, Breuer blamed psychological and ideational factors (Breuer and Freud 1955, 21–47). Physicians may have generally lost interest in the perils of music to the body in the interwar period, but the whole idea was instrumentalized in reactionary political circles, reaching a climax with the Nazi persecution of those connected to so-called degenerate music (Wenzel 2001; Dümling 1988; Ziegler 1939). Subsequent revivals in the medicalized debate on the dangers of music to the body have been based on a rejection of the psychological approach and an attempt to find basically somatic causation. For instance, the panic about musical brainwashing since the 1950s has drawn heavily on neo-Pavlovian thinking about the brain (Kennaway 2012b).

It is really only in the last twenty years or so that medical thinking on the body has made a dramatic comeback in discussions of music. In particular, contemporary neuroscience has brought the focus back to the impact of music on the nerves and lent its growing prestige to approaches to the subject based on that paradigm. The primary driving force behind the shift to a neuroscientific approach to music has been the development of brain scanning technology, from EEG to CT, PET, and especially fMRI. Previously, neuroscience had been limited to experiments involved measuring pulse, self-reported responses, and autopsies, but the ability to visualize activity in the living brain has revolutionized views of the subject that challenge long-standing ideas in aesthetics, therapy, and in culture more broadly. Today, the cognitive neuroscience of music has become a significant field in its own right, producing exciting results in terms of brain localization, neuroplasticity, memory, pitch, and aphasia.

Because of the prestige of hard science and the pressures of evidence-based medicine, music therapy is increasingly looking toward brain science, although there is no shortage of books even today that evoke cosmic harmony or parallel Eastern esoteric ideas instead (Gardner 1990). Some music therapy practitioners are basing their work on the direct nervous impact, more than at any time since the 1830s, when physicians wrote of doses of musical stimulation. Many observers have expressed hope that a cognitive neuroscience of music therapy would provide proof of the instinctive sense that music can benignly affect health, and refine what is for some the worryingly vague concept of well-being into quantifiable scientific evidence (Thaut and Hoemberg 2014; Koelsch 2009). To this end, neurologic music therapy (NMT) has been developed, focusing in particular on the rehabilitation of victims of stroke, Parkinson's, and Alzheimer's, drawing on the findings in neuroscience of evidence of brain plasticity among musicians (Pascual-Leone 2001).

The influence of neuroscience has also been partly responsible for something of a revival of the whole medicalized discourse of the mad genius, reflected in books like Kay Redfield Jamison's *Touched with Fire* (1994) and Nassir Ghaemi's *A First-Rate Madness* (2011), not to mention films such as *Shine* (1996) (see Kyaga 2015). Despite lists of mentally ill "geniuses," there are reasons to be skeptical. The idea of the mad genius is a curious mirror image of the stolid bourgeois. If the life of art is associated with a special breed of humanity, who pay for their talent with mental illness and an early death, then respectable citizens can continue a productive life, consuming art and worshipping dead artists, while despising living ones. Saul Bellow wittily alludes to this point in his novel *Humboldt's Gift* (1975): "But no, instead of being a poet he was merely the figure of a poet. He was enacting the "Agony of the American Artist." And it was not Humboldt, it was the United States that was making its point. "Fellow Americans, listen. If you abandon materialism and the normal pursuits of life you wind up at Bellevue like this poor kook" (154).

Neuroscience has also brought the focus onto the body in musical aesthetics. Although science has massively increased our understanding of the impact of music on the nervous system, in some ways the neuroaesthetics pioneered by the likes of Semir Zeki has much in common with the approach of late Enlightenment figures such as Johann Georg Sulzer, Nikolaus Forkel, or Daniel Webb, who sought to base thinking on music on a nervous stimulation paradigm drawn from medicine, and arguably has some of the same limitations (Van Horn 2004). While neuroscientific approaches no doubt have much to tell us about music, basing aesthetics on the body and "reward circuits" in the brain can struggle to deal with music as an aesthetic rather than a simply sensual phenomenon (Blood and Zatorre 2001). More theoretical work on embodied emotions is also perhaps necessary to evaluate the reported subjective experience and brain scan evidence and to refine questionable suggestions such as that the latter has "proved" that music has the ability to "induce authentic emotions" because of changes in limbic system (De L'Etoile and LaGasse 2013, 10). A historical approach might provide a starting point for a critical neuroscience of music that illuminates some of the potential theoretical and indeed political pitfalls of a medical paradigm for understanding music.

References

Barker, Andrew, ed. 1989. *Greek Musical Writings*. 2 vols. Cambridge: Cambridge University Press.

Barker, Andrew. 2000. "Timaeus on Music and the Liver." In *Reason and Necessity: Essays on Plato's Timaeus*, edited by M. R. Wright, 85–100. London: Gerald Duckworth.

Bastide, Roger. 1972. *Le reve, la transe et la folie*. Paris: Flammarion.

Bellow, Saul. 1975. *Humboldt's Gift*. Harmondsworth: Penguin.

Blood, Anne J., and Robert J. Zatorre. 2001. "Intensely Pleasurable Responses to Music Correlate with Activity in Brain Regions Implicated in Reward and Emotion." *Proceedings of the National Academy of Sciences of the United States of America* 98: 11818–11823.

Breuer, Josef, and Sigmund Freud. 1955. *Studies in Hysteria*. London: Hogarth.

Brocklesby, Richard. 1749. *Reflections on Ancient and Modern Music*. London: Cooper.

Brown, John. 1814. *Elements of Medicine*. Philadelphia, PA: Webster.

Browne, Richard. 1729. *Medicina Musica*. London: J. Cooke.

Burford, Mark. 2006. "Hanslick's Idealist Materialism." *Nineteenth-Century Music* 30 (2): 166–181.

Burney, Charles. 1775. *The Present State of Music in Germany, the Netherlands, and the United Provinces*. London: Printed for T. Becket.

Cabanes, Augustin. 1930. *Grands nevropathes*. 3 vols. Paris: Albin Michel.

Campbell, David. 1777. *Disquisitio inauguralis de musices effectu in doloribus leniendis aut fugandis*. Edinburgh: Balfour and Smellie.

Cartwright, Samuel. 1854. "Diseases and Peculiarities of the Negro." In *The Industrial Resources, etc. of the Southern and Western States*, edited by J. D. B. DeBow, 315–328. New York: Appleton.

Cohen, H. F. 1984. *Quantifying Music: The Science of Music at the First Stage of the Scientific Revolution, 1580–1650*. Dordrecht: D. Reidel.

Combe, George. 1839. *Lectures on Phrenology*. New York: Samuel Colman.

Cullerre, Alexandre. 1887. *Nervosisme et nervoses*. Paris: JB Bailliere.

De l'Etoile, Shannon K., and A. Blythe LaGasse. 2013. "Music Therapy and Neuroscience from Parallel Histories to Converging Paths." *Music Therapy Perspectives* 31 (1): 6–14.

Dümling, Albrecht. 1988. *Entartete Musik: Zur Düsseldorfer Ausstellung von 1938. Eine komentierte Rekonstruktion*. Düsseldorf: Düsseldorfer Symphoniker.

Dutton, G. H. J. 1892. *Music and Phrenology*. Skegness: G. H. J. Dutton.

Engel, Johann Jacob. 1802. *Über die musikalische Malerei*. Berlin: Myliuss.

Forkel, Johann Nikolas. 1788–1801. *Allgemeine Geschichte der Musik*. 2 vols. Leipzig: Schwickert.

Friedreich, Johannes Baptista. 1830. *Versuch einer Literaergeschichte der Pathologie und Therapie der psychischen Krankheiten*. Würzburg: Carl Strecker.

Gall, Franz Joseph. 1835. *Organology*. 6 vols. Boston: Marsh, Capen & Lyon.

Gardner, Kay. 1990. *Sounding the Inner Landscape: Music as Medicine*. Rockport: Element.

Ghaemi, Nassir. 2011. *A First-Rate Madness*. New York: Penguin.

Gouk, Penelope. 2004. "Raising Spirits and Restoring Souls: Early Modern Medical Explanations for Music's Effects." In *Hearing Cultures: Essays on Sound, Listening, and Modernity*, edited by Veit Erlmann, 87–105. Oxford: Berg.

Gouk, Penelope, and Ingrid Sykes. 2011. "Hearing Science in Mid-Eighteenth-Century Britain and France." *Journal of the History of Medicine and Allied Sciences* 64: 507–545.

Hanslick, Eduard. 1922. *Vom Musikalisch-Schönen*. Leipzig: Breitkopf und Härtel.

Hauge, Peter. 2011. *The Temple of Music by Robert Fludd*. Farnham: Ashgate.
Hecker, J. F. C. 1888. *The Black Death and the Dancing Mania*. London: Cassell & Co.
Herskovits, M. J. 1943. *Dahomey*. Evanston, IL: Northwestern University Press.
Hiller, Johann Adam. 1781. *Über die Musik und deren Wirkungen*. Leipzig: Jacobäer.
Hoffmann, E. T. A. 2004. "Councillor Krespel." In *Tales of Hoffmann*, translated by R. J. Hollingdale, 158–183. Harmondsworth: Penguin.
Hui, Alexandra. 2013. *The Psychophysical Ear: Musical Experiments, Experimental Sound, 1840–1910*. Cambridge, MA: MIT Press.
Hui, Alexandra. 2014. "Origin Stories of Listening, Melody and Survival at the End of the Nineteenth Century." In *Music and the Nerves, 1700–1900*, edited by James Kennaway, 170–190. Basingstoke: Palgrave.
Jamison, Kay Redfield. 1994. *Touched with Fire*. New York: Free Press.
Kant, Immanuel. [1790] 1951. *Critique of Judgment*. New York: Forgotten Books.
Kausch, Johann Joseph. 1782. *Psychologische Abhandlung über den Einfluß der Töne und ins besondere der Musik auf die Seele; nebst einem Anhang über den unmittelbaren Zweck der schönen Künste*. Breslau: Johann Friedrich Korn.
Kennaway, James. 2010. "From Sensibility to Pathology: The Origins of Nervous Music." *Journal for the History of Medicine and Allied Sciences* 65 (3): 396–426.
Kennaway, James. 2011. "The Piano Plague: The Nineteenth-Century Medical Critique of Female Musical Education." *Gesnerus* 68 (1): 26–40.
Kennaway, James. 2012a. *Bad Vibrations: The History of the Idea of Music as a Cause of Disease*. Basingstoke: Ashgate.
Kennaway, James. 2012b. "Musical Hypnosis: Sound and Self from Mesmer to Brainwashing." *Social History of Medicine* 25 (4): 271–289.
Koelsch, Stefan. 2009. "A Neuroscientific Perspective on Music Therapy." *Annals of the New York Academy of Sciences* 1169: 374–384.
Kramer, Cheryce. 2000. "Soul Music as Exemplified in Nineteenth Century German Psychiatry." In *Musical Healing in Cultural Contexts*, edited by Penelope Gouk, 137–148. Aldershot: Ashgate.
Kyaga, Simon. 2015. *Creativity and Mental Illness*. Basingstoke: Palgrave Macmillan.
Lasserre, Joannes-Luduvicus [Jean-Louis]. 1779. *Tentamen medicum de musica therapeutice considerata*. Montpellier: J. Martel.
Levillian, L. 1901. *Hygiene des Gens Nerveux*. Paris: Felix Alcan.
Loschi, Aeneas. 1844. *De musices in organismum humanum virtute*. Padua: Minerva.
Marx, Adolf Bernhard. 1855. *Die Musik des neunzehnten Jahrhunderts*. Leipzig: Breitkopf und Härtel.
Michaelis, Christian Friedrich. 1799. "Über den Rang der Tonkunst unter den schönen Künsten." *Allgemeine musikalische Zeitung* 2: 183–186.
Michaelis, Christian Friedrich. 1800. *Uber den Geist der Tonkunst, mit Rucksicht auf Kants Kritik der aesthetischen Urtheilskraft*. Leipzig, Schaefer.
Michaelis, Christian Friedrich. 1806. "Ein Versuch, das innere Wesen der Tonkunst zu entwickeln." *Allgemeine Musikalische Zeitung* 8: 673–683, 691–696.
Möbius, Paul Julius. 1906. *Über Robert Schumanns Krankheit*. Halle: Carl Marlhold.
Moreau de Tours, Jacques-Joseph. 1859. *La psychologie morbide das ses rapports avec la philosophie de l'histoire de l'influence des neuropathes sur la dynamisme intellectual*. Paris: Vicor Masson.

Neher, Andrew. 1962. "A Physiological Explanation of Unusual Behavior in Ceremonies Involving Drums." *Human Biology* 34: 151–160.

Niedt, Friedrich Erhardt. 1717. *Veritophili*. Hamburg: Benjamin Schiller.

Nordau, Max. 1892. *Entartung*. 2 vols. Berlin: Duncker.

Odier, Louis. 1770. *Epistola physiologica inauguralis de elementariis musicae sensationibus*. Edinburgh: Balfour, Auld and Smellie.

Ostwald, Peter. 1985. *Schumann: The Inner Voices of a Musical Genius*. Boston: Northeastern University Press.

Pascual-Leone, A. 2001. "The Brain That Plays Music and Is Changed by It." *Annals of the New York Academy of Sciences* 930: 315–329.

Pelosi, Francesco. 2010. *Plato on Music, Soul and Body*. Cambridge: Cambridge University Press.

Pilcz, Alexander. 1905. *Psychiatrie und Musik*. Vienna: Ambr. Opitz Nachfolger.

Puschmann, Theodor. 1872. *Richard Wagner: Eine psychiatrische Studie*. Berlin: Behr's Buchhandlung.

Richer, Paul. 1881. *Etudes cliniques sur l'hystéro-épilepsie ou grande hystérie*. Paris: Adrien Delahaye & Emile Lecrosnier.

Robinot de la Lande, A. 1835. *Dissertation sur les effets avantageux de la musique dans les maladies*. Paris: Didot le Jeune.

Rodrigues, Raimundo Nina. 1935. *O animismo fechista dos negros babianos*. Rio de Janeiro: Civiliazacao Brasiliera.

Roger, Joseph-Louis. 1803. *Traité des effets de la musique sur le corps humain*. Paris: Chez Brunot.

Rousseau, George. 2004. *Nervous Acts*. Basingstoke: Palgrave Macmillan.

Schmolznop, Wenceslaus R. 1836. *Dissertatio inauguralis medica de musae polyhymniae in Sanos et aegrostantes effectu*. Prague: Haase.

Schneider, Peter Joseph. 1835. *System einer medizinischen Musik*. 2 vols. Bonn: Carl Georgi.

Schöne, Carl. 1806. *Versuch eines systematischen Entwurfs der gesammten Medicin*. Berlin: C. Quien.

Sömmerring, Thomas. 1785. *Über die körperliche Verschiedenheit des Negers von Europaer*. Frankfurt: Varrentrapp Sohn und Wenner.

Sulzer, Johann Georg. 1777. *Allgemeine Theorie der schönen Künste*. 4 vols. Biel: Hellmannschen Buchhandlung.

Thaut, Michael H., and Volker Hoemberg. 2014. *Handbook of Neurologic Music Therapy*. Oxford: Oxford University Press.

Van Horn, J. D. 2004. "Cognitive Neuroimaging: History, Developments, and Directions." In *The Cognitive Neurosciences*, edited by M. S. Gazzaniga, 1281–1294. Cambridge, MA: MIT Press.

Vieussens, Raymond. [1716] 1998. *Neurographia Universalis*. Lyon: Joannes Certe.

Vila, Anne C. 1998. *Enlightenment and Pathology*. Baltimore, MD: Johns Hopkins University Press.

Virey, J. J. 1828. *Hygiene philosophique; ou, De la santé*. Paris: Crochard.

Voss, Angela. 2000. "Marsilio Ficino, the Second Orpheus." In *Music as Medicine*, edited by Horden, 154–172. Farnham: Ashgate.

Weber, F. A. 1802. "Von dem Einflusse der Musik auf den menschlichen Körper und ihrer medicinischen Anwendung." *Allgemeine Musikalische Zeitung* (26 May): 561–569, (2 June): 577–589, (9 June): 609–617.

Wenzel, Silke. 2001. "'Entartung' in der Musik: Aspekte eines Begriffssystems." In *Entartete Musik 1938: Teil I Weimar und die Ambivalenz*, edited by Hanns-Werner Heister. Saarbrücken: PFAU.
West, Martin. 2000. "Music Therapy in Antiquity." In *Music as Medicine: The History of Music Therapy since Antiquity*, edited by Peregrine Horden, 51–68. Farnham: Ashgate.
Ziegler, Hans Severus. 1939. *Entartete Musik: Eine Abrechnung*. Dusseldorf: Völkischer Verlag.

Further Readings

Feis, Oswald. 1910. *Studien über die Genealogie der Psychologie der Musiker*. Wiesbaden: Bergmann.
Stendhal. 1957. *Life of Rossini*. New York: Criterion.

CHAPTER 20

MUSIC IN BODY AND IMAGINATION

H. M. EVANS

Simplistic Approaches to Music, Therapy, and the Body

There can be few more compelling invitations to consider the relation between music and the body than Oliver Sacks's account of the musical epiphany that catalyzed his convalescence from a crippling injury to his leg; the familiar passage pivots thus: "And suddenly—into the silence, the silent twittering of motionless frozen images—came music, glorious music, Mendelssohn, *fortissimo*! Life, intoxicating movement!" (Sacks 1984, 108). This joyous declamation has, I daresay, been an inspirational starting-point for many who have thought about therapeutic engagements with music, standing in vibrant contrast to the utilitarian conception apparently underlying such titles as "The Effect of Music on the Neurohormonal Stress Response to Surgery under General Anesthesia" (Migneault et al. 2004). Those inspired by Sacks will probably regard the following as blindingly obvious: "The word *music* stands for a multiplicity of human events that are difficult to analyse from a reductionist point of view" (Hilleke, Nickel, and Bolay 2005, 275).

So obvious, in fact, as to go without saying—at best, a corrective that might be needed by others. It might be kept in reserve to be deployed against statements like this: "One of the great powers of music is to evoke hippocampal activity related to happiness" (Koelsch 2009, 378). Koelsch and Sacks need not, of course, be in actual disagreement, since they are scarcely talking about the same thing at all. Koelsch is telling us something about what the experience of listening to music *does*: what it may involve in the brain. By contrast Sacks is telling us something about what music *is* and *means*. Interestingly, both writers seem to attribute to music a kind of ontological independence of the listener: music comes into the silence; music is the possessor of great powers. Both seem to imply a kind of agency, inviting us to enquire into music's therapeutic potential.

How is that potential made manifest? For Koelsch, music stimulates felicific brain processes—simple enough, and as a story about beneficial effects it clearly could be regarded as reductionist, albeit perhaps defensibly so. Things are different for Sacks (after all he was himself a neurologist): music and our engagement with it are first and foremost existential matters, to be understood at the level of life-affirming (and life-changing) events. To appreciate his view properly, one must of course read the relevant parts of his work in full; this chapter only draws pertinent contrasts. One can think of "the therapeutic" in experiential, existential terms concerning thought, feeling, action, and purpose; and one can think of "the therapeutic" in material terms concerning tissues, cells, neurons, and physiological and neurological functioning. Broadly speaking, within the literatures exploring "music therapy," "musical therapy," and "music-as-therapy" one will encounter both approaches, characteristically taken by practitioners and neuroscientists respectively.

And here lies a great temptation. One might like to imagine that only the scientific literature is to be charged with a reductionist point of view, something from which the practitioners of the art of music-as-therapy and its variants remain free. But reductionism—the idea that a sophisticated and complex whole really boils down to some key feature or features that account for all the rest—can put in appearances in unexpected places; and even in the expected places it may not always be what it seems.

To begin with the latter, we may feel unsurprised to find neuroscientists enthusing about the role of central limbic structures in the musical "induction" of pleasurable experiences such as shivers down the spine (Koelsch 2009, 376), or neural activations that musical experience shares with "sexual activity, ingestion of chocolates, or consumption of drugs" (378–379). Perhaps more challenging is the conjectured antimicrobial benefit of music via its inducing "the peripheral production and release of nitric oxide" (Boso et al. 2006, 189). Moreover, anyone might wonder quite what to make of the alleged "Mozart effect" on epilepsy; the proffered explanation—increased coherence in the lower alpha bands of an EEG trace—seems like a comforting distraction from the troubling phenomenon it is meant to explain, namely that epilepsy sufferers apparently do better listening to Mozart than they do when listening to Haydn or Brahms (189–190).

It is too easy to assume that such studies are based on reductionist assumptions about music itself; they need not be. Koelsch cites modulation, reanimation, evocation, and activation among the modes of impact that listening to music can have on key structures and functions within the brain (2009, 378–380). Thaut emphasizes music as "a mediating stimulus . . . arousing, guiding, organising, focusing and modulating perception, attention and behaviour in the affective, cognitive and sensorimotor domains" (2005, 304). Such claims are concerned with music's therapeutic effects—effects that, while doubtless telling us something about what music is capable of, appear confined to the listener considered merely as an organism regulated by neurophysiology in apparent isolation from cultural influences on musical memory and imagination—in short, unburdened by any explanatory claims concerning music's nature qua music.

But perhaps that is the problem. Perhaps studies of experiences of music—including therapeutic experiences of music—ought to be grounded on, imbued with, a deep understanding of music such that the explanations are themselves contingent on music's nature, properly comprehensible only if we also understand something of what music is and does. Perhaps the satisfactory articulation of music's therapeutic power ought to be expressible only in terms that are themselves at least part-musical: "Mendelssohn, *fortissimo*!"

So we come to the other half of the temptation—to assume that practitioners of music-as-therapy are free of reductionism because they attend to the phenomena at the level of the experiential, the existential. Prominent here are Paul Nordoff and Clive Robbins, whose trademark "creative music therapy" with children emphasizes collaborative improvisation as a modus operandi that is "only partially communicable in words" and whose transactions are musical events that can be "apprehended" only through listening to recordings of them (2007, xxiii). Bunt (1994) and Ansdell (1995) usefully survey varieties of music-based therapeutic practices as preludes to expounding their own approaches; Ansdell confirms the importance of creative improvisation in giving expression to that which patients cannot otherwise express, or connections that could not otherwise be made. Confronting one form of scientific reductionism head-on, Ansdell identifies musical dialogue between patient and therapist as a "freely willed, intentional force," evidence of "the very inner, unconditional and dynamic forces that behaviourism would like to exclude from its picture of human functioning" (32–33).

Lurking even here, though, is the risk that there be some presumed "nub" of the therapeutic encounter to which the whole may be essentially reduced. To his credit, Ansdell defies the usual temptation that afflicts musicians and nonmusicians, scientists and nonscientists, practitioners and nonpractitioners alike—to suppose that music is essentially a language of the emotions, and that its therapeutic powers are essentially those of emotional release or catharsis (9–10). A simplistic emotivism afflicts the neuroscientific studies referred to above, if anything more than it does the therapeutic practitioners (Koelsch 2009, 380; Boso et al. 2006, 188; but see also among practitioners Hilleke et al. 2005, 278). The idea that emotion is the basis of our understanding music is intensified, though certainly not rescued, by an emphasis on affect as the prereflective, visceral bodily experience of emotion or feeling, prior to and independent of meaning or understanding. Such a view takes seriously the embodied nature of musical experience, but at the expense of dismissing the cognitive and imaginative interpretive structures of listening, within which alone can musical sounds be understood and enjoyed with mind or body (Evans 1990). The affective, of course, underlies all visceral enjoyment—but fails to account for the fact that we have to grasp what we are listening to in order to enjoy it at all! For a comprehensive demolition of the wider "turn to affect" and its implications, see Leys (2011). But what alternative, "unconditional and dynamic forces" has Ansdell in mind? He speaks of "grounding" and "alleviating" (literally, lightening) the patient (1995, 34 and note), and also of "entrain[ing]" (9–10). The problem is that without some credible theoretical grounding, such notions (perhaps still more so as "forces") will appear to be indefensibly foundational, and evidence of unconscious reductionism. Hilleke and colleagues put the point trenchantly: "An extensive corpus of modalities and treatment theories exist in the music therapy world. Some of them are psychoanalytic music therapy, humanistic music therapy, behavioural music therapy, Nordoff-Robbins music therapy, and music medicine. Often music therapists refer to one of them as the theoretical background; nevertheless, these traditional approaches can be considered as more or less belief systems" (2005, 276).

In summary, the position appears to be this: the heavily materialist approaches taken in much neuroscientific investigation of music-as-therapy may nonetheless avoid an objectionable reductionism if what they seek to explain is music's effects, rather than its nature. At the same time, the experiential or existential approaches taken by practitioners of music-as-therapy may rest on undeclared or undefended foundationalist notions that make them vulnerable to charges of an alternative form of reductionism.

Significantly, both neuroscientific and practice-led enquiries into music-as-therapy emphasize its practical applications and, thereby, music's own attributed effects—but with surprisingly little distinction between these effects and music as such. Hilleke and colleagues do recognize the distinction, but only to lament that amid the "therapeutic eclecticism" found among practitioners "a music-specific therapeutic paradigm does not exist" (276).

Very well, but equally puzzling is an apparent neglect, in the neuroscientific summaries, of the role of imagination in musical experience. Suffice it here to say that imagination is taken to be simply the playful ability of the mind to present ideas to itself—the ideas with which we are most concerned being musical ideas, of course. A contrast dominated instead by affect apparently presumes a somewhat vulgar emotivism that itself reflects a post-Romantic fixation with emotion for which musicians themselves are largely to blame. That is, a conception of music is presumed but only tacitly appealed to, and it is a simplistic and inadequate one at that. More interesting conceptions may be presumed elsewhere (by Thaut, for instance) but are not articulated. It seems hard to find an understanding of music-as-therapy that is grounded in an articulation of music qua music.

To understand why—as distinct from mechanically how—music can be experienced therapeutically, we seem to need an alternative conception of musical experience and its relation to ordinary embodiment. We also need to be skeptical about music's being "put to work" detached from its being experienced intrinsically as music. After all, we may not always be masters of music's power and agency, and we cannot lightly assume their benevolence: musical experience may on occasion be anything but therapeutic. In particular, as explored in what follows, music unbidden may be music unwelcome.

An Alternative Conception of Musical Experience and Ordinary Embodiment

We can start to sketch out such an alternative conception only if we first abandon the fixation with music's being the expression primarily of emotions (Koelsch 2009; Boso et al. 2006; Trainor and Schmidt 2003; Blood et al. 1999)—realized through a correspondingly limited subset of neurological functions more or less confined to the limbic system—in favor of a more inclusive understanding of music *in expressive relation to our bodies and our bodily self-experience.* Such a relation would embrace bodily rhythms and movements more generally, our inner awareness of ourselves and our movements (in proprioception and kinesthesia), our sense of willed action, and the role of an active imagination.

The first aspect—music's expressive relation to bodily movements and rhythms more generally—seems self-evident. Any music that we are capable of hearing and making must inevitably exhibit certain facts about our own embodiment and about the material world around us (Evans 2010). It speaks to the musculo-sensory backdrop of all experience, cashed out as those dynamic tensions and resolutions, repetitions, tonal resonances, and rhythms that are natural to us—the very material and matter of music's organized playfulness. This is as recognizable to a neuroscientist (consider Thaut's contention that "rhythmic sounds also entrain the timing of muscle activity" [2005, 304]) as to a therapeutic practitioner (such as Ansdell, who argues that "within the body our different physiological rhythms are 'musically' entrained" [1995, 9–10]).

The second aspect—music's relation to our inner self-awareness—underlay Sacks's brief, but glorious, Mendelssohnian epiphany: his lost kinesthesia was temporarily regained. Aside from such dramatic realizations, our inner self-awareness is perhaps less obvious a dimension of musical experience; yet nonetheless it is there, and if the urge to tap one's toes is something that one can resist in terms of "acting out," one certainly cannot resist feeling it.

However, this suggests a deeper connection between music and our sense of self. Damasio suggests that our inner "primordial feelings"—the "direct experience of one's own living body, wordless, unadorned and connected to nothing but sheer existence" (2010, 21)—arise from the core of the self, its earliest foundations. (His reasons include their constancy over a lifetime, offering a surer grounding for personal identity—the "simultaneous foundation of mind and self" [256]—than do our world-facing senses, for instance.) Damasio evidently feels the pull of a connection between self-experience and music. He compares the ineluctable feelings that are part of any subjective process to music, as a score that accompanies the *remainder* of the ongoing mental process, but noting that the performance is *within* the mental process too.

> When the main object in my consciousness is not the [Pacific] ocean but an actual music piece, then there are two musical tracks going in my mind, one with the Bach piece that is playing right now and another with the *music-like* track with which I react to the actual music in the language of emotion and feeling . . . call it music on music. Perhaps polyphonic music was inspired by an intuition of this accumulation of parallel "musical" lines in one's mind. (254)

As will be seen later, when turning to the idea of unwanted and intrusive musical experience, the ineluctable nature of these aspects of self-experience—particularly in kinesthetic terms—is confirmed by understanding how musical experience can be as wearisome when it is unbidden as it can be revitalizing when bidden.

The third aspect of inner experience to which music has an especially intimate relation is the experience of willed action. Now we can communicate those aspects of experience that can be conceptualized, put into words, but there is an irreducible remainder that cannot be put into words—the sheer "what it's like" to be experiencing, and to be an experiencing subject. The subjective reality of all experience has the ineffable at its core—the unnameable "what it's like"—whether this concerns the taste of chocolate, the darkness at the center of grief, the indecision prior to action, the sound of Bach's blazing majesty, or "the velvet diapason of the Wagnerian orchestra" (Gould 1966, 50).

Along with the inner sense of continuous identity and location (we are, after all, the spatial reference point for all our physical observations), we know directly, within ourselves, moods, memories, emotions, intentions, and—signally—acts of will. Among philosophers, Schopenhauer has most provocatively articulated the importance of our direct knowledge of our willed action (Schopenhauer 1966). We need not here follow him into the detail of his metaphysics, but it is striking that he thought that our immediate knowledge of our own willed actions—known by others through observable behavior, but by ourselves directly—gives us a crucial source of insight into our own nature that is otherwise largely hidden.

More strikingly still, he thought music—alone among the arts—offers us a glimpse of that hidden reality that underlies both our selves and our world. In his own words, music expresses "with the greatest distinction and truth, the inner being, the in-itself, of the world" (264). "Accordingly we could just as well call the world embodied music as embodied will" (262). In listening to music, we are directly glimpsing that underlying reality—or coming as close to it as we ever do, with this one exception: our direct immediate knowledge of our own willed actions, as those acts of will emerge "from the inaccessible depths of our unconscious selves," as Bryan Magee puts it (1983, 184). He goes on, "It can therefore be said that it is from these inscrutable regions that music speaks. In this sense what it articulates is insight into, and in *that* sense knowledge of, the hidden, inner nature of our own willing" (1983, 184).

This view—controversial in its wider metaphysics, but they are not needed here—importantly extends music's expression of our inner sense of embodiment beyond the passive (sensations, emotions) to the active and volitional. One fourth and final aspect remains, carrying the volitional forward into the playful (as music must at some level be): the embodied imagination.

The fairly general (and perhaps Humean) notion of imagination earlier proposed is more helpful than specific accounts. For instance, Warnock's view, that imagination supports our "thoughts about what is absent" (1976, 196) will fail for musical imagination; for unless we stipulate that music exists only in actual sound then its presence or absence in thought-as-such is moot: There is no reason to deny that a musical thought is any less musical than is that same thought realized in sound. In a related manner, Scruton's account (1974) of imagination as "unasserted" thought, the construction of a kind of narrative that goes beyond what is given or believed and that underpins a form of playful speculation, is one that works well for representational arts—where the distinction between an original and its imaginary re-creation or representation holds—but falls short in the case of music, where nothing is represented beyond the musical idea.

Imagination's role in composing music is obvious enough; but all musical experience including performing and listening involves active, musically thoughtful anticipation and, thus, imagination (Evans 1990). Musical thought is essentially playful and essentially imaginative: its manifestation in actual sound is at one level no more real—merely more public. In this sense, music is a form of imagination, whose activities are from time to time typically but not invariably shared in sound. Sound need not even constitute the majority of acts of musical imagination. As Sacks observes, alien visitors would be "stupefied when they realized that, even in the absence of external sources, most of us are incessantly playing music in our heads" (2007, 40). The musical imagination is to this extent music's primary domain.

Since our inner self-experience in proprioception, the kinesthesia of movement, and willed actions are inbound to the bodily experience of music, it follows that they are inbound to the imagination that is at play in listening to, performing, or composing music—and correspondingly that, in music, the imagination is intimately connected, inbound, to them. To imagine music is to imagine musical experience: and to imagine musical experience is inevitably also to imagine the inner self-experience, the kinesthesia, the willed action, that is fused with it (including, in Damasio's terms, the primordial feelings that run beneath and within the musical experience, the accompanying "music-like track" [2010, 254]). Where music reaches, so must the imagination. And perhaps, where the imagination can reach, so might music.

So far, this chapter has sketched out that alternative conception of musical experience that might allow us to consider why it is that music-as-therapy can be effective. The healing potential of music is made possible as much by the intensity of musical experience operating through the imagination and its inward reach into our inner self-experience—by *recalling, reimagining, and in a sense revitalizing the former sense of* our premorbid inner states—quite as much as it is by any more mechanistic attempt to reinstate them (through, for instance, neuromuscular "entrainment").

Here I suspect we are drafting the terms of reference for the instincts expressed by some advocates of "creative music therapy"—curiously redolent of Schopenhauer's partnering of music and willed action as offering glimpses into our true, but hidden, nature. As Ansdell puts it, "These [therapeutic musical practices] are possible because they engage aspects of our nature that are themselves in some sense musical manifestations. . . . 'Music' says Thomas Clifton 'is what I am when I experience it.' . . . David Aldridge suggests that we should begin to think of ourselves as 'symphonic' beings rather than the mechanical ones that reductionist scientific medicine would often have us be" (1995, 8). What is it to be such a "symphonic" being—what is it like to be one? The ubiquitous nature of musical engagement (Evans 2010) suggests that at one level we all know the answer in tacit, experiential, terms.

The remainder of this chapter concerns my own experiences of unbidden music, by no means all joyful or beneficent/benevolent. The unbidden music aligns well with, and hence offers confirmatory support to, the conception of embodied musical experience sketched out previously. The following autobiographical fragments will, I hope, illuminate this.

Musical Intrusions: A Sufferer's Perspective

If I can hear music of any kind, regardless of whether it is music that I know or music that I do not know, whether it is music that I like or that I dislike, or whether it has any merit or none, then I cannot help listening to it with most or all of my attention. Music grips me by the throat whether I want it to or not, and frequently disrupts my attention to anything else that might otherwise be occupying me.

The type of music involved does not make much difference. Even though loud, insistently rhythmic music (which often also has sexually charged lyrics and vocal delivery such as some soul music or hip-hop, etc.) is obviously intrusive to a conversation, even very restrained and abstract instrumental music equally grabs my attention in a different way, but to much the same effect of interrupting my ability to do anything else.

This is frequently a problem for me in attempting serious conversation with other people, where it is a real practical difficulty. (My sense of frustration is made worse by a sense of isolation, arising from the fact that I cannot understand why the other people are not finding the background music a problem in the way that I do.) If, instead of conversing, I am doing something on my own, it remains a problem but—provided I am not trying to read and concentrate on my reading—then it is usually annoying rather than disabling.

Perhaps this suggests that I am excessively, morbidly sensitized to music. I would never choose to have background music on while I am working, and I am completely unable to understand the wishes—or the experiences—of those who do, although I recognize that they might even be the majority.

Hearing and "Hearing"

Although external musical interruptions are a significant problem for me, their public sources make them seem more obviously like other irritants, such as loud and intrusive conversations nearby, particularly those conducted over a mobile phone (where to the interruption is added the torment of our irresistible urge to conjecture the unheard half of the conversation); these are things that I encounter along with everyone else in modern life and have to deal with.

But what I shall call "internal" musical intrusion—inner music—is different. This seems to be something that I somehow "do to myself." If I want help in dealing with external musical interruptions, I can ask another person such as the waiter or the manager to turn the music volume down, or off altogether. But there is no one other than myself whom I can ask in the case of inner music. And since I myself seem unable to turn the volume down, then it is a significant, sometimes even severe, problem that is unfortunately also very persistent.

Pathological Disruption of Thought and Concentration

It seems not to matter whether or not there is any possible connection or "cross-talk" between the music and the activity on which I am trying to concentrate. Presumably anyone would find it a problem to hear background music while they were trying to think about different music. But I find it a problem even to concentrate on buying groceries in the supermarket if there is background music. This is not so bad as, say, trying to have a meaningful conversation with someone in a restaurant while my attention is being pirated away by background music.

The problem of disruption of concentration by inner music is particularly obvious for me when reading, and it constitutes a significant handicap given my professional occupation as an academic. Some form of inner music is going on for a very large part of the time when I am not positively and successfully engaged in thinking about something else, something on which I can truly focus. If my attention wanders at all, then there is the inner music waiting to snap up my attention and hang on to it. When I am reading, the least lapse in concentration on my part will be hijacked by the inner music, and then I have to go back over and over the same paragraph or even sentence, trying to regain a sense of what I have missed. This makes reading in particular very laborious, time-consuming, and tiring.

Real and Invented Music

As a neurologist, Sacks is familiar with those people who present as patients with inner music that is inescapably disruptive and firmly pathological; he regards such music as morbidly continuous with the more plausibly standard phenomenon of "ear-worms," with which all are familiar. Moreover, he writes generally as though inner music is characteristically "ready-made." For example, he strongly contrasts visual memories and imaginings with musical ones:

> This [fundamentally different way in which the brain treats music and vision] may arise in part because we have to *construct* a visual world for ourselves, and a selective and personal character therefore infuses our visual memories from the start—whereas we are given pieces of music already constructed. A visual or social scene can be constructed or reconstructed in a hundred different ways, but the recall of a musical piece has to be close to the original. We do, of course, listen selectively, with differing interpretations and emotions, but the basic musical characteristics of a piece—its tempo, its rhythm, its melodic contours, even its timbre and pitch—tend to be preserved with remarkable accuracy. (2007, 47)

Sacks suggests, "the automatic or compulsive repetition of musical phrases is almost universal—the clearest sign of the overwhelming, and at times helpless, sensitivity of our brains to music" (45). Any relief I might feel on reading this suggestion that everyone shares my affliction is immediately dispelled by the realization that this cannot in fact be so, since the compulsive and repetitive elements are, for me, only half the story: the content of my musical imaginings is not confined to music remembered—actual tunes compulsively recalled—but extends also to music apparently invented by myself, for obscure purposes but with an all-too-clear effect, that of self-torment: the inventions are characteristically banal, hopelessly fragmentary, and, worse still, demonically aimless and unfulfilled.

When inner music is inflicting itself on me, then it is likely to take the form of a repeated sequence of notes and underlying harmonies such as we might often find in baroque music. It might be a debased fragment of some actual passage from (typically) Bach, but that scarcely seems to matter. I seem to cut it out of its context, then cycle and recycle it, often endlessly, following it through repeatedly unresolved key sequences, a sort of "Bach-nightmare" of a sequence that is perpetually trying to resolve itself but that never arrives at its destination—for, of course, it has no authentic destination. Throughout this, I am perfectly aware that I am "doing" it myself, but I cannot help it; the awareness in itself may add to the distress.

Sacks does mention patients whose inner music includes inventions: "Gordon B" endures "musical wallpaper, meaningless musical phrases and patterns" (64) and David Oppenheim mentions symptoms that include "new music that I was somehow making up" (75). Overwhelmingly, however, the inner music reported is firmly external in primary origin. For my own part, I have so far met only a handful of people who endure invented music of the kind familiar to me, and virtually no one who has less capacity than I to dismiss the intrusions of "ear-worms," regardless of their origin, be it in established, external music or in the musical imagination alone.

The Role of the Imagination

"What is more remarkable is [the finding] that listening to music or imagining, even without any overt movement or keeping time, activates motor cortex and subcortical motor systems, too. Thus the *imagination* of music, of rhythm, may be as potent, neurally, as actually listening to it" (Sacks 2007, 240–241, emphasis added; see also, for example, Hargreaves, Miell, and Macdonald 2011; Kraemer et al. 2005; Halpern 2001; Halpern and Zatorre 1999).

When I was about twelve, I memorized two entire symphonies of Sibelius, the third and seventh, and would "play" them through in my head at night while waiting to fall asleep. I would "hear" them (which means of course, imagine them) in intricate and satisfying detail including the sound of their wonderfully astringent instrumentation. This was the first time I can remember, or recognize, an ability that I have had ever since, to create an imagined musical experience, in detail, of my own choosing. At times I would do this in a more independent way as a part of making up my own music (something that in my earlier years seems to have had none of the futility or unreason that infects the compulsive repetitions described earlier). This is the ability, I think, that in adult life has come to take on the distorted form of my musical intrusions.

Latterly, when actually asleep, from time to time I dream established music that I know well, and it occasionally happens that I will dream it in the wrong key, realize in my dream that something is not right, and wake up, usually feeling nauseous—pathological music indeed! It may take some minutes for me to work out what has happened, what the music was, and in what way it was "wrong"—such as a passage being "heard" (imagined) in a key that is, for example, a perfect fourth away from its true key (for instance, heard in F major instead of C major). If this matters musically—if, for instance, it undoes the key relationships critical to a work's structure—then it seems that it matters personally as well; it is sufficient to produce actual physical distress to me upon waking until I can assure myself that it was "only a dream." Formally close to the original, the music has become a parody—the same, yet other.

The Role of Perfect Pitch

Awareness of the parody (and the distress to which it gave rise) stems largely from the phenomenon known as "perfect pitch" or "absolute pitch" (see Lennard's chapter for more on this subject).

At the age of six or seven, I was asked to play a piece that I knew perfectly well on a piano in the home of an elderly relative, a piano I had never seen or touched before. When I tried to begin playing the first couple of notes, I was quite unable to continue; indeed, I was wholly disorientated. I stopped, tried again, stopped again, and fell into a dismayed silence. The adults in the room were puzzled and asked why I was not playing.

"I can't play this piano," I replied. "It doesn't work properly." This was not well received, and I burst into tears and left the room.

I now know that the reason was that the piano had not been tuned for so long that the entire instrument was around a semitone below pitch. Every time I pressed a key, I heard the note that I should have expected from its immediate left-hand neighbor. I could not understand this, and was very disorientated and upset (at the time, unaccountably so). While any competent listener will hear the note we call A as being a certain "distance" higher than, say, the note we call F, someone with perfect pitch will hear it intrinsically as *being* an A—or an F, or a B—without reference to musical notation, the sound of other notes nearby, or visual cues from its place on an instrument. Indeed, to someone with perfect pitch an F cannot sound like anything else, and it is effectively impossible for such a person to imagine what it might be like to hear it without this pitch-attunement. It is difficult to convey to someone without perfect pitch what any of this means without resorting rather optimistically to analogy: for instance, that it might be rather like the impossibility of a native English-speaker imagining the sound of an English sentence for someone who does not speak English.

Even today, it causes me continuing problems; singing in a choir is extremely difficult and actually unpleasant (to the point almost of nausea) if we are singing in a key slightly different from what is written in the score—as sometimes happens intentionally if the conductor wants to bring two consecutive pieces into closer relationship, or (more usually) unintentionally because the choir is unaccompanied and cannot maintain secure pitch over the course of a long piece of music.

The Patterings of the Nervous System

In outlining my experiences of inner music, I have emphasized that its intrusiveness—its disruption of my own volitional processes of thought and concentration—is its most unwanted aspect. Even as I type these words, there is some sort of compulsive, unresolved, fragmentary inner music pattering away in the background, though fortunately at the present moment I notice that it does not entirely get between me and what I am trying to do. (Perhaps right now I even need it, in order to center my thinking on the phenomenon that I am trying to describe.) But why does it patter away at all?

Sacks considers parallels between involuntary music-hearing and involuntary movements, such as the observable tics that characterize choreas and other movement disorders. He broadly rejects the parallels, on the grounds that the content of musical compulsions has roots in experience and not simply in physiology:

> One may use the word "random" with regard to the effects of a low-level mishap in the basal ganglia—in the involuntary movement called chorea, for example. There is no personal element in chorea; it is wholly an automatism—it does not, for the most part, even make its way to consciousness and may be more visible to others than to the patient himself. But "random" is a word one would hesitate to use in regard

> to *experiences*, whether these are perceptual, imaginary, or hallucinatory. Musical hallucinations draw upon the musical experience and memories of a lifetime, and the importance that particular sorts of music have on the individual must surely play a major role. (2007, 81)

The prudent layman hesitates to differ from the specialist, but I wonder whether Sacks delimits experiences too tightly; it is hard to think that tics are never present to the sufferer's experience, and hard also to think that there is no connection (in either or both directions) between the "styles" of tics and the "styles" of the sufferer's learned or intentional behaviors. As for what is or is not "random," I cannot deny that the repertoire of my musical imaginings originates in active musical experiences and memories. However, given the existence of that repertoire, I do not see why the content of particular imaginings must always arise from the repertoire in some nonrandom, that is, directed or symbolic, fashion—why particular imaginings must always be in some sense meaningful, howsoever obscure.

On the other hand, the general, restless, and often parodic pattering of musical imagination as a whole may well constitute something that is purposive, as I shall shortly conjecture.

The Role of Kinesthesia—Its Specificity in My Case

One of Sacks's patients, the composer "John C.," experienced something that is very familiar to me:

> Though the music he hallucinated was usually vocal or orchestral in origin, it was instantly and automatically transcribed into piano music, often in a different key. He would find his hands physically "playing" these transcriptions "almost by themselves." He felt that there were two processes involved here: the refluxing of old songs, "musical information from the memory banks," and then an active reprocessing by his composer's (and pianist's) brain. (2007, 58–59)

Invariably, for me, inner music involves a muscular or kinesthetic component: primarily this component is what the exertions would feel like if I were physically to play the notes that I inwardly hear (which usually involve full harmony, so several notes simultaneously) on the keyboard and, often, the pedal board of an organ, regardless of the actual instrumental timbres involved. Just as for John C., there is no need for me to be imagining keyboard music; whatever instruments were involved in the original, then if one were to realize those same notes on the keys and pedals of a pipe organ, my imaginative "hearing" of the notes is inseparable from the sensation of the muscular exertions involved in producing them.

But in my case it is not the hands (or feet) alone. The sensations are also descriptively elusive, but experientially highly specific, forms of what I can only call visceral and "thoracic" kinesthesia. It is as though given harmonies—given chords, in given dispositions (roots or inversions)—are inwardly realized in more general kinesthetic or proprioceptive "dispositions" of my body as a whole. The directed "pull" of harmonic movement—the resolutions for which progressions of chords yearn musically—is inwardly experienced as bodily yearnings that are so inbound to my bodily self-experience as to be virtually automatic: certainly they are unopposable and inescapable.

When inner music prevents my falling asleep, as it often does, the intrusion consists both in imagined sound and in wholly actual kinesthesia. This, I think, must be true in sleep as well, thus contributing to the discomforts occasioned by my dreaming music in the wrong key, and hence—perhaps—with the "wrong" motoric sensations. Sacks notes motoric experiences in some of his patients but seems to me to pay curiously little attention to them, and he does not explore any one-to-one specificity in their relation to particular musical sounds, real, imagined, hallucinated, or dreamed.

Even so, a clue to their pressing nature may be discerned in what he terms "release" hallucinations as they are manifest in a deaf patient: "Given her deafness, the auditory part of the brain, deprived of its usual input, had started to generate a spontaneous activity of its own, and this took the form of musical hallucinations, mostly musical memories from her earlier life. The brain *needed to stay incessantly active*, and if it was not getting its usual stimulation, whether auditory or visual, *it would create its own stimulation* in the form of hallucinations" (52, emphasis added). This accords well with the conjecture that I have begun to frame in my own case, as follows.

Toward an Understanding: "Existential Reassurance"

It is perhaps not merely the brain that needs to "stay incessantly active" but also the embodied self as a whole, in a more general, existential sense: in kinesthetic and proprioceptive terms, the need for musico-motoric stimulation that confirms that one is still there, still oneself. It seems to me then that, in my own case, this pattering-away of music in some form is a kind of staving off, a keeping at bay, of an existential anxiety. The duple-time rhythms that predominate in my invented, parodic music align with a bodily connection, and my inner music is invariably felt as muscular exertions and proprioceptive changes, inseparably from its being "heard" in the imagination.

So perhaps this music is the form taken by a continuous stimulus to proprioception, to embodied self-presence—a way of making sure that in some weird sense the "I" that is my embodied self is "still there," a repeated checking that the general bodily disposition is what it usually is, and is not fading into dissolution. This aligns strongly, I think, with Damasio's suggestion that "primordial feelings" are, through their very lifelong qualitative continuity, the likeliest bodily ground of personal identity. If that is true, then perhaps it is not so surprising that in my own case it is music, the element of my greatest living joys, that both furnishes my most disruptive modes of existential restlessness and anxiety, and involuntarily placates them with bodily, kinesthetic, reassurance.

Can this restlessness be managed, tamed, redeemed in any way? Can the reassurance be summoned? Can it offer a clue as to *why* music-as-therapy can work? I believe it can.

"It is clear," says Sacks, "that music, above all else, can kickstart a damaged or inhibited motor system into action again" (2007, 236). He was, of course, speaking from personal experience. His kinesthesia and proprioception temporarily disrupted, he had forgotten how to walk; through music, Mendelssohn, *fortissimo*, he "remembered" again (1984, 108). But the music was inner music, recalled in the imagination. In music—"the very prototype of inwardness, being, soul" (167)—we have both the materials and the means by which, in imagination, we recall not merely the fluency of movement and agency but also the inner self-experience—the "primordial feelings" that Damasio identifies as kinesthetically inbound to all temporal experience that give continuity between our premorbid and our presently morbid selves. Our ordinary kinesthesia, our ordinary proprioception, our premorbid self-experience is temporarily (and sometimes cumulatively and lastingly) restored to our sense of who, and how, we are.

A feeling recalled is a feeling restored, for as long as the recalling is itself felt; and in music-as-therapy the recalling is nothing other than our musical imaginations at work. The musical imagination is, of course, an unruly thing, as my inner musical compulsions constantly remind me, but it can work for good as well as for ill, and when it does so it allows us both existential reassurance and a renewed glimpse of joy.

References

Ansdell, Gary. 1995. *Music for Life: Aspects of Creative Music Therapy and Adult Clients.* London: Jessica Kingsley.

Blood A. J., Robert J. Zatorre, P. Bermudez, and A. C. Evans. 1999. "Emotional Responses to Pleasant and Unpleasant Music Correlate with Activity in Paralimbic Brain Regions." *Nature Neuroscience* 2: 382–387.

Boso, Marianna, Pierluigi Politi, Francesco Barale, and Enzo Emanuele. 2006. "Neurophysiology and Neurobiology of the Musical Experience." *Functional Neurology* 21 (4): 187–191.

Bunt, Leslie. 1994. *Music Therapy: An Art beyond Words.* London: Routledge.

Damasio, Antonio. 2010. *Self Comes to Mind: Constructing the Conscious Brain.* London: William Heinemann.

Evans, Martyn. 1990. *Listening to Music.* London: Macmillan.

Evans, H. Martyn. 2010. "Music, Medicine and Embodiment." *Lancet* 375: 886–887.

Gould, Glenn. 1966. "The Prospects of Recording." *High Fidelity* 16 (4): 46–63.

Halpern, Andrea R. 2001. "Cerebral Substrates of Musical Imagery." *Annals of the New York Academy of Sciences* 930: 179–192.

Halpern, Andrea R., and Robert J. Zatorre. 1999. "When That Tune Runs through Your Head: A PET Investigation of Auditory Imagery for Familiar Melodies." *Cerebral Cortex* 9: 697–704.

Hargreaves, David, Dorothy Miell, and Raymond MacDonald, eds. 2011. *Musical Imaginations: Multidisciplinary Perspectives on Creativity, Performance, and Perception.* Oxford: Oxford University Press.

Hilleke, Thomas, Anne Nickel, and Hans Volker Bolay. 2005. "Scientific Perspectives on Music Therapy." *Annals of the New York Academic of Sciences* 1060: 271–282.

Koelsch, Stefan. 2009. "A Neuroscientific Perspective on Music Therapy." *Annals of the New York Academy of Science* 1169: 374–384.

Kraemer, David J. M., C. Neil Macrae, Adam E., Green, and William M. Kelley. 2005. "Sounds of Silence: Musical Imagery Activates Auditory Cortex." *Nature* 434: 158.

Leys, Ruth. 2011. "The Turn to Affect: A Critique." *Critical Enquiry* 37 (3): 434–472.

Magee, Bryan. 1983. *The Philosophy of Schopenhauer*. Oxford: Clarendon Press.

Migneault, Brigitte, Francois Girard, Caroline Albert, Philippe Chouinard, Daniel Boudreault, Diane Provencher, et al. 2004. "The Effect of Music on the Neurohormonal Stress Response to Surgery under General Anesthesia." *Anesthesia and Analgesia* 98 (2): 527–532.

Nordoff, Paul, and Clive Robbins. 2007. *Creative Music Therapy: A Guide to Fostering Clinical Musicianship.* 2nd ed. Gilsum, NH: Barcelona.

Sacks, Oliver. 1984. *A Leg to Stand On*. London: Duckworth.

Sacks, Oliver. 2007. *Musicophilia: Tales of Music and the Brain*. London: Picador.

Schopenhauer, Arthur. 1966. *The World as Will and Representation*. Translated by E. F. A. Payne. New York: Dover.

Scruton, Roger. 1974. *Art and Imagination*. New York: Harper and Row.

Thaut, Michael H. 2005. "The Future of Music in Therapy and Medicine." *Annals of the New York Academic of Sciences* 1060: 303–308.

Trainor, Laurel J., and Louis A. Schmidt. 2003. "Processing Emotions Induced by Music." In *The Cognitive Neuroscience of Music*, edited by I. Peretz and Robert J. Zatorre, 310–324. Oxford: Oxford University Press.

Warnock, Mary. 1976. *Imagination*. London: Faber and Faber.

PART VI

MUSIC AND THE MULTIMODAL BODY

CHAPTER 21

SPATIAL REPRESENTATIONS COMMON TO MUSIC AND BODILY EXPERIENCE

XUEJING LU AND WILLIAM FORDE THOMPSON

Cross-Modal Associations

To perceive the external environment accurately and efficiently, the human brain draws on multiple sources of sensory information—such as vision, audition, touch, and olfaction—and integrates these inputs to form a coherent multimodal percept (Ernst and Bulthoff 2004). Because objects in the world are perceived multimodally, strong cross-modal associations are formed, especially between vision and audition, which process most of the information that we perceive in everyday life. From an evolutionary standpoint, cross-modal associations are essential to survival. For example, when hearing a sound with a high frequency in a potentially dangerous environment, people tend to look up rather than look down to locate the sound source. In this case, pitch is a powerful localization cue because it is perceived regardless of the position of one's body or head. Audition can provide a rapid but imprecise cue about the location of objects, whereas vision can be used for more precise localization. As a result, people often use pitch to evaluate the likely spatial height of an object, and then draw on the vision system for more precise localization.

Pitch is not only associated with spatial height. Indeed, there is a body of evidence showing that pitch is associated with other visuospatial features such as shape, thickness, and brightness. High-pitched sounds usually correspond to sharp-edged, thin, or light-colored objects, whereas low-pitched sounds are associated with round, thick, or dim-colored objects. In addition to pitch, loudness that reflects the intensity of a sound is also associated with spatial location, brightness, and size (Dolscheid et al. 2013; Jeschonek, Pauen, and Babocsai 2013; Walker et al. 2010; Eitan, Schupak, and Marks 2008; Marks 1982, 1987, 1989; Wagner et al. 1981; Lewkowicz and Turkewitz 1980).

Importantly, all of these visual associations are related to "space," either in a literal sense (the location of a sound-emitting object in real three-dimensional space) or in a metaphorical sense (an association that does not reflect the actual spatial location of the sound-emitting object). Eitan and Granot (2006) discuss how physical and metaphorical uses of spatial concepts may be applied to several features of music. As examples, changes in pitch, loudness, or attack rate can be associated with visuospatial changes in verticality, distance, or velocity. In other words, a number of different spatial concepts can be used to make sense of music.

This chapter focuses on musical pitch and its association with space and bodily experience. Pitch is one of the most fundamental elements of music, and the tendency to map pitch onto space is one of the most robust cross-modal associations (for a review, see Spence 2011). Nonetheless, the implications drawn from discussions on musical pitch are applicable to other music attributes.

Spatial Representations of Pitch

Musical pitch is usually discussed, conceived, and represented spatially. The most obvious example is the Western music notation system, in which pitches are mapped to vertical locations: notes corresponding to higher pitches are presented at higher spatial positions while notes corresponding to lower pitches are presented at lower spatial positions on the staff. Another example is the layout of the piano keyboard, in which pitches are mapped to horizontal locations: high-pitched notes are on the right and low-pitched notes are on the left.

Formal theoretic and informal descriptions of pitch also employ spatial terms. One early geometric model depicts musical pitch as a helix with two psychological dimensions, referred to as pitch height and chroma, respectively. In language, most Western speakers use spatial metaphors such as "up-down," "ascending-descending," and "rising-falling" to describe pitch height and movement. However, spatial height is not the only metaphor for this attribute of sound. For Turkish, Farsi, and Zapotec speakers, pitch is referred to as "thin" or "thick." In Kpelle, a language spoken in Liberia, speakers use "large" or "heavy" to describe low-pitched sound and "small" or "light" to describe high-pitched sound (Dolscheid et al. 2013; Shayan, Ozturk, and Sicoli 2011; Stone 1981). These alternative characterizations of pitch challenge the notion that pitch is universally represented in terms of spatial height. Nonetheless, the association between pitch and spatial height is encoded in most languages of the world, and all conceptions of pitch appear to reflect some connection between pitch and space, suggesting that pitch is represented in an embodied manner across cultures.

Pitch-space associations have been extensively investigated since the late 1800s. Why are tones characterized as "high" or "low" in most languages? Stumpf (1883, as cited in Pratt 1930) argued that the association between pitch and vertical location might be accounted for by the frequent use of spatial metaphors derived from the experiences elicited by tones and other stimuli. For instance, low-pitched tones tend to give people impressions of objects that are massive, gloomy, or heavy. Creatures that emit low-pitched vocalizations are usually found at a lower location (i.e., close to the ground). Conversely, high-pitched tones tend to give people impressions of objects that are small, bright, and light. In turn, creatures that emit high-pitched vocalizations tend to be found at higher locations (i.e., up in the sky).

Pratt (1930) made the first empirical attempt to discover the mechanisms underlying pitch–space associations. He presented subjects with tones differing in pitch and asked them to locate the position of incoming tones on a scale running from the floor to the ceiling. Every subject placed tones in a top-to-bottom order corresponding to their pitch height from high to low. Based on this observation, Pratt reasoned that high- and low-pitched tones are phenomenologically higher and lower in space, respectively, suggesting a psychological bias in the perception of pitch. At the time, it was unknown whether the connection between pitch and space was learned or hardwired, but it was assumed that the association helps to facilitate sensory processing. For instance, when a visual object at a relatively high position is simultaneously presented with a high-pitched sound rather than with a low-pitched sound, both the visual object and the sound are more rapidly and accurately detected (Ben-Artzi and Marks 1995; Melara, 1989; Melara and O'Brien 1987; Bernstein and Edelstein 1971).

Other evidence showing pitch–space associations comes from studies adopting the stimulus-response compatibility (SRC) paradigm. The SRC effect can be observed in binary-choice reaction time tasks when one manipulates the assignment of the response and stimulus. The basic logic of the SRC paradigm is as follows: Participants are asked to classify a stimulus as rapidly and accurately as possible using response keys, whereby the spatial location of the response key for a correct classification is manipulated such that it is either congruent or incongruent with the implied (or metaphoric) spatial location of the stimulus. Subjects should respond faster and more accurately when the position of the (correct) response key is congruent with the implied position of the stimulus (Proctor and Vu 2006; Kornblum, Hasbroucq, and Osman 1990).

Adopting this paradigm, two research teams independently demonstrated an association between the pitch height of a tone and the spatial location of the response keys. This is known as the spatial–musical/pitch association of response code (SMARC/SPARC) effect (Lidji et al. 2007; Rusconi et al. 2006). Both teams found that subjects performed better when the pitch of the tone (higher or lower) was congruent with the position of the correct response key (higher or lower on the device) than when they were incongruent with each other. This effect occurred for both musically trained and untrained participants and hence cannot be explained as a learned association between pitch and actions required to produce tones on instruments such as the piano. Indeed, when response keys were aligned horizontally (i.e., left-right rather than top-bottom, as on a piano), the advantage for the combinations of high-pitched tone with right response key, and low-pitched tone with left response key, was only found in subjects with musical training but not in those who were musically naive (Lidji et al. 2007; Rusconi et al. 2006).

Spatial Metaphors of Pitch on the Basis of Bodily Experience

In an effort to understand the world, humans form mental representations through bodily and sensory experiences, and their mental representations reflect the nature of the world itself. As the world is three-dimensional, spatial coding is an efficient way to represent and interact with it. Spatial representations also appear to be universal and can be expressed using both language metaphors and bodily gestures (Nunez and Cooperrider 2013; Parkinson et al. 2012; Lakoff and Johnson 1980).

The examination of spatial metaphors can shed light on mental processes, as they mirror cross-modal experiences. Lakoff and Johnson (1980) argued that we conceptualize the external world through our own bodily experience in the physical environment, and bodily experiences are reflected in the metaphors we use. Orientational metaphors arise from spatial experience and assign nonspatial concepts a (metaphoric) spatial orientation. As an example, we often use the expression "wake up" to awaken somebody. This expression is formed on the basis that our body tends to be in an upright position when we are awake—standing, sitting, or walking, with the ground under our feet and the sky above our head. Therefore, the concept "wake" is oriented in the direction "up," leading to such an expression.

Similarly, we use spatial metaphors to describe the pitch of a tone and its movement on the basis of our experiences of environmental sounds that vary in average fundamental frequency. For example, bird song typically has a high average fundamental frequency and arises from up in the trees or in the sky; large animals that produce low-pitched sounds tend to be heavy and emit vocalizations at ground level. When we are singing high-pitched tones, the larynx rises, and when we are singing low-pitch tones, it descends. Such bodily experiences link pitch height to spatial position, which can lead to orientation metaphors of pitch. Analogously, musicians use movement metaphors to describe tempo changes in music such as "accelerando" (gradually accelerating) and "ritardando" (being slower), because they "feel" these movements in music. In short, we use bodily experiences as our most trusted way of representing, understanding, and interacting with the external world. As a consequence, our conceptual system and the metaphors we use are shaped and constrained by our bodily experiences. These embodied metaphors, in turn, allow us to map concrete experiences in the sensorimotor domain to subjective experience.

Music-Related Gesture and Its Affordances

Music performance is always accompanied by bodily movements. Such movements include actions required to produce sounds on musical instruments, ancillary bodily gestures and facial expressions, and physical responses to perceiving music. The analysis of such movements can provide considerable insight into mental representations of music.

In this chapter, the term "music-related movements" refer to movements required to produce sounds (i.e., music-producing gestures such as piano key press actions), and ancillary movements that often accompany music performance (e.g., emotive facial expressions, gestures that illustrate or reinforce a structural element) or music perception (e.g., dancing or tapping along with music). For performers, physical actions are required to make sounds. For example, percussionists must raise a mallet into the air before striking a drum. However, bodily movements are also a nonverbal means of communicating with audience members and with coperformers, and may function to express an emotional interpretation of the music, to reinforce metric and phrase boundaries, and to interact with and entertain audiences (Davidson and Correia 2002). For those perceiving the sounds, seeing performers' bodily movements can enrich their music experience. For instance, facial expressions shown by performers affect perceived interval size, dissonance, and sung syllables (Quinto et al. 2010; Thompson, Graham, and Russo 2005). In addition, audiences are likely to feel more engaged if their response to the music includes bodily movements and facial expressions, such as smiling.

Although a number of researchers have examined music-related gestures, this area of research is still in its infancy. In contrast, there is a large body of literature concerned with language-related gestures. Using gestures during talking is common in all cultural and linguistic backgrounds. Language-related gestures can even be observed in young children before the development of language, as well as in congenitally blind speakers who have never seen gestures and have no model for gesturing (Iverson and Goldin-Meadow 1998). Iverson and Thal (1998) define gestures as actions "produced with the intention to communicate . . . [they] typically involve fingers, hands, and arms (e.g., in pointing or showing) and also may employ facial features (e.g., lip smacking for cookies) or even the entire body (e.g., bouncing up and down for horse)" (59–60). Broadly speaking, gestures can be understood to consist of all facial expressions and body movements that have the potential to contribute to communication, whether such movements are intentional or not.

Iverson and Thal classify two primary categories of gesture: deictic and representational. Deictic gestures are used to indicate an object or event by directing listeners' attention (e.g., pointing or reaching for a toy). Representational gestures are used to indicate an object-related semantic content by signifying some feature of the referent (e.g., cupped hand to mouth to represent "drinking," and swinging arms rapidly to express "running"). Representational gestures can also be used in a more abstract or symbolic way (e.g., shrugging shoulders for "helpless").

Based on the categories of nonverbal behavior described by Ekman and Friesen (1969), McNeill (1992) provides another category—beat gestures—which typically do not convey information about concrete objects or semantic content, but are used rhythmically with speech to emphasize the significant or accented points in the accompanying speech. McNeil further divides representational gestures into three subcategories: iconic, metaphoric, and emblems. Iconic gestures indicate object-related semantic content by mirroring a specific feature of the referent, whereas metaphoric gestures represent abstract features of an object or event. Emblems are stereotypical gestures that are shared within a culture or a group of people, and can usually be translated into a word such as "hello," "ok," or "peace." For example, when greeting in Hindu culture, people usually bow slightly and put their palms together with fingers pointing upward and thumbs close to the chest (i.e., the gesture of Namaste).

According to Darwin (1871), music and language have a common evolutionary origin as a "musical protolanguage" that was used in courtship and territoriality (see S. Gilman's chapter for more on this subject). If so, then one might expect gestures used in music to have a similar communicative function as language-related gestures. For instance, conductors usually use deictic gestures with hands, arms, face, and head to direct instrument players' or performers' attention, thereby unifying them, setting the tempo, and so on. It is also common for performers to use gestures as a way of signaling musical changes in an iconic or metaphoric way. For example, singers often move their head or eyebrows in systematic ways, or display upward hand gestures when there is an increase in pitch or intensity. Such gestures contain secondary information about changes in music, supplementing and reinforcing the acoustic channel, and hence providing a safety net for breakdowns in acoustic communication.

Thompson, Graham, and Russo (2005) point out that music-related gestures are often used to signal "the timing of musical events, focusing listeners' attention to (or away from) critical acoustic information at specific moments in time" (203). In their view, "musical events" can be any significant melodic, harmonic, or rhythmic event based on performers' "understanding of segmentation (points of closure), intervallic information (whether a melodic interval is large or small), and points of expectancy fulfillment or violation" (204). In other words, performers can use bodily movements to convey their intentions and share their experience with listeners.

Although some music-evoked movements appear to be universal (e.g., tapping of feet to music), the same piece of music can induce different bodily responses from individual to individual. It is also common to observe differences in movements that people make to a single piece of music for the first and the second time. These variations in bodily responses to music may reflect unique interactions of individuals with music, in which music-related movements are made depending on several factors, such as individuals' background, their mood at a given time, and so on.

Godøy (2010) discusses the idea of "gestural affordances of musical sound" by adopting the notion of "affordance" from ecological psychology, which was first introduced by James J. Gibson in the 1970s. Gibson (1979) suggested, "the affordances of the environment are what it offers the animal, what it provides or furnishes either for good or ill" (127). Applying this to the gestural affordances of musical sounds, Godøy (2010) suggests, "music has very rich gestural affordance" and "gestures shape our perception of the music" (104). In his view, "musical sound is a transducer of source-information" (106), which is fulfilled by the potential gestures that music affords. People acquire knowledge of music-producing gestures through experience, such as seeing music performances. Once the association between musical sounds and gestures is established, people can mentally re-create these music-producing gestures even when they listen to music in the absence of visual images. By combining these mental images with the human nature of imitating others, listeners readily move to musical sounds on the basis of the gestures that they have seen or experienced previously (see Godøy's chapter for more on this subject).

Most features of musical sounds can be understood in relation to their gestural affordances. Therefore, Godøy considers these musical features as "trajectory shapes in time and space" (113). For instance, gestural responses such as moving hands up or down, left or right, near or far, and so on are elicited in response to musical changes. These spatial gestures, in turn, shape the music perception of listeners as well as performers, if we consider performers to be listeners of their own performance. From this perspective, music perception is not about perceiving music itself but also concerns how one interacts with its affordances. To support this idea, empirical evidence for the influence of bodily movements, such as hand gestures and facial expressions, on perceptual representations of music is reviewed, and then the connection between music perception and action planning is discussed.

The Influence of Music-Related Gestures on Representations of Music

A number of psychological studies on bodily movements and music experience have been conducted in the past few decades. In most of these studies, researchers record performers' bodily movements using video recorders or motion capture cameras and then conduct movement analysis of the gestures or facial expressions. Captured movements are also sometimes used as stimuli in perceptual experiments in which participants are asked to rate, judge, or reproduce these movements.

At the perceptual level, it has been well documented that the presence of visual information may enhance or alter one's auditory perception (e.g., van Wassenhove, Grant, and Poeppel 2005; McGurk and MacDonald 1976). Thus, music-related movements introduced intentionally by performers have the potential to enhance the listeners' music experience. Listeners with varied musical background can judge the size of sung intervals in the absence of acoustic information based on performers' facial expressions and head movements alone (Thompson, Russo, and Livingstone 2010). More generally, music-related movements greatly influence listeners' perception of musical structure. For instance, Connell, Cai, and Holler (2013) investigated how visuospatial movement biases pitch perception. They first presented a group of subjects with a synthesized note, and then showed them a video where an actress sang a target note and moved her hand either upward or downward at the same time. The task was to judge whether the first and the second note were the same or different in terms of the pitch height while ignoring the visual information. If the subject's answer was "different," the participant was asked whether the second note was higher or lower. Their results indicated an obvious bias on perceived pitch height caused by concurrent spatial gestures, such that downward gestures made notes seem lower in pitch than they really were, and upward gestures made notes seem higher in pitch.

The same research team conducted another study to examine the influence of gesture on perceived tone duration (Cai, Connell, and Holler 2013). This time, researchers asked a group of subjects to view a video of an actress singing a musical note while making a long- or short-distance gesture. After watching the video, subjects were asked to reproduce the duration of the sung note by holding down the spacebar on a computer keyboard. As expected, the perceived duration of the sung note was longer when it was accompanied by a long-distance gesture than when it was accompanied by a short-distance gesture.

In a related study, Schutz and Kubovy (2006) presented subjects with a video showing a world-renowned percussionist performing a single percussive note. They observed that the visual stroke type, that is, the gesture employed for producing a long note or a short note, affected subjects' perceived tone duration, even when they were informed that the visual and auditory information might be mismatched. More importantly, such influences of bodily movements on perceived music structure were not dependent on the level of musical training (Schutz 2007; Schutz and Lipscomb 2007).

A Link between Music Perception and Action Planning

William James (1890) stated, "every representation of a movement awakens in some degree the actual movement which is its object" (2:526), suggesting a direct perception–action link. More recently, this idea was developed into a formal model known as "common coding theory" (Hommel et al. 2001; Prinz 1984, 1997). Common coding theory holds that perception and action modulate each other because they share a common representational coding system. Hommel and colleagues (2001) emphasize, "the stimulus representations underlying perception and the action representations underlying action planning are coded and stored not separately, but together in a common representational medium" (849). Hommel (2011) further explains, "the human brain seems to code all sorts of spatial aspects of a stimulus event and there is evidence that all these codes can interact with the spatial representation of an action directed to, or at least triggered by a given stimulus" (193). In other words, using common coding allows people to combine information from different sensory sources into an amodal spatial representation and also to transfer the mental representation to a behavioral level. From this perspective, the common coding theory also reflects an embodied view of cognition, where the subjective experience and physical or biological mechanisms are connected as tightly as possible (Leman 2008).

In the music domain, the perception-action link is well supported by evidence from behavioral studies, such as the priming effect of perception on action that we mentioned earlier (the SMARC/SPARC effect). Alternatively, both active movements (adults bouncing themselves in synchrony with the experimenter) and passive movements (babies being bounced in the experimenter's arms) elicit a biased hearing of an ambiguous rhythmic pattern (Phillips-Silver and Trainor 2005, 2007, 2008). In addition, Ammirante and Thompson (2012) also found that motor plans could be incorporated into nonmusicians' perception of melodic motion. In this study, when subjects were asked to tap continuously to trigger successive tones in a melody, their finger movement was systematically affected by the auditory feedback that they were instructed to ignore. These findings suggest that actions such as bodily movements are not merely the outcome of high-level cognitive functions. Instead, they have great potential to exert a reciprocal influence on perception.

Further evidence of a perception-action link comes from neuroimaging studies (Baumann et al. 2007; Bangert et al. 2006; Popescu, Otsuka, and Ioannides 2004; Lotze et al. 2003; Grezes and Decety 2001; for a review, see Zatorre et al. 2007). Such studies indicate that listening to music without playing it, or even just imagining a music performance, can lead to activation in motor areas of the brain in both musicians and nonmusicians. Conversely, when playing music, even with no auditory feedback, the right primary auditory cortex can be activated.

The Role of Musical Skill

On the one hand, based on the current empirical evidence summarized previously, it is clear that people with different levels of expertise in music make gestures in response to music, and the influences of gestures on music perception do not seem to depend on musical training. On the other hand, musical training may increase the stability of spatial representations of music, thereby affecting music perception. Indeed, in addition to a vertical representation of pitch, a horizontal representation of pitch tends to occur in musicians but not in nonmusicians (Küssner et al. 2014; Lega et al. 2014; Taylor and Witt 2014; Stewart et al. 2012; Lidji et al. 2007; Rusconi et al. 2006). A possible explanation is that musical training allows people to establish new associations by linking musical sounds and corresponding actions via a shared spatial representation. In general, the association between pitch and vertical space is more straightforward than the association between pitch and horizontal space for most people with no musical training on the basis of their daily experience. Therefore, when faced with situations that require associating pitch with horizontal space, musically untrained individuals may need extra effort to mentally project horizontal space to vertical space. In contrast, musicians, such as pianists, are exposed to associative learning processes due to the immense amount of practice accompanying their training, where they can progressively learn the horizontal representation of pitch. In particular, pianists move toward the left of the keyboard to correspond to decreasing pitch and to the right of the keyboard to correspond to increasing pitch. As a consequence, the horizontal spatial representation of musical pitch is strongly established without involving extra mental spatial manipulations.

On the other end of the spectrum of musical ability, people with congenital amusia show unstable vertical spatial representations of pitch when compared with normal listeners in the SRC task, suggesting that spatial conceptions of musical pitch might be disrupted and cause difficulties in associating pitch height with spatial action among these people (Lu et al. 2014; Douglas and Bilkey 2007). Systematic investigations into congenital amusia started only a dozen years ago (Foxton et al. 2004; Hyde and Peretz 2004; Ayotte, Peretz, and Hyde 2002). Based on the current understanding of this disorder, some researchers believe amusia to be a disorder of pitch awareness caused by a disconnection between auditory perceptual and high-level processing that involves conscious awareness (Lu et al. 2015; Zendel et al. 2015; Hyde, Zatorre, and Peretz 2010; Loui, Alsop, and Schlaug 2009; Peretz et al. 2009; Loui et al. 2008). In other words, the musical input fails to be cognitively represented for further processing. The unstable representation of musical pitch can, of course, lead to other musical impairments at higher levels, such as melodic contour processing, timbre perception, and pitch memory (Albouy et al. 2013; Marin, Gingras, and Stewart 2012; Williamson and Stewart 2010; Gosselin, Jolicoeur, and Peretz 2009; Tillmann, Schulze, and Foxton 2009).

Drawing from findings of musician and amusia studies, the possibility that spatial representations of pitch normally play a valuable function for listeners and performers, bridging music perception and actions and reinforcing the ability to encode and retain auditory information, is considered.

A Cognitive-Motor Framework

This chapter reviewed theories related to music perception and music-related gestures from embodied accounts and provided an overview of recent evidence in support of these ideas. In essence, embodied music cognition stresses the human body as a biological transducer that decodes music and translates it into intended expressive movements. Music perception, which is shaped by bodily experience, encourages action tendencies due to the rich gestural affordances of music (Godøy 2010; Leman 2008). More recently, combining the new evidence showing the functional role of the motor system in auditory perception, Maes and colleagues (2014) provide a theoretical framework to emphasize that action can modulate music perception reciprocally. This framework draws on the common coding theory, which suggests a representational overlap between perception and action planning.

Here, the focus is on spatial representation, and a cognitive-motor framework is proposed to capture its important role in linking music perception and bodily movements. Given the fact that the human body is the only medium between the mental world and the physical world, it makes sense to assume that spatial representation is an efficient and universal way to exchange information between these two worlds. Both concrete and abstract information can be represented in a kind of virtual space. Indeed, spatial representation of music is not specific to pitch. Other attributes of music—loudness, timbral brightness, tempo, rhythm, note duration, temporal order, and melodic accent—can also be represented spatially. The shared spatial representation allows people to link musical sounds to the corresponding gestures by means of progressively associative learning, and also allows people to execute actions in response to music based on these learned associations. Based on the evidence from studies on musical expertise and congenital amusia, this framework emphasizes that the degree of stability of spatial representations in music relies on the level of music skill. Musical training can reinforce associations between music sounds and corresponding actions. Furthermore, the representation of musical pitch may be disrupted in individuals with congenital amusia. The unstable representations may lead to impairments at both low and high levels of music processing.

Conclusion

Music is not purely acoustic. Rather, it is a multisensory expressed and processed activity that involves both perception and motion systems (see Trehub's chapter for more on this subject). Indeed, the affordances of musical sound allow the body to interact with it. In other words, bodily gestures can be incorporated into music perception and be shaped by it. This chapter proposed that spatial representations on the basis of bodily experience can account for the connections between musical sounds and music-related movements. Furthermore, these representations might be reinforced in musicians but disrupted in individuals with musical impairments.

References

Albouy, Philippe, Jeremie Mattout, Romain Bouet, Emmanuel Maby, Gaetan Sanchez, Pierre-Emmanuel Aguera, et al. 2013. "Impaired Pitch Perception and Memory in Congenital Amusia: The Deficit Starts in the Auditory Cortex." *Brain* 136: 1639–1661.

Ammirante, Paolo, and William F. Thompson. 2012. "Continuation Tapping to Triggered Melodies: Motor Resonance Effects of Melodic Motion." *Experimental Brain Research* 216: 51–60.

Ayotte, Julie, Isabelle Peretz, and Krista L. Hyde. 2002. "Congenital Amusia: A Group Study of Adults Afflicted with a Music-Specific Disorder." *Brain* 125: 238–251.

Bangert, Marc, Thomas Peschel, Gottfried Schlaug, Michael Rotte, Dieter Drescher, Hermann Hinrichs, et al. 2006. "Shared Networks for Auditory and Motor Processing in Professional Pianists: Evidence from fMRI Conjunction." *Neuroimage* 30: 917–926.

Baumann, Simon, Susan Koeneke, Conny F. Schmidt, Martin Meyer, Kai Lutz, and Lutz Jancke. 2007. "A Network for Audio-Motor Coordination in Skilled Pianists and Non-Musicians." *Brain Research* 1161: 65–78.

Ben-Artzi, Elisheva, and Lawrence E. Marks. 1995. "Visual-Auditory Interaction in Speeded Classification: Role of Stimulus Difference." *Perception and Psychophysics* 57: 1151–1162.

Bernstein, Ira H., and Barry A. Edelstein. 1971. "Effects of Some Variations in Auditory Input upon Visual Choice Reaction Time." *Journal of Experimental Psychology* 87: 241–247.

Cai, Zhenguang G., Louise Connell, and Judith Holler. 2013. "Time Does Not Flow without Language: Spatial Distance Affects Temporal Duration Regardless of Movement or Direction." *Psychonomic Bulletin and Review* 20: 973–980.

Connell, Louise, Zhenguang G. Cai, and Judith Holler. 2013. "Do You See What I'm Singing? Visuospatial Movement Biases Pitch Perception." *Brain and Cognition* 81: 124–130.

Darwin, Charles. 1871. *The Descent of Man, and Selection in Relation to Sex*. London: John Murray.

Davidson, Jane W., and Jorge S. Correia. 2002. "Body Movement." In *The Science and Psychology of Music Performance: Creative Strategies for Teaching and Learning*, edited by Richard Parncutt and Gary McPherson, 237–251. New York: Oxford University Press.

Dolscheid, Sarah, Shakila Shayan, Asifa Majid, and Daniel Casasanto. 2013. "The Thickness of Musical Pitch: Psychophysical Evidence for Linguistic Relativity." *Psychological Science* 24: 613–621.

Douglas, Katie M., and David K. Bilkey. 2007. "Amusia Is Associated with Deficits in Spatial Processing." *Nature Neuroscience* 10: 915–921.

Eitan, Zohar, and Roni Y. Granot. 2006. "How Music Moves: Musical Parameters and Listeners' Images of Motion." *Music Perception* 23: 221–248.

Eitan, Zohar, Assi Schupak, and Lawrence E. Marks. 2008. "Louder Is Higher: Cross-Modal Interaction of Loudness Change and Vertical Motion in Speeded Classification." Paper presented at the 10th International Conference on Music Perception and Cognition, Adelaide, Australia. August 25–29.

Ekman, Paul, and Wallace V. Friesen. 1969. "The Repertoire of Nonverbal Behavior: Categories, Origins, Usage, and Coding." *Semiotica* 1: 49–98.

Ernst, Marc O., and Heinrich H. Bulthoff. 2004. "Merging the Senses into a Robust Percept." *Trends in Cognitive Sciences* 8: 162–169.

Foxton, Jessica M., Jennifer L. Dean, Rosemary Gee, Isabelle Peretz, and Timothy D. Griffiths. 2004. "Characterization of Deficits in Pitch Perception Underlying 'Tone Deafness.'" *Brain* 127: 801–810.

Gibson, James J. 1979. *The Ecological Approach to Visual Perception*. Boston: Houghton Mifflin.

Godøy, Rolf I. 2010. "Gestural Affordances of Musical Sound." In *Musical Gestures: Sound, Movement, and Meaning*, edited by Rolf I. Godøy and Marc Leman, 103–125. New York: Routledge.

Gosselin, Nathalie, Pierre Jolicoeur, and Isabelle Peretz. 2009. "Impaired Memory for Pitch in Congenital Amusia." *Annals of the New York Academy of Sciences* 1169: 270–272.

Grezes, Julie, and Jean Decety. 2001. "Functional Anatomy of Execution, Mental Simulation, Observation, and Verb Generation of Actions: A Meta-Analysis." *Human Brain Mapping* 12: 1–19.

Hommel, Bernhard. 2011. "The Simon Effect as Tool and Heuristic." *Acta Psychologica* 136: 189–202.

Hommel, Bernhard, Jochen Müsseler, Gisa Aschersleben, and Wolfgang Prinz. 2001. "The Theory of Event Coding (TEC): A Framework for Perception and Action Planning." *Behavioral and Brain Sciences* 24: 849–937.

Hyde, Krista L., and Isabelle Peretz. 2004. "Brains That Are Out of Tune but in Time." *Psychological Science* 15: 356–360.

Hyde, Krista L., Robert J. Zatorre, and Isabelle Peretz. 2010. "Functional MRI Evidence of an Abnormal Neural Network for Pitch Processing in Congenital Amusia." *Cerebral Cortex* 21: 292–299.

Iverson, Jana M., and Susan Goldin-Meadow. 1998. "Why People Gesture When They Speak." *Nature* 396: 228.

Iverson, Jana M., and Donna J. Thal. 1998. "Communicative Transitions: There's More to the Hand Than Meets the Eye." In *Transitions in Prelinguistic Communication: Preintentional to Intentional and Presymbolic to Symbolic*, edited by Amy M. Wetherby, Steven F. Warren, and Joe Reichle, 59–86. Baltimore, MD: Paul H. Brookes.

James, William. 1890. *The Principles of Psychology*. Vol. 2. New York: Holt.

Jeschonek, Susanna, Sabina Pauen, and Lysett Babocsai. 2013. "Cross-Modal Mapping of Visual and Acoustic Displays in Infants: The Effect of Dynamic and Static Components." *European Journal of Developmental Psychology* 10: 337–358.

Kornblum, Sylvan, Thierry Hasbroucq, and Allen Osman. 1990. "Dimensional Overlap: Cognitive Basis for Stimulus-Response Compatibility—a Model and Taxonomy." *Psychological Review* 97: 253–270.

Küssner, Mats B., Dan Tidhar, Helen M. Prior, and Daniel Leech-Wilkinson. 2014. "Musicians Are More Consistent: Gestural Cross-Modal Mappings of Pitch, Loudness and Tempo in Real-Time." *Frontiers in Psychology* 5: 789.

Lakoff, George, and Mark Johnson. 1980. *Metaphors We Live By*. Chicago: University of Chicago Press.

Lega, Carlotta, Zaira Cattaneo, Lotfi B. Merabet, Tomaso Vecchi, and Silvia Cucchi. 2014. "The Effect of Musical Expertise on the Representation of Space." *Frontiers in Human Neuroscience* 8: 250.

Leman, Marc. 2008. *Embodied Music Cognition and Mediation Technology*. Cambridge, MA: MIT Press.

Lewkowicz, David J., and Gerald Turkewitz. 1980. "Cross-Modal Equivalence in Early Infancy: Auditory-Visual Intensity Matching." *Developmental Psychology* 16: 597–607.

Lidji, Pascale, Régine Kolinsky, Aliette Lochy, and José Morais. 2007. "Spatial Associations for Musical Stimuli: A Piano in the Head?" *Journal of Experimental Psychology: Human Perception and Performance* 33: 1189–1207.

Lotze, M., G. Scheler, H. R. Tan, C. Braun, and N. Birbaumer. 2003. "The Musician's Brain: Functional Imaging of Amateurs and Professionals during Performance and Imagery." *Neuroimage* 20: 1817–1829.

Loui, Psyche, David Alsop, and Gottfried Schlaug. 2009. "Tone Deafness: A New Disconnection Syndrome?" *Journal of Neuroscience* 29: 10215–10220.

Loui, Psyche, Frank H. Guenther, Christoph Mathys, and Gottfried Schlaug. 2008. "Action-Perception Mismatch in Tone-Deafness." *Current Biology* 18: R331–332.

Lu, Xuejing, Pin Zhang, Yanan Sun, and William F. Thompson. 2014. "Spatial Representations of Pitch among Individuals with Congenital Amusia." Paper presented at the 13th International Conference on Music Perception and Cognition, Seoul, South Korea, August 4–8.

Lu, Xuejing, Hao Tam Ho, Fang Liu, Daxing Wu, and William F. Thompson. 2015. "Intonation Processing Deficits of Emotional Words among Mandarin Chinese Speakers with Congenital Amusia: An ERP Study." *Frontiers in Psychology* 6: 385.

Maes, Pieter-Jan, Marc Leman, Caroline Palmer, and Marcelo M. Wanderley. 2014. "Action-Based Effects on Music Perception." *Frontiers in Psychology* 4: 1008.

Marin, Manuela M., Bruno Gingras, and Lauren Stewart. 2012. "Perception of Musical Timbre in Congenital Amusia: Categorization, Discrimination and Short-Term Memory." *Neuropsychologia* 50: 367–378.

Marks, Lawrence E. 1982. "Bright Sneezes and Dark Coughs, Loud Sunlight and Soft Moonlight." *Journal of Experimental Psychology: Human Perception and Performance* 8: 177–193.

Marks, Lawrence E. 1987. "On Cross-Modal Similarity: Auditory-Visual Interactions in Speeded Discrimination." *Journal of Experimental Psychology: Human Perception and Performance* 13: 384–394.

Marks, Lawrence E. 1989. "On Cross-Modal Similarity: The Perceptual Structure of Pitch, Loudness, and Brightness." *Journal of Experimental Psychology: Human Perception and Performance* 15: 586–602.

McGurk, Harry, and John MacDonald. 1976. "Hearing Lips and Seeing Voices." *Nature* 264: 746–748.

McNeill, David. 1992. *Hand and Mind: What Gestures Reveal about Thought*. Chicago: University of Chicago Press.

Melara, Robert D. 1989. "Dimensional Interaction between Color and Pitch." *Journal of Experimental Psychology: Human Perception and Performance* 15: 69–79.

Melara, Robert D., and Thomas P. O'Brien. 1987. "Interaction between Synesthetically Corresponding Dimensions." *Journal of Experimental Psychology: General* 116: 323–336.

Nunez, Rafael, and Kensy Cooperrider. 2013. "The Tangle of Space and Time in Human Cognition." *Trends in Cognitive Sciences* 17: 220–229.

Parkinson, Carolyn, Peter Jes Kohler, Beau Sievers, and Thalia Wheatley. 2012. "Associations between Auditory Pitch and Visual Elevation Do Not Depend on Language: Evidence from a Remote Population." *Perception* 41: 854–861.

Peretz, Isabelle, Elvira Brattico, Miika Jarvenpaa, and Mari Tervaniemi. 2009. "The Amusic Brain: In Tune, Out of Key, and Unaware." *Brain* 132: 1277–1286.

Phillips-Silver, Jessica, and Laurel. J. Trainor. 2005. "Feeling the Beat: Movement Influences Infant Rhythm Perception." *Science* 308: 1430.

Phillips-Silver, Jessica, and Laurel J. Trainor. 2007. "Hearing What the Body Feels: Auditory Encoding of Rhythmic Movement." *Cognition* 105: 533–546.

Phillips-Silver, Jessica, and Laurel J. Trainor. 2008. "Vestibular Influence on Auditory Metrical Interpretation." *Brain and Cognition* 67: 94–102.

Popescu, Mihai, Asuka Otsuka, and Andreas A. Ioannides. 2004. "Dynamics of Brain Activity in Motor and Frontal Cortical Areas during Music Listening: A Magnetoencephalographic Study." *Neuroimage* 21: 1622–1638.

Pratt, Carroll C. 1930. "The Spatial Character of High and Low Tones." *Journal of Experimental Psychology* 13: 278–285.

Prinz, Wolfgang. 1984. "Modes of Linkage between Perception and Action." In *Cognition and Motor Processes*, edited by Wolfgang Prinz and Andries F. Sanders, 185–193. Berlin: Springer.

Prinz, Wolfgang. 1997. "Perception and Action Planning." *European Journal of Cognitive Psychology* 9: 129–154.

Proctor, Robert W., and Kim-Phuong L. Vu. 2006. *Stimulus-Response Compatibility Principles: Data, Theory, and Application*. Boca Raton, FL: CRC Press.

Quinto, Lena, William F. Thompson, Frank A. Russo, and Sandra E. Trehub. 2010. "A Comparison of the McGurk Effect for Spoken and Sung Syllables." *Attention, Perception, and Psychophysics* 72: 1450–1454.

Rusconi, Elena, Bonnie Kwan, Bruno L. Giordano, Carlo Umilta, and Brian Butterworth. 2006. "Spatial Representation of Pitch Height: The SMARC Effect." *Cognition* 99: 113–129.

Schutz, Michael. 2007. "Do We Hear Silent Gestures?" Paper presented at the 153rd Meeting of the Acoustical Society of America, Salt Lake City, Utah, June 4–8.

Schutz, Michael, and Michael Kubovy. 2006. "Seeing Music: The Percussionists' Strategic Use of Gesture in Live Performance." Paper presented at the 2nd International Conference on Music and Gesture, Royal Northern College of Music, Manchester, UK, July 20–23.

Schutz, Michael, and Scott Lipscomb. 2007. "Hearing Gestures, Seeing Music: Vision Influences Perceived Tone Duration." *Perception* 36: 888–897.

Shayan, Shakila, Ozge Ozturk, and Mark A. Sicoli. 2011. "The Thickness of Pitch: Crossmodal Metaphors in Farsi, Turkish, and Zapotec." *The Senses and Society* 6: 96–105.

Spence, Charles. 2011. "Crossmodal Correspondences: A Tutorial Review." *Attention, Perception and Psychophysics* 73: 971–995.

Stewart, Lauren, Rinus G. Verdonschot, Patrick Nasralla, and Jennifer Lanipekun. 2012. "Action-Perception Coupling in Pianists: Learned Mappings or Spatial Musical Association of Response Codes (SMARC) Effect?" *Quarterly Journal of Experimental Psychology* 66: 37–50.

Stone, Ruth M. 1981. "Toward a Kpelle Conceptualization of Music Performance." *Journal of American Folklore* 94: 188–206.

Stumpf, Carl. 1883. *Tonpsychologie*. Vol. 1. Leipzig: S. Hirzel.

Taylor, J. Eric T., and Jessica K. Witt. 2014. "Listening to Music Primes Space: Pianists, but Not Novices, Simulate Heard Actions." *Psychological Research* 79: 175–182.

Thompson, William F., Philip W. Graham, and Frank A. Russo. 2005. "Seeing Music Performance: Visual Influences on Perception and Experience." *Semiotica* 156: 203–227.

Thompson, William F., Frank A. Russo, and Steven R. Livingstone. 2010. "Facial Expressions of Singers Influence Perceived Pitch Relations." *Psychonomic Bulletin and Review* 17: 317–322.

Tillmann, Barbara, Katrin Schulze, and Jessica M. Foxton. 2009. "Congenital Amusia: A Short-Term Memory Deficit for Non-Verbal, but Not Verbal Sounds." *Brain and Cognition* 71: 259–264.

van Wassenhove, Virginie, Ken W. Grant, and David Poeppel. 2005. "Visual Speech Speeds up the Neural Processing of Auditory Speech." *Proceedings of the National Academy of Sciences of the United States of America* 102: 1181–1186.

Wagner, Sheldon, Ellen Winner, Dante Cicchetti, and Howard Gardner. 1981. " 'Metaphorical' Mapping in Human Infants." *Child Development* 52: 728–731.

Walker, Peter, J. Gavin Bremner, Uschi Mason, Jo Spring, Karen Mattock, Alan Slater, and Scott P. Johnson. 2010. "Preverbal Infants' Sensitivity to Synaesthetic Cross-Modality Correspondences." *Psychological Science* 21: 21–25.

Williamson, Victoria J., and Lauren Stewart. 2010. "Memory for Pitch in Congenital Amusia: Beyond a Fine-Grained Pitch Discrimination Problem." *Memory* 18: 657–669.

Zatorre, Robert J., Joyce L. Chen, and Virginia B. Penhune. 2007. "When the Brain Plays Music: Auditory-Motor Interactions in Music Perception and Production." *Nature Reviews Neuroscience* 8: 547–558.

Zendel, Benjamin R., Marie-Elaine Lagrois, Nicolas Robitaille, and Isabelle Peretz. 2015. "Attending to Pitch Information Inhibits Processing of Pitch Information: The Curious Case of Amusia." *Journal of Neuroscience* 35: 3815–3824.

CHAPTER 22

MULTIMODAL MUSIC IN INFANCY AND EARLY CHILDHOOD

SANDRA E. TREHUB

Music is invariably difficult to define even though we recognize it when we hear it, at least for examples from our own culture. Brandt, Gebrian, and Slevc (2012) offer an intriguing definition of music as "creative play with sound" (12). The definition is reasonable, but it does not go far enough in recognizing the centrality of the body to musical experiences. An altered and more nuanced definition, especially for the early years, is "creative play with sound and body." Accordingly, we recognize something as music or not when we see and hear it, but often when we feel it as well. In this chapter, music in infancy and early childhood is examined through this lens, that is, from the experiences of the central characters—infants, preschoolers, and children in the early school years—and the consequences for those characters at home, at school, and in the laboratory.

Infants

The period of infancy, extending from birth to two years of age, has much to reveal about musical exposure, predispositions, and learning as well as developmental changes in the perception and production of music and movement. Musical engagement in infancy also has implications for social and emotional regulation.

Exposure to Music

Infants from developing countries are generally carried in slings or other devices throughout the day, and they sleep alongside their mother at night. One consequence of ubiquitous contact between mother and infant is that the infant hears whatever the mother hears, in other words, the full range of musical genres in their community. These mothers also sing directly to infants, like mothers everywhere, except that they typically do so without maintaining face-to-face contact. Nevertheless, the infants in question experience maternal singing in conjunction with physical contact, warmth, and maternal movement, which jointly increase their composure and reduce their crying relative to infants in the developed world. In some communities, cradles are passed from one generation to the next, so that infants experience simultaneous rocking and lulling. In these developing societies, caregivers typically sing lullabies to infants (Trehub and Trainor 1998), and these lullabies are readily distinguished from other song forms even when the musical culture in unfamiliar (Mehr et al., 2018; Trehub, Unyk, and Trainor, 1993).

Although infants in these circumstances experience considerable passive movement from maternal walking, working, vocalizing, or rocking, their self-generated movement is constrained by the holding or carrying devices in which they spend much of their time. Once they begin walking, however, they typically have more freedom of movement than infants from developed countries. In any case, infants' movement, whether active or passive, influences their perception of music (Phillips-Silver and Trainor 2005), as it does for adults.

Our knowledge of infants' musical experiences and responsiveness comes largely from developed countries. In contrast to their peers in developing countries, infants in developed countries typically have limited exposure to the general community and therefore to the adult world of music. Their waking time is spent primarily in face-to-face interactions with mother, often without touch. In general, they sleep in their own bed (crib), often in a separate room. Although infants are dependent on mothers (or primary caregivers) in most respects, mothers in developed countries, especially in North America, promote infants' independence in a variety of ways. For example, they take pride in infants who can fall asleep without maternal interventions like rocking, patting, or singing. Infants' failure to achieve such milestones of independence leads some mothers to consult infant sleep specialists for guidance in promoting self-soothing and independent sleep. Nevertheless, parents often engage in presleep routines that include lullaby singing to mark the transition from play to rest.

During infants' waking hours, mothers talk to them in a highly expressive manner that incorporates numerous musical elements—distinctive pitch contours and rhythms, sustained vowels, and a large dynamic range (Fernald 1991). When mothers sing to infants, they typically select lively play songs rather than lullabies, performing them at a slightly higher pitch level, slightly slower tempo, and with greater temporal regularity than their singing in other contexts (Trehub and Trainor 1998; Trehub et al. 1997). Mothers also smile almost continuously when singing to infants but only intermittently when talking to them (Trehub, Plantinga, and Russo 2016). At times, they sway from side to side or bob their head in time to the music, enabling infants to observe coordinated movement to music even if they are not experiencing active or passive movement. At other times, mothers grasp infants' hands or waist and move or bounce them to the beat. Remarkably, mothers' multimodal performances of specific songs are repeated almost identically on different occasions (Bergeson and Trehub 2002), so that they become much like rituals. These stereotyped performances allow infants to forge links among the auditory, facial, and gestural aspects of each song.

Infants contribute indirectly to the expressiveness of maternal performances. Mothers' affective state and vocal tone are influenced by the infant's presence and degree of engagement. For example, when an opaque screen is inserted between mother and infant, maternal singing and speech become less expressive (Trehub, Plantinga, and Russo 2016). Perhaps it is not surprising, then, that mothers and fathers are unable to produce fully realized simulations of infant-directed singing in the infant's absence (Trehub et al. 1997). Such infant-absent performances lack the warmth and expressiveness of face-to-face performances.

In addition to live maternal singing, which is infants' predominant musical experience in the developed world, they also hear children's recordings, usually songs with instrumental accompaniment (Young 2008). Mothers may dance or sing along with these recordings, perhaps while holding infants, transforming a potentially passive, unimodal event into an active, multimodal event.

Responsiveness to Infant-Directed Singing

Infants are more attentive to speech and singing in the maternal style (i.e., infant-directed) than to conventional speech and singing, even with auditory-only stimuli (Trainor 1996; Fernald 1985). The heightened expressiveness and positive affect of maternal vocalizations are thought to underlie their attention-eliciting effects (Singh, Morgan, and Best 2002). Interestingly, audio-only versions of infant-directed singing are no more effective than infant-directed speech in capturing infants' attention (Corbeil, Trehub, and Peretz 2013). With audiovisual renditions (videos), however, infants are considerably more engaged by maternal singing than by maternal speech (Nakata and Trehub 2004). Their rapt attention to singing is evident from their extended fixation on the mother's face as well as their reduced body movement as they watch and listen. Not surprisingly, live, multimodal singing is even more effective than bimodal singing at capturing infants' attention and regulating their arousal levels (Ghazban 2013; Shenfield, Trehub, and Nakata 2003).

Although audio-recordings of infant-directed singing are no more effective than infant-directed speech for *capturing* infant attention, they are considerably more effective than infant-directed speech for *maintaining* infants' attention and composure. In one study, English-learning infants who listened to recordings of an unfamiliar Turkish play song remained calm and content for over twice as long (9 minutes) as infants who listened to the lyrics of the song spoken in a playful, infant-directed manner (4 minutes) (Corbeil, Trehub, and Peretz 2016). In another study, live, multimodal maternal singing was much more effective than multimodal maternal speech in reducing the arousal level and negative affect of infants who had been exposed to a distress-induction procedure

(Ghazban 2013). The salutary effects of infant-directed singing become increasingly evident as the listening time is extended.

When music is captivating, as is the case for live maternal singing or audiovisual depictions of maternal singing, infant movement is often suppressed, at least initially, by absorption in the performance (Nakata and Trehub 2004). From about five months of age, audio-recordings of rhythmic music, whether vocal or instrumental, tend to provoke rhythmic movement, but that movement is not synchronized with the music (Zentner and Eerola 2010). Although infants move, to a limited extent, to Baby Einstein recordings, their movement in such circumstances is not rhythmic, perhaps because of the inexpressiveness of the synthesized renditions. Fortunately, the once popular Baby Einstein materials fell out of favor after scientists debunked the Disney Corporation's promise of enhanced cognitive development. In a rare acknowledgment of their misleading claims, the corporation provided a public apology and refunds.

From the time infants are six or seven months of age, mothers focus increasingly on action songs such as "Itsy Bitsy Spider," "The Wheels on the Bus," and "Two Little Dickie Birds." Infants reproduce some of the actions of familiar maternal songs (e.g., the corresponding hand movements for "Fly Away Peter, Fly Away Paul") well before they can reproduce any vocal features. After infants' first birthday, mothers often encourage them to participate vocally by pausing at the end of each line (e.g., *Twinkle, Twinkle, Little* ___) so that infants can add the missing word or some approximation to it (e.g., *star*).

Social and Emotional Implications

Multimodal maternal singing is a means of emotion-sharing, regulating infant arousal, and intensifying dyadic bonds. Because mothers use a small repertoire of children's songs and perform them nearly identically on different occasions (Bergeson and Trehub 2002), each song becomes identified with the mother, specific contexts of singing, and positive affect. As a result, her singing is particularly effective in ameliorating distress, perhaps because of the comfort afforded by highly predictable performances of highly familiar songs (Ghazban 2013). Even when unfamiliar women sing to five-month-old infants, they are regarded more favorably if they sing a song that the mother performed repeatedly in the immediately preceding two weeks than if they sing an unfamiliar song (Mehr, Song, and Spelke 2016). Song familiarity is not the critical factor. Instead, it is the special significance accorded to songs sung regularly by the mother. When infants are familiarized with the same song for the same period by other means, such as a woman singing and interacting via Skype, they are no more interested in a woman who sings the familiar song than in another woman who sings an unfamiliar song. There is no failure of learning on the part of infants because they exhibit long-term memory for the song sung by the Skype singer.

After their first birthday, infants increasingly engage in prosocial behavior, offering help to others who need assistance. For example, fourteen-month-olds retrieve more "accidentally" dropped objects for an unfamiliar woman if she sings a song from the mother's repertoire than if she sings another song (Cirelli and Trehub, 2018).

Synchronous movement to music also increases prosocial behavior in adults and in preschool children (Kirschner and Tomasello 2009). Although infants move to music, as noted, they are unable to synchronize their movements with the music (Zentner and Eerola 2010) or with the movements of others. Nevertheless, when fourteen-month-olds are bounced to music in synchrony with a visible, bouncing adult, they subsequently offer more help to the adult than do infants whose bouncing was out of synchrony with the adult (Cirelli, Einarson, and Trainor 2014). Thus, synchronous movement to music has positive social consequences even when the movement is not self-initiated.

Other Cross-Modal Links

Adults commonly move to music, and their movement influences their perception of its rhythmic structure. That is also the case for infants. When infants are bounced on every second beat to a drumming pattern with an ambiguous rhythm (i.e., no accented notes), they perceive the pattern in duple meter (Phillips-Silver and Trainor 2005). When bounced on every third beat of the same pattern, they perceive it in triple meter. Moving to music, especially lively music, is a joyous activity at all ages. Although rudimentary dancing is evident before infants' first birthday (Zentner and Eerola 2010), dancing increases in regularity and vigor in subsequent months. As is the case for singing, dancing is an important self-regulatory activity.

Like adults, infants perceive links between auditory and visual cues to the identity of unfamiliar singers. After listening to an audio-only sample of infant-directed singing, infants look longer at a silent video of the previously heard singer than that of another singer, which parallels adults' ability to match the sounds of singing to silent videos of singers (Trehub et al. 2013).

Musical Activities Outside the Home

In recent years there has been a proliferation of parent-infant music programs that feature musical games, group singing, movement to music, and, at times, percussion instruments. These classes are enjoyable for parents and infants, but the more controversial claim is that they enhance infants' cognitive, social, and musical development. Such claims must be considered with caution because the families who participate in musical or other enrichment programs generally have higher levels of education and income compared to families who do not choose to participate. It is impossible to establish that these programs *cause* gains in musical or other skills unless infants are randomly assigned to the programs and, ideally, compared with infants who participate in alternative programs.

In one of the few infant intervention studies with random assignment and an alternative intervention, six-month-olds participated in a six-month program with multimodal musical experiences or an alternative program with nonmusical play and passive musical exposure (i.e., children's music as background). Infants who completed the multimodal musical program showed enhanced perception of Western tonality as well as enhanced social interaction and communication relative to those who completed the program with play and passive listening (Gerry, Unrau, and Trainor 2012). In another study, nine-month-olds were randomly assigned to a four-week musical intervention featuring synchronous movement to music or a four-week play intervention without music (Zhao and Kuhl 2016). Infants who participated in the multimodal musical intervention exhibited enhanced temporal processing of speech and music relative to infants who participated in the nonmusical intervention.

Laboratory Perspectives

With the exception of the small body of research on maternal singing and infants' responsiveness to such singing, the laboratory study of music in infancy has focused largely on pattern discrimination. Despite the fact that infants' everyday life is dominated by vocal music in multimodal contexts, the stimuli in most laboratory settings are nonvocal and unimodal, usually involving synthesized instrumental patterns or even sine-wave patterns.

Despite its lack of ecological validity, such laboratory research has revealed infants' sensitivity to subtle changes in pitch, timing, and timbre and to relational aspects of pitch and timing patterns (Trehub 2003). Like adults, infants perceive the invariance of a melody across changes in key or pitch level. As a result, they recognize "Twinkle, Twinkle, Little Star" or any other familiar tune at different pitch levels and tempos even though they also remember the pitch level at which they usually hear it (Volkova, Trehub, and Schellenberg 2006). Similarly, they recognize specific rhythmic patterns or groupings even when the tempo changes. They also exhibit long-term memory for recordings of vocal or instrumental melodies heard at home, although they remember more details from vocal than from instrumental recordings (Volkova et al. 2006). Infants' memory is especially enduring for songs experienced multimodally (Mehr et al. 2016).

Infants are sensitive to various culture-general aspects of music, but knowledge of culture-specific aspects depends on exposure and learning. At times, the absence of such knowledge leads infants to outperform their adult counterparts on melody and rhythm discrimination tasks. For example, unlike adults, who more readily detect an altered note in a melody if the note goes outside the key rather than remaining within it, infants are equally adept at detecting in-key and out-of-key changes (Trainor and Trehub 1992). In fact, musically untrained adults fail to notice a four-semitone change when the change remains within the key and implied harmony of the melody, but infants readily notice such a change. Similarly, although adults detect subtle tuning changes in the context of the familiar major scale, they fail to notice comparable changes in the context of an unfamiliar, invented scale. Infants perform equally well in both contexts (Trehub, Schellenberg, and Kamenetsky 1999).

In the temporal domain, Western adults detect timing changes in the context of music with simple meters, which prevail in Western music, but not in the context of music with complex meters, which are present in other cultures. Infants perform equally well in both contexts (Hannon and Trehub 2005a). By twelve months of age, however, Western infants experience difficulty with complex meters, which implies attunement to culture-specific rhythms (Hannon and Trehub 2005b). Limited exposure to music with complex meter reverses infants' difficulty but not that of adults (Hannon and Trehub 2005b), which implies that infants have greater perceptual flexibility than adults. It is possible that adults could overcome their difficulty with complex meters if they moved to the music while listening.

Preschoolers

During the preschool period, which extends from two to five years of age, children experience considerable expansion in their musical and social world. They also exhibit notable gains in singing and dancing proficiency.

The Home Front

Mothers continue to sing to preschoolers, but their singing is less frequent than it was during infancy. Instead of performing *for* children, as before, they perform *with* them, often engaging in exuberant action songs such as "If You're Happy and You Know It," with joyful hand-clapping and foot-stomping. Preschoolers initiate considerable independent singing, often while playing with toys, moving around a room, and in the solitary moments before they fall asleep (e.g., Sole, 2017). At such times, they sing conventional songs, often with improvised words or lyrics, or invented songs with real or nonsense words. Their rhythm and tempo are typically coordinated with the movement patterns of their ongoing activities.

Preschoolers' live, multimodal musical experiences are supplemented by recorded music from CDs (intended for them or overheard), children's television programs, and music emanating from their parents' cell phones or tablets, the latter two sources often providing visual as well as auditory input. Listening to recorded music, with or without visuals, accounts for about a third of children's musical episodes and is generally an active rather than passive experience (Lamont 2008). Preschoolers tend to sing along, jump, dance, clap, or engage in other forms of playful activity while listening to familiar recordings. They develop enduring affection for the cartoon characters from their favorite television programs (e.g., *Dora the Explorer, Backyardigans*, and *Bob the Builder*), singing along with the theme songs and singing those songs at other times as well.

Nursery School and Day Care Settings

Informal musical activities play an important role in nursery schools and day care centers. In addition to group singing, music is usually part of movement activities and games in such group settings. Songs are also used to scaffold learning about seasons, holidays, emotions, letters of the alphabet, numbers, and so on. The lyrics of these songs provide preschoolers with a repository of knowledge that they can access, as needed, in future months or years (e.g., *What letter comes after L?*).

In principle, children's frequent bouts of singing in preschool settings could provide rich information about the early developmental course of singing. In practice, however, their improvisations and song inventions preclude accurate assessment of their abilities. According to some scholars, preschoolers' renditions of conventional songs focus on words rather than tunes and exhibit a compressed pitch range (Welch, Sergeant, and White 1998). Other scholars contend, however, that preschoolers produce recognizable tunes with appropriate pitch range, contours, and rhythm (Gudmundsdottir and Trehub 2018; Trehub and Gudmundsdottir 2015). Interestingly, preschoolers' singing seems most "deficient" when they sing songs on request, especially songs learned at school, and they perform them in unfamiliar contexts (e.g., a quiet room near their classroom). By contrast, their performances are most proficient when they sing spontaneously at home. In contrast to the former "test" context, the latter context is familiar and comfortable, with children choosing what to sing, when to sing it, and being free to move and play while they sing.

In preschool classrooms, children's joint music and movement activities enhance mood, social bonds, and prosocial behavior (Kirschner and Tomasello 2010). Social contexts also enhance preschoolers' ability to synchronize their movements to music. For example, when preschoolers engage in drumming to instrumental music, their drum beats exhibit greater synchronization with the music in the presence of a social partner (Kirschner and Tomasello 2009).

Beyond teacher-directed musical activities in preschool settings, preschoolers spontaneously produce solos, usually with improvised or invented tunes and lyrics, and often with the goal of capturing the attention of their classmates. At times, their lyrics include taboo words (e.g., "poo"), much to the amusement of playmates and the chagrin of teachers. Teachers are inclined to ignore these scatological songs for fear of drawing further attention to them and emboldening the singer.

Music preferences influence young adults' patterns of socialization, especially in relation to new acquaintances. For example, they are more inclined to pursue friendships with those who share their musical tastes, as revealed by playlists on cell phones. In a similar vein, preschoolers indicate greater interest in socializing with children whose favorite songs are familiar rather than unfamiliar (Soley and Spelke 2016).

Music Training

Some children begin music training in the preschool period, with lessons commonly occurring in small groups rather than individually. The social context makes learning more palatable and perhaps more efficient than it might be otherwise. The coordinated sounds and movements of participants also strengthen interpersonal connections. Suzuki and Yamaha programs feature highly structured routines and daily at-home practice to supplement in-class learning. These programs focus as much on ear training as on instrument mastery and often use a fixed-*do* rather than movable-*do* system. Early onset of training, in conjunction with a fixed-do system, increases the likelihood that children will acquire absolute pitch (Wilson et al. 2012). There is no evidence, however, that early onset of training (i.e., in the preschool period) yields better long-range outcomes in terms of musical proficiency or overall duration of training. Nevertheless, there is suggestive evidence that early music lessons have long-term consequences for fine-motor skills (see Lennard's chapter for more on this subject).

Parents seek music training for preschool children for a variety of reasons, including their own or their child's keen interest in music, their interest in arts education more generally, or the presumed nonmusical benefits of training. Unfortunately, studies that reveal nonmusical benefits of music training tend to receive more publicity than those that fail to find such benefits. In any case, *informal* musical activities at home may be as effective as *formal* activities in enhancing musical knowledge and generating nonmusical benefits (Putkinen, Saarikivi, and Tervaniemi 2013). In non-Western countries, musical skills are often acquired informally by peer mentoring. In Bali, for example, preschool girls learn to dance, and boys learn to play musical instruments, not from experts or parents but from older children (Dunbar-Hall 2011). They learn holistically by copying the behaviors of others.

Laboratory Perspectives

Laboratory music studies with preschoolers, like those with older children and adults, focus largely on perception, usually with unimodal stimuli (i.e., synthesized auditory patterns) presented in nonsocial environments. In such situations, it is unclear whether the observed limitations stem from preschoolers' insensitivity to musical features or from stimuli and contexts that are not ecologically valid. In one study, for example, adults judged the happiness or sadness of synthesized classical excerpts varying in tempo, mode (major, minor), or both (Dalla Bella et al. 2001). Not surprisingly, adults' judgments were influenced by tempo and mode, with tempo playing a greater role. When children were required to respond to the musical excerpts by pointing to one of two schematic faces depicting happiness (laughing) or sadness (crying), four-year-olds responded randomly, five-year-olds were influenced by tempo but not mode, and six-year-olds were influenced by tempo and mode. A subsequent study that replaced the classical music excerpts with children's songs (instrumental renditions) found that four-year-olds were indeed capable of judging music with fast tempo as happy and with slow tempo as sad (Mote 2011). In other words, four-year-olds succeeded in discerning the emotional intentions of music only in the context of music that was meaningful for them, whether that music was familiar or unfamiliar.

The association of the major mode with happiness and the minor mode with sadness is not universal, so like other musical conventions, it must be acquired. Nevertheless, there is considerable variability in adults' interpretation of the expressive intentions of music. For example, musical pieces considered sad by some may be considered soothing or peaceful by others. In the case of four-, five-, or six-year-olds, it is unlikely that they would label music as sad in a task that involved open-ended responding (e.g., *What feeling does the music have? How does it make you feel?*). Children's choice of "sad" in two-choice contexts may be akin to judgments of "not happy." In live, multimodal contexts, gestures of face, limbs, and torso influence listeners' judgments of musical emotions.

Children's performances of songs provide insight into their understanding of emotional expression in music. For example, children as young as four use a range of vocal and gestural devices when they sing a familiar song to make their listeners feel happy or sad (Adachi and Trehub 1998). Specifically, they sing louder, faster, and at a higher pitch level for their happy versions than for their sad versions. They also smile and have an upright bearing during their happy renditions, in contrast to frowns and a slumping posture during their sad renditions. These expressive differences are not unique to music, being relevant to the expression of happiness or sadness in speech. However, preschoolers are much more likely to move their torso and limbs when singing than while speaking, which underlines the importance of multimodal influences in song and other forms of music.

Preschoolers' preferred tempo for music is faster than that of older children and adults. Their preferred tempo influences their perception of various musical features as well as the accuracy with which they can synchronize their movements to music. Older children and adults differ not only in having a slower preferred tempo but also in synchronizing their movements over a wider tempo range (McAuley et al. 2006). Unfortunately, entrainment to music is typically assessed in laboratory settings by finger-tapping to musical patterns. It is possible that young children's entrainment skills are underestimated by the use of stimuli (e.g., synthesized patterns) and tasks (e.g., finger-tapping) that lack ecological validity. Even for older children and adults, toe-tapping, clapping, dancing, and walking are much more natural than finger-tapping. As noted, moreover, preschoolers exhibit better synchrony in social contexts than in nonsocial contexts.

The Early School Years

Important shifts in musical and social influences occur in the early school years, with peers playing an increasingly important role.

The Home Front

Once children begin primary school, their lives become more autonomous than they were previously. Although they participate in family musical activities, their musical tastes are increasingly influenced by their peers. In general, second-graders do not have personal listening devices, but they exhibit considerable interest in the pop idols of older children, and they may have online access to such recordings. Among their valued activities with friends are listening, singing, and dancing to their favorite pop and hip-hop artists.

Classroom and Schoolyard

In general, young schoolchildren are profoundly interested in the music of their peer group, but they are often bored by classroom musical selections (e.g., classical recordings) and by passive activities such as lessons in music theory or music appreciation. Group singing, which engages children's bodies, minds, and ears, evokes considerably greater enthusiasm.

In the schoolyard—before school begins, at recess, and during lunch breaks—children engage in group musical activities that are joyful and creative (Countryman 2014; Marsh 2009). For example, girls improvise traditional playground chants and songs in conjunction with various games (e.g., *My name is A___, and my sister's name is A___, and we come from A___ with a basket full of___; My name is B___, etc.*). They also participate in clapping games that feature intricate rhythms and recitations. Their expertise in these rhythmic clapping games is at odds with many music educators' claims about children's limitations in rhythm production or comprehension.

Boys and girls also use counting-out rhymes, traditional and novel, when they choose someone to join or leave a group activity or decide who will be "it" in a game of tag (e.g., *Eeny, Meeny, Miney, Mo; Duck, Duck, Goose; Tinker, Tailor*). These chants, songs, and rhymes are belted out cheerfully, with dramatic gestures. Children jump for joy if the consequences are favorable (e.g., chosen to be "it"), and they groan if the consequences are unfavorable (e.g., eliminated from a game). The stark contrast between children's enthusiasm and skill level on the playground and in the classroom should give pause to music educators or those who develop music curricula.

Laboratory Perspectives

In the early school years, children exhibit further gains in sensitivity to Western tonality and rhythm and to conventions of emotional expressiveness. Even without training, they are sensitive to harmony as well as key structure (Corrigall and Trainor 2014). Just as twelve-month-old infants learn to discern foreign rhythms more readily than adults (Hannon and Trehub 2005b), young school-age children do so as well (Hannon, Vanden Bosch der Nederlanden, and Tichko 2012), indicating their continuing perceptual flexibility. They consider mode as well as tempo when judging musical emotions even though they are less consistent than adults in their judgments (Dalla Bella et al. 2001). Perhaps adults are less disadvantaged than children by the synthesized stimuli in these laboratory studies. With samples of children's singing as stimuli, children are more accurate than adults at identifying the intended emotion (Adachi, Trehub, and Abe 2004). Children's greater accuracy may be attributable to the engaging and ecologically valid stimuli and to their greater insight into same-age children's manner of expression.

With few exceptions, like the aforementioned studies of children's singing, research with children has used synthesized instrumental stimuli. It is clear, however, that children seven and older remember melodies better when they are presented vocally (sung to *la la*) rather than instrumentally (Weiss et al. 2015). Younger children do not exhibit a comparable vocal memory advantage, but they tend to judge vocal melodies as familiar even when hearing them for the first time.

Cross-cultural research has the potential to shed light on the impact of divergent cultural practices and values on children's perception and production of music. To date, however, few studies have compared the musical abilities of children across cultures. In one such comparison, Canadian listeners were more accurate at identifying children's happy renditions of songs than sad renditions, and Japanese listeners were more accurate at identifying children's sad renditions than happy renditions (Adachi et al. 2004). These cross-cultural differences may be explained, in part, by the Japanese emphasis on sensitivity to the suffering of others. In many other respects, one would expect differences in music perception and production in cultures in which participatory and inclusive musical activities are dominant as opposed to the Western demarcation between performers (a small minority) and listeners (the overwhelming majority).

Conclusion

The definition of music as creative play with sound and body, building on an earlier definition centered on sound alone (Brandt, Gebrain, and Slevc 2012), seems particularly appropriate for infants and young children. Music is at the foreground of children's lives, in contrast to the background role that it plays for Western adults most of the time. Although music and movement are linked for adults, they are inseparable for infants and young children. Mothers smile and move as they sing to infants, forging indelible connections across modalities as well as enduring emotional and social connections.

The affect-regulatory aspects of maternal music, arousing at times and soothing at other times, lay the foundation for lifelong, self-regulation strategies in which music plays a central role. Similarly, the social-regulatory aspects of maternal music pave the way for musical bonding and social cohesion among preschoolers and young school-age children.

Although mothers produce stereotyped renditions of songs for infants and toddlers, music-making in young children is marked by innovation in sound and movement along with boundless enthusiasm. Their playful and creative approach to music remains evident in the early school years, mainly in playgrounds rather than classrooms, where music often sheds its natural multimodality.

Acknowledgments

The preparation of this chapter was assisted by the Social Sciences and Humanities Research Council, AIRS (Advanced Interdisciplinary Research in Singing) and the Natural Sciences and Engineering Research Council of Canada.

References

Adachi, Mayumi, and Sandra E. Trehub. 1998. "Children's Expression of Emotion in Song." *Psychology of Music* 26: 133–153.

Adachi, Mayumi, Sandra E. Trehub, and Jun-Ichi Abe. 2004. "Decoding Emotion in Children's Songs across Age and Culture." *Japanese Psychological Research* 46: 322–336.

Bergeson, Tonya R., and Sandra E. Trehub. 2002. "Absolute Pitch and Tempo in Mothers' Songs to Infants." *Psychological Science* 13: 71–74.

Brandt, Anthony, Molly Gebrian, and Robert L. Slevc. 2012. "Music and Early Language Acquisition." *Frontiers in Psychology* 3: 327.

Cirelli, Laura K., Kathleen M. Einarson, and Laurel J. Trainor. 2014. "Interpersonal Synchrony Increases Prosocial Behavior in Infants." *Developmental Science* 17: 1003–1011.

Cirelli, Laura K., and Sandra E. Trehub. 2018. "Infants Help Singers of Familiar Songs." *Music & Science* 1: 1–11.

Corbeil, Mariève, Sandra E. Trehub, and Isabelle Peretz. 2013. "Infants Prefer Speech to Singing Only When Speech Sounds Happier." *Frontiers in Psychology* 4: 372.

Corbeil, Mariève, Sandra E. Trehub, and Isabelle Peretz. 2016. "Singing Delays the Onset of Infant Distress." *Infancy* 21: 373–391.

Corrigall, Kathleen, and Laurel J. Trainor. 2014. "Enculturation to Musical Pitch Structure in Young Children: Evidence from Behavioral and Electrophysiological Methods." *Developmental Science* 17: 142–158.

Countryman, June. 2014. "Start-up Games on School Playgrounds: Instances of Ceremonial Rituals." *Bulletin of the Council on Research in Music Education* 202: 7–27.

Dalla Bella, Simone, Isabelle Peretz, Luc Rousseau, and Nathalie Gosselin. 2001. "A Developmental Study of the Affective Value of Tempo and Mode in Music." *Cognition* 80: B1–B10.

Dunbar-Hall, Peter. 2011. "Children's Learning of Music and Dance in Bali: An Ethnomusicological View of the Cultural Psychology of Music Education." In *A Cultural Psychology of Music*, edited by Margaret S. Barrett, 17–40. New York: Oxford University Press.

Fernald, Anne. 1985. "Four-Month-Old Infants Prefer to Listen to Motherese." *Infant Behavior and Development* 8: 181–195.

Fernald, Anne. 1991. "Prosody in Speech to Children: Prelinguistic and Linguistic Functions." *Annals of Child Development* 8: 43–80.

Gerry, David, Andrea Unrau, and Laurel J. Trainor (2012). "Active Music Classes in Infancy Enhance Musical, Communicative and Social Development." *Developmental Science* 15: 398–407.

Ghazban, Niusha. 2013. "Emotion Regulation in Infants Using Maternal Singing and Speech." PhD diss., Ryerson University.

Gudmundsdottir, Helga, and Sandra E. Trehub. 2018. "Adults Recognize Toddlers' Song Renditions." *Psychology of Music* 46: 281–291.

Hannon, Erin E., and Sandra E. Trehub. 2005a. "Metrical Categories in Infancy and Adulthood." *Psychological Science* 16: 48–55.

Hannon, Erin E., and Sandra E. Trehub. 2005b. "Tuning in to Musical Rhythms: Infants Learn More Readily Than Adults." *Proceedings of the National Academy of Sciences USA* 102: 12639–12643.

Hannon, Erin E., Christina M. Vanden Bosch der Nederlanden, and Parker Tichko. 2012. "Effects of Perceptual Experience on Children's and Adults' Perception of Unfamiliar Rhythms." *Annals of the New York Academy of Sciences* 1252: 92–99.

Kirschner, Sebastian, and Michael Tomasello. 2009. "Joint Drumming: Social Context Facilitates Synchronization in Preschool Children." *Journal of Experimental Child Psychology* 102: 299–314.

Kirschner, Sebastian, and Michael Tomasello. 2010. "Joint Music Making Promotes Prosocial Behavior in 4-Year-Old Children." *Evolution and Human Behavior* 31: 354–364.

Lamont, Alexandra. 2008. "Young Children's Musical Worlds." *Journal of Early Childhood Research* 6: 247–261.

Marsh, Kathryn. 2009. *The Musical Playground: Global Tradition and Change in Children's Songs and Games*. New York: Oxford University Press.

McAuley, J. Devin, Mari R. Jones, Shayla Holub, Heather M. Johnston, and Nathaniel S. Miller. 2006. "The Time of Our Lives: Life Span Development of Timing and Event Tracking." *Journal of Experimental Psychology: General* 135: 348–367.

Mehr, Samuel A., Manvir Singh, Hunter York, Luke Glowacki, and Max M. Krasnow. 2018. "Form and Function in Human Song." *Current Biology* 28: 356–368.

Mehr, Samuel A., Lee A. Song, and Elizabeth S. Spelke. 2016. "For Five-Month-Old Infants, Melodies Are Social." *Psychological Science* 27: 486–501.

Mote, Jasmine. 2011. "The Effects of Tempo and Familiarity on Children's Affective Interpretation of Music." *Emotion* 11: 618–622.

Nakata, Takayuki, and Sandra E. Trehub. 2004. "Infants' Responsiveness to Maternal Speech and Singing." *Infant Behavior and Development* 27: 455–464.

Phillips-Silver, Jessica, and Laurel J. Trainor. 2005. "Feeling the Beat: Movement Influences Infant Rhythm Perception." *Science* 308: 1430.

Putkinen, Vesa, Katri Saarikivi, and Mari Tervaniemi. 2013. "Do Informal Musical Activities Shape Auditory Skill Development in Preschool-Age Children?" *Frontiers in Psychology* 4: 572.

Shenfield, Tali, Sandra E. Trehub, and Takayuki Nakata. 2003. "Maternal Singing Modulates Infant Arousal." *Psychology of Music* 31: 365–375.

Singh, Leher, James L. Morgan, and Catherine T. Best. 2002. "Infants' Listening Preferences: Baby Talk or Happy Talk?" *Infancy* 3: 365–394.

Sole, Meryl. 2017. "Crib Song: Insights into Functions of Toddlers' Spontaneous Singing." *Psychology of Music* 45: 172–192.

Soley, Gaye, and Elizabeth S. Spelke. 2016. "Shared Cultural Knowledge: Effects of Music on Young Children's Social Preferences." *Cognition* 148: 106–116.

Trainor, Laurel J. 1996. "Infant Preferences for Infant-Directed Versus Noninfant-Directed Playsongs and Lullabies." *Infant Behavior and Development* 19: 83–92.

Trainor, Laurel J., and Sandra E. Trehub. 1992. "A Comparison of Infants' and Adults' Sensitivity to Western Musical Structure." *Journal of Experimental Psychology: Human Perception and Performance* 18: 394–402.

Trehub, Sandra E. 2003. "The Developmental Origins of Musicality." *Nature Neuroscience* 6: 669–673.

Trehub, Sandra E., and Helga R. Gudmundsdottir. 2015. "Mothers as Singing Mentors for Infants." In *Oxford Handbook of Singing*, edited by G. F. Welch, J. Nix, and D. Howard. Oxford: Oxford University Press.

Trehub, Sandra E., Judy Plantinga, Jelena Brcic, and Magda Nowicki. 2013. "Cross-Modal Signatures in Maternal Speech and Singing." *Frontiers in Psychology* 4: 811.

Trehub, Sandra E., Judy Plantinga, and Frank A. Russo. 2016. "Maternal Vocal Interactions with Infants: Reciprocal Visual Influences." *Social Development* 25: 665–683.

Trehub, Sandra E., E. Glenn Schellenberg, and Stuart B. Kamenetsky. 1999. "Infants' and Adults' Perception of Scale Structure." *Journal of Experimental Psychology: Human Perception and Performance* 25: 965–975.

Trehub, Sandra E., and Laurel J. Trainor. 1998. "Singing to Infants: Lullabies and Play Songs." *Advances in Infancy Research* 12: 43–77.

Trehub, Sandra E., Anna M. Unyk, Stuart B. Kamenetsky, David S. Hill, Laurel J. Trainor, Joanna L. Henderson, and Myra Saraza. 1997. "Mothers' and Fathers' Singing to Infants." *Developmental Psychology* 33: 500–507.

Trehub, Sandra E., Anna M. Unyk, and Laurel J. Trainor. 1993. Adults Identify Infant-Directed Music across Cultures. *Infant Behavior and Development* 16: 193–211.

Volkova, Anna, Sandra E. Trehub, and E. Glenn Schellenberg. 2006. "Infants' Memory for Musical Performances." *Developmental Science* 9: 584–590.

Weiss, Michael W., E. Glenn Schellenberg, Sandra E. Trehub, and Emily Dawber. 2015. "Enhanced Processing of Vocal Melodies in Childhood." *Developmental Psychology* 51: 370–377.

Welch, Graham. F., Desmond C. Sergeant, and Peta J. White. 1998. "The Role of Linguistic Dominance in the Acquisition of Song." *Research Studies in Music Education* 10: 67–74.

Wilson, Sarah J., Dean Lusher, Catherine L. Martin, Genevieve Rayner, and Neil McLachlan. 2012. "Intersecting Factors Lead to Absolute Pitch Acquisition That Is Maintained in a Fixed *Do* Environment." *Music Perception* 29: 285–296.

Young, Susan. 2008. "Lullaby Light Shows: Everyday Musical Experience among Under-Two-Year-Olds." *International Journal of Music Education* 26: 33–46.

Zentner, Marcel, and Tuomas Eerola. 2010. "Rhythmic Engagement with Music in Infancy." *Proceedings of the National Academy of Sciences USA* 107: 5768–5773.

Zhao, T. Christina, and Patricia K. Kuhl. 2016. "Musical Intervention Enhances Infants' Neural Processing of Temporal Structure in Music and Speech." *Proceedings of the National Academy of Sciences USA* 113: 5212–5217.

CHAPTER 23

OPERA AS FILM

Multimodal Narrative and Embodiment

YAYOI U. EVERETT

Introduction

A successful opera draws on the synergy established between music, libretto, and staging. Yet a serious opera fan expects to be enthralled by the singing voice above and beyond the visual glamor of the production. This chapter explores how the advent of high-definition (HD) broadcasting—offering close-up shots of singers from multiple visual angles—has transformed the ways in which we attend to opera as film. How does the filmic technology—the camera angles, frames, and shot sequences—alter the viewer's experience of opera? How about the physical environment in which we view the opera, be it live transmission in a theater with other audience or in the privacy of our own home? Drawing on recent studies on film and multimedia, this chapter examines audio-visual congruence as a key to understanding the viewer's embodied response to staged opera in its remediated form. To this end, it refers to the concept of multimodality, Anabel Cohen's hierarchical model of narrative construction, and Mark Johnson's conceptual metaphor theory. Analytical case studies include selected scenes from Richard Wagner's *Die Walküre* (directed by Holten 2008 and LePage 2011), John Adams's *Doctor Atomic* (directed by Sellars 2006 and Woolcock 2007), and Kaija Saariaho's *Adriana Mater* (directed by Sellars 2006). The conclusion provides a summary of the distinctive types of audio-visual congruence established in these works and a discussion of further avenues for exploring opera and film through the lens of hermeneutical and cognitive research.

Multimodal Narrative and Theoretical Framework for Analysis

Productions of operas, both old and new, are continually shaped and transformed by new media and aesthetic trends. David Levin claims that opera has emerged as "an unsettled site of signification," requiring the audience to attend to a surfeit of competing systems for this reason (2007, 3). Each operatic production is unique in the sense that the multimedial elements—music, libretto, film, objects, lighting, mime, and/or dance—operate as interdependent structural and semantic elements that shape the narrative production of the whole. Linda and Michael Hutcheon introduce the concept of "multimodal narrative" to refer to the embodied action of viewing opera as follows: "the physical move from what theatre semiotician Keir Elam calls the 'dramatic texts'—or the *design*—to live performance with actual material bodies in actual physical space is an act of narrative *production* of both the verbal libretto and the musical score" (2009, 66). Hutcheon and Hutcheon argue that production adds to design the accumulation of meaning(s) and helps "organize the interpretive strategies of the audience members" (67). Furthermore, the DVD format or HD broadcast of a given production establishes a narrative angle through filmic devices (e.g., camera angle, shot sequence) that shape the viewer's multimodal experience in an entirely different way from viewing opera live in a theater.

From a cognitive perspective, psychologists have focused on cross-modal congruence between audio and visual domains as key to understanding the viewer's embodied response to film.[1] Scott D. Lipscomb, who has investigated accentual alignment of audio and visual domains in film, observes that when the audio and visual components are simple (e.g., coordination achieved via Mickey-mousing technique[2]), accentual alignment emerged as a key factor in perceived effectiveness, while in more complex multimedia contexts, the alignment between film and soundtrack could be out-of-phase without having a significant impact on the viewer's perception of effectiveness (2013, 206). Audio-visual congruence refers to the extent to which visual patterning of images lines up with auditory patterning of sounds. Lipscomb's study lays down an important foundation for modeling perceptual salience in film by distinguishing the semantic (associative) from structural (alignment) dimensions of audio-visual congruence. Building on these tenets, Anabel Cohen has developed a hierarchical framework for modeling the viewer's construction of narrative based on the interplay of multiple sensory domains.

As replicated in Figure 23.1, her congruence-association model (CAM) comprises five levels: level A lists the separate media sources that come into play (i.e., text, speech, visual, music, sound effects, and kinesthetic[3]); level B differentiates structural from semantic features associated with each domain; at level C, the composite information enters a region of short-term memory (called "working narrative"); at levels D and E, the viewer retrieves relevant information from long-term memory to shape the narrative.[4]

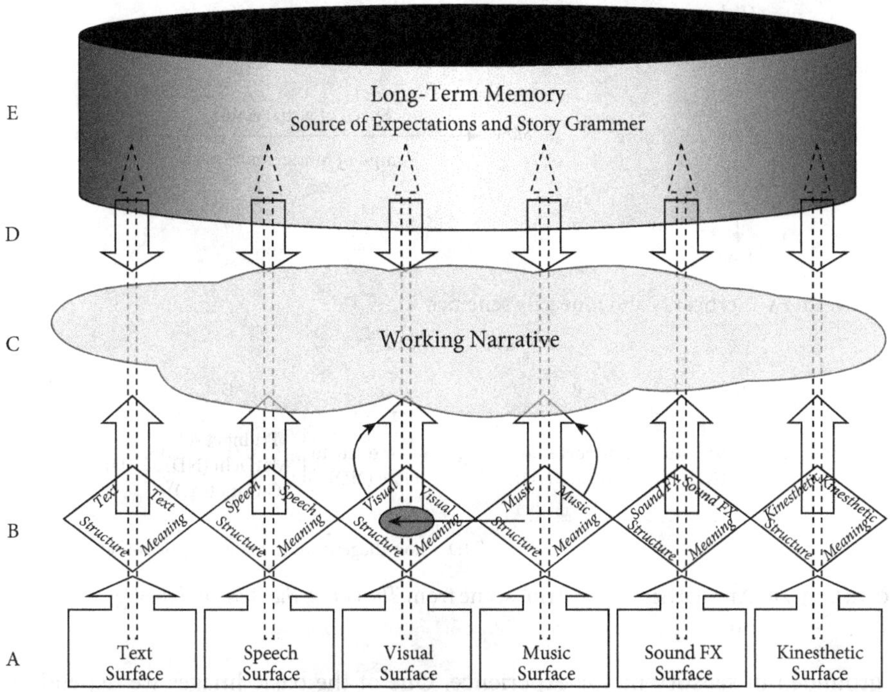

FIGURE 23.1 Cohen's Congruence-Association Model (CAM)

Reprinted from Anabel J. Cohen, "Film Music from the Perspective of Cognitive Science," in *Oxford Handbook of Film Music Studies*, edited by David Neumeyer (2014), Figure 5.3 (p. 120). By permission of Oxford University Press.

The lateral arrow in level B illustrates a case whereby the musical structure brings the visual domain into greater focus, as the dual arrows pointing from levels B to C enable this event to enter into the short-term memory of the viewer's working narrative. Cohen also remarks that mood congruencies established between the visual-kinesthetic and sonic domains enhance absorption and long-term memory (2009, 445). As a prime example, she proposes that a musical leitmotif can shape the viewer's long-term memory of narrative through audio-visual congruence established at the lower levels. In her analysis of *The Red Violin* (1998) she reveals how similarities established between film and music with respect to three kinds of musical structure—tonality, large-scale form, and motivic interconnections—shape the viewer's construction of narrative (2002, 215).

Another perspective is afforded by scholars who adopt Mark Johnson's conceptual metaphor theory (CMT) to illustrate analogical correspondences established between the visual and sonic domains in film and multimedia. Johnson claims that abstract concepts are defined by systematic mappings from bodily-based domains onto abstract target domains. Integral to CMT is the image schema, which Johnson defines as "a dynamic cognitive construct that functions somewhat like an abstract structure of an image and thereby connects together a vast range of different experiences that manifest this same recurring experience" (1987, 2). Maarten Coëgnarts and Peter Kravanja advance the notion that image schemas provide evidence for embodied cognition

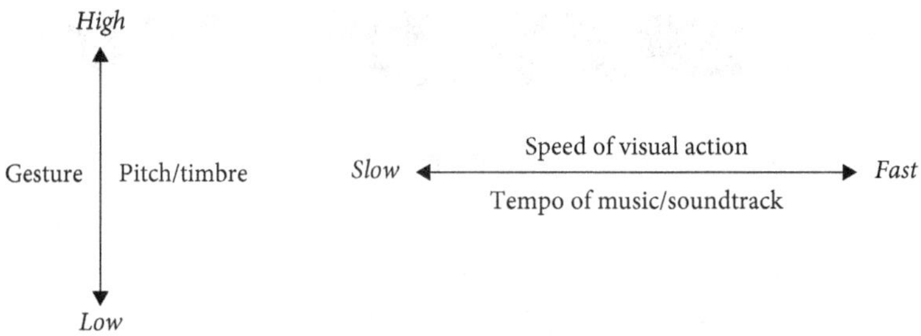

FIGURE 23.2A Verticality and Linearity schemas

A	B	C	A′	B′	C′
Guitar ostinato Flute (ND)	Trumpet solo Mariachi (ND)	(D)	Guitar ostinato Flute (ND)	Trumpet solo Mariachi (ND); Gunshots (D)	(D)

(ND = non-diegetic music; D = diegetic sound)

FIGURE 23.2B An analysis of the "Trio" scene from *The Good, the Bad, and the Ugly*

as grounded in sensory motor-experience. One of the basic images we depend on is conceiving time in terms of space: following the spatial logic of the SOURCE-PATH-GOAL schema, we constitute time as manifesting itself in movement with a starting point and an end (Coëgnarts and Kravanja 2015, 66). Other image schemas (e.g., CENTER-PERIPHERY) are similarly conceived as intrinsically dynamic, grounding us in basic experiential patterns.

Because both music and filmic images are organized hierarchically, an analogy between the visual and sonic domains can be drawn in multiple ways; for example, under the VERTICALITY schema, the pitch contour of high and low finds analogical correspondences with images that move up or down on the screen, while under the LINEARITY schema, the speed of visual or kinesthetic movement may be correlated with tempo of soundtrack (slow to fast), as shown in Figure 23.2a. Basic Mickey-mousing techniques are founded on such analogic correspondences to establish alignment and mood congruence. Lawrence Zbikowski, in his work on musical and physical gestures in film, remarks that "both music and gesture draw on a common pool of cognitive resources to create analogical representations of dynamic processes" (2011, 97). Additionally, Juan Chattah illustrates how psychological tension (low/high) finds analogical correspondences in pitch height (low/high), volume (soft/loud), and consonance and dissonance in David Shire's film scores for *Return to Oz* (1985) and *Short Circuit* (1986) (2015, 104–105).

Chattah also refers to Johnson's CONTAINER schema as a commonplace technique for filmmakers to structure scenes; he argues, for example, that Shire maps the transition between the two worlds in which the narrative takes place in *Return to Oz* by compartmentalizing audio and visual elements into alternating blocks (108–109). The three-way duel scene at the conclusion of Sergio Leone's *The Good, the Bad, and the Ugly* (1966) illustrates this type of audio-visual congruence par excellence. As shown in Figure 2c, while Enrico Morricone skillfully alternates between a sequence of rising guitar solo (A) and trumpet solo in mariachi style (B), Leone alternates between close-up shots and long shots of the three men in rhythmic alignment with the soundtrack. In the middle of this scene, the nondiegetic music stops and gives way to a diegetic sound of a crow croaking (C). Then the opening sequence recurs with heightened tension. The pacing of shots accelerates in the penultimate B′ section as the music speeds up and incorporates gunshot sounds in the nondiegetic music in anticipation of the shooting that takes place. The actual gunshots signal the end of the duel (C′). The audio-visual congruence based on the CONTAINER and LINEARITY schemas build suspense and tension in this concluding scene and in a manner that is distinctive to film.

The ensuing analysis of operatic scenes uses Johnson's image schemas to illustrate specific ways in which audio-visual congruence elicits the viewer's attention and embodied response. While Cohen maintains the position that music/soundtrack brings focus to visual images through cross-modal congruence, here it is proposed that music has the capacity to generate its own "sonic" imagery independent of the visual domain. In the contexts of opera and music theater, instrumental overtures and postludes are capable of shaping the meaning and expression of a particular scene for the viewers in absentia of visual information. Michel Chion calls this acousmatic music, which he defines as music without a visible source, and this chapter illustrates how the absence of visual elements allows music to generate its own imagistic content that shapes narrative (1994, 72). In such instances, the absence or erasure of diegetic sounds and dialogue plays a key role in eliciting an embodied response from the viewer. Let us now turn to three case studies to illustrate how these analytical tools explain the interpretive processes at work.

Case Study #1: Wagner's *Die Walküre* by Lepage (2011) and Holten (2008)

Robert LePage's staging of Richard Wagner's *Der Ring des Nibelungen* garnered attention from the press in 2012, primarily for the breakthrough technology of using a forty-five-ton set of twenty-four rotating planks with video projections. Inspired by the imagery of volcanic plates in Iceland, LePage (and his company Ex Machina) developed this machine so that, when used in combination with video projection, it could transform the stage into anything from a tangle of trees in a forest to a desolate mountain peak. In spite of critics who complained about the excess of technology as distractive, the rhythm by which the machine shifted its form in synchrony with Wagner's music in the HD broadcasting was captivating; in particular, during the interludes that separated Siegfried's journey down the Rhine river in *Götterdämmerung*, the movement of the machine and the projection of pulsating and amorphous colors provided a riveting visual counterpoint to Wagner's meandering soundscape.

Suffice to say, leitmotifs play a significant role in steering the audience's attention to the underlying themes in the Ring cycle. Amid over a hundred leitmotifs that represent anything from Valhalla to the forsaken ring, the "sword" motif stands out for its bright timbre (trumpet) and its sweeping arpeggiation of the C major triad. When this motif is sounded toward the end of Act 1 in *Die Walküre*, Wagner provides the following instruction: "With a violent effort [Siegmund] pulls the sword from the tree and shows it to Sieglinde, who is seized by astonishment and ecstasy" (Spencer and Millington 2010, 139). As shown in Figure 23.3a, Wagner's music prolongs the triumphant moment by drawing out the C major chord via upward and downward arpeggiations in the harps and strings and the trumpets' sounding of the motif across five bars. The harmony then abruptly shifts to E major, as Siegmund acknowledges his new identity: a Wälsung hero and bridegroom to Sieglinde.

The HD broadcast of LePage's production for the Metropolitan Opera renders the dramatic moment salient through the deft use of camera angles and audio-visual "downbeat."[5]

Here, the center shot of Siegfried holding the sword (Figure 23.3b) coincides with the registral peak in the music (see the arrow in Figure 23.3a)—creating a strong audio-visual alignment by synchronizing the physical gesture of raising the sword with the melodic peak. This is followed by a long shot that captures the slow descent of the upper panel of the machine as Sieglinde watches Siegmund in awe (Sieglinde is placed in the periphery). The camera then shifts to a bottom-up shot angle and slowly pans in circular motion from right to left to capture the joyous union of lovers—a common filmic technique used to convey an enraptured state, inducing bodily empathy from viewers.[6] The VERTICAL and CENTER-PERIPHERY schemas work in synchrony to heighten the euphoric sensation that accompanies this closing scene from Act 2. This filmic rendering of the scene corroborates Cohen's theory with its perceptually salient moment established via cross-modal congruence. Not every sounding of the "sword" motif receives this kind of audio-visual reinforcement, although an attentive listener will pick up other nuanced expressions; for instance, when the motif is sounded in the postlude of *Das*

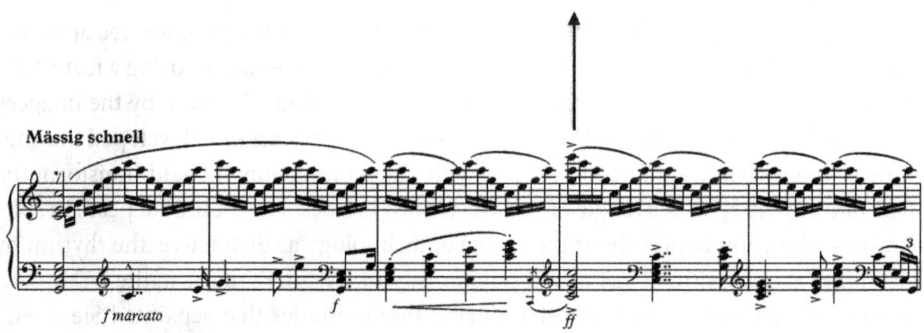

FIGURE 23.3A Wagner's music featuring the "sword" leitmotif (reduced score)

FIGURE 23.3B LePage's production of *Die Walküre*, Act I (DVD: 1:03:31–1:05:50)

FIGURE 23.3C Holten's production of *Die Walküre*, Act I (DVD 1:00:35–1:01:15)

Rheingold in D♭ major, it foreshadows the coming of a mortal hero as the gods enter the ill-begotten Valhalla.

Now compare LePage's mise-en-scène with Kasper Bech Holten's 2008 production for the Royal Danish Opera. Titled *The Copenhagen Ring*, Holten's production privileges Brünnhilde as the heroine who saves the gods from ultimate ruin. Unlike the common portrayal of the demigoddess as a pitiable victim, Holten sees her as a heroine "whose infinite visionary powers enable her to liberate Wotan after so many years"

(Wagner 2008, 13). And women, aside from Brünnhilde, are featured in the Copenhagen Ring cycle as the movers and shakers of the narrative. At the comparable scene where Siegmund and Sieglinde acknowledge their union in *Die Walküre*, the lovers are situated in a dark forest with red lighting cast on the ash tree (see Figure 23.3c). In the moment when they profess their love for each other, Siegmund urges Sieglinde to pull the sword from the ash tree; as the motif sounds, it is Sieglinde who pulls the sword and holds it up in the air in sheer amazement. The camera angle shifts back and forth between the two lovers without notable panning effects, but focuses on Sieglinde as the heroine who will take hold of the fate of the Wälsung twins. This production fails to establish a strong audio-visual congruence around the sounding of the leitmotif because the shot sequence and the music do not exactly line up.

The camera's focus on Sieglinde is not just a localized effect, but speaks to Holten's reconfiguration of the narrative on the whole. In *The Copenhagen Ring*, Brünnhilde tells the whole saga in flashbacks and is shown to be holding a child at the end of the cycle. Instead of the Gods of Valhalla perishing at the end of the narrative, the image of the child symbolizes rebirth, one made possible through Brünnhilde's bold actions. In short, Wagner's original tale is turned into a feminist heroine's quest. There are numerous other filmic productions of the Ring cycle, but the two versions discussed here could not be more different with respect to the filmic rendering of this critical scene in *Die Walküre*.

Case Study #2: John Adams's *Doctor Atomic* by Sellars (2007) and Woolcock (2008)

A contemporary opera that has garnered attention through the two contrasting productions directed by Peter Sellars and Penny Woolcock, *Doctor Atomic* (2006) is the last in the trilogy of Adams's operas based on a political subject from postwar American history. *Nixon in China* (1985–1987) revolves around Nixon's historical visit to China and his meeting with Mao Zedong and then-prime-minister Zhou Enlai; the opera cleverly satirizes the "disconnect" between the American and Chinese officials as Nixon and Kissinger engage in a one-sided conversation with the aging tyrant of communism (see S. Johnson's chapter for more on this subject). On a more controversial note, *Death of Klinghoffer* (1992) foregrounds the murder of an American Jewish civilian at the hands of Palestinian terrorists aboard the Italian cruise ship *Achille Lauro*, which has led critics and scholars to debate the moral authority behind the violence of one ethnic and political group against another.[7] *Doctor Atomic*, in turn, takes us back in American history to the Manhattan Project; in July 1945, a group of scientists led by Robert Oppenheimer conducted the first atomic testing in New Mexico in anticipation of its use to bring an end to World War II. Here, the scientists were faced with the ethical dilemma of whether or not to carry out the test explosion of the thermonuclear bomb in the Los Alamos desert prior to dropping the bomb in Japan. The "pact with the devil" implication and Oppenheimer's subsequent fall from grace in his political career lead us to view the opera as a Faustian parable set in the modern age.

The opera's narrative takes place during the 48 hours leading up to the countdown of the test explosion in New Mexico. By interpolating sounds of screaming children and the disembodied voice of a Japanese woman asking for water, the concluding scene conjures up the actual horror inflicted on the victims of Hiroshima and Nagasaki in virtual space and time. In this respect, the ending catapults the audience from the once-now (Los Alamos) to soon-to-be-now (Hiroshima) and casts the operatic narrative within a broader historical framework, enabling them to participate in the construction of its meaning.

Peter Sellars's mise-en-scène is striking because of the physicality of actions that he demands from the singers. In fact, his complex stage design, lighting design, and choreography pay respect to post-Brechtian theater in terms of (1) the simultaneous appropriation and revision of Brechtian political theater on the part of feminists and non-Western writers; (2) experiments in decentering the function of dramatic/theatrical authorship; and (3) the obsessive interest in the body as a political unit, and its function within the play of political forces (Garner 1990, 146). Sellars's staging of *Doctor Atomic* embodies nearly all of these traits: the construction of the libretto itself departs from a monolithic model of authorship; his inclusion of women's poetic voices (Rukeyser, Tewa Indian folklore) infuses the narrative with perspectives of feminist and oppressed others, and most importantly, Sellars makes use of the human body as a political site of engagement.

In this respect, post-Brechtian theater disrupts Brecht's concept of alienation (*Verfremdungseffekt*) because it intends to evoke a cathartic response from the viewer. In an ensemble number called "At the Sight of This," the chorus sings the text from the Bhagavad Gita (Discourse XI), imploring the Hindu God Vishnu for protection. In the filmic production of Sellars's staging of this scene, close-up shots and a lurid combination of colors (orange, green, and red) call attention to individual faces and bodies as the site of political unrest (see Figure 23.4a). Sellars uses close-up shots to capture the distorted facial expressions of the chorus, while zooming in on Pasqualita's face and placing it in the center of the frame, as if to call attention to her strength and profound connection to nature. The writhing bodies of the chorus members visually "mirror" the images of human casualties in World War II, shown in the documentary film that precedes the opera proper (see Figure 23.4b). The twisted bodies and hand gestures convey physical casualties inflicted on the American military workers in a manner that suggests that they are not necessarily victors, but are portrayed as the imagined victims of the atomic explosion.

Adams's music for this choral number is based on a modified binary form (ABA′B′A), with the opening section in C minor. It appears to be conspicuously modeled after the

FIGURE 23.4A Sellars's mise-en-scène of "At the sight of this" (Act II/scene 3)

FIGURE 23.4B Documentary film that precedes the opera (DVD1: 1′18″)

FIGURE 23.4C Woolcock's mise-en-scène of "At the sight of this" (Act II/scene 3)

dramatic refrain "O Fortuna" that opens and closes Carl Orff's *Carmina Burana*. The text that speaks of "fate" being monstrous, changeable, and villainous from the opening number of *Carmina Burana* parallels the stupendous and terrifying image of Vishnu from the Bhagavad Gita text. The A section, cast in a march topic with strong timpani beats driving the pulse, is contrasted by the B section, in which the texture shifts to a harmonic progression built on a falling, chromatically descending bass; an uncanny piccolo motive hovers in the high register. As the music descends, the chorus surrenders their hands in the air in deference to Vishnu. The close-up shots of their faces communicate fear and terror to the viewer.

Woolcock brings an entirely different political perspective to her staging of *Doctor Atomic* for the Metropolitan Opera (2008). In her previous collaboration with Adams, she remediated *The Death of Klinghoffer* as a film that intersperses images from different time periods and cultural vistas to comment on the Jewish and Palestinian identities. Ruth Longobardi argues that the visual domain is made primary in Woolcock's filmic opera; it provides for "a singular position from which a viewer follows events and characters, while libretto and music sustain its representations" (2009, 294–295). In Woolcock's theatrical production of *Doctor Atomic*, one could argue similarly for the primacy of the visual domain. In her staging of the opera, she breaks up the theatrical space by placing multileveled panels with cubicles to segregate the scientists from the other workers and domestic servants. The scientists (all men) congregate around the table in front of the stage; the female chorus provides commentary but does not participate in dialogue; and the native American Indians who work as servants are "seen" but not "heard." Kitty is portrayed as a glamorous white woman, poised for the cover of fashion magazines from the time period, in contrast to Pasqualita, who is portrayed as a servant clad in her Tewa costume.

The choral number "At the Sight of This" evinces the most striking departure Woolcock makes from Sellars's mise-en-scène (see Figure 23.4c). In lieu of choreography, Woolcock has the chorus stand erect and sing the number devoid of any physical gestures. Behind the chorus, photographs of scientists are projected onto the middle row of cubicles, while Tewa Indians clad in full battle gear stand in the top row of cubicles. The camera zooms in and out of the photographs of scientists, then slowly pans from left to right in capturing the chorus. A strong audio-visual synchronization occurs when the camera shifts to the projection of rain-like drizzle at the beginning of the B section; the panels that hold the Tewa Indians open from the middle to reveal projected images of a thunderstorm over the machinery positioned in the back of the stage. The whole effect is abstract and kinesthetically charged, but the psychological impact of terror (associated with the text from the Bhagavad Gita) is significantly undermined. The filmic rendering of this number delineates the musical form via rhythmic alignment with the projected images. In favoring long shots and linear panning, Woolcock distances the viewer from the physicality of the human body in the interest of calling attention to the ideological collision between nature and science.

Case Study #3: Saariaho's *Adriana Mater* (Directed by Peter Sellars)

The final case study features a contemporary opera about war by the Finnish composer, Kaija Saariaho. Working with the librettist Amin Maalouf and stage director Peter Sellars, Saariaho quickly developed the theme for her second opera (2005). For Maalouf, the Latin title *Adriana Mater* evoked the image of a modern incarnation of a Mater Dolorosa, the mother of atoning sacrifice through whose suffering the world is saved. Drawing on his first hand experience of war as a journalist, it was important for Maalouf to place the opera's narrative within a quasi-mythical framework. That is to say, the narrative takes place during and after a nameless war, without a specific time and place; in the libretto, he indicates that the diegetic time and location has to do with the "present" and the place a "country at war."

The opera, in fact, is about coping with the memory of war: the characters in the opera are traumatized by the memory of a recent war, which has corrupted people and undermined their optimism for living. At the beginning of the story, all we are given is an unknown village in a war-torn country where another war is about to begin. One evening, the protagonist Adriana is raped and impregnated by Tsargo, a soldier from her own community. She gives birth to the child (Yonas) despite her sister Refka's protest, and she decides to raise the illegitimate baby on her own. Much of the psychological drama is focused on the women's struggle over the fate of Adriana's son, as to whether he will grow up to be like his mother or his father: "would he be Cain or Abel?" she asks herself. She nonetheless conceals the truth from Yonas by telling him that his father had died during the war. Seventeen years later, Yonas discovers that his father is still alive, confronts Adriana and Refka about the truth of his identity, and vows to kill his father. Yet once he encounters Tsargo, who is now blind and helpless, Yonas is unable to carry out the revenge. Adriana tells her son that his action has proven that he is truly her son and that the curse is finally broken. The nightmare is over.

While most of the opera takes place within a constructed reality of a war zone, the plot formation is decisively nonlinear, marked by interpolated dream sequences that provide a window into the characters' unconscious desires and repressed emotions. With Saariaho's music, we enter into a mythical narrative of trauma and ambivalence where inner expression presides over spoken words.[8] Dream sequences are interpolated into the narrative progression of the opera as a way for the characters to come to terms with the "belatedness" and "incomprehensibility" that accompany their experiences of traumatic repetition, as defined by Cathy Caruth (1996, 27). Saariaho endows music with a capacity to narrate independently of the spoken words and makes acousmatic commentary on violence and trauma in lieu of actions. Cyclical repetition and stasis in her music correspond with Caruth's concept of repetitive seeing. The present study focuses on the role of the overture and the orchestral interludes that accompany the traumatic scenes of rape and the encounter between father and son seventeen years later.

The somatic quality of the overture and the expressive oppositions established in the opening gestures are especially striking. As shown in Figure 23.5, the solo trumpet sounds the principle dyadic motif (S1) in three parts over the sustained chords in the brass and the undulating bass drum rolls; here expressive markings alternate between *Disperato* and *Dolce*, creating an expressive opposition between the alternating gestures. As the intensity of the bass drum dies down, the strings echo the chord in the brass with a softer chord, played *sul tasto*; the descending glissando (D–C♯) in the cello, which inverts the ascending contour of S1 in the solo trumpet, conveys a sense of melancholic resignation. I refer to the septachord played by strings, labeled S2, as the *Urklang* for the opera as it alludes to Adriana's maternal yearning and suffering throughout the rest. As the tremolo in the bass drum grows louder at m. 3, this three-bar unit repeats again. Accompanied by the intensity that ebbs and flows through the bass drum rolls, the chords exchanged between the brass and strings resemble a cycle of breath—loud exhalation followed by a soft inhalation.

In its somatic and organic quality, it thus captures Adriana's maternal yearning pitted against the unknown fate of the child. In its third and final reiteration, the trumpet motif sounds again, this time marked by an interrogatory gesture (angular leap from A to G♯), as if to pose the question "what will become of this child?"[9]

Toward the end of scene 2, the first transgressive act is conveyed exclusively through music and changes in lighting; the audience sees Adriana's house and the surrounding domes turning bright red, but is not privy to the presumed action that takes place. Following Adriana's last cry of refusal ("Non! Non! Non!"), Tsargo is seen entering her house. As the house lights up, the instrumental postlude proceeds with a massive texture

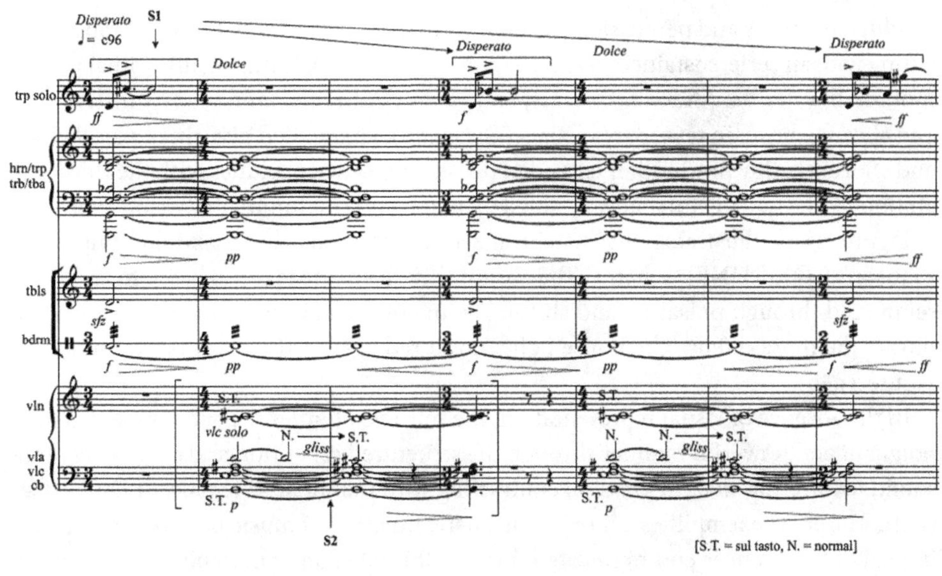

FIGURE 23.5 Instrumental overture (mm. 1–7)

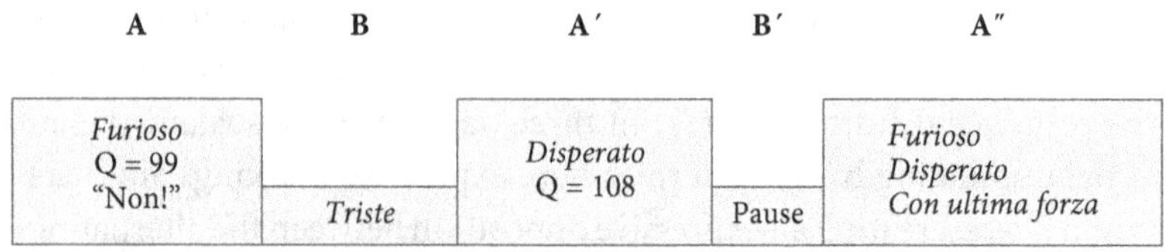

FIGURE 23.6A Schematic illustration of Saariaho's music for the "rape" scene

FIGURE 23.6B Sellars's mise-en-scène for Scene II (Helsinki production, 2006)

in which the brass and percussion pound out rhythms that convey the act of rape, alternating with an eerie, sustained passage (*Triste molto calmo*), featuring the primary "sigh" motif in the chorus. After another desperate cry by Adriana, reinforced by instruments and chorus, the main texture reappears (*Disperato*), augmented by strings, woodwinds, and chorus. Following another extended pause, the scene concludes with the third reiteration of the main texture, now doubled in tempo and rhythmic subdivision (*Furioso*).

Figure 23.6a illustrates the synchronized changes in music and lighting using Johnson's CONTAINER schema. The abrupt shift from one textural block to another is reinforced through pulsating and shifting colors of the lighting—the more violent the music, the more intense the house lights up in red against the black background (see Figure 23.6b).

In the absence of visual representation, the music embodies the act of rape in a most poignant and powerful manner through the structured repetition and gendering of the sound texture—as male aggression contrasting with the silenced female, or female passivity. The effect exemplifies Chion's acousmatic function of music in film: the dramatic import of this scene is communicated through the collision of masculine and feminine sound textures, disembodied from the presumed action.[10] The bright red lighting on the house and domes plays a complementary role in signifying trauma and violence.

FIGURE 23.7 Saariaho's postlude for Scene VI (reduced score)

The second act of transgression occurs in the later scene where Tsargo, now a decrepit old man, is reunited with his son Yonas seventeen years later. Saariaho brings back the acousmatic function of music, enabling music to convey the affect of trauma in lieu of actions on stage. Just as in the scene of rape, the trauma of their encounter is made visceral and psychological through the musical enactment of "repetition" compulsion.[11]

As shown in Figure 23.7, the slow and static music eventually builds to the volatile instrumental postlude. In spite of his revulsion, Yonas is unable to resist his father's desire to touch him. Against the dark red background, the characters stand facing each other, and Yonas is unable to pull the trigger. As the two depart from the stage and the music builds to an explosive end, the viewer is fully absorbed into the inner expression of trauma that accompanies this reunion. Here, mood congruence for each scene is established through the alignment between music and lighting in Sellars's minimal staging, which in turn articulates the dramatic turns in the narrative.[12]

Conclusion and Further Considerations

Current research on film and multimedia studies emphasizes alignment and associative dimensions in audio-visual congruence as important determinant of salience—memorable moments that induce an embodied response from the viewer. The case studies presented here attest to the importance of audio-visual congruence in viewing staged opera as film. When accompanied by accentual alignment and schematic congruence in the audio and visual domains, Wagner's *Die Walküre* enhances the viewer's absorption of the leitmotif that heralds the birth of a mortal hero and the euphoric sensation. The two productions of *Doctor Atomic* illustrate how the directors are able to manipulate the audio-visual congruence through camera angles and shot sequences toward radically different aesthetic ends. Lastly, Sellars's staging of Saariaho's *Adriana Mater* illustrates the music's potential to generate its own imagistic content in lieu of kinesthetic actions on stage. While these observations are preliminary at best, here are further considerations for investigating audio-visual congruence in multimodal narrative:

1. The camera angles, framing, and shot sequence are integral to preserving or disrupting Brechtian alienation—the distancing effect that accompanies theatrical viewing of opera. Sellars's deployment of close-up shots creates an intimacy that disrupts this condition by inducing empathy in the viewer, while Woolcock's filming technique preserves the objective distance through the medium- to long-shot takes. At the next stage, it would be important to compare filmic opera (where the opera is conceived as film) and HD broadcasting of staged opera from this perspective.[13]
2. Johnson's image schemas provide a useful paradigm for tracking analogical correspondences established between the visual, kinesthetic, and/or musical dimensions in film. While further empirical research is needed for verification, it can be speculated with reasonable certainty that an audio-visual congruence established through the combination of image schemas articulates salient moments in the viewer's experience of opera and film.
3. Cohen provides an elegant model for integrating bottom-up and top-down processes involved in constructing narrative, but it requires context-specific refinement. How are specific connections established across sensory domains at the preattentive levels (A and B) before the composite meaning can be constructed at level C? Notwithstanding her general claim that music/soundtrack brings the visual domain into focus, operatic music can undermine or cover over the visual by generating its own imagistic content, as demonstrated by the acousmatic function of music in Saariaho's *Adriana Mater*. The acousmatic use of music is hardly limited to opera. A prime example in film can be found in Oliver Stone's *Platoon* (1986), in which Samuel Barber's "Adagio for Strings" covers over the diegetic sounds of battle. Instead of underscoring the exterior "noises" one hears during a battle, Barber's music directs the viewer's attention to the interior emotion of the characters and by extension provides a larger-scale commentary on the subject of war in a manner that transcends words. In a similar vein, in the climatic battle scene in Akira Kurosawa's *Ran* (1985), Toru Takemitsu composes an orchestral score, which is Mahlerian in its expression of sheer anguish; the images and

actions unfold in slow motion in structural conformance to the elegiac tone of Takemitsu's music.

Hermeneutic and cognitive research on film and multimedia still has a long way to go in explaining the complex manners by which media sources lend meaning and expression to our understanding of multimodal narrative. However, uncovering such nuanced techniques for manipulating the visual and sonic domains in film provides the first step in understanding the interpretive processes that connect music, body, and the mind.

Notes

1. In the last two decades alone, Siu-lan Tan, Annabel J. Cohen, Scott D. Lipscomb, and Roger A. Kendall, among other psychologists, have spearheaded cognitive research on audio-visual congruence in film and multimedia (Lipscomb 2013).
2. Mickey-mousing refers to a film technique developed during the early days of animation in which the actions on the screen are synchronized with the soundtrack.
3. Text, speech, visual, music, and sound fx are Christian Metz's literary-critical categories for analyzing film. To this list, Cohen added the kinesthetic domain.
4. Cohen claims that while music transports a variety of information, "the brain selects what is useful for the goal at hand with the help of all the information stored in long-term memory at Level E, including story grammar" (2014, 121).
5. The idea of "downbeat" is taken from Nicholas Cook; in his analysis of Stravinsky's *The Rite of Spring* in Disney's *Fantasia*, Cook claims that coincidences established between image sequence and music articulate a "large-scale downbeat-afterbeat pattern" (1998, 184).
6. In the film version of *The Phantom of Opera* directed by Joel Schumacher (2004), a technique of circular panning (360 degrees) is used to transform Christina's rise from a chorus girl to a star in her stage debut.
7. See Renihan (2014) and Longobardi (2009).
8. During my interview with Saariaho in 2009, she told me that she was physically in pain for the duration of three years while she composed this work.
9. For a more detailed analysis of the opera, see chapter 3 from Everett (2015).
10. Where this concept differs from film is that the viewer is denied the opportunity to see the action, but may imagine it through sound. In filmic contexts, Chion claims, "acousmatization is a process whereby we are made to hear without seeing after having first been allowed to see and hear at the same time" (2014).
11. Saariaho has gone so far as to eliminate parts of the dialogue in Maalouf's original libretto in order to communicate the dramatic tension of the scene in purely musical terms.
12. The mood for each scene corresponds with Sellars's choice of color in the lighting; for example, dark red is used throughout the traumatic scene of rape, while blue and gold are featured in the subsequent scene that announces Adriana's hopeful prospect in raising the child to be like her.

13. Jeongwon Joe discusses the disruptive effects brought on by the exploitation of aesthetic and technical conflicts between opera and film (Joe and Theresa 2002, xii).

References

Caruth, Cathy. 1996. *Unclaimed Experience: Trauma, Narrative, and History*. Baltimore, MD: John Hopkins University Press.

Chattah, Juan. 2015. *David Shire's The Conversation: A Film Score Guide*. Lanham, MD: Rowman & Littlefield.

Chion, Michel. 1994. *Audio-Vision: Sound on Screen*. New York: Columbia University Press.

Chion, Michel. 2014. "100 Concepts pour penser et décrirer le cinema sonore." Accessed June 10, 2014. http://www.michelchion.com/v1/index.php?option=com_content&task=view&id=45&Itemid=60.

Coëgnarts, Maarten, and Peter Kravanja, eds. 2015. *Embodied Cognition and Cinema*. Leuven: Leuven University Press.

Cohen, Annabel J. 2002. "Music Cognition and the Cognitive Psychology of Film Structure." *Canadian Psychology* 43: 215–232.

Cohen, Annabel J. 2009. "Music in Performance Arts: Film, Theatre and Dance." In *The Oxford Handbook of Music Psychology*, edited by Susan Hallam, Ian Cross, and Michael Thaut, 441–451. New York: Oxford University Press.

Cohen, Annabel J. 2014. "Film Music from the Perspective of Cognitive Science." In *The Oxford Handbook of Film Music Studies*, edited by David Neumeyer, 96–130. New York: Oxford University Press.

Cook, Nicholas. 1998. *Analysing Musical Multimedia*. Oxford: Clarendon Press.

Everett, Yayoi Uno. 2015. *Reconfiguring Myth and Narrative in Contemporary Opera: Osvaldo Golijov, Kaija Saariaho, John Adams, and Tan Dun*. Bloomington: Indiana University Press.

Garner, Stanton B., Jr. 1990. "Post-Brechtian Anatomies: Weiss, Bond, and the Politics of Embodiment." *Theatre Journal* 42 (2): 145–164.

Hutcheon, Michael, and Linda Hutcheon. 2009. "Opera: Forever and Always Multimodal." In *New Perspectives on Narrative and Multimodality*, edited by Ruth Page, 65–77. New York: Routledge.

Joe, Jeongwon, and Rose Theresa, eds. 2002. *Between Opera and Cinema*. New York: Routledge.

Johnson, Mark. 1987. *The Body in the Mind: The Bodily Basis of Meaning, Imagination, and Reason*. Chicago: University of Chicago Press.

Levin, David. 2007. *Unsettling Opera: Staging Mozart, Verdi, Wagner, and Zemlinsky*. Chicago: University of Chicago Press.

Lipscomb, Scott. 2013. "Cross-Modal Alignment of Accent Structures in Multimedia." *The Psychology of Music in Multimedia*, edited by Siu-Lan Tan, Annabel Cohen, Scott David Lipscomb, and Robert Allen Kendall, 192–213. New York: Oxford University Press.

Longobardi, Ruth. 2009. "Re-producing *Klinghoffer*: Opera and Arab Identity before and after 9/11." *Journal of the Society for American Music* 3 (3): 273–310.

Renihan, Colleen. 2014. "Gesture, Temporality, and the Politics of Engagement in Opera on Film: Penny Woolcock's *The Death of Klinghoffer*." *Music, Sound, and Moving Image* 8 (1): 57–85.

Spencer, Stewart, and Barry Millington. 2010. *Wagner's Ring of the Nibelung: A Companion*. New York: Thames and Hudson.
Zbikowski, Lawrence. 2011. "Musical Gesture and Musical Grammar: A Cognitive Approach." In *New Perspectives on Music and Gesture*, edited by Anthony Gritten and Elaine King, 83–98. Farnham: Ashgate.

Videography

Adams, John. 2007. *Doctor Atomic*. Netherlands Philharmonic Orchestra and Chorus. Opus Arte. OA-0998-D. DVD.
Adams, John. 2008. *Doctor Atomic*. The Metropolitan Opera Live on Demand. New York: The Metropolitan Opera. https://www.metoperafamily.org/ondemand/.
Wagner, Richard. 2008. *Die Walküre: The Copenhagen Ring*. Production by Kasper Holten. Royal Danish Opera. Decca. 074 3266. DVD.
Wagner, Richard. 2011. *Die Walküre*. Production by Robert Lepage. The Metropolitan Opera HD Series. Deutsche Grammaphon. 00440 073 4848.
Webber, Andrew Lloyd. 2004. *The Phantom of the Opera*. Directed by Joel Schumacher. Warner Bros. Pictures. 079 079 5396. DVD.

Further Readings

Eitan, Z., and R. Y. Granot. 2006. "How Music Moves: Musical Parameters and Images of Motion." *Music Perception* 23: 221–248.
Simonson, Mary. 2013. *Body Knowledge: Performance, Intermediality, and American Entertainment at the Turn of the Twentieth Century*. New York: Oxford University Press.

CHAPTER 24

LISTENING TO THE MUSICKING BODY

A Cross-Disciplinary and Historical Perspective

YOUN KIM

It is through my body that I understand other people; just as it is through my body that I perceive "things." The meaning of a gesture thus "understood" is not behind it, it is intermingled with the structure of the world, outlined by the gesture, and which I take up on my own account. It is arrayed all over the gesture itself.

—Maurice Merleau-Ponty (1962, 186)

As Maurice Merleau-Ponty emphasizes, the human body is connected to the external world through sensation and perception. How we become aware of the environment is one of the oldest questions central to both the sciences and humanities. The topic is generally discussed in terms of different types of modes. Most contemporary textbooks on physiology and psychology teach about the sense of seeing, hearing, taste, smell, and touch as separate units (see, for example, Hall 2015; Boring 1942). Such prototypical differentiation of sensorial modes can be traced all the way back to Aristotle (Murphy and Kovach 1972, 331). Despite the distinction between sensory modalities, however, these senses are not independent of each other, and in fact, they influence each other. The interactions between different sensory modalities such as synesthesia (see Haverkamp 2012) are examples of such influence. More importantly, despite the qualitative distinctions between modalities, some commonalities exist between them. This shared set of characteristics can be considered as a framework through which we sense, perceive, and ultimately understand the world. Sensation and perception, therefore, not only are the physiological and psychological processes of each bodily receptor picking up environmental stimuli but also constitute a matter of general epistemology, subject to contextual and cultural conditions.

Listening is one of the mediating pathways through which we get to know the external world. Traditionally, listening has been considered to be related to the ear: the auditory system is regarded as the primary organ for listening, and these approaches have mostly been grounded in the natural scientific approach to hearing. Recently, however, the study of listening has witnessed significant growth, not only in terms of quantity but also, and more importantly, in terms of the range of disciplines examining listening. A more holistic conceptualization of listening that goes beyond the ear and addresses the whole human body is emerging. The cultural and historical spaces in which body acts occur are becoming increasingly significant to the study of listening (see, for example, Szendy 2008; Nancy 2007; Johnson 1996; and the special issue, "Listening," of the *Journal of the Royal Musical Association*, Bacht 2010).

This chapter examines the plasticity of the notion of listening across the modal and/or disciplinary dimension and locates it in its historical contexts. It closely examines the late nineteenth and early twentieth centuries. This period witnessed significant advances in the biological and medical sciences and is often referred to as "the time of the body": the human body and its physiology and psychology were completely reconceptualized in this era (Cowan 2011). Musicology and acoustics also witnessed significant changes and advances during the period, bringing about a new understanding of both the body and music. Initially, this chapter discusses several theories of seeing, hearing, and kinesthetics from this period and explores how listening has been understood based on a holistic understanding of the human body across different sensory modes. In these theories, certain core ideas first permeated the study of vision and auditory perception, which were then extended to motor functions as common traits across various modes. These ideas appeared across various and apparently remote fields, forming the discursive space of music and the body. This complex web of ideas led to the formation of the comprehensive notion of musical listening. Exploring how the mode of listening had been discussed and constructed in these different theories in the past will shed light on the current understanding of listening in the context of the holistic human body. In what follows, this chapter examines the journey of these ideas as they appear in the writings of scientific psychologists, philosophers, and music theorists, all of whom were direct or indirect interlocutors to each other. Considering scientific texts as a fluctuating cultural product instead of a stable and solid system, this chapter suggests one possible way of connecting psychology and cultural history, the two major approaches to listening.

Afterimages and Aftersound

The most apparent bodily organ to consider—although certainly not the only one—for a discussion on hearing and listening would be the ear, but the current discussion begins with the eye. Most early psychologists of the late nineteenth century had rather general and comprehensive research interests that were not confined to a specific sensory modality. It is not surprising, therefore, that these scholars attempted to find parallels and counterparts in visual and auditory perception. In this regard, an examination of the notions of "afterimage" (*Nachbild*) and its equivalent in hearing, "aftersound" (*Nachklingen*), is relevant here.

Afterimage refers to the visual perceptions experienced even after the removal of the stimulus. For example, if you stare at a negative photo for a while with your eyes fixed on a spot without blinking and then move your eyes to a plain surface, you should be able to see a full-color image of the photo for a moment (Colman 2009; Matsumoto 2009; Roeckelein 2004). Many scholars across time were engaged in studying the phenomenon of the afterimage. Isaac Newton, as well as some of his predecessors in the seventeenth century, knew about the afterimage (Roeckelein 2004, 28; Boring 1957, 103), and Goethe discussed the phenomenon in his *Theory of Colours* (1810). It was, however, only after the mid-nineteenth century that afterimages began to be investigated more substantially and empirically. Physiological works by Jan Evangelista Purkyně (1787–1869) and later by Ewald Hering (1834–1918) are the most well-known works that began to unravel the afterimage. Hermann von Helmholtz (1821–1894) discussed afterimages in two lectures in 1858 (1858a, 1858b; cited in Cahan 1994, 241) and in the second volume of his *Treatise on Physiological Optics* (1860).

Although the afterimage is the most readily observable example of "aftersensation" (*Nachempfindung*), parallel phenomena have been thought to exist in most other sense-modalities as well. Helmholtz, who made significant contributions to various fields, including optics and acoustics, substantially discussed both the optical afterimage and the auditory "aftersound" (*Nachklingen*) (Helmholtz 1863, 212–218; 1954, 142–145). In fact, his well-known explanation of the hearing mechanism effectively explains aftersound. According to his theory, the transverse fibers on the basilar membrane are "tuned" to sympathetically vibrate at different frequencies, similar to the strings of a piano. Similar to how any other elastic sounding body set into sympathetic vibration goes on sounding even after the original tone ceases, Helmholtz postulated that the fibers in the ear continue vibrating for a perceptible duration, and that this continuing vibration of fibers results in the phenomenon of aftersound.[1] An analogical strategy of comparing hearing and seeing features heavily in Helmholtz's *On the Sensation of Tone* (1863). Visual phenomena such as "floaters," "blind spots," and "double vision" are discussed at length only in order to explain the perceptual process of hearing.[2]

In addition to Helmholtz, many other scholars became involved in finding a parallel for the phenomenon of afterimage in the realm of hearing. The otologist Victor Urbantschitsch (1847–1921) studied residual hearing (Urbantschitsch 1881), and the physicist Alfred M. Mayer (1836–1897) took an acoustical interest in the subject and attempted to calculate the duration of the residual sonorous sensations based on the number of vibrations per second (Mayer 1874).

Although their explanations of aftersound differed from each other in their specifics, a few significant commonalities arise in these discussions. First, the phenomena of aftersound and afterimage are intrinsically entangled in terms of *temporality*: these sensations unfold in time, and moreover, they do not exactly correspond to the outer stimulus in terms of time. As Jonathan Crary phrases it in his study of the visual culture of the nineteenth century, "as observation is increasingly tied to the body, in the early nineteenth century, temporality and vision become inseparable" (1992, 98). The discussion on aftersound is much more than a mere parallelism of afterimage in hearing, for the phenomenon points to the temporality of perception itself.

The proof for the objective existence of aftersound had to wait until 1964, when the Zwicker tone was discovered (Zwicker 1964): a faint pure tone can be heard for a few seconds immediately after the presentation of a broadband noise with a spectral gap. After the "real" noise ceases, an apparent sound can be heard whose frequency falls within this frequency gap in the original signal. Although this auditory illusion that is comparable to the afterimage in vision was proven to exist only in the late twentieth century, the increased discussion on the aftersound in the late nineteenth century signifies a critical shift in the psychology of music. The focus shifted from identifying individual and discrete mental states or representations (such as visual images or a single sonic event) to the temporal relations between them.

Similar to aftersensations, Helmholtz's notion of "reaction time" deserves a mention following the discussion on temporality. The phrase "lost time" (*temps perdu*) is best known to us through the title of Marcel Proust's masterpiece novel *In Search of Lost Time* (1913). However, an earlier reference to the phrase is found much earlier, and quite surprisingly, in Helmholtz's letter to the French Academy of Sciences in 1851 (about sixty years earlier than Proust), in which he described his famous nerve-muscle experiments on frogs.[3] Earlier, it was assumed that nerves transmitted sensory information almost instantaneously and immediately. Contrary to this belief, Helmholtz found that when a recently dissected frog's leg was electrically stimulated, it took a quite substantial and measurable amount of time for the muscle to contract in response to the nervous shock. To denote this lag between a stimulus and the response to it, Helmholtz used the French phrase meaning "lost time" and later extended this experiment to humans and measured what psychologists now call the "simple reaction time" (see Piccolino 2003; Schmidgen 2002, 2014). This observation points to the delay between the stimulus and the response and, consequently, the fleeting temporality of perception, just as the phenomena of afterimage and aftersound do.

The second significant issue in the context of afterimage and aftersound is the *subjectivity* of perception. Afterimage, above all, is the "presence of sensation *in the absence of* stimulus" (Doane 2002, 79; emphasis added). In other words, the experience is generated and produced within the human body. One of the earliest to explicate on afterimage, Goethe had noted the implications of the human body being the "active producer of optical experience" and mentions the idea of "subjective vision" (Crary 1990, 69). Attempts were made in the earlier explanations of aftersound to ground the phenomenon physiologically by considering it as a result of the continuation of the vibration (*Nachschwingen*) of the nerve fibers of the basilar membrane for a short amount of time after the acoustical sounding body stops its vibration. Nevertheless, the focus of the investigation is not so much the acoustics of sound but the physiological properties of the human action, that is, listening.

The idea of the afterimage continued to be further developed in relation to subjective states of mind, such as imagination, fantasy, and memory. Among the nineteenth-century psychologists who discussed afterimages, Gustav Fechner (1801–1887) distinguished ordinary afterimages from images of imagination (*Elements of Psychophysics*, 1860). William James (1842–1910) discusses afterimages in chapters on imagination, memory, and hypnotism (*The Principle of Psychology*, 1890; see Roeckelein 2004, 25). In addition to the afterimage, these scholars proposed several other types of mental imageries, such as the eidetic (or perceptual) image (*Anschauungsbild*) and memory-images (*Vorstellungsbild*), which are closer to imagination rather than sensation (see Roecklin 2004, 172).

The idea was also extended to the realm of exploration of sound and music in the early twentieth century. Psychologists such as Erich Jaensch (1883–1940) and Rudolf Kochmann considered the "afterimages of tones" mostly as eidetic imagery and studied them by locating them in the context of memorizing and imagining tonal sequences, particularly for children (see also Godøy and Jørgensen 2001). A similar line of approach is also found in philosophical discussions: the Austrian philosopher Alexius Meinong (1853–1920) recounts this residual auditory sensation as imagination, "I hear the note c which is produced by a tuning fork. It becomes less and less audible. Then there is a range where the tone itself is not there anymore, that is, the tuning fork ceases to vibrate, but internally we keep hearing it for a while. That is fantasy (*Phantasie*)" (as explained in Kalsi 1978, 67).

In addition to the signification of temporality and subjectivity, afterimage and aftersound carry considerable implication for aesthetic experiences. Goethe's account of afterimage is noteworthy here:

> I had entered an inn towards evening, and, as a well-favoured girl, with a brilliantly fair complexion, black hair, and a scarlet bodice, came into the room, I looked attentively at her as she stood before me at some distance in half shadow. As she presently afterwards turned away, I saw on the white wall, which was now before me, a black face surrounded with a bright light, while the dress of the perfectly distinct figure appeared of a beautiful sea-green. (Goethe 1810, cited in Doane 2002, 72–73)

The film theorist Mary Ann Doane notes the eroticism of vision indicated here: Goethe explains afterimages as something "accessible only through an experience of intensity, of dazzlement" (2002, 73). Given this extension of the physiological theory to the aesthetic theory of vision, the parallelism found in listening is not surprising. The phenomenon of aftersound, previously explained on a physiological basis, is found in the writings of music theorists. The Czech composer/theorist Leoš Janáček (1854–1928), for example, found scientific support for his theory of "chord connection" in the physiological phenomenon of aftersound. However, his remarks sometimes reveal an ambivalent attitude, as Michael Beckerman rightly pointed out. Janáček wrote, "(a tone) sounds from an instrument, but also floats in our mind when it has really faded out" (cited in Beckerman 1994, 61). Janáček speaks of the "ghost" of a tone, which continues to sound after the actual tone ceases, and sees the foundation for the chord connection in this illusionary moment. The ostensible basis he wanted to use to substantiate his theory was physiological, but with expressions such as "floats in the mind," Janáček appears to refer to "a process more closely related to psychology" (Beckerman 1994, 61). The music theorist Ernst Kurth (1886–1946) also discussed aftersounds as the aftereffect of a tonal impression and designated them as "conceptual images" (*Vorstellungsbild*) (1931, 88). To Kurth, the aftereffect (*Nachwirken*) of the tone is neither acoustical nor physiological but psychological in nature: "Without doubt, this after-effect of the tonal stimulus promotes, despite its short duration, continuity of tonal connection: however, this is mainly based upon a completely different [kind of continuum], namely, *energetic* continuum, which shows broader independence from a purely sonic [continuum]; it is conserved even across rests and staccato-tones" (89; emphasis in original).

Imagining the Gap between Tones

The implications of these specific physiological phenomena soon began to be extended to seeing and hearing more comprehensively and even to aesthetics. The French philosopher Paul Souriau (1852–1926), for example, writes:

> Any luminous point displacing itself in the visual field leaves behind it a visible wake . . . when the object displaces itself, each of its points leaves behind it a wake into which enters the next point, so that all of them tend to melt into a continuous line. (1983, 116)

Afterimages thus connote continuity. What then is happening during the gap between two stimuli? Even Helmholtz, whose initial concern was the delayed response to a stimulus, later adopted the term "Zwischenzeit," subtly shifting the focus to the "interim" between two events (Schmidgen 2014, 18). The aforementioned psychologist Meinong made a more explicit contention regarding the continuity between tones. In the case of the spatial continuum, such as a line connecting two points, continuity is passively experienced because it is imparted externally. On the other hand, he argues, a continuum between two tones is an artificial one, constructed by an active imagination, and not one imparted by the external stimuli (Meinong 1889).[4] Meinong's fellow psychologist Carl Stumpf provides an account along the following lines: the X between the two tones of m and n is a tone sui generis, a "tonal motion" (*Tonbewegung*) (1883, 184). It is we who fill the empty space between two boundary tones by imagining the between-tones (*Zwischentöne*) as a whole and as a unit (*Einer*). It is suggested here that aftersound, the phenomenon previously considered as purely physiological, is actually a psychological phenomenon that relates itself to the mind rather than to the ear. Similar to how afterimage migrated toward memory and fantasy as discussed earlier, aftersound then no longer belongs to the realm of physiological sensation and has come to be regarded as a product of the imagination instead.

It should be noted that it is not only the holistic continuity between two events but also the motion, as a result of our imagination, that is emphasized here. The previous quotation from Souriau is in fact taken from his monograph on the perception and aesthetics of moving objects, and the final pages are devoted to a discussion of the perception of melodic movement. He speaks of the "persistence of auditory images," which enable us to hear melodic movement. Our ears follow the "sonorous trails" of the melody in the sonic space, similar to how our eyes follow "those motile ephemeral images traced by a moving light" in the visual space (1983, 153–154). Even when discussing the visual afterimages, Souriau refers to an example related to music, that is, the "electric measure beater" (*le batteur de mesure électrique*) constructed by Jules Carpentier (1851–1921) for the Paris Opera.

> In a blackened panel there are two ridges in the shape of a V. In each ridge is placed a square ruler, the faces of which are painted alternately black and white. A simple mechanism turns at will each ruler a quarter turn, so that the first shows its white face and the other its black face which becomes invisible. You think you are seeing a wand oscillating like the baton of an orchestra leader. The illusion is striking. (1983, 116)

Additionally, early Gestalt psychologists often used melody as an example when explaining such "apparent motions,"[5] whose experience, in Roget Shepard's words, falls "somewhere between perception and imagery" (Shepard 1984, 423; also see Burtt 1917).

Imagining and Producing the Motion

The works of the French scientist Étienne-Jules Marey (1830–1904) can be used to understand this wider sense of listening, including imagining and moving, especially his explorations of perception, particularly in the context of body movements. Marey is known for his photographs and his earlier films emphasizing the continuous trajectory of the movement of human bodies. What makes him more relevant to this discussion is his concern for afterimages and his genuine interest in the acoustical study of sound. The afterimage effect was central to Marey's descriptions of human movement. In fact, Marey was a serious advocate of Helmholtz, who, as discussed previously, made substantial contributions to the theory of both afterimage and aftersound.[6]

As a physiologist whose primary concern was moving bodies, Marey was interested in expressing the "rhythms" of human progressions.

By attaching a device to the body of a runner/walker, Marey measured the duration for which the right and the left foot apply pressure on the ground, indicated with white and gray bars respectively, and called this a type of "staff notation" (*la portée*). More interesting is the curve line above that represents the bodily movement of the right and left foot and the curve of the vertical reactions of the body. The two representations underscore the gap between the end of one bar and the beginning of the next one (marked in Figure 24.1). This moment "between the pressure of the two feet" constitutes the "silence in the rhythm," which is composed of "that instant of the course when the body is suspended above the ground" (Marey 1874, 133).

The significance of this restoration of the lost moment is not limited to the human gait. Including commenting on Carpentier's invention of the melograph, Marey discussed sound and music quite extensively, subjects that had not received much scholarly attention so far. In his *Movement* (1894), for example, Marey mentions a "musical notation," the recording of Bach's airs played on a special harmonium with tracing needles attached to its keyboard,[7] which recorded of the movements of a pianist's fingers on the keyboard (Figure 24.2).

Here, the durations of the sounding notes are represented by the length of the stroke, instead of the conventional half, quarter, and eighth notes. This may resemble, at first, some of the later graphic representations of music, such as the ones by Cogan and Escot (1976). However, when located in the context of Marey's emphasis on the moving body, they are entirely different. These modern notations represent the sound itself (whether descriptively or prescriptively). However, while Marey's "surface" rhythm of music is more or less consistent with the articulatory movements of the performer, they are not always "one and the same" (Shove and Repp 1995, 60). Marey's human gait and foot pressure represent this difference: the transitional in-between moment, the lost moment that is indicated simply as blank in "staff notation," is restored in the representation of the bodily actions as continuing curves.[8] The same reasoning can be applied to musical notation. Mareyan notation endeavors to register the fingerings on the keyboard. The

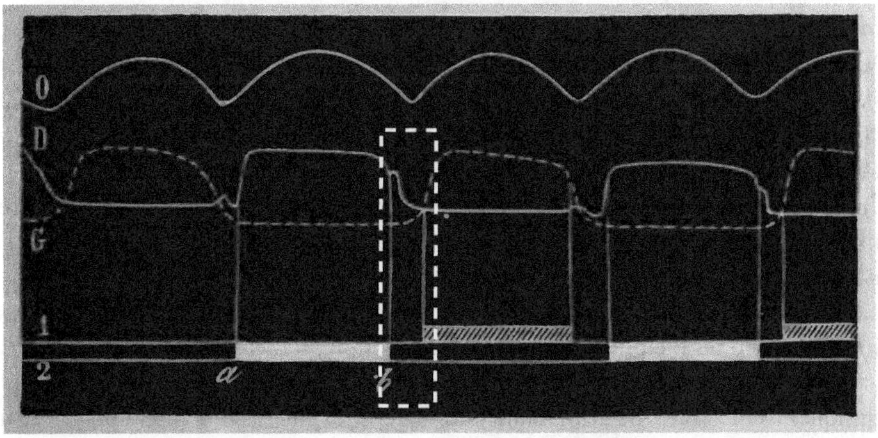

FIGURE 24.1 The rhythm of walking in Marey's notation: the moment marked with the square is "the instant when the body is suspended"

(from Marey 1874, 133, fig. 34).

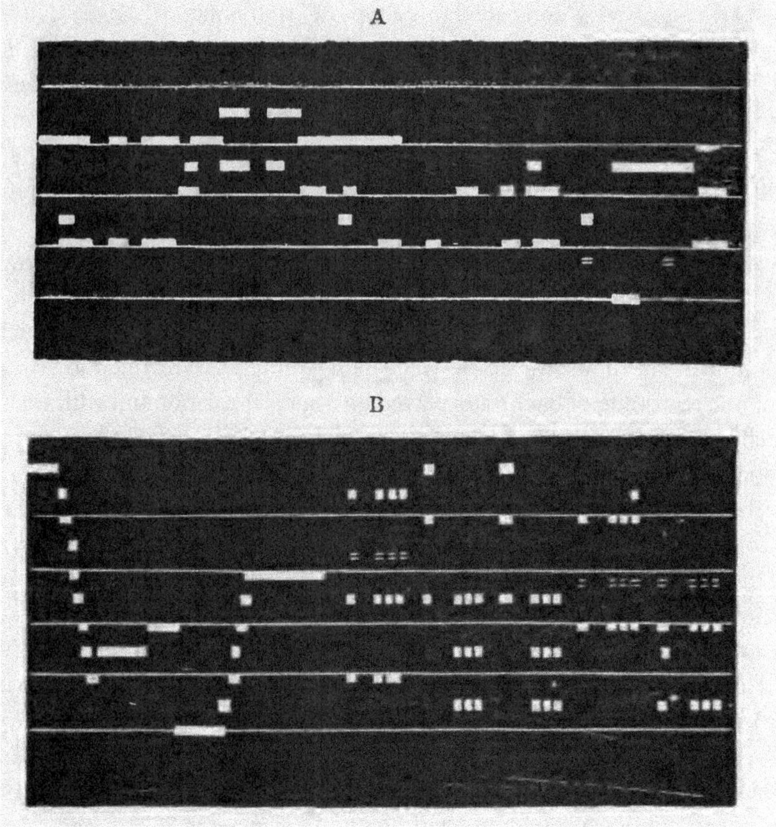

FIGURE 24.2 "Record of two airs played on the keyboard of a harmonium"
(from Marey 1895, 13, fig. 10).

moving body of the performer producing the sound is awarded more prominent consideration here. The lost moment in notational representation is not empty time, as it is restored in the curve as the moment of the suspended body. It is in this sense that Marey emphasizes the significance of the ear as a perceptual organ: "The ear is more competent than the eye to perceive these rhythms" (1878, 289).

In the realm of music-making, this agency of motion, corresponding to Marey's walking body, is the musician or the performer's body. Marey's photographic method was in fact adopted in some performance studies. A notable example is the study by the American music educator Otto Ortmann (1889–1979), who proposed pedagogical principles based on the physical understanding of piano mechanics and an analysis of human anatomy and physiology (Ortmann [1929] 1962). He adopted these photographic techniques to illustrate the arm, hand, and finger movements of a piano player. A tiny electric lamp was attached to the moving part of the pianist's hands or arm (Figure 24.3).

Figure 24.3 shows Ortmann's photograph of hand movement (Figure 24.3B) playing the octave figure with a crescendo (Figure 24.3A). As the distance increases, the

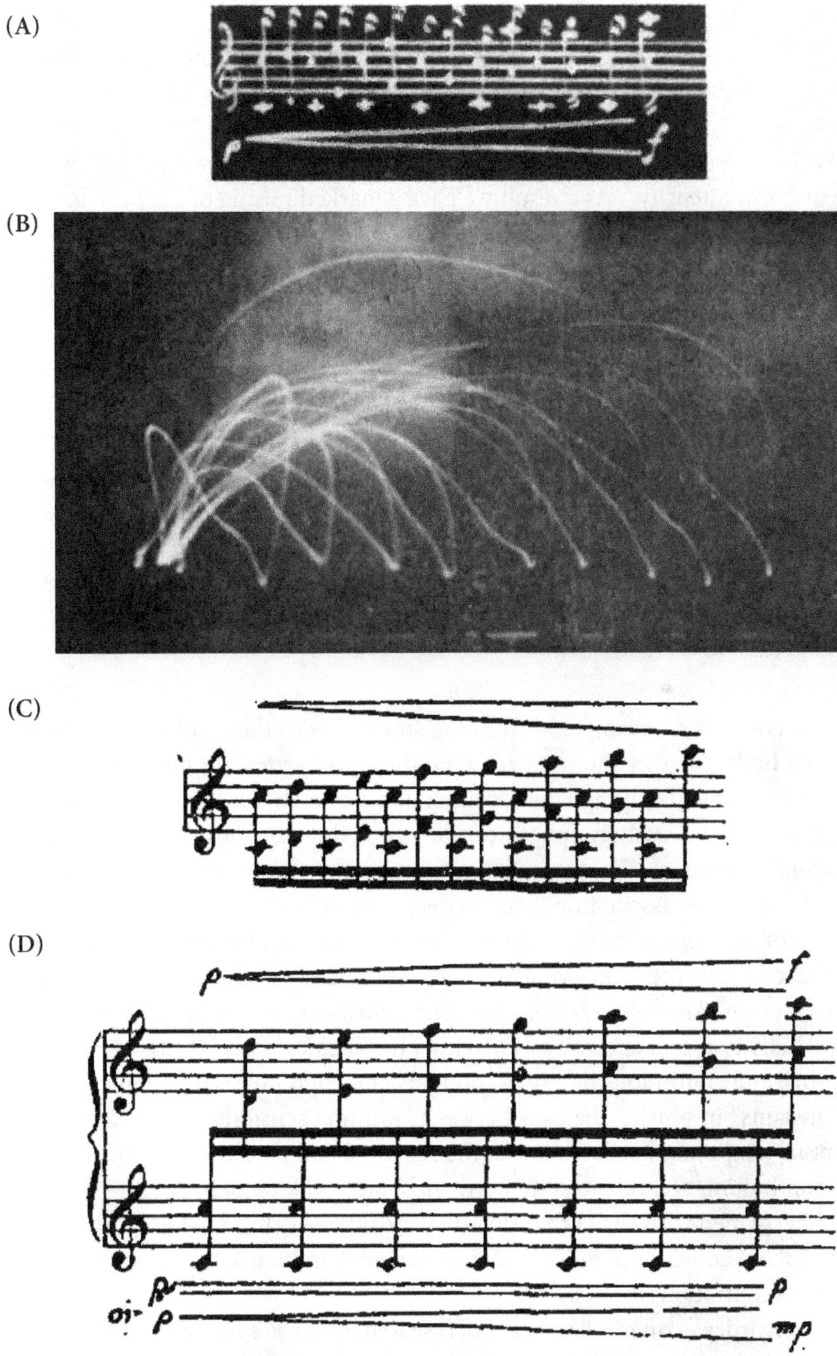

FIGURE 24.3 Hand movement in the octave figure with a crescendo: A) notated octave figure with a crescendo; B) photograph of hand movement playing the notated figure; C) printed version of the octave figure with crescendo evenly distributed across the passage; D) the figure with the crescendo as often performed.

(from Ortmann [1929] 1962, 170–171).

amplitude of the curves increases too. This increase in height of curve is not found in playing the same octave figure with the same intensity. According to his explanation, "the greater distance travelled results in a greater hand velocity, which causes the crescendo" (ibid., 171). This particular photograph also reveals that "the crescendo applied primarily to the upper tones, those forming the scale" and the "repeated C was kept at a more uniform intensity." "As a result we have a marked raising of the peak of the curve to the right and no similar displacement for the return strokes to the left. . . .The printed page marks the crescendo as extending uniformly over high and low tones as at [Figure 24.3C] but in reality it is often played, for artistic reasons resulting from the melodic relationship of the upper tones as at [Figure 24.3D]" (ibid., 171).

The adoption of this technology allowed us to see what happens "in reality," which is different from the score indication. In other words, it reveals that which is hidden from the natural ears. As Marey did in his physiological study of the moving body, Ortmann also sought "the true form of a movement" in piano playing, a phenomenon that may be beyond the thresholds of our perception.

Listening with and to the Body

Ortmann's book is an example of "training studies," with the emphasis on how the coordinated body parts of the performer produce and articulate musical motion. This suggests an immediately apparent connection between the new conception of the moving body in the early twentieth century and "the most obvious source of musical movement" (Shove and Repp 1995, 58), the human's—the performer's—bodily movement. However, the association is much deeper than literal bodily movement. The exploration of aftersound, various searches for lost time, and the attention to the moving body that connects the sonic events are intimately linked to the notion of music as continuous motion. After all, it is not only the performer's body that is moving. Once the moving body is approached by a "vivifying physiological gaze" (Cartwright 1995, xiii), the mobile and continuous nature of thoughts and cognition is highlighted, leaving no "dead instants" in music. The gaps between "minima sensibilia" (Capek 2013, 62), the perceptual temporal units, are filled at the sensorial and psychological level and constitute a continuum. The relationship between sonic and kinetic musical representations are so intertwined with each other that they are intrinsically embedded. Studies on the music-making activities of human bodies inevitably influence the sonic representation of music as continuous motion.

In this regard, we can recall a couple of researchers in the early twentieth century who worked with this connection between real body movement and the motion expressed by music itself. Gustav Becking (1894–1945), for example, is known for "Becking curves," graphic representations of "the conducting shapes of various composers" (for a table with these curves, see Becking [1928] 2011, 278–281), which is the type of motion characterizing each composer's musical personality. He argues that "one should let oneself be conducted by this work rather than trying to conduct the music according to one's own preferences" (Becking [1928] 2011, 19; Schneider 2009, 88). While Becking was concerned with the composer's "human attitude to the world" (*Erkenntnisquelle*, Nettheim's translation, 2011), as revealed in music rhythm, Alexander Truslit (1889–1971) attempted to visualize musical motions, being more concerned with the performer's body (Truslit 1938; see also Brandner 2012; Shove and Repp 1995; Repp 1993).

Here, in these discussions in the early twentieth century, we already begin to see that the music-body discourse extends beyond the apparent fact that music is created by the physical movements of performers. The idea that our perception and cognition of music are grounded in bodily action appeared in the psychological studies on music in the early twentieth century. Raymond Herbert Stetson (1872–1950), an American speech scientist who studied with William James, for example, argued for the "motor theory of musical rhythm," which states that rhythm is neither a purely affective nor a purely mental experience, but a dynamic experience that consists of "actual movements" (Stetson 1905, 257). A continuous and dynamic "rhythm-movement," he argues, is not only produced during the performance but also "induced in the act of perceiving rhythms" (307). To paraphrase Stetson's definition of speech, music is "a set of movements made audible rather than a set of sounds produced by movements" (cited in Maconie 2005, 77).

The notion of such embodied music cognition can also be found in the music-theoretical discourse of the early twentieth century, particularly in the writings of the theorists who conceived of music in dynamic terms. A group of music theorists called "energeticists" particularly thematized force as the central metaphor of music theory and promoted the new dynamic aural mode of perception in the early twentieth century (see Rothfarb 2002). This is a rather natural consequence of developments since the late nineteenth century, particularly that of the symphonic works by Brahms and Bruckner. The essential experience of these works was dynamic flow and trajectories (Schneider 2009, 84). In dealing with this "psychological, subjective" experience, such theories appear rather abstract and transcendental. However, the connection between this seemingly nonconcrete energeticist conception of music and the contemporaneous discussions of the more tangible and concrete kind of bodily movement in psychology and performance discourse may be closer than it appears.

In theorizing music in a dynamic mode, the energeticist music theorists may have considered music as the "sonic product of *action*" (Baily 1985, 237; emphasis added). For instance, Heinrich Schenker (1868–1935), whose interests in practical performance and the dynamic conception of music are relatively less acknowledged, speaks of "piano singing (*Klaviersingen*), applying principles of the voice and string instruments to the piano" (2002, 8). Citing the beginning of Mozart's piano trio in G major as an example, Schenker writes that this phrase "gives the pianist an opportunity to use a technique learned by violinists." A violinist will continue the motion of the bow through rest. "By using the hand's impulse," the pianist can produce a similar effect. "The pianist thus has a device, heretofore unrecognized, for giving the piano similarity to the orchestra" (22). What constitutes the melodic single impulse here is the motion of the hand (Figure 24.4).

FIGURE 24.4 Mozart, Piano Trio, K. 564, III, mm. 1–4

(from Schenker 2002, 66).

Evoking string instruments and ultimately singing are not unusual in the piano performance discourse of the early twentieth century. Here, Schenker is invoking strings and voice as a model to emphasize the continuous, flowing movement in piano playing. The Polish pianist Theodor Leschetizky (1830–1915) once wrote, "there is more rhythm between the notes than in the notes themselves" (cited in Newcomb 1967, 40). For Schenker, what fills the gap between the notes is this more comprehensive sense of continuity, signifying "something to do with continuity of physical and emotional commitment to a line" (Crutchfield 1991, 54).

This emphasis on continuity allows for a kind of kinesthetic hearing. As Roland Barthes asserts, we hear the performing and sounding body in between the tones (1992). The consciousness of the way our body moves when producing music allows such listening to and with the body. The musical motion is heard through our awareness of the body.

Concluding Remarks

We can sum up our excursion into the complex relationship between the science of perception and the meanings attached to music in the body as follows: (1) We began with how the visual concept of afterimage traveled from the eye to the ear and appeared as aftersound, (2) which in turn navigated from the ear to the mind by being reinterpreted as a psychological phenomenon instead of a physical one. The discussion on aftersound led to the concerns about the transition between tones, (3) which then engendered the performer's motion linking the notes and (4) also the listening not only to the sonic entity but also to the body implied in the music.

The overarching trajectory described in this chapter, however, should not be perceived as suggesting either neatly divisible states or the linear progression from sensation to perception and from cognition to motor movement. The whole journey leads to the close intertwining of perception/cognition and bodily movement. It illustrates the increasing concerns for "transitive moments" in the temporal dimension that connect individual elements. Apparently referring to the early development of photography, the philosopher Walter Benjamin once remarked that photography reveals the secret of "what happens during the fraction of a second when a person actually takes a step" and therefore the existence of the "optical unconscious" (1999, 510). We may think of a parallel phenomenon in the auditory domain, rediscovering "an aspect of human experience that remains largely 'hidden in plain view'" (Stern 2010, 3). With the new growing emphasis on the trajectory of the agent and on transitional moments, the musical mind has been reconceptualized with what may be phrased as the "auditory unconscious." Such restoration of the lost moment is substantiated by the recent theory of the continuity of the mind, which is "constantly in motion" (Spivey 2008, 8). Additionally, it captures the moment lost in the staff notation and renders listening more expansively so as to include listening to music and the performing body.

The notion that music is not just a sonic entity but also a "gestural-sonorous object" is more prominent than ever in today's discourse on music (see Godøy's chapter for more on this subject). Research topics such as "musical gestures" (Cox 2011), "embodied music cognition" (Godøy and Leman 2009; Jensenius 2007), "shaping music in performance" (Leech-Wilkinson and Küssner 2013), and "corporeal signification" (Leman 2008) are all themes that are being advanced. Various empirical methodologies are being adopted in these studies, including not only interviews, questionnaires, and experiments but also some of the most advanced motion capture system and recording technologies. Recent neuroscientific studies also support the idea of "action-listening" (Jensenius 2007, 19). When we listen to music, we mentally simulate how sounds are produced and in doing so, make sense of what we hear. Mirror neurons are activated by auditory stimuli, supporting the idea of "action-listening" (see Overy and Molnar-Szakacs 2009; Lahav, Saltzman, and Schlaug 2007).

Nevertheless, it has not been always like this, and these topics have not been popular for a very long time. For instance, in the 1980s, when the psychology of music began to thrive, the ethnomusicologist John Bailey had to argue for the necessity of recognizing music not just as a sonic event but also as a "motion event." The present study is an attempt to examine the traveling concepts across various domains of perception, movement, and music and to restore the lost moments in the history of music psychology, when the notion of the perception-action coupling was formed and prognosticated the dynamic approach to music as motion. Listening to music not only is an activity of the ear but also includes the rest of the body, which needs to be extended beyond the actual body and contextualized to include its environment and the working space that it is embedded in (Kirsch 1995; Gibson 1986).

NOTES

1. As for the duration of this aftersound, Helmholtz provided only a rather vague answer:

> Now, although we are not able to discover how long the ear and its individual parts, when set in motion, will continue to sound, yet well-known experiments allow us to form some sort of judgment . . . when making a shake, we can readily strike 8 or 10 notes in a second, so that each tone separately is struck from 4 to 5 times. If then, the sound of the first tone had not died off in our ear before the end of the second sound, at least to such an extent as not to be sensible when the latter was sounding, the tones of the shake, instead of being individually distinct, would merge into a continuous mixture of both. (Helmholtz 1954, 143; 1863, 233–234)

Thus, Helmholtz appears to suggest approximately 1/10 of a second as the duration of the aftersound of a single tone.

2. Interestingly, this part (chapter 4) was significantly revised between the third (1870) and the fourth (1877) editions. The perceptual process, which had been described in terms of "sensations" and "perceptions," was later rearticulated in terms of "perception" and "apperception." Nevertheless, the analogy to vision was retained, although much reduced in length and expressed in plainer terms, indicating its weighty role in explaining the perceptual process of hearing a tone.
3. The assumption of the connection between Helmholtz and Proust is plausible because, above all, Helmholtz's work was well known in the French scientific community to which his letter was addressed, and Proust's father, Achille Adrien Proust (1834–1903), was an eminent medical doctor.
4. Therefore, a melody is an "object of higher order" (*Gegenstände höherer Ordnung*), based on the relations among individual tones. (For a more detailed discussion of Meinong, see Schramm 2005 and Schubert 2012).
5. Gjerdingen notes that the motions perceived in music are not real but apparent: despite that, there is no continuous movement in frequency but we still perceive motion in music (1994, 35). He applies the Grossberg-Rudd model of visual perception to the study of Gestalt phenomenon in music. While the perceived motion in his study is mainly concerned with that in the pitch dimension, the latter parts of the present chapter deals with a more comprehensive sense of motion expanded from a sonic object to "gestural-sonorous object" (Godøy 2006).
6. It can be even presupposed that Marey may have played a key role in more immediately mediating Helmholtz's and Proust's lost time. Marey discussed Helmholtz's device of measuring reaction time in detail in his book *Movement*, which played an important role in propagating Helmholtz's ideas in France (see Piccolino 2003). A letter from Marey to Helmholtz written in 1868 is available at http://echo.mpiwg-berlin.mpg.de/ECHOdocuView?url=/permanent/vlp/lit16129/index.meta.
7. The same figures first appeared in his article "Moteurs animés," in *La Nature*, 1878. He did not specify the authorship of these figures, but simply made a reference to the recording of a performance by "a celebrated organist" during a conference on animal movement at one of the scientific soirees at the Sorbonne (Marey 1895, 13).
8. Marc Leman presents an interesting comparison between the player's movement, the sonogram, and the listener's corporeal articulations in a case study on the performance of *Guqin*, a traditional Chinese instrument. The difference between gestures and the sonic output is more striking in the case of Marey because in the latter's work, the footsteps and the sounds generated by the keyboard were represented using a succession of bars (see Leman 2008, 148–156).

References

Bacht, Nikolaus, ed. 2010. "Listening: Interdisciplinary Perspectives." *Special issue, Journal of the Royal Musical Association* 135, supplement 1.

Baily, John. 1985. "Music Structure and Human Movement." In *Musical Structure and Cognition*, edited by Peter Howell, Ian Cross, and Robert West, 237–258. London: Academic Press.

Barthes, Roland. 1992. *The Responsibility of Forms: Critical Essays on Music, Art, and Representation*. New ed. Berkeley: University of California Press.

Beckerman, Michael. 1994. *Janacek as Theorist*. Stuyvesant, NY: Pendragon.

Becking, Gustav. [1928] 2011. *How Musical Rhythm Reveals Human Attitudes*. Translated by Nigel Nettheim. New York: Peter Lang.

Benjamin, Walter. 1999. *Walter Benjamin: Selected Writings, Volume 2: 1927–1934*. Cambridge, MA: Belknap Press of Harvard University Press.

Boring, Edwin G. 1942. *Sensation and Perception in the History of Experimental Psychology*. New York: Appleton-Century-Crofts.

Boring, Edwin G. 1957. *A History of Experimental Psychology*. 2nd ed. New York: Appleton-Century-Crofts.

Brandner, Hans. 2012. *Bewegungslinien der Musik: Alexander Truslit und seine Lehre der Körpermusikalität, der Kinästhesie der Musik*. Augsburg: Wissner.

Burtt, Harold E. 1917. "Auditory Illusions of Movement—A Preliminary Study." *Journal of Experimental Psychology* 2: 63–75.

Cahan, David, ed. 1994. *Hermann von Helmholtz and the Foundations of Nineteenth-Century Science*. Berkeley: University of California Press.

Capek, Milič. 2013. *The New Aspects of Time: Its Continuity and Novelties*. N.p.: Springer.

Cartwright, Lisa. 1995. *Screening the Body: Tracing Medicine's Visual Culture*. Minneapolis: University of Minnesota Press.

Cogan, Robert, and Pozzi Escot. 1976. *Sonic Design: The Nature of Sound and Music*. Englewood Cliffs, NJ: Prentice-Hall.

Colman, Andrew M. 2009. *A Dictionary of Psychology*. 3rd ed. New York: Oxford University Press.

Cowan, Michael. 2011. *Technology's Pulse: Essays on Rhythm in German Modernism*. London: IGRS, University of London.

Cox, Arnie. 2011. "Embodying Music: Principles of the Mimetic Hypothesis." *Music Theory Online* 17. http://www.mtosmt.org/issues/mto.11.17.2/mto.11.17.2.cox.html.

Crary, Jonathan. 1990. *Techniques of the Observer: On Vision and Modernity in the 19th Century*. Cambridge, MA: MIT Press.

Crutchfield, Will. 1991. "Crutchfield at Large." *Opera News* 56: 54.

Doane, Mary Ann. 2002. *The Emergence of Cinematic Time: Modernity, Contingency, the Archive*. Cambridge, MA: Harvard University Press.

Gibson, James J. 1986. *The Ecological Approach to Visual Perception*. Boston: Psychology Press.

Gjerdingen, Robert O. 1994. "Apparent Motion in Music?" *Music Perception* 11: 335–370.

Godøy, Rolf Inge. 2006. "Gestural-Sonorous Objects: Embodied Extensions of Schaeffer's Conceptual Apparatus." *Organised Sound* 11 (2): 149–157.

Godøy, Rolf Inge, and Harald Jørgensen, eds. 2001. *Musical Imagery*. New York: Routledge.

Godøy, Rolf Inge, and Marc Leman, eds. 2009. *Musical Gestures: Sound, Movement, and Meaning*. New York: Routledge.

Goethe, Johann Wolfgang von. [1810] 1994. *Theory of Colours*. Translated by Charles Lock Eastlake. Cambridge, MA: MIT Press.

Hall, John E. 2015. *Guyton and Hall Textbook of Medical Physiology*. 13th ed. Philadelphia, PA: Saunders.

Haverkamp, Michael. 2012. *Synesthetic Design*. Basel: Birkhaeuser.

Helmholtz, Hermann von. 1858a. "Ueber die subjectiven Nachbilder im Auge." *Sitzungsberichte der Niederrheinischen Gesellschaft für Natur—und Heilkunde zu Bonn* 15: xcviii–c.

Helmholtz, Hermann von. 1858b. "Ueber Nachbilder." *Amtlicher Bericht über die Versammlung deutscher Naturforscher und Aerzte* 34: 225–226.

Helmholtz, Hermann von. 1863. *Die Lehre von den Tonempfindungen als physiologische Grundlage für die Theorie der Musik*. Braunschweig: F. Vieweg.

Helmholtz, Hermann von. 1954. *On the Sensations of Tone*. 2nd ed. New York: Dover.

Jensenius, Alexander Refsum. 2007. "Action—Sound: Developing Methods and Tools to Study Music-Related Body Movement." PhD diss., University of Oslo. http://urn.nb.no/URN:NBN:no-18922.

Johnson, James H. 1996. *Listening in Paris: A Cultural History*. Berkeley: University of California Press.

Kalsi, Marie-Luise Schubert. 1978. *Alexius Meinong: On Objects of Higher Order and Husserl's Phenomenology*. Boston: M. Nijhoff.

Kalsi, Marie-Luise Schubert. 2012. *Alexius Meinong: On Objects of Higher Order and Husserl's Phenomenology*. Springer Science and Business Media.

Kirsch, David. 1995. "The Intelligent Use of Space." *Artificial Intelligence* 73: 31–68.

Kurth, Ernst. 1931. *Musikpsychologie*. Berlin: Max Hasse.

Lahav, Amir, Elliot Saltzman, and Gottfried Schlaug. 2007. "Action Representation of Sound: Audiomotor Recognition Network While Listening to Newly Acquired Actions." *Journal of Neuroscience* 27: 308–314.

Leech-Wilkinson, Daniel, and Mats B. Küssner. 2013. "Introduction to Special Issue on Music and Shape." *Empirical Musicology Review* 8 (1): 1.

Leman, Marc. 2008. *Embodied Music Cognition and Mediation Technology*. Cambridge, MA: MIT Press.

Maconie, Robin. 2005. *Other Planets: The Music of Karlheinz Stockhausen*. Lanham, MD: Scarecrow.

Marey, Étienne-Jules. 1874. *Animal Mechanism: A Treatise on Terrestrial and Aerial Locomotion*. New York: D. Appleton.

Marey, Étienne-Jules. 1878. "Moteurs animés: Expériences de physiologie graphique." *La Nature* 278, 279: 273–278, 289–295.

Marey, Étienne-Jules. 1894. *Le mouvement*. Paris: Libraire de l'Académie de Médecine.

Matsumoto, David, ed. 2009. *The Cambridge Dictionary of Psychology*. Cambridge: Cambridge University Press.

Mayer, Alfred. 1874. "Researches in Acoustics, Paper Nos. 5, 6, and 7." *American Journal of Science* 3: 81–109, 170–182, 241–255, 362–366.

Meinong, Alexius. 1889. "Phantasie-Vorstellung und Phantasie." *Zeitschrift fur Philosophie und Philosophische Kritik* 95: 161–244.

Merleau-Ponty, Maurice. 1962. *Phenomenology of Perception*. London: Routledge & Kegan Paul.

Murphy, Gardner, and Joseph K. Kovach. 1972. *Historical Introduction to Modern Psychology*. New York: Harcourt Brace Jovanovich.

Nancy, Jean-Luc. 2007. *Listening*. New York: Fordham University Press.

Newcomb, Ethel. 1967. *Leschetizky as I Knew Him*. New York: Da Capo.

Ortmann, Otto. (1929) 1962. *The Physiological Mechanics of Piano Technique: An Experimental Study of the Nature of Muscular Action as Used in Piano Playing and of the Effects Thereof upon the Piano Key and the Piano Tone*. New York: E. P. Dutton.

Overy, Katie, and Istvan Molnar-Szakacs. 2009. "Being Together in Time: Musical Experience and the Mirror Neuron System." *Music Perception: An Interdisciplinary Journal* 26: 489–504.

Piccolino, Marco. 2003. "A 'Lost Time' between Science and Literature: The 'Temps Perdu' from Hermann von Helmholtz to Marcel Proust." *Audiological Medicine* 1: 261–270.

Repp, Bruno H. 1993. "Music as Motion: A Synopsis of Alexander Truslit's (1938) Gestaltung und Bewegung in Der Musik." *Psychology of Music* 21 (1): 48–72.

Roeckelein, Jon. 2004. *Imagery in Psychology: A Reference Guide*. Westport, CT: Praeger.

Rothfarb, Lee. 2002. "Energetics." In *Cambridge History of Western Music Theory*, edited by Thomas Christensen, 927–955. Cambridge: Cambridge University Press.

Schenker, Heinrich. 2002. *The Art of Performance*. Edited by Heribert Esser. Translated by Irene Schreier Scott. New York: Oxford University Press.

Schmidgen, Henning. 2002. "Of Frogs and Men: The Origins of Psychophysiological Time Experiments, 1850–1865." *Endeavour* 26 (4): 142–148.

Schmidgen, Henning. 2014. *The Helmholtz Curves: Tracing Lost Time*. Translated by Nils F. Schott. New York: Fordham University Press.

Schneider, Albrecht. 2009. "Music and Gestures: A Historical Introduction and Survey of Earlier Research." In *Musical Gestures: Sound, Movement, and Meaning*, edited by Rolf Inge Godøy and Marc Leman, 69–100. New York: Routledge.

Schramm, Alfred. 2005. *Meinong Studies/Meinong Studien*. Frankfurt: Ontos.

Shepard, R. N. 1984. "Ecological Constraints on Internal Representation: Resonant Kinematics of Perceiving, Imagining, Thinking, and Dreaming." *Psychological Review* 91: 417–447.

Shove, Patrick, and Bruno Repp. 1995. "Musical Motion and Performance: Theoretical and Empirical Perspectives." In *The Practice of Performance: Studies in Musical Interpretation*, edited by John Rink, 55–83. Cambridge: Cambridge University Press.

Souriau, Paul. 1983. *The Aesthetics of Movement*. Translated by Manon Souriau. Amherst: University of Massachusetts Press.

Spivey, Michael. 2008. *The Continuity of Mind*. New York: Oxford University Press.

Stern, Daniel N. 2010. *Forms of Vitality: Exploring Dynamic Experience in Psychology and the Arts*. New York: Oxford University Press.

Stetson, Raymond Herbert. 1905. "A Motor Theory of Rhythm and Discrete Succession." *Psychological Review* 12: 250–270, 293–330.

Stumpf, Carl. 1883. *Tonpsychologie*. Vol. 1. Leipzig: S. Hirzel.

Szendy, Peter. 2008. *Listen: A History of Our Ears*. New York: Fordham University Press.

Truslit, Alexander. 1938. *Gestaltung und Bewegung in der Musik: Ein tönendes Buch vom musikalischen Vortrag und seinem bewegungserlebten Gestalten und Hören*. Berlin-Lichterfelde: Vieweg.

Urbantschitsch, Victor. 1881. "Über das An—und Ab-klingen akustischer Empfindungen." *Pflüger's Archiv* 25: 323–342.

Zwicker, Eberhard. 1964. "'Negative Afterimage' in Hearing." *Journal of the Acoustical Society of America* 36: 2413–2415.

Index